ACOUSTIC AND ELASTIC WAVE FIELDS IN GEOPHYSICS, II

Methods in Geochemistry and Geophysics
(Volumes 1–28 are out of print)

1. A.S. RITCHIE — CHROMATOGRAPHY IN GEOLOGY
2. R. BOWEN — PALEOTEMPERATURE ANALYSIS
3. D.S. PARASNIS — MINING GEOPHYSICS
4. I. ADLER — X-RAY EMISSION SPECTROGRAPHY IN GEOLOGY
5. THE LORD ENERGLYN AND L. BREADLY — ANALYTICAL GEOCHEMISTRY
6. A.J. EASTON — CHEMICAL ANALYSIS OF SILICATE ROCKS
7. E.E. ANGINO AND G.K. BILLINGS — ATOMIC ABSORPTION SPECTROMETRY IN GEOLOGY
8. A. VOLBORTH — ELEMENTAL ANALYSIS IN GEOCHEMISTRY, A: MAJOR ELEMENTS
9. P.K. BHATTACHARYA AND H.P. PATRA — DIRECT CURRENT GEOELECTRIC SOUNDING
10. J.A.S. ADAMS AND P. GASPARINI — GAMMA-RAY SPECTROMETRY OF ROCKS
11. W. ERNST — GEOCHEMICAL FACIES ANALYSIS
12. P.V. SHARMA — GEOPHYSICAL METHODS IN GEOLOGY
13. C.H. CHEN (Editor) — COMPUTER-AIDED SEISMIC ANALYSIS AND DISCRIMINATION

14A. O. KOEFOED — GEOSOUNDING PRINCIPLES, 1. RESISTIVITY SOUNDING MEASUREMENTS

14B. H.P. PATRA AND K. MALLICK — GEOSOUNDING PRINCIPLES, 2. TIME-VARYING GEOELECTRIC SOUNDINGS

15. A.A. KAUFMAN AND G.V. KELLER — THE MAGNETOTELLURIC SOUNDING METHOD
16. A.A. KAUFMAN AND G.V. KELLER — FREQUENCY AND TRANSIENT SOUNDINGS
17. C.H. CHEN (Editor) — SEISMIC SIGNAL ANALYSIS AND DISCRIMINATION
18. J.E. WHITE — UNDERGROUND SOUND APPLICATION OF SEISMIC WAVES
19. M.N. BERDICHEVSKY AND M.S. ZHDANOV — ADVANCED THEORY OF DEEP GEOMAGNETIC SOUNDINGS

20A. A.A. KAUFMAN AND G.V. KELLER — INDUCTIVE MINING PROSPECTING, PART I: THEORY

21. A.W. WYLIE — NUCLEAR ASSAYING OF MINING BOREHOLES - AN INTRODUCTION
22. C.H. CHEN (Editor) — SEISMIC SIGNAL ANALYSIS AND DISCRIMINATION III
23. R.P. PHILP — FOSSIL FUEL BIOMARKERS
24. R.B. JOHNS (Editor) — BIOLOGICAL MARKERS IN THE SEDIMENTARY RECORD
25. J.C. D' ARNAUD GERKENS — FOUNDATION OF EXPLORATION GEOPHYSICS
26. P. TYGEL AND P. HUBRAL — TRANSIENT WAVES IN LAYERED MEDIA
27. A.A. KAUFMAN AND GV. KELLER — INDUCTION LOGGING
28. J.G. NEGI AND P.D. SARAF — ANISOTROPY IN GEOELECTROMAGNETISM
29. V.P. DIMRI — DECONVOLUTION AND INVERSE THEORY - APPLICATION TO GEOPHYSICAL PROBLEMS
30. K.-M. STRACK — EXPLORATION WITH DEEP TRANSIENT ELECTROMAGNETICS
31. M.S. ZHDANOV AND G.V. KELLER — THE GEOELECTRICAL METHODS IN GEOPHYSICAL EXPLORATION
32. A.A. KAUFMAN AND A.L. LEVSHIN — ACOUSTIC AND ELASTIC WAVE FIELDS IN GEOPHYSICS, I
33. A.A. KAUFMAN AND P.A. EATON — THE THEORY OF INDUCTIVE PROSPECTING
34. A.A. KAUFMAN AND P. HOEKSTRA — ELECTROMAGNETIC SOUNDINGS
35. M.S. ZHDANOV AND P.E. WANNAMAKER — THREE-DIMENSIONAL ELECTROMAGNETICS
36. M.S. ZHDANOV — GEOPHYSICAL INVERSE THEORY AND REGULARIZATION PROBLEMS
37. A.A. KAUFMAN, A.L. LEVSHIN AND K.L. LARNER — ACOUSTIC AND ELASTIC WAVE FIELDS IN GEOPHYSICS, II

Methods in Geochemistry and Geophysics, 37

ACOUSTIC AND ELASTIC WAVE FIELDS IN GEOPHYSICS, II

A.A. KAUFMAN
Colorado School of Mines
Golden, CO, USA

A.L. LEVSHIN
University of Colorado
Boulder, CO, USA

K.L. LARNER
Colorado School of Mines
Golden, CO, USA

2002

ELSEVIER

Amsterdam - Boston - London - New York - Oxford - Paris
Tokyo - San Diego - San Francisco - Singapore - Sydney

ELSEVIER SCIENCE B.V.
Sara Burgerhartstraat 25
P.O. Box 211, 1000 AE Amsterdam, The Netherlands

First edition 2002

Library of Congress Cataloging-in-Publication Data
A catalog record from the Library of Congress has been applied for.

British Library Cataloguing in Publication Data
A catalogue record from the British Library has been applied for.

ISBN: 0 444 50642 X
ISSN: 0076-6895

∞ The paper used in this publication meets the requirements of ANSI/NISO Z39.48-1992 (Permanence of Paper).
Printed in The Netherlands.

Contents

Introduction

This volume is a continuation of the Part 1, Acoustic and Elastic Wave Fields in Geophysics, I (Kaufman and Levshin, 2000). As in the first volume, we consider here acoustic media in which only longitudinal waves are present. We focus in this media because compressional waves play a dominant role in exploration geophysics and because so many aspects of wave propagation theory apply to both acoustic and elastic wave fields.

The first chapter is devoted entirely to principles of geometric acoustics, when a wave can be approximated by a plane wave. It is essential that its amplitude and phase (eikonal) in general depend on coordinates of a point. First, we represent the complex amplitude of the potential in the form of an asymptotic series. This allows us to derive the eikonal and transport equations. Proceeding from the eikonal and the concept of the ray, we obtain the system of equations characterizing ray behavior in different media. As an example, we study the reflected and head waves in media with plane interfaces. We focus especially on ray tracing and wavefront construction. Several examples demonstrate the geometry of rays in simple as well as rather complicated cases, when, for instance, shadow zones and a caustic are observed. A discussion of the relationship between ray equations and Fermat's principle concludes Chapter 1.

In the second chapter, we turn our attention to dynamic properties of wave fields. Using the Poynting vector, we investigate the behavior of pressure and displacement amplitudes in the zero approximation along ray tubes. The same result is obtained from solution of the transport equation. In this light, we discuss determination of a cross-section of elementary ray tubes. As an example, we study in detail amplitudes of reflected and transmitted waves, including points on the boundary. In particular, we derive Fresnel coefficients. This chapter contains several examples related to shadow zones and caustics. Applying the Kirchhoff integral, we consider a diffraction phenomenon near the caustic, as well as the distribution of pressure. We also discuss the concepts of ray, ray tube, and Fresnel volume.

In the third chapter, our study of the kinematic and dynamic properties of waves in the high-frequency spectrum allows us to describe some aspects of seismic migration, on the basis of use of the Kirchhoff integral. To begin, we consider an approach, that does not require knowledge of the wave amplitude beneath the surface of observation. It is based on such concepts as the line of equal reflection time, the envelope of these lines, the diffraction point, and the diffraction curve. In the second section, we discuss basic principles of Kirchhoff migration. For example, we show that an envelope of diffraction curves characterizes a time, when constructive interference of wavelets takes place. In this example, the locus of apexes of diffraction curves corresponds to the reflector. Also we demonstrate that the envelope of curves of equal reflection is the reflector. Besides a relationship between ray tracing and Kirchhoff migration is considered.

In the next chapter we study the behavior of a plane wave in a horizontally layered medium, when normal incidence takes occurs. Chapter 5 describes the behavior of plane waves in a layered medium with an oblique incidence. First we study reflected and transmitted waves in the presence of a single interface. We pay special attention to the case in which the incident angle exceeds the critical angle and the evanescent wave arises. In addition, we consider reflection and transmission of inhomogeneous plane waves. The study of oblique incidence of plane waves in media with two interfaces allows us to investigate propagation of homogeneous and evanescent waves in different parts of a medium. Also in chapter 5, we consider formation of normal modes in wave guides. The chapter concludes with some interesting features of waves in a media with continuously changing velocity.

Chapter 6 is devoted completely to a spherical wave in the presence of a horizontal interface. First, two limiting cases are considered, when one part of a medium is either a free space or a rigid space. Then, expressions for the field in a general case are derived. Using these equations and applying the stationary phase method, we obtain for reflected and transmitted waves, asymptotic formulas that are valid in the wave zone. Applying integration along branch cuts, we derive approximate formulas for the head waves. In addition, we present some information about inhomogeneous waves. Using the steepest descent method, we improve to some extent the asymptotical formulas and describe the behavior of the reflected, head, transmitted, and evanescent waves.

In Chapter 7 we investigate propagation of sinusoidal and transient waves in a layer with a relatively small velocity of propagation. We derive expressions for the complex amplitude of the acoustic potential first. Then we represent the potential as a sum of two parts, one associated with poles and the other with branch points of the integrand, respectively. The main attention is paid to the first one, which describes normal modes. Correspondingly, we derive the dispersion equation and study frequency responses of the phase velocity for different modes. The results allow us to investigate propagation of transient waves and to distinguish wavelet elements moving with different group velocities.

Chapter 8 describes propagation of waves inside a cylinder when both media are acoustic. As in the previous chapter, we consider mainly normal modes, whose behavior is similar to that in the layer bounded by two plane interfaces. At the end of this volume there are several appendixes that can help the reader to understand different aspects of the mathematics used in this monograph.

Acknowledgments

We express our thanks to Drs. N. Bleinstein, I. Fuks, D. Hale, K. Elleffson, P. Martin, K. Naugolnykh, K. Osipov, L. Ostrovsky, R. Snieder, and L. Tenorio for very useful comments and suggestions. Numerous discussions with Drs. B. Schneider and H. Jaramillo in preparing the second section of Chapter 3 were extremely instructive. Their involvement was so strong that we consider them coauthors of this section. We gratefully acknowledge their generous contribution. We are also grateful to Dr. O. Godin, who helped us to describe wave behavior using the steepest descent method. We express our sincere thanks to Dr. T. Young for the permanent moral support of this work.
Albena Mateeva carried the major burden of technical work in preparing this manuscript. We also thank Andres Pech, Rosa Bernal and Morrakot Khebcharon for their technical assistance. We appreciate editorial help of Faith Rogers. We are grateful for funding of our efforts by Drs. A. Saito (Japan), and Y. Shoham (Shell International) and Mr. M. Bostic (Geonics). We thank Blackwell Publishing for permission to reproduce figures from I.G. Hagedorn's article "A process of seismic reflection interpretation", published in *Geophysical Prospecting.*

List of Symbols

a	acceleration
a_n, b_n	coefficients of Fourier's series
A	work
A_n, B_n	amplitudes of normal modes, coefficients of Debye expansion
A_R	resultant amplitude
c	phase velocity
c_a	apparent velocity
c_{gn}, c_{pn}	group and phase velocities of the nth mode
c_n	velocity in the n-th layer
C	compressbility
ds	elementary displacement
e_k	density of kinetic energy
E_k	kinetic energy
$\mathbf{F_s}$	spreading factor
f	frequency
G	Green's function
$H_n^{(1)}(mr)$, $H_n^{(2)}(mr)$	Hankel functions of the nth order
$i = \sqrt{-1}$	imaginary unit
$\mathbf{i}$,$\mathbf{j}$	unit vector
$I_n(mr)$, $K_n(mr)$	modified Bessel functions of the nth order
$J_n(mr)$, $Y_n(mr)$	Bessel functions of the nth order
J	acoustic length, Jacobian
k	wavenumber
k_r	wavenumbers of normal modes
K	curvature, inclination factor
l	radius of Fresnel zone
L	eikonal
M	total mass, bulk modulus
$\mathbf{M}$	vector
n	refraction index
$\mathbf{n}$	unit normal
$\mathbf{p}$	slowness vector
p	slowness, observation point

P	pressure
P_a	additional pressure
$\mathbf{P}$	momentum of motion
$\mathcal{P}$	complex amplitude of pressure
r	distance from the z-axis
r	radius-vector
R	distance from the source
Rij	reflection coefficient
s	displacement
$\mathbf{s}_0$	unit vector along ray
t	time
T	period, travel time
U	acoustic potential
$\mathcal{U}$	complex amplitude of potential
$\mathbf{v}$	velocity
V_0	elementary volume
W	total energy
x, y, z	cartesian coordinates
$\mathbf{Y}$	Poynting vector
z	complex number
Z	impedance
δS	elementary area
γ	Euler's constant
ϵ	small number
φ	dip angle
θ	incidence angle
θ_r	angle of reflection
θ_2	angle of refraction
θ_c	critical angle
λ	wavelength
$\boldsymbol{\nu}$	unit vector
Φ	Airy function
τ	time constant, time delay
ρ	density of mass
ρ_a	additional density
Σ	surface
ω	circular frequency

Chapter 1

Principles of geometrical acoustics

In the last chapter of Part I we described the theoretical basis of the geometrical acoustics, and we now continue discussion of this subject. To facilitate this study, it seems reasonable to start with reminding the most important elements of the theory. Such an approach only slightly increases volume of this monograph, but makes it is rather independent of the previous one; in other words, some portions of this chapter serve as the bridge between two books.

1.1 Transition to high-frequency acoustics

As before (Part I), we proceed from the physical laws which are

(a) the principle of conservation of mass

$$\operatorname{div} \mathbf{s} = \frac{\Delta V}{V} = -\frac{\rho_a}{\rho} \tag{1.1}$$

(b) the relationship between pressure and density

$$P_a = \alpha\, \rho_a \tag{1.2}$$

(c) Newton's second and third laws

$$\rho \frac{\partial^2 \mathbf{s}}{\partial t^2} = -\operatorname{grad} P \tag{1.3}$$

In these equations, $\mathbf{s}$ is the particle displacement produced by the wave, $\Delta V/V$ is the relative change of the elementary volume (the dilatation), ρ is the medium density before a wave arrival; ρ_a is the density change caused by the wave, α is the coefficient equal to

$$\alpha = c^2, \tag{1.4}$$

and c is the wave velocity. Also P_a is the additional pressure due to the wave action, whereas P_0 is its value at an equilibrium. The total pressure is

$$P = P_0 + P_a \tag{1.5}$$

Note that we consider propagation of an acoustic wave, provided that

$$\rho_a \ll \rho_0 \,, \qquad P_a \ll P_0 \,, \qquad \Delta V \ll V \tag{1.6}$$

Because eqs. 1.1 and 1.3 contain derivatives with respect to the coordinates, they are applied at regular points only. At such places, the medium density and the propagation velocity are continuous functions.

Bearing in mind that the density change ρ_a is not subject to measurements, it is natural to eliminate it from the set 1.1–1.3. Then we obtain

$$\operatorname{div} \mathbf{s} = -\frac{P_a}{M} \qquad \text{and} \qquad \rho \frac{\partial^2 \mathbf{s}}{\partial t^2} = -\operatorname{grad} P_a \tag{1.7}$$

Here

$$M = \rho c^2 \tag{1.8}$$

is the bulk modulus. The system of equations contains two unknowns, P_a and $\mathbf{s}$, and describes their behavior at regular points. To further simplify eqs. 1.1–1.3, we consider some important features of the displacement $\mathbf{s}$ as a function of spatial coordinates.

Vector field s

As is well known, (Part I) any vector field is characterized by its curl and divergence. For the field $\mathbf{s}$, the latter is given by the first equation of set 1.7:

$$\operatorname{div} \mathbf{s} = -\frac{P_a}{M}$$

In order to obtain the second equation, curl $\mathbf{s}$, we take the curl from both sides of eq. 1.3 and change the order of differentiation. This yields

$$\frac{\partial}{\partial t} (\operatorname{curl} \rho \, \mathbf{v}) = 0, \tag{1.9}$$

because the equality

$$\operatorname{curl} \operatorname{grad} P_a = 0$$

always takes place. Here

$$\mathbf{v} = \frac{\partial \mathbf{s}}{\partial t}$$

is the particle velocity. As follows from eq. 1.9

$$\operatorname{curl} \rho \mathbf{v} = \mathbf{N}_1 , \tag{1.10}$$

where $\mathbf{N}_1$ is a vector independent of time t.

Taking into account the fact that at the beginning, before the wave arrives, particles of the medium were at rest, $\mathbf{v} = 0$, we conclude that

$$\mathbf{N}_1 = 0 \qquad \text{or} \qquad \frac{\partial}{\partial t} (\operatorname{curl} \rho\, \mathbf{s}) = 0$$

This gives

$$\operatorname{curl} \rho\, \mathbf{s} = \mathbf{N}_2$$

Applying the same initial condition, we have

$$\operatorname{curl} \rho\, \mathbf{s} = 0, \tag{1.11}$$

and it shows that the vector

$$\mathbf{N} = \rho\, \mathbf{s}$$

does not have vortexes and, correspondingly, its circulation is zero

$$\oint_L \mathbf{N} \cdot d\mathbf{l} = 0 \tag{1.12}$$

Here $d\mathbf{l}$ is an elementary vector tangential to the line L. Both vectors $\mathbf{N}$ and $\mathbf{s}$ have the same direction but, in general, differ in magnitude.

Our goal is to study the field $\mathbf{s}$ and, for this reason, let us write eq.1.11 as

$$\operatorname{curl} \rho\, \mathbf{s} \;=\; \rho \operatorname{curl} \mathbf{s} + \operatorname{grad} \rho \times \mathbf{s} \;=\; 0$$

or

$$\operatorname{curl} \mathbf{s} \;=\; -\frac{1}{\rho} \, (\nabla \rho \times \mathbf{s}) \tag{1.13}$$

Thus, within an elementary volume, which is in equilibrium, the particle displacement $\mathbf{s}$ obeys the system of equations

$$\operatorname{curl} \mathbf{s} = \frac{\mathbf{s} \times \nabla \rho}{\rho} \qquad \text{and} \qquad \operatorname{div} \mathbf{s} = -\frac{P_a}{M} \tag{1.14}$$

This set of equations can be interpreted two ways. First, we see that the field $\mathbf{s}$ has both vortexes and sources. The former are present at the points, where

$$\mathbf{s} \times \nabla\rho \neq 0\,,$$

while the sources are directly proportional to the additional pressure P_a.

The system 1.14 also demonstrates that, as in the case of a solid medium, any elementary volume can be subjected to different types of motion and deformation, (Part I). These include a translation and rotation, as well as a compression (expansion), and simple and pure shears. Of course, the elementary volume undergoes neither deformation nor rotation in the vicinity of points, where

$$\mathbf{s} \times \nabla\rho = 0 \qquad \text{and} \qquad P_a = 0$$

As follows from vector analysis, Part I, the field $\mathbf{s}$, which obeys the system 1.14, cannot be described by either a scalar or vector potential. In accordance with Helmholtz's theorem, it is necessary to use both of them. However, at the beginning, we assume that with respect to density the medium is homogeneous in the vicinity of regular points, that is

$$\nabla\rho = 0 \tag{1.15}$$

Taking into account this condition, our system of equations is greatly simplified to

$$\operatorname{curl}\mathbf{s} = 0 \qquad \text{and} \qquad \operatorname{div}\mathbf{s} = -\frac{P_a}{M} \tag{1.16}$$

Thus, we arrived at the source field where wave propagation can cause only the translation of elementary volume, its compression or expansion, and pure shear, (Part I).

Scalar potential, wave and Helmholtz equations

From the mathematical point of view, the absence of vortexes allows us to express both the pressure and the displacement in terms of a scalar potential U only. In fact, it follows from the first equation of the set 1.16 that

$$\mathbf{s} = \operatorname{grad} U\,, \tag{1.17}$$

because

$$\operatorname{curl}\operatorname{grad} U \equiv 0\,,$$

and the function U is called the acoustic potential. Certainly, there is an infinite number of functions U describing the same displacement field $\mathbf{s}$. Substitution of eq. 1.17 into equations of the set 1.7 yields

$$\operatorname{div}\operatorname{grad} U = \nabla^2 U = -\frac{P_a}{M},$$

and (1.18)

$$\rho\operatorname{grad}\frac{\partial^2 U}{\partial t^2} = -\operatorname{grad} P_a$$

Eliminating P_a from the latter, we have

$$\operatorname{grad}\left(\rho\frac{\partial^2 U}{\partial t^2} - M\nabla^2 U\right) = 0,$$

that is,

$$\rho\frac{\partial^2 U}{\partial t^2} - M\nabla^2 U = \text{const} \tag{1.19}$$

Inasmuch as a constant is independent of the point position, and the wave field is absent at infinity, we conclude that this constant is zero.

Therefore, we arrive at the wave equation for the potential

$$\nabla^2 U = \frac{1}{c^2}\frac{\partial^2 U}{\partial t^2}, \tag{1.20}$$

because

$$M = \rho c^2$$

It is a simple matter to derive the same equation for the displacement and the pressure, (Part I).

As follows from the first equation of the set 1.18,

$$P_a = -M\nabla^2 U$$

Then, making use of the wave equation, we have

$$P_a = -\rho\frac{\partial^2 U}{\partial t^2}$$

Thus, we find that at regular points of homogeneous medium

$$\nabla^2 U = \frac{1}{c^2}\frac{\partial^2 U}{\partial t^2}, \qquad P_a = -\rho\frac{\partial^2 U}{\partial t^2}, \qquad \mathbf{s} = \operatorname{grad} U \tag{1.21}$$

It may be proper to notice that the acoustic potential is an auxiliary function that hardly has any physical meaning. For this reason, we should not be surprised to observe unexpected behavior of this function in some cases. At the same time, introduction of the potential U greatly facilitates the study of wave fields.

Next, assume that the waves are sinusoidal, that is

$$U = \text{Re}\, \mathcal{U} e^{-i\omega t}, \tag{1.22}$$

where $\mathcal{U}$ is the complex amplitude of the potential. Substitution of eq. 1.22 into eq. 1.20 gives the Helmholtz equation

$$\nabla^2 \mathcal{U} + k^2 \mathcal{U} = 0 \tag{1.23}$$

Here

$$k = \frac{\omega}{c}$$

is the wavenumber. Correspondingly, we obtain for complex amplitudes of the displacement and the pressure

$$\mathcal{S} = \text{grad}\, \mathcal{U} \qquad \text{and} \qquad \mathcal{P} = \rho \omega^2 \mathcal{U}, \tag{1.24}$$

where $\mathcal{P}$ is the additional pressure.

As noted previously, eqs. 1.23–1.24 are valid at regular points only. In particular, they are not applied at points of boundaries between media with different parameters ρ and c.

Boundary conditions

Behavior of the potential at such interfaces is dictated by the fact that wave propagation can cause neither a gap between different parts of the medium nor overlap of them as well as infinite acceleration of the particles. This means that the normal components of the displacement and the pressure are continuous functions

$$s_{1n} = s_{2n} \qquad \text{and} \qquad P_1 = P_2$$

Respectively from eqs. 1.21 we have

$$\frac{\partial U_1}{\partial n} = \frac{\partial U_2}{\partial n} \qquad \text{and} \qquad \rho_1 \frac{\partial^2 U_1}{\partial t^2} = \rho_2 \frac{\partial^2 U_2}{\partial t^2}$$

In particular, in the frequency domain

$$\frac{\partial \mathcal{U}_1}{\partial n} = \frac{\partial \mathcal{U}_2}{\partial n} \qquad \text{and} \qquad \rho_1 \mathcal{U}_1 = \rho_2 \mathcal{U}_2$$

Now we are ready to derive the equations describing waves in the high-frequency range.

Basic assumptions of high-frequency asymptotics

If a point source is located in a homogeneous medium, then with an increase of frequency the near and intermediate zones occupy a relatively small volume around the source, (Part I), and in the limit $\omega \to \infty$ the wave zone is observed almost everywhere. In such a case, the wave field is described by the function

$$A(k,\, r)\, e^{i\,k\,r}\,, \tag{1.25}$$

where the amplitude A, in general, depends on the distance r from the source and the wavenumber k, and the phase is equal to

$$\varphi = k\,r = \frac{\omega}{c}\,r = \frac{2\,\pi\,r}{\lambda}\,, \tag{1.26}$$

where λ is the wavelength.

In the case of several sources, in place of eq. 1.25 we have

$$A_1\, e^{i\,k\,r_1} + A_2\, e^{i\,k\,r_2} + \ldots + A_n\, e^{i\,k\,r_n}\,, \tag{1.27}$$

and the total wave field is the result of interference of these waves.

Next, suppose that the medium is inhomogeneous with respect to velocity c and the wavelength tends to zero

$$\lambda \to 0 \qquad \text{or} \qquad k \to \infty$$

To derive a high-frequency asymptotics of the Helmholtz equation, P. Debye assumed that its solution has the same form

$$\mathcal{U} = A\, e^{i\,k_0\,L(r)} \qquad \text{if} \qquad \omega \to \infty \tag{1.28}$$

even though the medium is inhomogeneous, (Part I). Here k_0 is the wavenumber at some point of the medium, A is the amplitude, and L, the function that has the dimension of the length, is called the eikonal. Both of them depend on the parameters of the medium and the coordinates of the observation point.

Since $k_0 \to \infty$, the scalar potential $\mathcal{U}$ is a rapidly varying function of position. In contrast, it is assumed that the amplitude A and the eikonal L change relatively slowly from one point to another, and they do not grow unlimitedly when $k_0 \to \infty$. This means that the variations of A, as well as the medium parameters, are small within any interval approximately equal to the wavelength. Certainly, such an assumption is natural because $\lambda \to 0$. In other words, eq. 1.28 suggests that the wave can be treated locally as a plane wave if these conditions are met. This hypothesis is a foundation of the high-frequency (geometrical) acoustics.

Note that, in accordance with eq. 1.28, the surfaces

$$L = \text{const}$$

are the surfaces of equal phases, that is, they are the wavefronts. As was shown in Part I, normals to them characterize the ray directions. In general, if the propagation velocity varies, the rays are curved.

Derivation of the eikonal and transport equations

The right-hand side of eq. 1.28 contains two unknown functions $A(r)$ and $L(r)$. In order to find them, we substitute the complex amplitude $\mathcal{U}$, given by eq. 1.28, into the Helmholtz equation

$$\frac{\partial^2 \mathcal{U}}{\partial x^2} + \frac{\partial^2 \mathcal{U}}{\partial y^2} + \frac{\partial^2 \mathcal{U}}{\partial z^2} + k^2 \mathcal{U} = 0 \tag{1.29}$$

Differentiation with respect to x yields

$$\frac{\partial \mathcal{U}}{\partial x} = \frac{\partial A}{\partial x} e^{i k_0 L} + i k_0 A \frac{\partial L}{\partial x} e^{i k_0 L}$$

or

$$\frac{\partial \mathcal{U}}{\partial x} = \frac{\partial A}{\partial x} e^{i k_0 L} + i k_0 \mathcal{U} \frac{\partial L}{\partial x} \tag{1.30}$$

Correspondingly, the second derivative is

$$\begin{aligned} \frac{\partial^2 \mathcal{U}}{\partial x^2} &= \frac{\partial^2 A}{\partial x^2} e^{i k_0 L} + i k_0 \frac{\partial A}{\partial x}\frac{\partial L}{\partial x} e^{i k_0 L} + \\ &+ i k_0 \mathcal{U} \frac{\partial^2 L}{\partial x^2} + i k_0 \frac{\partial L}{\partial x}\left(\frac{\partial A}{\partial x} e^{i k_0 L} + i k_0 A \frac{\partial L}{\partial x} e^{i k_0 L}\right) \end{aligned} \tag{1.31}$$

Since it is assumed that $k_0 \to \infty$, the right-hand side of this equation contains terms of different orders. Ignoring the smallest one,

$$\frac{\partial^2 A}{\partial x^2} e^{i k_0 L},$$

we obtain

$$\frac{\partial^2 \mathcal{U}}{\partial x^2} = 2\,i\,k_0\,\frac{\partial A}{\partial x}\,\frac{\partial L}{\partial x}\,e^{i\,k_0\,L} + i\,k_0 \mathcal{U}\,\frac{\partial^2 L}{\partial x^2} - k_0^2 \left(\frac{\partial L}{\partial x}\right)^2 \mathcal{U}$$

or

$$\frac{\partial^2 \mathcal{U}}{\partial x^2} = -\,k_0^2 \left(\frac{\partial L}{\partial x}\right)^2 \mathcal{U} + 2\,i\,k_0 \mathcal{U} \left(\frac{\partial L}{\partial x}\,\frac{\partial \log A}{\partial x} + \frac{1}{2}\,\frac{\partial^2 L}{\partial x^2}\right)$$

The expressions for $\partial^2 \mathcal{U}/\partial y^2$ and $\partial^2 \mathcal{U}/\partial z^2$ are similar, we just replace x with y and z, respectively. Therefore, substitution of the second-order derivatives into the Laplacian gives

$$\nabla^2 \mathcal{U} = -\,k_0^2 \mathcal{U}\,D(L) + 2\,i\,k_0 \mathcal{U} \left(\nabla L \cdot \nabla \log A + \frac{1}{2}\,\nabla^2 L\right), \tag{1.32}$$

where

$$D(L) = \left(\frac{\partial L}{\partial x}\right)^2 + \left(\frac{\partial L}{\partial y}\right)^2 + \left(\frac{\partial L}{\partial z}\right)^2, \tag{1.33}$$

$$\nabla^2 L = \frac{\partial^2 L}{\partial x^2} + \frac{\partial^2 L}{\partial y^2} + \frac{\partial^2 L}{\partial z^2}, \qquad \nabla L = \frac{\partial L}{\partial x}\,\mathbf{i} + \frac{\partial L}{\partial y}\,\mathbf{j} + \frac{\partial L}{\partial z}\,\mathbf{k} \tag{1.34}$$

and

$$\nabla \log A = \frac{1}{A}\left(\frac{\partial A}{\partial x}\,\mathbf{i} + \frac{\partial A}{\partial y}\,\mathbf{j} + \frac{\partial A}{\partial z}\,\mathbf{k}\right) \tag{1.35}$$

Correspondingly, approximation to the Helmholtz equation is

$$\begin{aligned} \nabla^2 \mathcal{U} + k^2\,\mathcal{U} = -k_0^2\,\mathcal{U}\left[D(L) - \frac{k^2}{k_0^2}\right] \\ + 2\,i\,k_0 \mathcal{U}\left(\operatorname{grad} L \cdot \operatorname{grad}\,\log A + \frac{1}{2}\,\nabla^2 L\right) = 0 \end{aligned} \tag{1.36}$$

Here k is the wavenumber at the point, at which eq. 1.36 is considered.

In deriving eq. 1.36, we ignored the product

$$\nabla^2 A\;e^{i\,k_0\,L}, \tag{1.37}$$

while the retained terms have different orders: one of them is proportional to k_0^2, the other to k_0. Respectively, the Helmholtz equation in our approximation is satisfied if the functions L and A obey two equalities:

$$\left(\frac{\partial L}{\partial x}\right)^2 + \left(\frac{\partial L}{\partial y}\right)^2 + \left(\frac{\partial L}{\partial z}\right)^2 = n^2\,, \tag{1.38}$$

where

$$n = \frac{k}{k_0} = \frac{c_0}{c}\,, \tag{1.39}$$

and

$$\operatorname{grad} L \cdot \operatorname{grad} \log A = -\frac{\nabla^2 L}{2} \tag{1.40}$$

Thus, we arrive at two equations for the unknown functions L and A, that are simpler than the original Helmholtz equation. In fact, they are the first-order partial differential equations and, therefore, can be reduced to the systems of ordinary ones.

Equations 1.38 and 1.40 are called the eikonal and transport equations, respectively. After solving the eikonal equation, the function $L(r)$ becomes known allowing us to calculate ∇L and $\nabla^2 L$. Then the amplitude $A(L)$ is determined from the transport equation. Both the eikonal and transport equations are the theoretical basis of geometrical acoustics. Let us note that: **(a)** These equations imply that the first derivatives of the eikonal and the amplitude are defined at each point p, and **(b)** We assumed that a medium was homogeneous with respect to ρ. However, later it will be demonstrated that, unlike eq. 1.40, the eikonal equation is applied at regular points where density varies. Taking into account this fact, we will make use of eq. 1.38 to study ray behavior even when $\rho = \rho(p)$.

Asymptotic series of potential $\mathcal{U}$

Inasmuch as we study the high-frequency acoustics, $k \to \infty$, it is natural to represent the potential $\mathcal{U}$ as a series with respect to inverse powers of k. Such a series, called the Debye expansion, has the form, (Part I)

$$\mathcal{U}(k,\, r) = e^{i\,k_0\,L(r)} \sum_{m=0}^{\infty} \frac{A_m(r)}{(i\,k_0)^m}\,, \tag{1.41}$$

and comparison with eq. 1.28 shows that this sum represents the expansion of the amplitude A. As follows from eq. 1.41, each term of the series is characterized by the same eikonal $L(r)$.

Prior to deriving the equations for L and A_m, let us preserve the term given by eq. 1.37 and write Helmholtz equation as

$$[\,n^2 - (\nabla L)^2\,]\,A + \frac{i}{k_0}\,(2\,\nabla A \cdot \nabla L + A\,\nabla^2 L) + \frac{\nabla^2 A}{k_0^2} = 0 \tag{1.42}$$

It is obvious that if in place of $A\,e^{i\,k_0\,L}$ we substitute the series

$$\sum_{m=0}^{\infty} \frac{1}{(i\,k_0)^m} A_m\, e^{i\,k_0\,L} ,$$

eq. 1.42 becomes

$$\sum_{m=0}^{\infty} \frac{[\,n^2 - (\nabla L)^2\,]\,A_m}{(i\,k_0)^m} - \sum_{m=0}^{\infty} \frac{2\,\nabla A_m \cdot \nabla L + A_m\,\nabla^2 L}{(i\,k_0)^{m+1}} - \sum_{m=0}^{\infty} \frac{\nabla^2 A_m}{(i\,k_0)^{m+2}} = 0 \qquad (1.43)$$

This equality, representing the Helmholtz equation, takes place if the sums of terms with the same powers of k_0 vanish. For $m = 0$, it gives

$$(\nabla L)^2 = n^2 \qquad (1.44)$$

and

$$2\,\nabla A_0 \cdot \nabla L + A_0\,\nabla^2 L = 0\,, \qquad (1.45)$$

for $m = 1$,

$$2\,\nabla A_1 \cdot \nabla L + A_1\,\nabla^2 L = -\nabla^2 A_0 \qquad (1.46)$$

and, in general,

$$2\,\nabla A_m \cdot \nabla L + A_m\,\nabla^2 L = -\nabla^2 A_{m-1} \qquad (1.47)$$

In principle, this system allows one to obtain the function L and the coefficients A_m. Let us make several comments:

(1) The Debye series, describing the high-frequency spectrum of the potential $\mathcal{U}$, is the asymptotic one for the potential amplitude A, and it begins to diverge as the number of terms increases.

(2) The potential $\mathcal{U}$ can represent different types of waves. It may happen that their high-frequency asymptotics does not contain, for instance, the term $m = 0$. As will be shown later, this occurs in the case of head waves.

(3) Each term is characterized by the same eikonal function, which depends on the type of the wave.

(4) In essence, the Debye expansion is the series with respect to the non-dimensional parameter that characterizes the change of the propagation velocity and the wave amplitude within an interval comparable with the wavelength, (Part I).

From now on to the end of this chapter we restrict ourselves to the geometrical acoustics. Correspondingly, we begin with reviewing the eikonal and ray equations and then, in the following chapter, discuss transport equations and a behavior of a wave amplitude in a medium, which is inhomogeneous with respect to density and velocity of propagation.

Eikonal equation

As was already pointed out, the eikonal equation

$$\left(\frac{\partial L}{\partial x}\right)^2 + \left(\frac{\partial L}{\partial y}\right)^2 + \left(\frac{\partial L}{\partial z}\right)^2 = n^2 \tag{1.48}$$

is the inhomogeneous partial differential equation of the first order and the second degree. Its integration allows us to determine the geometry of the wavefronts. In fact, by definition, eq. 1.28, the surfaces of equal eikonal and phase

$$L = \text{const}\,, \qquad \varphi = k_0 L = \text{const}$$

coincide with each other.

Usually it is necessary to solve eq. 1.48 numerically in order to find the function $L(r)$. However, in several well-known cases the integration gives analytical expressions for the eikonal. For instance, the simplest solution for homogeneous media is

$$L = n\,(\alpha\, x + \beta\, y + \gamma\, z)\,, \tag{1.49}$$

where $n = \text{const}$. Substitution of eq. 1.49 into eq. 1.48 yields

$$\alpha^2 + \beta^2 + \gamma^2 = 1$$

It is clear that the phase surfaces are planes, and the parameters α, β, and γ characterize the direction of wave propagation.

Consider two more examples when $n = \text{const}$. With this purpose in mind, it is convenient to represent the eikonal equation as

$$(\text{grad}\, L)^2 = n^2\,, \tag{1.50}$$

because

$$\text{grad}\, L = \frac{\partial L}{\partial x}\mathbf{i} + \frac{\partial L}{\partial y}\mathbf{j} + \frac{\partial L}{\partial z}\mathbf{k}$$

Bearing in mind that the gradient is invariant with respect to the choice of coordinates, (Part I), eq. 1.50 can be used in any orthogonal system of coordinates. Suppose that the function L depends on the spherical coordinate R only. Then, eq. 1.50 becomes

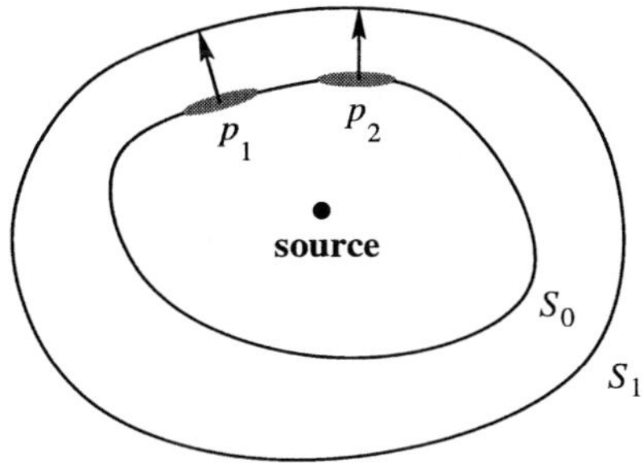

Figure 1.1: Construction of the eikonal surface.

$$\frac{\partial L}{\partial R} = n \qquad \text{or} \qquad L = n\, R\,, \tag{1.51}$$

and we deal with the spherical wave. In the case when L is the function of the cylindrical coordinate r only, we have

$$\frac{\partial L}{\partial r} = n \qquad \text{and} \qquad L = n\, r\,, \tag{1.52}$$

and the wavefronts are the cylinders. Of course, those examples are extremely simple and along with others, which are more complicated, were studied in detail. Some of them are described in Part I.

As was mentioned earlier, in general, solution of the eikonal equation can be found only numerically. Let us visualize the process of constructing the eikonal surfaces in the following way. Suppose that the eikonal

$$L = L_0$$

is known and constant at the surface S_0, surrounding the primary source, Fig. 1.1. Our goal is to find the function L in the volume V outside this surface, where the parameter n is given. We also assume that the wave travels away from S_0 and, correspondingly, its phase, as well as the eikonal L, increases with the distance from the source. Let us divide the surface S_0 into a large number of sufficiently small elements, so that they can be treated as pieces of planes and draw perpendiculars to them. By definition of the gradient, the change of the eikonal along each normal is equal to

$$\Delta L(p) = |\,\text{grad}\, L\,|\, \Delta l(p)$$

or

$$\Delta L(p) = n\, \Delta l(p)\,, \tag{1.53}$$

where Δl is the elementary displacement along the perpendicular.

If the lengths of those perpendiculars are chosen in such a way that the change of the eikonal ΔL remains the same, we obtain

$$L_1 = L_0 + \Delta L = \text{const}$$

The latter defines a new eikonal surface. Repeating this procedure, we can, in principle, determine the field L and construct the eikonal surfaces. Certainly, with a decrease of the intervals $\Delta l(p)$ the accuracy of calculations becomes higher. This is only an illustration, based on Huygens principle, and in one of the next sections we describe a conventional method for calculation of eikonal surfaces.

Now consider a different approach to studying the wave fields using the concept of rays. By definition they are lines that are orthogonal to the eikonal (phase) surfaces, (Part I).

Ray equation

Prior to deriving two forms of the ray equation, let us return to the eikonal surfaces $L = \text{const}$. At each point of those surfaces we have

$$\frac{\partial L}{\partial t} = 0\,, \tag{1.54}$$

where Δt is the elementary displacement tangent to this surface, Fig. 1.2. On the other hand, the derivative $\partial L/\partial l$ along an arbitrary direction $\mathbf{i}_l$ relates to $\operatorname{grad} L$ as, (Part I),

$$\frac{\partial L}{\partial l} = \mathbf{i}_l \cdot \operatorname{grad} L \tag{1.55}$$

Here $\mathbf{i}_l$ is the unit vector tangent to line l. Inasmuch as the function L is constant at the eikonal surface, we have

$$\mathbf{i}_t \cdot \operatorname{grad} L = 0\,, \tag{1.56}$$

where $\mathbf{i}_t$ is the unit vector tangent to this surface. Therefore, we conclude that the vector

$$\mathbf{N} = \operatorname{grad} L \tag{1.57}$$

is perpendicular to the eikonal surface. This is an important result because it shows that $\operatorname{grad} L$ is tangent to the rays.

In order to derive the ray equation, let us express the vector $\mathbf{N}$ in terms of the radius-vector $\mathbf{r}$ and the elementary displacement ds along the ray, Fig. 1.3. As is seen from this figure, we have

$$|\,d\mathbf{r}\,| \to d\,s \qquad \text{if} \qquad d\,r \to 0$$

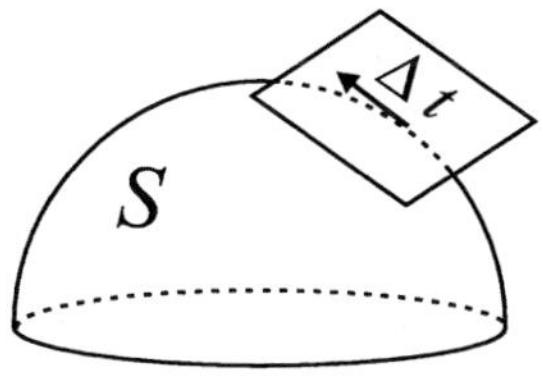

Figure 1.2: Displacement in the plane tangent to the eikonal surface.

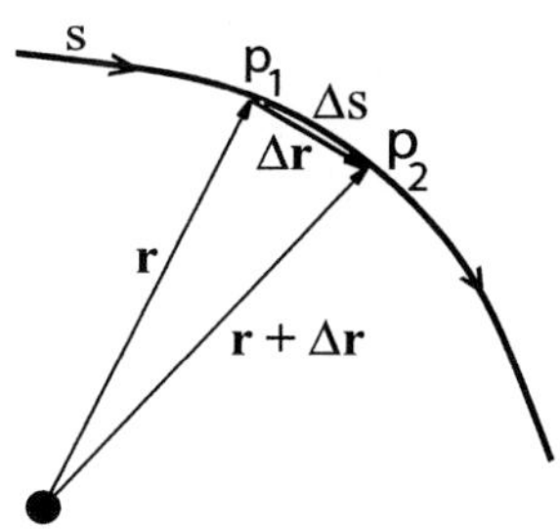

Figure 1.3: Illustration of eq. 1.60.

Hence,

$$\frac{d\,\mathbf{r}}{d\,s} = \mathbf{s}_0\,, \tag{1.58}$$

where $\mathbf{s}_0$ is the unit vector tangent to the ray and

$$d\,\mathbf{r} = \mathbf{r}_2 - \mathbf{r}_1$$

Since the vectors $\mathbf{s}_0$ and $\mathbf{N}$ are parallel to each other, we obtain

$$\frac{d\,\mathbf{r}}{d\,s} = \mathbf{s}_0 = \frac{\operatorname{grad} L}{|\operatorname{grad} L\,|} \tag{1.59}$$

Finally, substitution of eq. 1.50 into the right-hand side of eq. 1.59 gives

$$n\,\mathbf{s}_0 = \operatorname{grad} L \qquad \text{or} \qquad n\,\frac{d\,\mathbf{r}}{d\,s} = \operatorname{grad} L \tag{1.60}$$

Thus, we arrived at the differential equation for rays which allows us to find their position, provided that the function L and the propagation velocity are known. Of course, in such a simple case this task can be easily accomplished without solving eq. 1.60. In fact, the rays can be obtained by drawing the lines normal to the eikonal surfaces. However, the function L is usually unknown and, therefore, it is natural to eliminate L from eq. 1.60 and find a relation between the unit vector $\mathbf{s}_0$, characterizing the ray direction, and the refraction index n, that is, the velocity c, because $n = c_0/c$. To achieve this goal, we recall that, by definition, the magnitude of the gradient L is equal to the maximal derivative of this function, eq. 1.55. As we know, the latter is observed in the direction normal to the eikonal surfaces

$$\frac{dL}{ds} = |\operatorname{grad} L|$$

Here ds is the elementary displacement along the ray.

Then, making use of eq. 1.50, we have

$$\frac{dL}{ds} = n \tag{1.61}$$

This equality illustrates another important feature of the eikonal, namely, its rate of change along the ray is equal to n. In order to derive the ray equation, which does not contain L, it is necessary to form the gradients of both sides of eq. 1.61. This gives

$$\operatorname{grad} \frac{dL}{ds} = \operatorname{grad} n$$

or

$$\frac{d}{ds}(\operatorname{grad} L) = \operatorname{grad} n \tag{1.62}$$

To make use of this equality, we differentiate eq. 1.60 with respect to s

$$\frac{d}{ds}\left(n\frac{d\mathbf{r}}{ds}\right) = \frac{d}{ds}\operatorname{grad} L$$

Therefore,

$$\frac{d}{ds}\left(n\frac{d\mathbf{r}}{ds}\right) = \operatorname{grad} n \tag{1.63}$$

This is the second form of the differential equation of rays, and it demonstrates that they are defined by the velocity of propagation only.

Equation 1.63 is written in the vector form which is equivalent to three equations in any system of coordinates. For example, in the Cartesian system, we have

$$\mathbf{r} = x\,\mathbf{i} + y\,\mathbf{j} + z\,\mathbf{k}$$

and, therefore, eq. 1.63 gives

$$\begin{aligned} \frac{d}{ds}\left(n\,\frac{dx}{ds}\right) &= \frac{\partial n}{\partial x}\,, \\ \frac{d}{ds}\left(n\,\frac{dy}{ds}\right) &= \frac{\partial n}{\partial y}\,, \\ \frac{d}{ds}\left(n\,\frac{dz}{ds}\right) &= \frac{\partial n}{\partial z} \end{aligned} \tag{1.64}$$

Thus, a ray behavior is described by a linear system of inhomogeneous partial differential equations of the second order. The solution of those equations is the set of three functions

$$x = x(s)\,, \qquad y = y(s) \qquad \text{and} \qquad z = z(s) \tag{1.65}$$

that characterize the ray in parametric form, and s is the distance along the ray from some reference point p_0.

In general, we have to solve eq. 1.63 numerically in order to find the ray. At the same time, there are few exceptions, when it is possible to obtain the solution explicitly, Part I. As an illustration of eq. 1.63 consider two models.

Example one: Homogeneous medium Suppose that the medium is homogeneous ($n = \text{const}$) inside some volume V. Then, eq. 1.63 is greatly simplified, and it becomes

$$\frac{d^2\mathbf{r}}{ds^2} = 0 \tag{1.66}$$

The solution of this equation is

$$\mathbf{r} = \mathbf{a}\,s + \mathbf{b} \tag{1.67}$$

Here $\mathbf{a}$ and $\mathbf{b}$ are constant vectors. The latter means that rays in homogeneous media are such that radius-vectors of their points satisfy eq. 1.67. Consider three arbitrary points of a ray with the radius vectors $\mathbf{r}_1$, $\mathbf{r}_2$ and $\mathbf{r}_3$, respectively, Fig. 1.4, and form their differences

$$\Delta\mathbf{r}_1 = \mathbf{r}_2 - \mathbf{r}_1 = \mathbf{a}\,\Delta s_1 \qquad \text{and} \qquad \Delta\mathbf{r}_2 = \mathbf{r}_3 - \mathbf{r}_2 = \mathbf{a}\,\Delta s_2\,,$$

where Δs_1 and Δs_2 are the separations between the points. Because both vectors $\mathbf{r}_1$ and $\mathbf{r}_2$ have a common point and are parallel to each other, we conclude that the rays are the straight lines. Of course, the same conclusion follows from behavior of the eikonal surfaces.

Example two: Vertically inhomogeneous medium Now we assume that the propagation velocity c is a continuous function, which varies only along the z-axis, and that

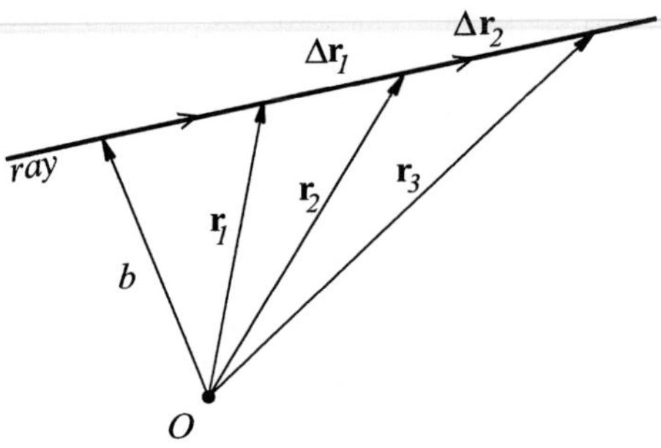

Figure 1.4: Ray in homogeneous medium.

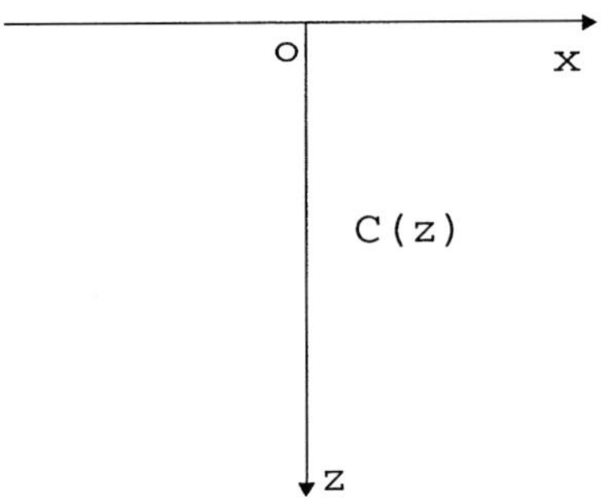

Figure 1.5: Velocity model and the source location.

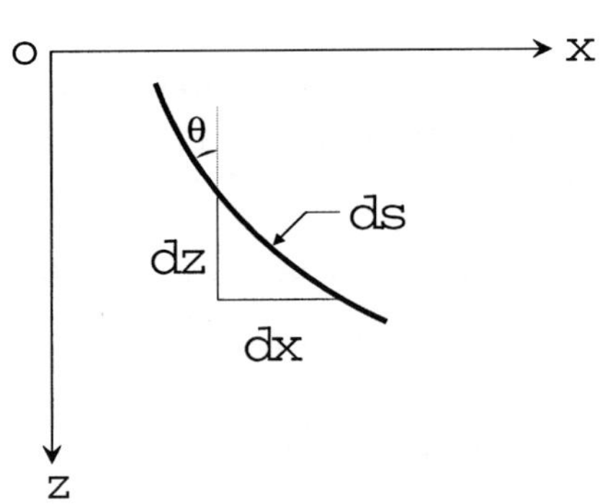

Figure 1.6: Illustration of eqs. 1.69.

the source of the wave is situated at the coordinate origin, Fig. 1.5. Since such a medium has axial symmetry with respect to the vertical axis, it is sufficient to discuss ray behavior in any plane containing the z-axis, (Part I). For example, suppose that the rays are located in the plane xOz, Fig. 1.6. Bearing in mind that

$$\frac{\partial n}{\partial x} = 0 \qquad \text{and} \qquad \frac{\partial n}{\partial y} = 0\,,$$

eq. 1.64 gives

$$\frac{d}{ds}\left(n\,\frac{dx}{ds}\right) = 0 \qquad \text{and} \qquad \frac{d}{ds}\left(n\,\frac{dz}{ds}\right) = \frac{\partial n}{\partial z}$$

or

$$n\,\frac{dx}{ds} = C \qquad \text{and} \qquad \frac{d}{ds}\left(n\,\frac{dz}{ds}\right) = \frac{\partial n}{\partial z}\,, \tag{1.68}$$

where C is some constant characterizing the ray. As is seen from Fig. 1.6,

$$\frac{dx}{ds} = \sin\theta \qquad \text{and} \qquad \frac{dz}{ds} = \cos\theta \tag{1.69}$$

Here θ is the angle between the z-axis and the ray direction. Thus, the first equation of the set 1.68 becomes

$$\sin\theta = \frac{C}{n} = C\,\frac{c(z)}{c_0} = p\,c(z)$$

or

$$\frac{\sin\theta}{c(z)} = p\,, \tag{1.70}$$

where p is the ray parameter, often called the slowness. This equation describes Snell's law in a medium with continuous velocity change along the z-axis. It shows that the ratio of the sine of the incidence angle θ to the velocity $c(z)$ remains constant at each point of the ray, but it varies from one ray path to another.

Suppose that in the vicinity of the source the propagation velocity is c_0 and the incidence angle is θ_0. Then Snell's law, eq. 1.70, can be written as

$$\frac{\sin\theta}{c} = \frac{\sin\theta_0}{c_0} \tag{1.71}$$

The slowness of the ray is

$$p = \frac{\sin\theta_0}{c_0} > 0\,,$$

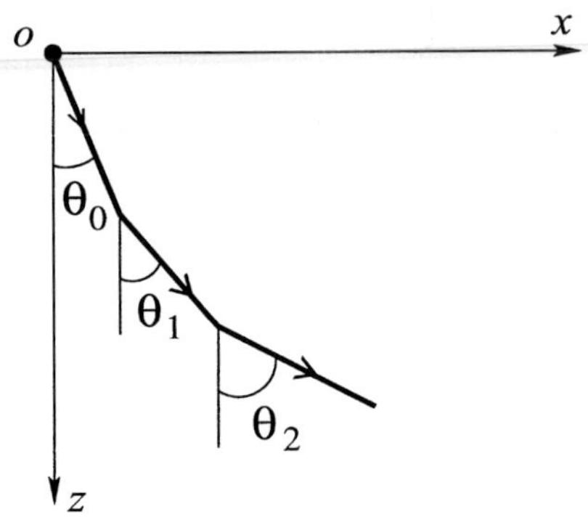

Figure 1.7: Construction of the ray path.

and it changes within the range

$$0 \leq p \leq \frac{1}{c_0}$$

Its dimension is

$$[p] = \mathsf{s\,m}^{-1}$$

The relationship 1.70 represents the ray equation. If the parameter p and the function $c(z)$ are known, the ray can be found numerically, and this procedure is called the ray tracing. For example, it can be visualized in the following way. Let us choose the ray with the incidence angle θ_0 at the source. By definition, the first element of the ray makes the angle θ_0 with the z-axis. At the end of this element we draw the next one. Its angle θ_1 is defined by eq. 1.71:

$$\sin\theta_1 = \frac{c(z)}{c_0}\sin\theta_0\,,$$

where z is the vertical coordinate of the end of the first segment. In the same manner, we plot other ray elements, Fig. 1.7. It is clear that the accuracy of this procedure increases as the length of the segments decreases. Later we will outline some conventional methods for solving this problem.

Now it is natural to express the second equation of the set 1.68

$$\frac{d}{ds}\left(n\frac{dz}{ds}\right) = \frac{\partial n}{\partial z}$$

in terms of θ and $c(z)$. Differentiation at its left-hand side yields

$$\frac{d}{ds}\left(n\frac{dz}{ds}\right) = \frac{d}{ds}(n\cos\theta) = -n\sin\theta\frac{\partial\theta}{\partial s} + \cos\theta\frac{\partial n}{\partial s} = \frac{\partial n}{\partial z}$$

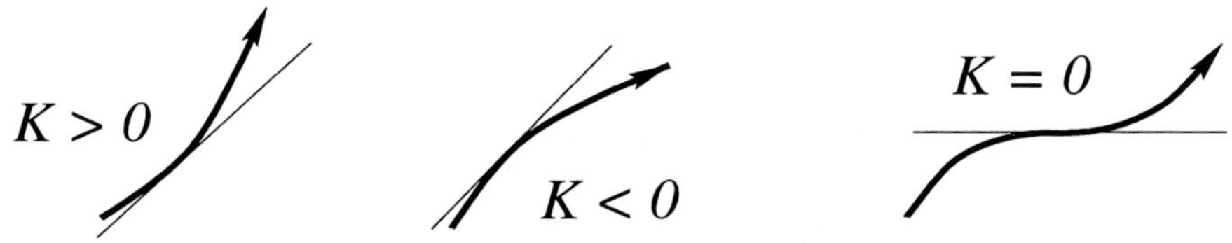

Figure 1.8: Ray curvature.

or

$$\frac{\partial n}{\partial z} = -n \sin\theta \frac{\partial \theta}{\partial s} + \cos\theta \frac{\partial n}{\partial z} \frac{\partial z}{\partial s} = -n \sin\theta \frac{\partial \theta}{\partial s} + \cos^2\theta \frac{\partial n}{\partial z}$$

or

$$\frac{\partial \theta}{\partial s} = -\frac{\sin\theta}{n} \frac{\partial n}{\partial z} \tag{1.72}$$

Making use of the equality

$$\frac{\partial n}{\partial z} = \frac{\partial}{\partial z}\left(\frac{c_0}{c}\right) = -\frac{c_0}{c^2} \frac{\partial c}{\partial z},$$

in place of eq. 1.72 we obtain

$$\frac{\partial \theta}{\partial s} = \frac{\sin\theta}{c} \frac{\partial c}{\partial z} \qquad \text{or} \qquad \frac{\partial \theta}{\partial s} = p \frac{\partial c}{\partial z} \tag{1.73}$$

This form of the ray equation is also useful for understanding of ray behavior.

As mentioned previously, the incidence angle θ between the z-axis and the ray is measured counter-clockwise. By definition, (Part I), the ray curvature K is

$$K = \frac{\partial \theta}{\partial s}, \tag{1.74}$$

and it can be positive, negative, or equal to zero, Fig. 1.8. Then, eq. 1.73 is written as

$$K = p \frac{\partial c}{\partial z} \tag{1.75}$$

and it shows that the ray curvature is directly proportional to the slowness and rate of a change of the velocity. It may be proper to notice that eq. 1.75 also follows from a more general relationship between the ray curvature and the velocity gradient, (Part I).

Thus, we arrived at two forms of the ray equation which relate θ and $c(z)$:

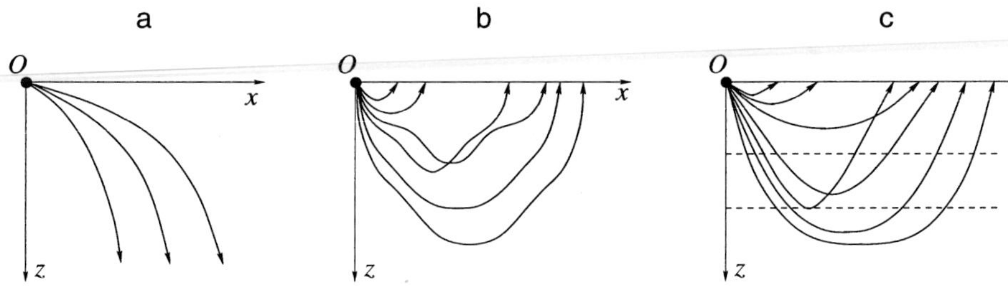

Figure 1.9: Ray behavior in media with different velocity-depth laws.

$$\frac{\sin\theta}{c} = p \qquad \text{and} \qquad \frac{\partial\,\theta}{\partial\,s} = p\,\frac{\partial\,c}{\partial\,z} \tag{1.76}$$

Examples given in Fig. 1.9 illustrate possible ray behavior in media where the propagation velocity is a continuous function of the coordinate z, (Part I). Certainly, the ray geometry also essentially depends on the source position. In general, rays are curved in inhomogeneous media, and under certain conditions they have turning points, Fig. 1.9b, c. They may intersect each other and form "shadow" zones. All those features were described in some detail in Part I. Bearing in mind that we consider the high-frequency acoustics, it is natural to think that all properties of rays result from diffraction of elementary waves.

Having considered rays in media with continuously changing velocity, we now turn to ray behavior near interfaces in piece-wise homogeneous media.

Snell's law for reflected and transmitted rays

Suppose that an incident wave propagates through a medium and its upper and lower parts have the velocities c_1 and c_2, respectively, Fig. 1.10. The boundary between them can be almost an arbitrary surface, provided that its normal $\mathbf{n}$ is defined everywhere. In accordance with the main assumption of the high-frequency acoustics, the complex amplitude of the incident wave is represented as

$$\mathcal{U}_i(\mathbf{r}) = A_i(\mathbf{r})\, e^{i\,k_1\,L(\mathbf{r})}\,, \tag{1.77}$$

where $\mathbf{r}$ is the radius vector characterizing the point position and $L(\mathbf{r})$ is the eikonal defining the wave phase. Inasmuch as both portions of the medium are homogeneous, $n = 1$, it follows from eq. 1.48 that

$$|\,\text{grad}\,L\,| = 1$$

By definition, $\text{grad}\,L$ is normal to the phase surface.

It is natural to think that secondary sources appear at the interface due to the incident wave, and they cause both reflected and transmitted waves, as well as the others. This

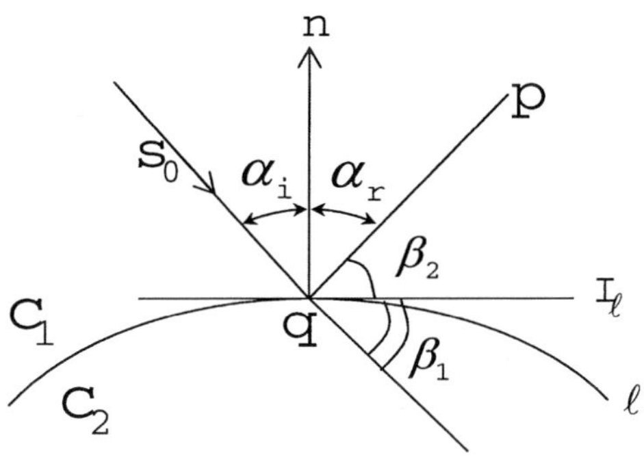

Figure 1.10: Illustration of Snell's law.

means that the total wave field is the result of interference (diffraction) of elementary waves produced by these sources. Following the Kirchhoff approximation, (Part I), let us assume that the phase of the source at some point q on the boundary coincides with that of the incident wave. As an example, we discuss a two-dimensional case, in which the wave field and the medium parameters do not vary in the y direction, Fig. 1.10. Correspondingly, the rays can be studied in any plane perpendicular to this axis. Then, as was shown in Part I, the potential of each elementary source is expressed through the Hankel function $H_0^{(1)}(kr)$. At the same time, if observations are performed at distances exceeding the wavelength, this function behaves as

$$H_0^{(1)}(kr) = \sqrt{\frac{2}{\pi\, k\, r}}\, e^{-i\,\pi/4}\, e^{i\,k\,r} \tag{1.78}$$

First, consider the reflected wave. Applying the principle of superposition, (Part I), the potential of this wave at some point p is written as

$$\mathcal{U}_r(p) = \int_{-\infty}^{\infty} F_r\, e^{i\,k_1\,[\,r(q,\,p) + L(q)\,]}\, dl \tag{1.79}$$

Here q is the point of the boundary l and dl is elementary displacement along this boundary. The function F_r is proportional to the amplitude of the incident wave and depends on the distance between the points q and p. The presence of the term $e^{i\,k_1\,L(q)}$ shows that the phase of the secondary source near point q and that of the incident wave are the same.

In order to determine the locations of such sources which give the main contribution to the reflected wave at some point p, we make use of the stationary phase method, (Part I). As is well known, the stationary point, that characterizes the position of such sources, is defined by the equality

$$\frac{d}{d\,l}\,[\,r(q,p) + L(q)\,] = 0 \tag{1.80}$$

The latter can be rewritten in the following form, (eq. 1.55):

$$\mathbf{i}_l \cdot \operatorname{grad} r + \mathbf{i}_l \cdot \operatorname{grad} L = 0\,, \tag{1.81}$$

where $\mathbf{i}_l$ is the unit vector tangent to the line l. Bearing in mind that

$$\operatorname{grad} r = \frac{\mathbf{r}}{r} = -\,\mathbf{r}_0 \qquad \text{and} \qquad \operatorname{grad} L = \mathbf{s}_0$$

are the unit vectors along rays, in place of eq. 1.81 we have

$$-\mathbf{i}_l \cdot \mathbf{r}_0 + \mathbf{i}_l \cdot \mathbf{s}_0 = 0 \tag{1.82}$$

As is seen from Fig. 1.10, this equality means that

$$\cos\beta_2 = \cos\beta_1 \qquad \text{or} \qquad \sin\alpha_r = \sin\alpha_i\,,$$

that is,

$$\alpha_r = \alpha_i\,, \tag{1.83}$$

where α_r is the reflection angle formed by normals to the boundary and the phase surface of the reflected wave. Thus, we arrive at Snell's law of reflection.

Next, suppose that the observation point is located in the lower medium with the velocity c_2. By analogy, the potential of the transmitted wave can be represented in the form

$$\mathcal{U}_t(p) = \int\limits_{-\infty}^{\infty} F_t\, e^{i\,(k_2\,r + k_1\,L\,)}\, dl \tag{1.84}$$

Respectively, the stationary point is defined by equation

$$\frac{\partial}{\partial\,l}\,(k_2\,r + k_1 L\,) = 0 \qquad \text{or} \qquad -\,k_2\,\mathbf{i}_l \cdot \mathbf{r}_0 + k_1\,\mathbf{i}_l \cdot \mathbf{s}_0 = 0$$

or

$$\frac{\cos\beta_1}{c_1} = \frac{\cos\beta_2}{c_2}$$

Since

$$\cos(\mathbf{s}_0,\,\mathbf{i}_l) = \sin\alpha_i \qquad \text{and} \qquad \cos(\mathbf{r}_0,\,\mathbf{i}_l) = \sin\beta\,,$$

we obtain Snell's law of transmission

$$\frac{\sin\alpha_i}{c_1} = \frac{\sin\beta}{c_2}\,, \tag{1.85}$$

where α_i and β are the incident and transmission angles.

Later in the same manner this law will be studied in the three-dimensional case. We have investigated behavior of sinusoidal waves at high-frequency spectrum. Next, consider transient waves and with this purpose in mind let us discuss the main concept of time fields.

Time fields

As is well known, the arrival time of a wave at the observation point is one of the main parameters measured in seismology. Assuming that the wavelength is sufficiently small, we again introduce the concept of time fields, (Part I), since both the eikonal and time fields represent two equivalent approaches allowing one to describe kinematics of waves when $\lambda \rightarrow 0$. Suppose that a non-stationary wave propagates with the velocity c through the medium and we want to describe the position of its front. Any other phase surface of the wave can be considered in the same manner. Inasmuch as the wavefront appears at a given point of the medium at some instant t, it is natural to treat the time t as a function of spatial coordinates. For example, in the Cartesian coordinate system we have

$$t = t(x,\, y,\, z) \tag{1.86}$$

As well known, the latter represents the scalar field that characterizes arrival time of the wavefront or any other phase surface at the observation point $p(x,\, y,\, z)$. To study behavior of this scalar field, that is its change in different directions, we use the gradient

$$M = \operatorname{grad} t\,, \tag{1.87}$$

which describes the magnitude and the direction of the maximum rate of increase of the field. In a general curvilinear orthogonal system of coordinates we have

$$\operatorname{grad} t = \frac{1}{h_1}\frac{\partial t}{\partial x_1}\mathbf{i}_1 + \frac{1}{h_2}\frac{\partial t}{\partial x_2}\mathbf{i}_2 + \frac{1}{h_3}\frac{\partial t}{\partial x_3}\mathbf{i}_3\,, \tag{1.88}$$

where $\mathbf{i}_1$, $\mathbf{i}_2$ and $\mathbf{i}_3$ are the unit vectors along the coordinate axes and h_1, h_2 and h_3 are the corresponding metric coefficients, (Part I).

Certainly, a change of the coordinate system results in a change of the components of the gradient, however, its magnitude and direction remain the same since they depend on the field only.

The derivative of the time field in the direction l relates to the gradient as

$$\frac{\partial t}{\partial l} = \mathbf{i}_0 \cdot \operatorname{grad} t = \operatorname{grad}_l t$$

or

$$\frac{\partial t}{\partial l} = |\,\mathrm{grad}\, t\,| \cos(\nabla t,\, \mathbf{i}_0)\,, \tag{1.89}$$

where $\mathbf{i}_0$ is the unit vector along the line l. Two obvious features follow from eq. 1.89, namely, the derivative $\partial t/\partial l$ is maximal along the gradient because

$$\cos(\nabla t,\, \mathbf{i}_0) = 1\,,$$

while in the perpendicular direction the field does not change. In order to visualize the time field it is convenient to introduce the surfaces, where $t = \mathrm{const}$, which are called the isochrones and it is a conventional way to study scalar fields. The physical meaning of such surfaces is clear. They show the positions of phase surfaces at different times. Thus, knowing the location of isochrones for a sequence of times $t_1,\, t_2,\, \ldots,\, t_n$, we can observe behavior of the wavefronts. In general, the wavefronts may have a complicated shape, and even intersect. Consider two isochrones with close values of times t_1 and t_2, Fig. 1.11, and suppose that the time t between them varies linearly. Then, the time change

$$\Delta t = t_2 - t_1$$

along the normal $\mathbf{s}_0$ and any direction $\mathbf{i}_0$ can be written as

$$\Delta t = \frac{\partial t}{\partial s}\,\Delta s \qquad \text{and} \qquad \Delta t = \frac{\partial t}{\partial l}\,\Delta l\,,$$

which gives

$$\frac{\partial t}{\partial s}\,\Delta s = \frac{\partial t}{\partial l}\,\Delta l$$

Bearing in mind that

$$\Delta s = \Delta l\, \cos\alpha\,,$$

we obtain

$$\frac{\partial t}{\partial l} = \frac{\partial t}{\partial s}\, \cos\alpha\,, \tag{1.90}$$

where α is the angle between the normal $\mathbf{s}_0$ and the direction of the displacement Δl.

This equality clearly shows that the maximal rate of the time change occurs along the direction normal to the isochrone, and, therefore, according to the definition of gradient

$$\mathrm{grad}\, t = |\,\mathrm{grad}\, t\,|\, \mathbf{s}_0 \tag{1.91}$$

Correspondingly, lines normal to the isochrones characterize the directions along which

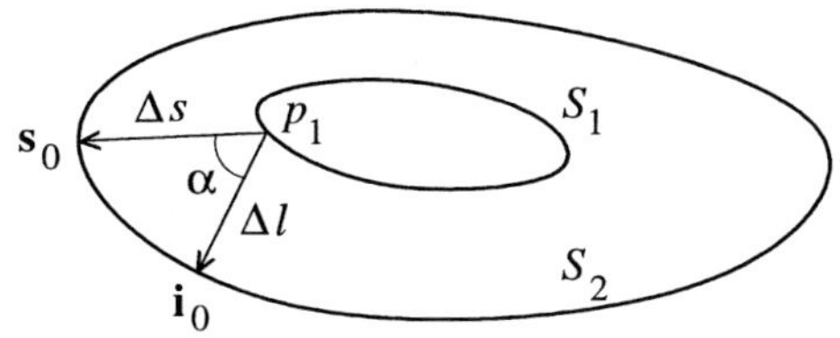

Figure 1.11: Isochrones.

time increases more rapidly than along any other paths. Taking into account that isochrones have the same shape as phase surfaces, we can say that these lines coincide with the rays along which the acoustic energy propagates through the medium. Therefore, the vector

$$\mathbf{M} = \operatorname{grad} t$$

is tangential to them. It is obvious that if the family of rays and time of propagation along them is known, the time field is defined. In fact, drawing the surfaces that are normal to the rays, we obtain the set of isochrones, and their parameter t is determined from the intersections with the rays. Thus, both isochrones and rays describe the time field.

Next, taking into account the equivalence between the phase surface and the isochrone, we derive an equation for the time field. Suppose that the wavefront coincides with the isochrone S_1 at the instant t_1. To determine its position at the moment

$$t_2 = t_1 + \Delta t\,,$$

we apply the Huygens principle. Then, each point of the surface S_1 is considered as the center of the elementary spherical surface with the radius Δs being equal to

$$\Delta s = c\,\Delta t \tag{1.92}$$

Correspondingly, the envelope of these surfaces represents the isochrone S_2, where $t = t_2$. By definition, the gradient t is directed along the normal $\mathbf{s}_0$ and, therefore,

$$\operatorname{grad} t = \frac{\partial t}{\partial s}\,\mathbf{s}_0 = \frac{\mathbf{s}_0}{c} \qquad \text{or} \qquad |\operatorname{grad} t\,| = \frac{1}{c} \tag{1.93}$$

since

$$c(p) = \frac{\partial s}{\partial t}$$

in the vicinity of each point p of the isochrone. Thus, in a curvilinear orthogonal coordinate system, eq. 1.93 yields

$$\sqrt{\frac{1}{h_1^2}\left(\frac{\partial t}{\partial x_1}\right)^2 + \frac{1}{h_2^2}\left(\frac{\partial t}{\partial x_2}\right)^2 + \frac{1}{h_3^2}\left(\frac{\partial t}{\partial x_3}\right)^2} = \frac{1}{c(p)}$$

or

$$\frac{1}{h_1^2}\left(\frac{\partial t}{\partial x_1}\right)^2 + \frac{1}{h_2^2}\left(\frac{\partial t}{\partial x_2}\right)^2 + \frac{1}{h_3^2}\left(\frac{\partial t}{\partial x_3}\right)^2 = \frac{1}{c^2(p)} \tag{1.94}$$

The latter is called the equation of the time field and, in essence, coincides with the eikonal equation if t and c^2 are replaced with L and c^2/c_0^2, respectively. In particular, in Cartesian coordinates eq. 1.94 becomes

$$\left(\frac{\partial t}{\partial x}\right)^2 + \left(\frac{\partial t}{\partial y}\right)^2 + \left(\frac{\partial t}{\partial z}\right)^2 = \frac{1}{c^2(p)} \tag{1.95}$$

This equation, as well as the eikonal one, has an infinite number of solutions, and it is necessary to know the boundary and initial conditions in order to find the time field. At the same time, the solution of the forward problem for the eikonal, characterizing the sinusoidal waves, does not require the initial conditions. Having described some general features of time fields, we next consider the simplest special case.

Isochrones of direct wave

Suppose that the wave propagates with the velocity c from the source, Fig. 1.12. The origin O of the Cartesian coordinate system coincides with the position of the source and the z-axis is directed downward. Taking into account that time t is a function of the distance R from the source only, eq. 1.94 is greatly simplified in the spherical coordinate system:

$$\frac{\partial t}{\partial R} = \frac{1}{c}$$

Then

$$t = \frac{R}{c} + C$$

Assuming that the source starts generating the wave at the instant $t = 0$, we can set $C = 0$ and then

$$t = \frac{R}{c} = \frac{1}{c}\sqrt{x^2 + y^2 + z^2} \tag{1.96}$$

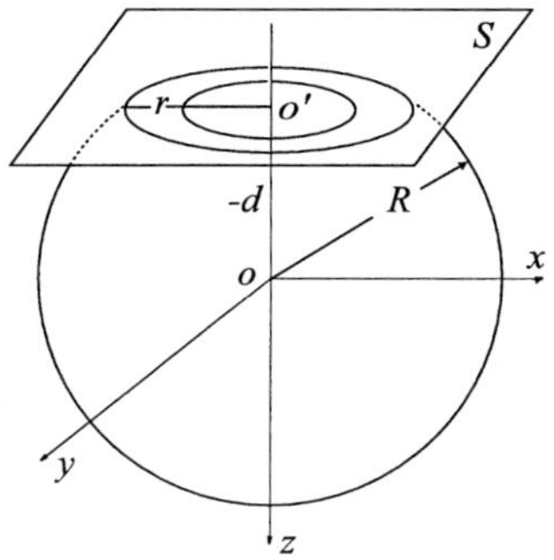

Figure 1.12: Isochrones in the plane S.

Thus, the isochrones of the direct wave are concentric spheres with the radii $R = ct$ and their centers at the source. Correspondingly, the rays are straight lines leaving the source in the radial directions. Certainly, this is not a novel result. Suppose the observations are performed in the horizontal plane S located at $z = -d$, Fig. 1.12. Then, substituting this value of z in eq. 1.96, we obtain the equation for the time field in the plane

$$t = \frac{\sqrt{x^2 + y^2 + d^2}}{c} = \frac{\sqrt{r^2 + d^2}}{c}, \tag{1.97}$$

where r is the distance from point O' in Fig. 1.12.

As we see, the map of isochrones in the plane consists of concentric circles and, therefore, the function $t(x,\, y,\, -d)$ is described by the hyperboloid with the apex at the t-axis, Fig. 1.13a. Consider behavior of the time along a line, for instance, $y = b$ on the plane S. From eq. 1.97 we have

$$t = \frac{\sqrt{x^2 + b^2 + d^2}}{c} = \frac{\sqrt{x^2 + d_1^2}}{c}, \tag{1.98}$$

where d_1 is the distance from the source to the line. The function 1.98 is a hyperbola symmetric with respect to the coordinate x, Fig. 1.13b.

To determine the apparent velocity of the wave on plane S, Part I, we take the derivative from both sides of eq. 1.98 with respect to x. This gives

$$\frac{\partial t}{\partial x} = \frac{1}{c} \frac{x}{\sqrt{x^2 + d_1^2}} \qquad \text{or} \qquad c_a = c\sqrt{1 + \frac{d_1^2}{x^2}} \tag{1.99}$$

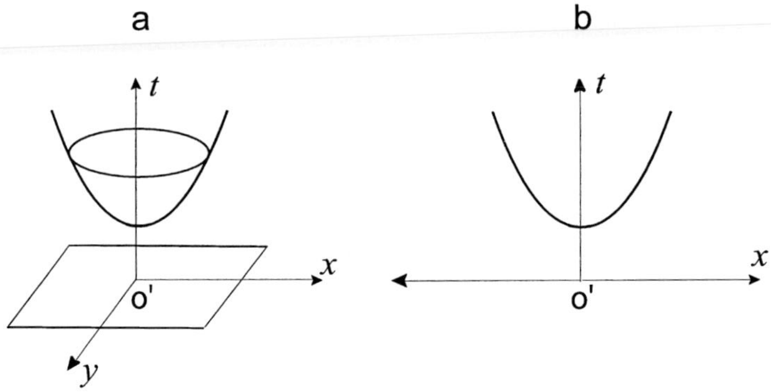

Figure 1.13: (a) Function $t(x,\, y,\, -d)$ (b) Illustration of eq. 1.98.

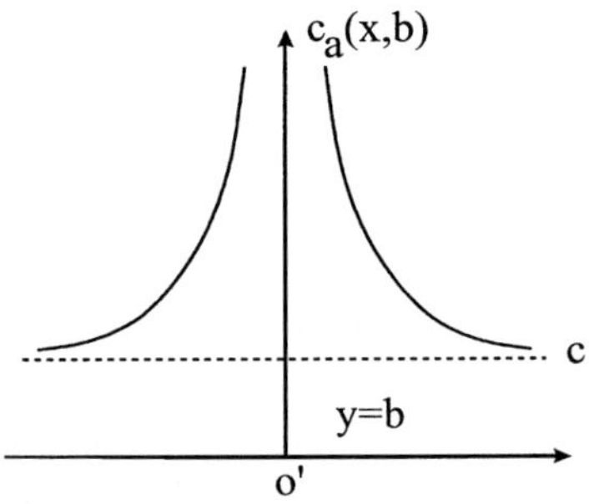

Figure 1.14: Behavior of apparent velocity, eq. 1.99.

Thus, as distance x increases, the apparent velocity gradually changes from infinity at point $x = 0$, $y = b$ to the medium velocity c, Fig. 1.14. Let us remind that, by definition, the function $c_a(x)$ characterizes the velocity of propagation of a wave along the x-axis. If the source is located on the plane $d = 0$, then eq. 1.97 becomes

$$t = \frac{\sqrt{x^2 + y^2}}{c} = \frac{r}{c},$$

and the isochrone is described by a conical surface, Fig. 1.15a. In this case, the dependence of time along the line, that is, the moveout or the traveltime curve, is

$$t = \pm \frac{x}{c} \qquad \text{if} \qquad y = 0$$

and consists of two straight lines, Fig. 1.15b.

As an illustration of time fields it may be useful to study their behavior in layered media. This will be a subject of two next sections. Taking into account the fact that

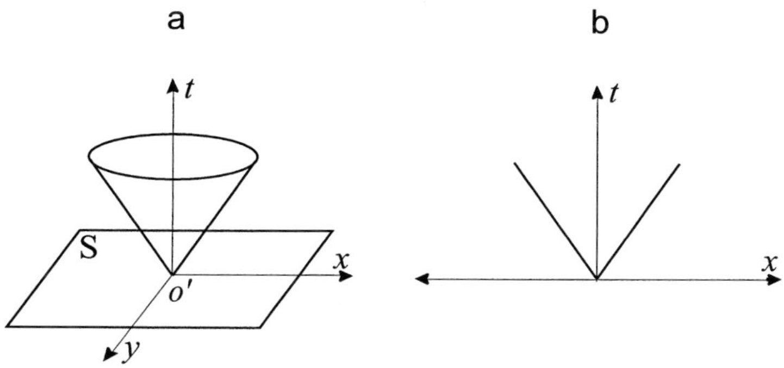

Figure 1.15: Dependence on time, $d = 0$.

those models are of a considerable practical importance in exploration seismology, special attention will be paid to the relationships between moveouts and the medium parameters. Sometimes we will derive formulas which allow one to estimate those parameters, but, these should not be treated as methods for solving kinematic inverse problems.

1.2 Reflected waves in media with plane interfaces

Case one: Horizontal reflector At the beginning let us assume that a source of the wave is situated in the vicinity of Earth's surface, S_0, and beneath there is a single horizontal interface, S, Fig. 1.16a. The velocity of propagation in the upper layer is equal to c, and its thickness is H. In order to describe the time field, we choose the Cartesian system of coordinates with the origin at point o, where the source is located and the z-axis is directed downward. As we know, when the wave, caused by the source, reaches the boundary, $z = H$, it is reflected simultaneously at all points of the interface located at the same distance r from the z-axis, Fig. 1.16b. To determine the time field of this wave, that is, the time of arrival of its front, we mentally place a fictitious source at the point o^*, which is a mirror reflection of the point o with respect to the plane $z = H$. This approach was proved in deriving Snell's law of reflection (Part I). Therefore, the arrival time of the reflected wave at points of Earth's surface S_0 is

$$t = \frac{R}{c} \qquad \text{or} \qquad t = \frac{\sqrt{r^2 + 4\,H^2}}{c}$$

or

$$t = \frac{\sqrt{x^2 + y^2 + 4\,H^2}}{c} \tag{1.100}$$

Before we investigate the behavior of this function, it is appropriate to notice that the

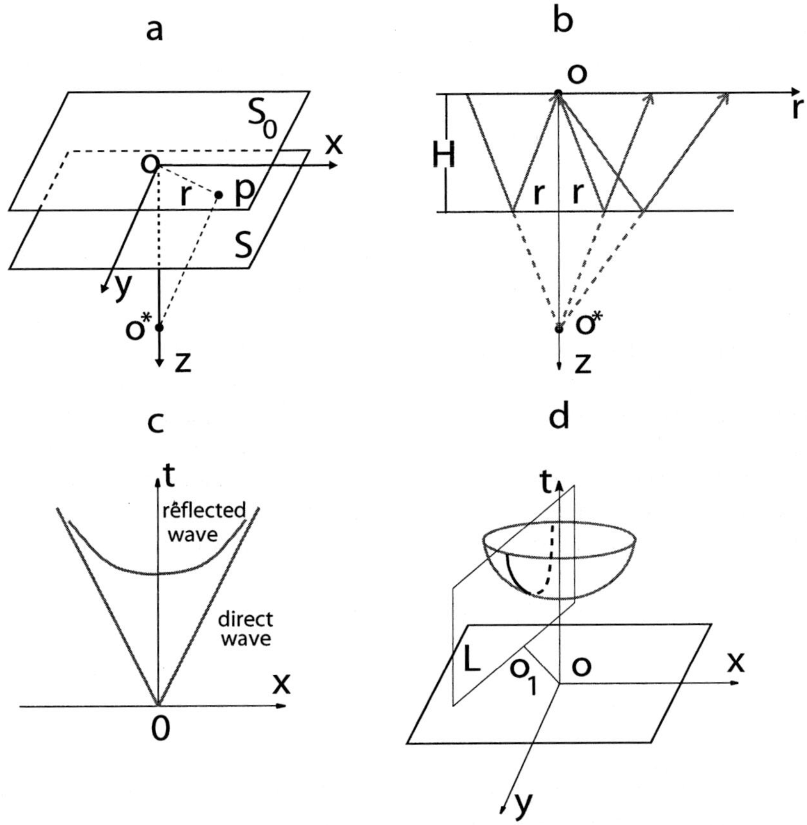

Figure 1.16: (a) Model geometry. (b) Reflected rays. (c) Traveltime curves of direct and reflected waves. (d) Function $t(x,\ y)$.

wave, approaching the boundary: $z = 0$, also experiences the reflection. It gives rise to the wave moving toward the interface: $z = H$, that causes the second and higher order multiple waves. But in this case, as well as the next one, we pay attention to the reflected wave of the first order.

In accordance with eq. 1.100 the function

$$t = t(x, y)$$

is described by the hyperboloid of rotation which is symmetrical with respect to the t-axis. Respectively, the time as a function of the distance r from the source is characterized by hyperbola. In particular, along the x-axis we have

$$t = \frac{1}{c}\sqrt{x^2 + 4H^2} \tag{1.101}$$

This dependence is shown in Fig. 1.16c together with that of the direct wave. Note that in accordance with Snell's law of reflection all paths, formed by the reflected rays, that intersect the x-line, are situated on the vertical plane, $y = 0$. The minimum of the function: $t = t(x)$, eq. 1.101, observed at the origin o, is understandable since in this case a length of the ray path is the smallest one. With the increase of the x-coordinate the time also increases because the corresponding ray path becomes longer. At the beginning, however, when $|x| < 2H$, this increase is relatively small. In contrast, at sufficiently large distances from the source, ($x \gg H$), lengths of the direct and reflected ray paths almost coincide, and time t is almost directly proportional to x.

As follows from eq. 1.100 along the line parallel to the x-axis, ($y = d$), we have

$$t = \frac{1}{c}\sqrt{x^2 + d_1^2}\,, \tag{1.102}$$

where

$$d_1^2 = 4H^2 + d^2$$

Again, the function $t = t(x, d)$ is described by the hyperbola, which is symmetrical with respect to the x-axis and has a minimum at the point $x = 0$ and $y = d$.

In general, for a time field along some straight line, L,

$$y = ax + b, \tag{1.103}$$

the function $t = t(x, y)$ is still characterized by a hyperbola. This is demonstrated by drawing a vertical plane containing the line L, since its intersection with the hyperboloid, $t = t(x, y)$, gives rise to a hyperbola, Fig. 1.16d. The minimum of this curve takes place at point o_1, where a perpendicular drawn from the source crosses the line L, and the total length of the incident and reflected rays, arriving at the line is the smallest one. As is seen from Fig. 1.17a all rays which intersect L are located in the same plane formed

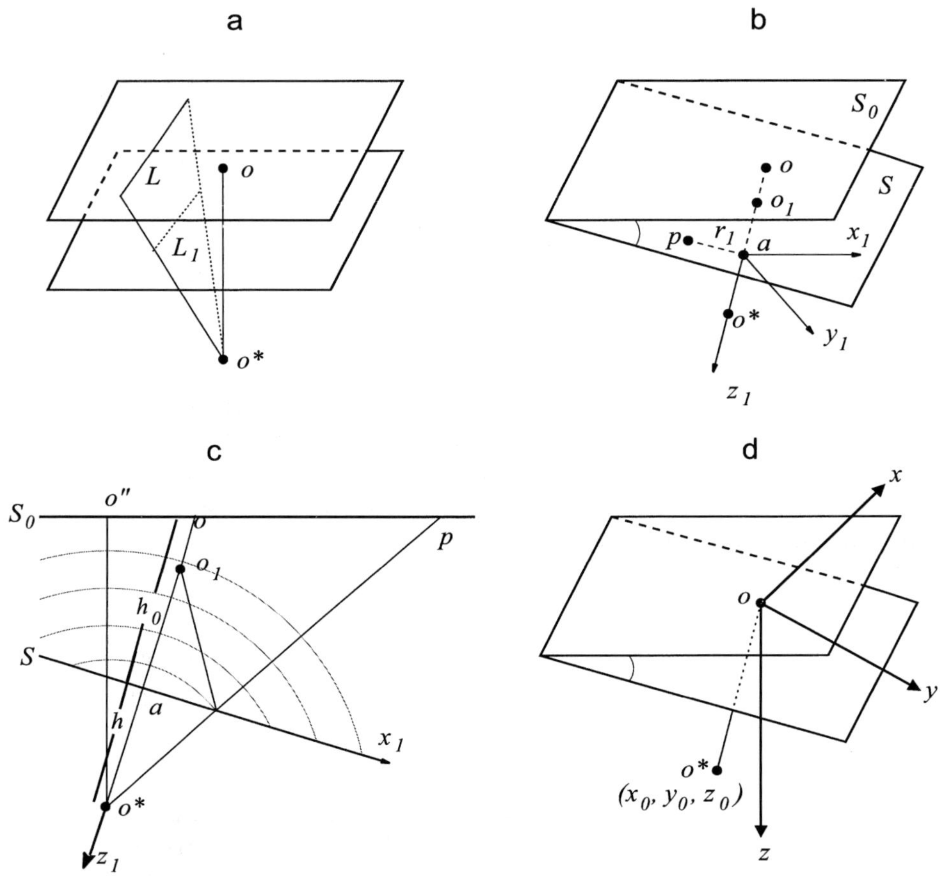

Figure 1.17: (a) Location of rays intersecting line L. (b, d) Model with a dipping reflector. (c) Cross-section of isochrones.

by this line and the imaginary source, o^*. The intersection of this plane with the surface S is also the straight line, L_1, usually called the reflection line.

Now we show how to estimate the medium parameters in this simplest case, assuming that the arrival time is known along some line, for example, the x-axis. Consider two points located at distances: x_1 and x_2 from the source. In accordance with eq. 1.101 we have

$$t_1 = \frac{1}{c}\sqrt{x_1^2 + 4\,H^2} \qquad \text{and} \qquad t_2 = \frac{1}{c}\sqrt{x_2^2 + 4\,H^2} \tag{1.104}$$

Whence

$$\frac{t_2^2}{t_1^2} = \frac{x_2^2 + 4\,H^2}{x_1^2 + 4\,H^2} \qquad \text{and} \qquad H = \frac{1}{2}\sqrt{\frac{x_2^2\,t_1^2 - x_1^2\,t_2^2}{t_2^2 - t_1^2}} \tag{1.105}$$

If one of the observations is performed in the vicinity of the source,

$$x_1 = 0, \qquad t_1 = t_0$$

we obtain

$$H = \frac{1}{2}\,\frac{x_2\,t_0}{\sqrt{t_2^2 - t_0^2}} \tag{1.106}$$

On the other hand, as it follows from eq. 1.101

$$H = \frac{c\,t_0}{2}\,, \tag{1.107}$$

where t_0 is the time of propagation of the wave along the z-axis from the source to the interface and back. Then, making use of eq. 1.106 we arrive at the expression for the velocity of the wave in the upper layer

$$c = \frac{2\,H}{t_0} = \frac{x_2}{\sqrt{t_2^2 - t_0^2}} \tag{1.108}$$

This procedure can be repeated for receivers situated at different distances from the source, thus increasing the accuracy of determination of H and c.

Let us compare the time of arrival t at the point located at the distance x with that near the source. From eqs. 1.101 and 1.107 we have

$$\Delta t = t - t_0 = \frac{1}{c}\left(\sqrt{x^2 + 4\,H^2} - 2\,H\right) \tag{1.109}$$

and it is used for a transformation called the normal moveout. The latter, Δt, can be also represented in a different form. In fact we have

$$t = \frac{\sqrt{x^2 + c^2\, t_0^2}}{c} \qquad \text{or} \qquad t^2 = \frac{x^2 + c^2\, t_0^2}{c^2}$$

Thus

$$t^2 - t_0^2 = \frac{x^2}{c^2} \qquad \text{or} \qquad \Delta t = \frac{x^2}{c^2\,(t + t_0)} \tag{1.110}$$

Case two: Dipping reflector Consider a more general case in which surfaces S_0 and S are not parallel to each other and the source of the incident wave is situated beneath Earth's surface at point o_1, Fig. 1.17b. To determine the time field of the wave reflected from S, we draw from the point o_1 the line perpendicular to this plane and find the mirror reflection, o^*. At the beginning it is convenient to choose the Cartesian system of coordinates, x_1, y_1, z_1 with the origin at point a where the line oo^*, (z_1-axis) crosses the plane S while x_1 and y_1-axes are arbitrarily oriented in this plane. Since the direct wave arrives simultaneously at points of the boundary S, which are located at the same distance, r_1, from the origin a, the time field of the reflected wave is the axially symmetrical with respect to the z_1-axis. As we know, this wave is equivalent to that caused by a fictitious source situated at point o^*, and, correspondingly, its isochrones are concentric spherical surfaces which have a common center at point o^*. Therefore, at any point of a medium bounded by planes S_0 and S the time of arrival of the wave is defined as

$$t = \frac{1}{c}\sqrt{r_1^2 + (z_1 - h)^2} \tag{1.111}$$

where

$$r_1 = \sqrt{x_1^2 + y_1^2} \qquad \text{and} \qquad h = o_1\, a = o^*\, a$$

As an illustration, a cross-section of these isochrones by the plane $y_1 = 0$ is shown in Fig. 1.17c.

Before we begin to study a time field on the plane S_0, it is useful to define the dip of the surface S with respect to S_0. These planes have a common line l, directed along the strike of the boundary S, and an intersection of them by the vertical plane which is perpendicular to l, forms dip angle, φ, Fig. 1.17d. Let us introduce a new Cartesian system of coordinates: x, y, z so that the z-axis is directed downward, but the x and y-axes are arbitrary oriented on the plane S_0. Its origin o is an intersection of this plane by the line o^*o_1, Fig. 1.17c. Suppose that x_0, y_0 and z_0 are coordinates of the point o^*. Then, the time of arrival of the reflected wave at any point p of Earth's surface, $z = 0$ is

$$t = \frac{o^*\, p}{c} = \frac{1}{c}\sqrt{(x - x_0)^2 + (y - y_0)^2 + z_0^2} \tag{1.112}$$

As is seen from Fig. 1.17c

$$o^* o = \sqrt{x_0^2 + y_0^2 + z_0^2} = h + h_0 \tag{1.113}$$

where

$$h = o^* a, \qquad h_0 = o\,a,$$

h_0 is the slant depth of the surface S beneath the origin o. Correspondingly, eq. 1.112 may be rewritten in the form

$$t = \frac{1}{c}\sqrt{x^2 + y^2 - 2\,x_0\,x - 2\,y_0\,y + (h + h_0)^2} \tag{1.114}$$

Comparison of eq. 1.112 or eq. 1.114 with eq. 1.111 shows that they are similar and with a proper change of coordinates they may coincide. Thus, eq. 1.114 describes the hyperboloid of rotation: $t = t\,(x, y)$ and its axis is parallel to that of time and passes through the point o'. This point is the projection of the imaginary source on the plane S_0, Fig. 1.18a. Because the point o^* is always shifted toward up-dip, the surface $t = t\,(x, y)$ is also displaced in the same direction with respect to the origin o. It is obvious that isochrones on the plane S_0 are concentric circles with the common origin at the point o', where the time of arrival is minimal, Fig. 1.18b. Given that the apparent velocity c_a decreases with an increase of the distance from o', the surface gradient

$$\text{grad}^S\, t$$

becomes larger and approaches the value $1/c$.

To evaluate the displacement of the point o' with respect to o consider the triangle oo^*o' Fig. 1.18a. By definition, it is located in the vertical plane and perpendicular to the line l. Taking into account the fact that pairs of segments, $o'o^*$ and oo^* and c_1d_1 and c_1d_2, are situated in parallel planes and are mutually perpendicular to each other, we can say that the angle between $o'o^*$ and o^*o is equal to the dip:

$$\angle\, o'\, o^*\, o = \varphi,$$

and, therefore,

$$o'\, o = d = (h + h_0)\,\sin\varphi \tag{1.115}$$

The source of the incident wave is usually located on Earth's surface. In this case

$$h = h_0 \tag{1.116}$$

and correspondingly, eq. 1.114 becomes

$$t = \frac{1}{c}\sqrt{x^2 + y^2 - 2\,x_0\,x - 2\,y_0\,y + 4\,h^2} \tag{1.117}$$

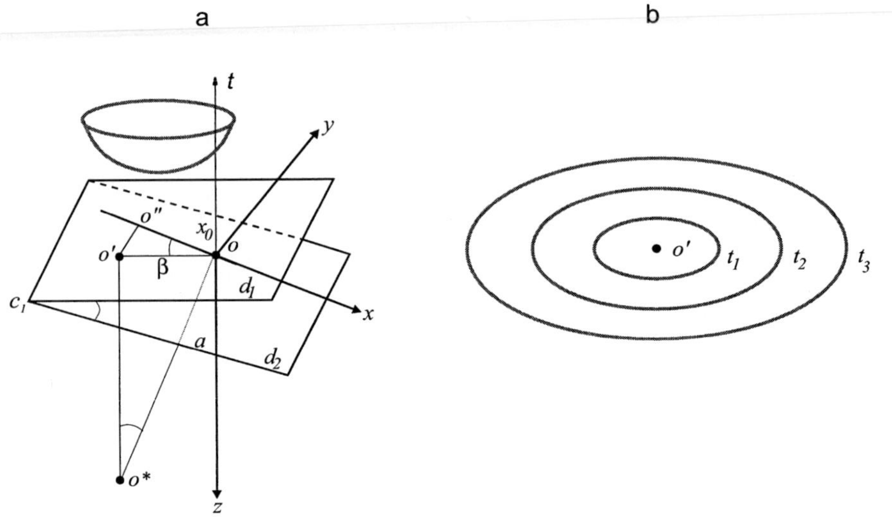

Figure 1.18: (a) Projection of imaginary source o^* on plane S_0. (b) Isochrones on the plane S_0.

and

$$d = 2\,h\,\sin\,\varphi \tag{1.118}$$

Next consider the behavior of an arrival time along a straight line passing through the origin o and located in the plane S_0. Since an orientation of horizontal axes is chosen arbitrarily, assume that the profile coincides with the x-axis. Then, letting $y = 0$ in eq. 1.117 we have

$$t = \frac{1}{c}\sqrt{4\,h^2 + x^2 - 2\,x_0\,x} \tag{1.119}$$

where the coordinate x_0 depends on the geometry of a medium.

Let us imagine the plane of rays of the reflected wave which cross points of the x-line and merge at point o^*. This plane intersects boundary S along line R, and it represents the reflection line, Fig. 1.19a. Note, that as in the case of the horizontal interface, S, this line is rectilinear. The angle between the profile, $y = 0$, and the reflection line, R, is named the apparent dip, φ_a. If the x-axis is perpendicular to the line l, the apparent and real dips are equal to each other

$$\varphi = \varphi_a$$

Suppose that o'' is the projection of the point o^* on the x-axis. It means that

$$x_0 = o\,o''$$

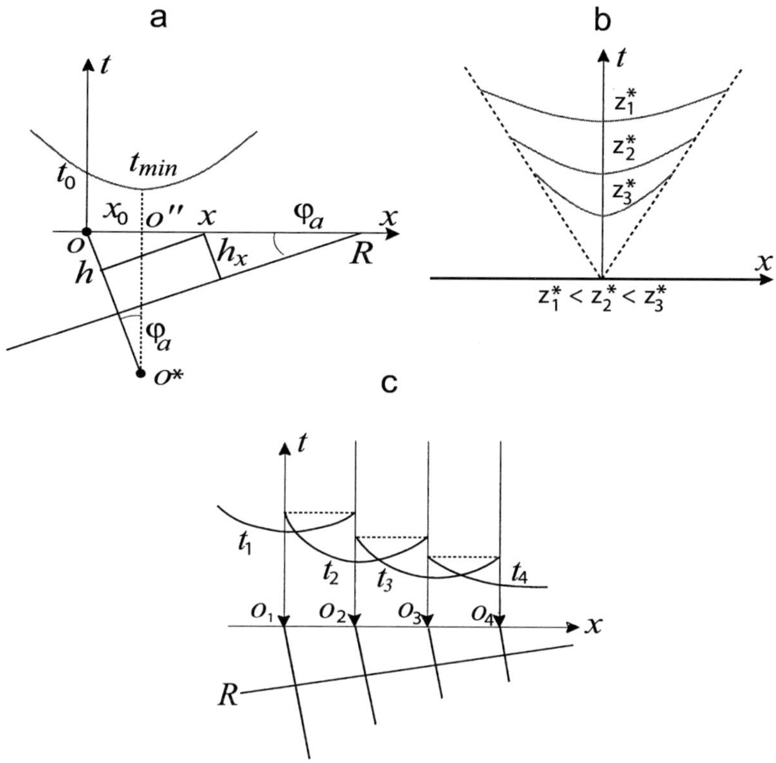

Figure 1.19: (a) Illustration of eq. 1.124. (b) Family of traveltime curves for different reflector depths. (c) Principle of reciprocity.

and from the triangle oo^*o'', it follows that

$$x_0 = 2\,h\,\sin\,\varphi_a \tag{1.120}$$

Thus, the coordinate x_0 of the fictitious source is expressed in terms of the slant depth of the source and the apparent dip. As seen in eq. 1.120 the angle φ_a is positive if the x -axis is directed toward up-dip.

Now we establish the relation between the real and apparent dips. From the triangle $oo'o''$, Fig. 1.18a, we have

$$x_0 = o'\,o\,\cos\,\beta \tag{1.121}$$

where β is the angle between the x-axis and the element oo', which is perpendicular to the line l, (the strike of the surface S).

Making use of eqs. 1.115 and 1.120 we obtain

$$2\,h\,\sin\,\varphi_a = 2\,h\,\sin\,\varphi\,\cos\,\beta$$

or

$$\sin\varphi_a = \cos\beta\,\sin\varphi \tag{1.122}$$

Therefore, the apparent dip angle φ_a varies within the following limits

$$0 \le \varphi_a \le \varphi \tag{1.123}$$

Correspondingly, eq. 1.119 can be rewritten as

$$t = \frac{1}{c}\sqrt{4\,h^2 + x^2 - 4\,h\,x\,\sin\varphi_a} \tag{1.124}$$

and it represents the hyperbola, symmetrical with respect to the vertical axis that crosses the point o'', Fig. 1.19a. Its minimum is shifted toward the up-dip on the distance equal to x_0. As follows from eq. 1.124, the time of arrival at the origin, t_0, and at the point of the minimum, t_{min}, are defined as

$$t_0 = \frac{2\,h}{c}, \qquad t_{min} = \frac{2\,h\,\cos\varphi_a}{c} = \frac{z_*}{c} \tag{1.125}$$

Here

$$z_* = o''o^* = 2\,h\,\cos\varphi_a \tag{1.126}$$

is the distance from the imaginary source to the profile.

Let us introduce the coordinate x_2 as

$$x = x_2 + x_0,$$

Then eq. 1.124 becomes

$$t = \frac{1}{c}\sqrt{4\,h^2 + x_2^2 + 2\,x_2\,x_0 + x_0^2 - 4\,h\,(x_0 + x_2)\,\sin\varphi_a}$$

Substitution of eq. 1.120 into the latter gives

$$t = \frac{1}{c}\sqrt{4\,h^2 + x_2^2 + 4\,h\,x_2\,\sin\varphi_a + 4\,h^2\,\sin^2\varphi_a - 4\,h\,x_2\,\sin\varphi_a - 8\,h^2\,\sin^2\varphi_a}$$

or

$$t = \frac{1}{c}\sqrt{4\,h^2 + x_2^2 - 4\,h^2\,\sin^2\varphi_a} = \frac{1}{c}\sqrt{x_2^2 + z_*^2} \tag{1.127}$$

Taking the derivative with respect to time from both sides of this equation we obtain for the apparent velocity

$$1 = \frac{x_2}{c\sqrt{x_2^2 + z_*^2}}\,c_a$$

or

$$c_a = c\sqrt{1+\frac{z_*^2}{x_2^2}} = c^2\,\frac{t}{x_2} \tag{1.128}$$

In particular, in the vicinity of the source, eq. 1.125, we have

$$c_a = c^2\,\frac{t_0}{2\,h\,\sin\,\varphi_a} = c^2\,\frac{2\,h}{c\,2\,h\,\sin\,\varphi_a}$$

or

$$c_a = \frac{c}{\sin\,\varphi_a} \tag{1.129}$$

At the same, time near the minimum, $x_2 = 0$, the apparent velocity becomes infinitely large but with an increase of the distance it tends to c

$$\lim_{x\to\infty} c_a = c$$

The family of hyperbolae, $t = t\,(x, z_*)$, for different values of z_* is shown in Fig. 1.19b. With an increase of the depth z_*, the curvature of hyperbolae near the minimum becomes smaller, which is related to the fact that rather small changes of x,$(x \ll z_*)$, have almost no affect on the length of ray paths. As follows from eq. 1.128 with an increase of the distance from the minimum the slope $\partial t/\partial x$ tends to the value $1/c$ and all curves approach each other, while their asymptote being the function $t = t\,(x)$, corresponding to the direct wave.

Sometimes it is useful to represent eq. 1.124 in a different form. As can be seen in Fig. 1.19a, the slant depth near the source, h, and that at the point with the coordinate x are related and we have

$$h_x = h - x\,\sin\,\varphi_a = h - \Delta h_x \tag{1.130}$$

where

$$\Delta h_x = x\,\sin\,\varphi_a \tag{1.131}$$

Respectively, eq. 1.124 becomes

$$t(x) = \frac{1}{c}\sqrt{x^2 + 4\,h^2 - 4\,h\,\Delta h_x}$$

$$= \frac{1}{c}\sqrt{x^2 + 4\,h^2 - 4\,h\,(h - h_x)} = \frac{1}{c}\sqrt{4\,h\,h_x + x^2} \tag{1.132}$$

By analogy the function $t\,(x, y)$ may be rewritten in a similar manner.

From eq. 1.132 it follows that if the source occupies the position of a receiver and vice-versa, then the time of arrival remains the same. In fact, this procedure does not change a value of the product: hh_x in this equation. In other words, we have demonstrated the principle of reciprocity. This principle is, of course, applicable to observations along a profile with different positions of sources: $o_1(x)$, $o_2(x)$, ..., Fig. 1.19c.

Thus, in accordance with eq. 1.132, the relationship between the traveltime curves for waves reflected from the boundary is

$$t_i(x_k) = t_k(x_i) \tag{1.133}$$

Let us make one more comment. Since observations are often performed at relatively small distances from the source ($x \ll 2h$), the right hand side of eq. 1.124 may be expanded in a power series

$$t = t_0 - \frac{t_0 \sin\varphi_a}{2}\frac{x}{h} + \frac{t_0 \cos^2\varphi_a}{8}\left(\frac{x}{h}\right)^2 + \frac{t_0 \sin\varphi_a \cos^2\varphi_a}{16}\left(\frac{x}{h}\right)^3 + \ldots \tag{1.134}$$

Usually, it is possible to keep only the first three terms.

As follows from eq. 1.124 three parameters define the behavior of the travel time curve, namely

$$c, \quad h, \quad \text{and} \quad \varphi_a,$$

which can be determined in different ways. For example, as shown previously

$$t_0 = \frac{2h}{c}, \qquad t_{min} = \frac{2h\cos\varphi_a}{c} = \frac{z_*}{c} \tag{1.135}$$

and

$$x_{min} = 2h\sin\varphi_a \tag{1.136}$$

Thus, the apparent dip can be calculated as

$$\frac{t_{min}}{t_0} = \cos\varphi_a \qquad \text{or} \qquad \sin\varphi_a = \sqrt{1 - \frac{t_{min}^2}{t_0^2}} \tag{1.137}$$

while the slant depth beneath the source and the velocity of propagation are

$$h = \frac{x_{min}}{2\sqrt{1 - \dfrac{t_{min}^2}{t_0^2}}}, \qquad c = \frac{2h}{t_0} = \frac{x_{min}}{\sqrt{t_0^2 - t_{min}^2}} \tag{1.138}$$

Knowing the values of h and φ_a we can draw the reflection line R. In accordance with eq. 1.129 the angle φ_a may also be determined from the apparent velocity near the source.

Until now we have studied time of arrival of the wave along a profile crossing the origin, for example, the x-axis. Next suppose that the line of observations is parallel to the y-axis and $x = d_0$. Then eq. 1.117 becomes

$$t = \frac{1}{c}\sqrt{d_0^2 + y^2 - 2\,d_0\,x_0 - 2\,y_0\,y + 4\,h^2}\,, \tag{1.139}$$

or, taking into account eq. 1.120, we have:

$$t = \frac{1}{c}\sqrt{d_0^2 + y^2 - 4\,d_0\,h\,\sin\varphi_a - 2\,y_0\,y + 4\,h^2} \tag{1.140}$$

As is seen from Fig. 1.20a the line of reflection, as before, is the rectilinear one and it is situated in the same plane as the profile and the fictitious source.

Case three: Multiple reflections So far we have considered the wave which is reflected on the surface S only once. At the same time when this wave arrives at the Earth's surface it also experiences the reflection, and, correspondingly, multiple reflected waves arise. In order to determine their time field, we proceed from the fact that each reflected wave is the spherical one regardless of how many reflections occur. First, we assume that the interface S is the horizontal plane and the source is situated at point o, Fig. 1.20b. Taking into account the axial symmetry, consider a distribution of fictitious sources along the z-axis, directed downward, Fig. 1.20b. As we know, the imaginary source of the first reflected wave is located at the distance $z_1 = 2h$ from the origin. Bearing in mind the similarity of triangles $o_1^* o p_1$ and $o_2^* o p_2$, we conclude that the imaginary source of the second reflected wave o_2^*, has the coordinate: $z_2 = 4h$. Therefore, the time field of this wave on Earth's surface is

$$t = \frac{1}{c}\sqrt{r^2 + (4\,h)^2}$$

or

$$t = \frac{1}{c}\sqrt{x^2 + y^2 + 16\,h^2} \tag{1.141}$$

Generalizing this result for the reflected wave of the order n, we obtain

$$t = \frac{1}{c}\sqrt{x^2 + y^2 + 4\,n^2\,h^2} \tag{1.142}$$

This equation was derived for a ray with an incident angle equal to θ. Correspondingly, waves of different order but with the same incident angle θ arrive at different points of Earth's surface.

Now suppose that boundary S has a dip, and consider the reflected wave of the second order. As is seen from Fig. 1.20c,

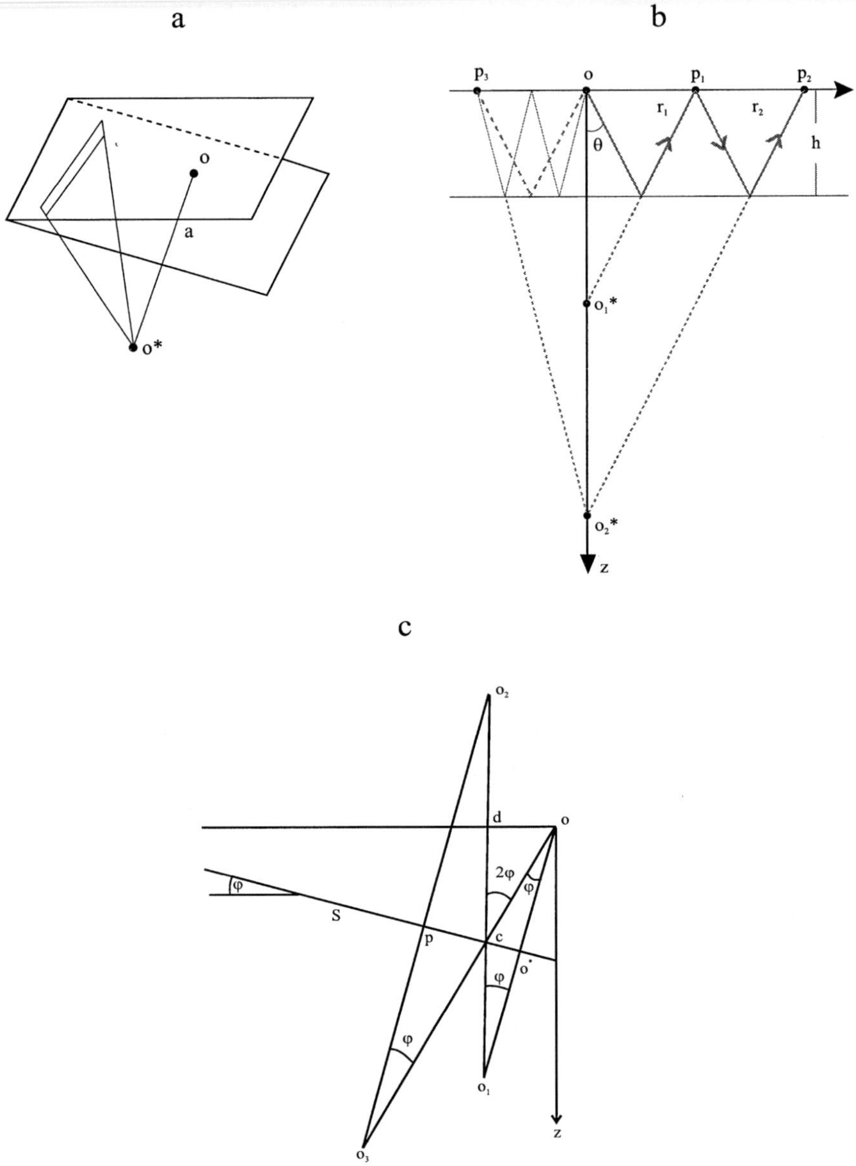

Figure 1.20: (a) Illustration of eq. 1.140. (b) Multiple reflections. (c) Second-order multiple from a dipping reflector.

$$o_2\, d = o_1\, d = 2\, h\, \cos\varphi$$

and

$$d\, c = c\, o\, \sin\left(\frac{\pi}{2} - 2\,\varphi\right) = c\, o\, \cos\, 2\,\varphi$$

From the triangle oo^*c, we have

$$c\, o = \frac{h}{\cos\, \varphi}$$

Thus,

$$d\, c = h\, \frac{\cos\, 2\,\varphi}{\cos\, \varphi}$$

Whence

$$o_2\, c = o_3\, c = 2\, h\, \cos\, \varphi + \frac{\cos\, 2\,\varphi}{\cos\, \varphi}\, h$$

and therefore

$$o\, o_3 = h\, \left[2\, \cos\, \varphi + \frac{\cos\, 2\,\varphi}{\cos\, \varphi} + \frac{1}{\cos\, \varphi}\right] = 4\, h\, \cos\, \varphi$$

The last expression allows us to determine coordinates of the fictitious source of the reflected wave of the second order. In fact, we have

$$x\,(o_3) = 4\, h\, \cos\, \varphi\, \sin\, 2\,\varphi$$

$$z\,(o_3) = 4\, h\, \cos\, \varphi\, \cos\, 2\,\varphi$$

Then, letting in eq. 1.112

$$x_0 = x\,(o_3)\,, \qquad y_0 = 0, \qquad \text{and} \qquad z_0 = x\,(o_3)$$

we obtain

$$t = \frac{1}{c}\, \sqrt{x^2 + y^2 - 4\, h_1\, x \sin\, 2\,\varphi + 4\, h_1^2}\,, \tag{1.143}$$

where

$$h_1 = 2\, h\, \cos\, \varphi$$

and, as is seen from Fig. 1.20c, the wave of this order can be observed if $\varphi < \dfrac{\pi}{4}$.

1.3 Head waves in media with plane interfaces

Considering in Part I a behavior of rays near an interface and applying Fermat's principle we found that the head wave may exist along with reflected and transmitted waves. It appears when an underlying medium has a higher velocity than the upper one, $c_2 > c_1$, Fig. 1.21a. The ray of this wave consists of three elements,

(1) The segment oa along which the wave propagates with the velocity c_1. The incidence angle of this ray element on the boundary is equal to the critical angle θ_c

$$\sin \theta_c = \frac{c_1}{c_2} \tag{1.144}$$

(2) The segment ab is situated in the second medium, in the vicinity of the boundary, where the velocity of propagation is equal to c_2.

(3) The ray segment, bd which passes through an observation point and its angle with the normal to the boundary coincides with the critical one as in the case of the segment oa.

As was demonstrated earlier this type of ray satisfies Fermat's principle, and experiments prove the existence of such waves. Also it is useful to explain the existence of head waves proceeding from the Huygens principle which was already applied to derive Snell's laws of reflection and refraction. Let us consider phase surfaces of the direct and refracted (transmitted) waves in the presence of the horizontal interface, Fig. 1.21b. Taking into account the axial symmetry with respect to the z-axis, it is sufficient to study a wave field in any vertical plane containing the origin o. As we see lines with indexes: "1", "2" and "3" describe phase surfaces of the direct (incident) wave, which is the spherical one. The phase surface with the index "4" has already reached the boundary S and gave rise to the refracted wave in the second medium. Its wave front has index "$4'$". Of course, the incident wave also generates the reflected wave. The angle between the front of the transmitted wave at the point a of the interface is equal to $\pi/2$. Starting from the point a the front of this wave remains perpendicular to the boundary regardless of the distance. Certainly beneath the interface the phase surfaces of the refracted wave have a different orientation. Since in an underlying medium a wave propagates faster than in the upper layer, we may expect that at points of the interface where $r > o_1a$, the wave front of the transmitted wave is located at greater distances from the source than that of the direct and reflected waves, Fig. 1.21b. The propagation of this wave is accompanied by a motion of particles of a medium, particularly, in the vicinity of the interface. Respectively, vibrations arise in the upper medium and the head wave appears.

It is a simple matter to describe approximately the behavior of this wave using Huygens principle. Suppose that at instances t_1, t_2, t_3 the wave front reaches points: b_1, b_2, b_3 of the boundary, Fig. 1.21c. Therefore, distances between them can be represented as

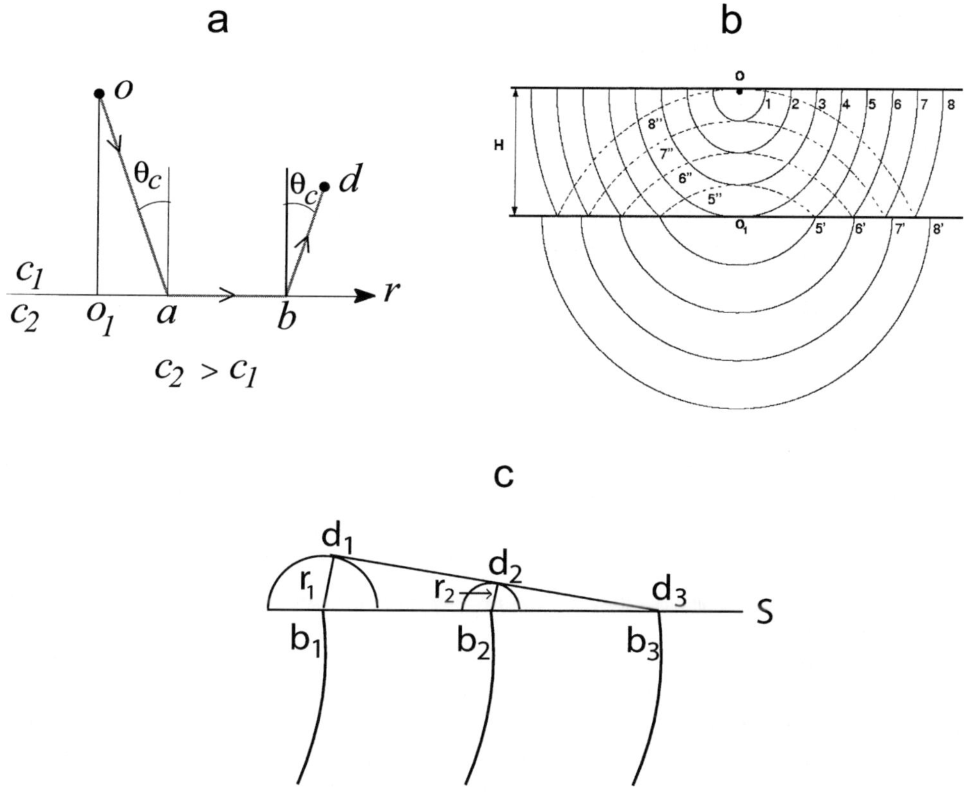

Figure 1.21: Head wave: (a) Ray path of the head wave. (b) Phase surfaces of direct, transmitted waves are solid lines and reflected waves are dashed lines. (c) Huygens principle and phase surfaces of the head wave.

$$b_1 b_2 = c_2 \, (t_2 - t_1) \, , \qquad b_1 b_3 = c_2 \, (t_3 - t_1) \, , \qquad b_2 b_3 = c_2 \, (t_3 - t_2) \tag{1.145}$$

At each point of the interface where the refraction wave arrives, it generates an elementary spherical wave in the upper medium, and an envelope of these elementary waves describes the phase surface of the head wave. It is clear that at the moment: $t = t_3$ the radii of wavelets at points: b_1, b_2 and b_3 are

$$r_1 = c_1 \, (t_3 - t_1) \, , \qquad r_2 = c_1 \, (t_3 - t_2) \, , \qquad r_3 = 0 \tag{1.146}$$

As is seen from Fig. 1.21c the line $d_1 d_3$ is tangential to the elementary wave front and therefore lines: $d_2 d_3$ and $d_1 d_3$ are perpendicular to the radii r_2 and r_1, respectively.

As follows from eqs. 1.145–1.146 two ratios

$$\frac{b_2 \, d_2}{b_2 \, b_3} \qquad \text{and} \qquad \frac{b_1 \, d_1}{b_1 \, b_3}$$

are equal to each other. In fact, it gives

$$\frac{b_2 \, d_2}{b_2 \, b_3} = \frac{c_1 \, (t_3 - t_2)}{c_2 \, (t_3 - t_2)} = \frac{c_1}{c_2}$$

and

$$\frac{b_1 \, d_1}{b_1 \, b_3} = \frac{c_1 \, (t_3 - t_1)}{c_2 \, (t_3 - t_1)} = \frac{c_1}{c_2} \tag{1.147}$$

On the other hand, these ratios characterize angles between lines $d_2 d_3$ and $d_1 d_3$ and the interface S. Since it is the same for both lines, we conclude that points d_1, d_2 and d_3 belong to one straight line, which forms with the horizontal axis r the angle equal to the critical one, θ_c, Fig. 1.21c. Thus, it was demonstrated that on the vertical planes lines of a constant phase of the head wave are rectilinear and parallel to each other. Taking into account the axial symmetry, we see that the phase surfaces are conical, Fig. 1.22a. For this reason head waves are sometimes called the conical waves. Let us notice that from Fermat's principle we arrived at the same conclusion about the geometry of head waves. Next consider several examples that illustrate the time field of this wave on the surface S_0 of a layered medium.

Case one: Two-layer medium Suppose that beneath Earth's surface there is one horizontal interface and the thickness of the upper layer is equal to h. Because in this case the time field possesses by the axial symmetry with respect to the z-axis, it is sufficient to study a function $t = t(x)$ along an arbitrary oriented x-axis, when the source

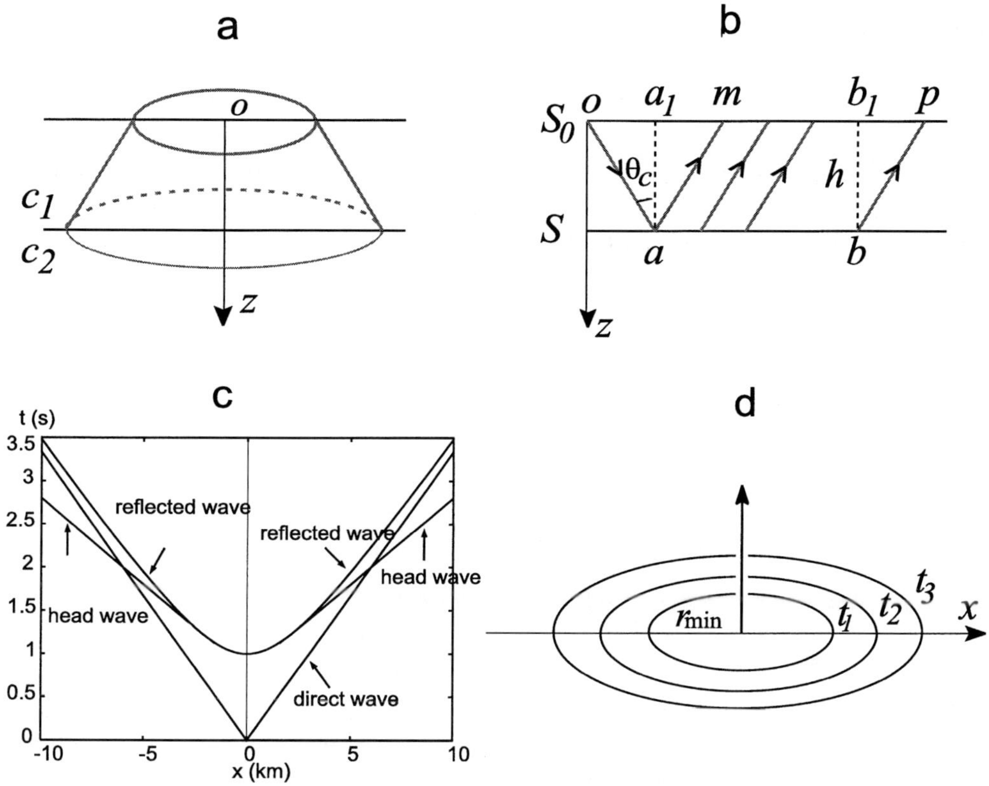

Figure 1.22: Head wave in a two-layer medium: (a) Phase surface of the head wave. (b) Medium with one horizontal interface. (c) Traveltime curves. (d) Isochrones in the horizontal plane.

is located at the origin o, Fig. 1.22b. The time of arrival of the head wave, traveling along the path $oabp$ from the source to an observation point p can be represented as a sum

$$t = t_1 + t_2 + t_3, \tag{1.148}$$

where t_1, t_2 and t_3 are the time of propagation along segments: oa, ab and bp, respectively.

As is seen from Fig. 1.22b

$$t_1 = \frac{o\,a}{c_1} = \frac{b\,p}{c_1} = \frac{h}{c_1 \cos\theta_c} = t_3 \tag{1.149}$$

Since

$$a\,b = a_1\,b_1 = o\,p - o\,a_1 - b_1\,p \qquad \text{or} \qquad a\,b = x - 2\,h\tan\theta_c\,,$$

we have

$$t_2 = \frac{a\,b}{c_2} = \frac{x - 2\,h\tan\theta_c}{c_2} \tag{1.150}$$

Thus, the total time, t, is equal to

$$t = \frac{2\,h}{c_1 \cos\theta_c} + \frac{x - 2\,h\tan\theta_c}{c_2} \tag{1.151}$$

The latter describes the dependence of the time of arrival of the head wave on the distance, x, from the source. The function $t = t(x)$ can also be written as

$$t = \frac{2\,h}{\cos\theta_c}\left(\frac{1}{c_1} - \frac{\sin\theta_c}{c_2}\right) + \frac{x}{c_2},$$

or, taking into account that

$$\sin\theta_c = \frac{c_1}{c_2}, \qquad \cos\theta_c = \sqrt{1 - \frac{c_1^2}{c_2^2}}\,,$$

we have

$$t = \frac{2\,h}{c_1 \cos\theta_c}\left(1 - \sin^2\theta_c\right) + \frac{x}{c_2} = \frac{2\,h}{c_1}\cos\theta_c + \frac{x}{c_2} \tag{1.152}$$

or

$$t = \frac{2\,h}{c_1\,c_2}\sqrt{c_2^2 - c_1^2} + \frac{x}{c_2} \tag{1.153}$$

These relations clearly show that the traveltime curve is the straight line with the slope, equal to $1/c_2$, Fig. 1.22c. Therefore, as was already mentioned, the function $t = t(x, y)$, which characterizes a time field on the surface S_0, is described by a cone. For this reason isochrones on Earth's surface are concentric circles with the center at point o, Fig. 1.22d. The range of change of their radii is

$$r_{min} \leq r < \infty \qquad \text{where} \qquad r_{min} = Om\,, \tag{1.154}$$

and m is the point of the profile, corresponding to arrival of the first ray of the head wave.

It is obvious that if t_2=0, eq. 1.150 yields

$$r_{min} = x_{min} = 2\,h\,\tan\,\theta_c \tag{1.155}$$

As we know, within the interval

$$0 < r < r_{min}$$

the head wave is absent and for this reason this interval is sometimes called the "dead" zone. By definition, at the point m the times of arrival of the head and reflected waves coincide, since the ray am belongs simultaneously to both waves. At a greater distance from the source there is a point n with the coordinate x_c, where the direct and head waves arrive at the same instant and x_c is termed; the crossover distance, Fig. 1.22c. At distances that exceed x_c the head wave arrives earlier than the direct one, while within the interval

$$x_{min} \leq x < x_c$$

the first arrival corresponds to the direct wave. Since

$$x_c = c_1\,t_c \tag{1.156}$$

eq. 1.153 gives

$$t_c = \frac{2\,h}{c_1\,c_2}\,\sqrt{c_2^2 - c_1^2} + \frac{c_1}{c_2}\,t_c$$

or

$$t_c = \frac{2\,h}{c_1\,c_2}\,\frac{c_2\sqrt{1-\dfrac{c_1^2}{c_2^2}}}{1-\dfrac{c_1}{c_2}} \qquad \text{or} \qquad t_c = \frac{2\,h}{c_1}\sqrt{\frac{1+\dfrac{c_1}{c_2}}{1-\dfrac{c_1}{c_2}}} \tag{1.157}$$

Let us mentally draw the straight line $t = t(x)$ and find the ordinate of the point where it crosses the vertical axis of time, Fig. 1.22c. Since $x = 0$, eq. 1.153 becomes

$$t_i = \frac{2\,h}{c_1\,c_2}\sqrt{c_2^2 - c_1^2} \tag{1.158}$$

and it is called the intercept time. Now, assuming that the velocity of the upper layer c_1 is known, from observations of the time of arrival of the direct wave, for example, it is a simple matter to determine the thickness h and the velocity c_2. In particular, the slope of the function $t = t(x)$ allows one to evaluate c_2 and then, making use of eq. 1.158 we have

$$h = \frac{c_1\,c_2\,t_i}{2\sqrt{c_2^2 - c_1^2}} \tag{1.159}$$

The thickness h can also be found from eq. 1.157

$$h = \frac{c_1\,t_c}{2}\sqrt{\frac{c_2 - c_1}{c_2 + c_1}} = \frac{x_c}{2}\sqrt{\frac{c_2 - c_1}{c_2 + c_1}} \tag{1.160}$$

Before we consider other examples, it is appropriate to make one comment. Our study is based on principles of the geometrical acoustics, and therefore, as in the case of the reflected waves, we assume that a wave length λ is much smaller than a layer thickness.

Case two: Three-layer medium Now consider the travel time curves of the head waves in a three-layer medium, Fig. 1.23a, assuming that

$$c_1 < c_2 < c_3 \tag{1.161}$$

Taking into account this behavior of the velocity with the depth the head wave arises at each interface beneath S_0, but we pay attention to that which appears at the lower interface

$$z = h_1 + h_2$$

Suppose that the segment a_1a_2 of the ray $oa_1a_2b_2b_1p$ forms with the z-axis the critical angle θ_c, where

$$\sin\,\theta_c = \frac{c_2}{c_3} \tag{1.162}$$

As follows from Snell's law the ratio

$$\frac{\sin\,\theta}{c}$$

remains constant along the same ray and, therefore, the inclination angle θ at point a_1 of the boundary S_1, is defined from the condition

$$\frac{\sin\,\theta}{c_1} = \frac{\sin\,\theta_c}{c_2}$$

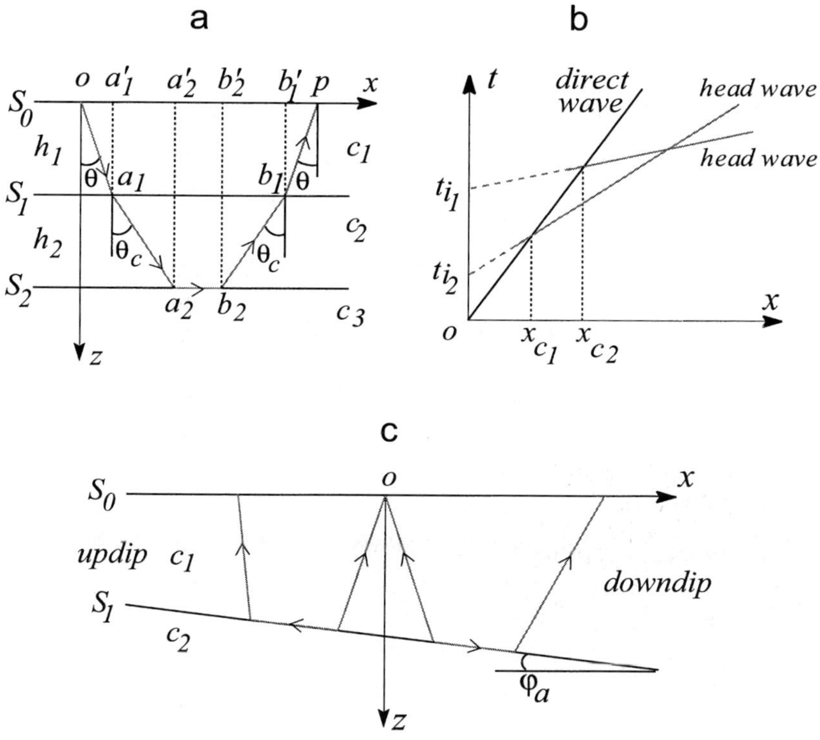

Figure 1.23: Head waves: ray scheme and traveltime curves (a) Medium with two horizontal interfaces. (b) Traveltime curves. (c) Medium with a single dipping interface.

Then, applying eq. 1.162 we obtain

$$\sin\theta = \frac{c_1}{c_2}\sin\theta_c = \frac{c_1}{c_3} \tag{1.163}$$

By definition the critical angle, $\theta_c^{(1)}$, at the boundary S_1, is equal to

$$\sin\theta_c^{(1)} = \frac{c_1}{c_2}$$

and, since $c_3 > c_2$, we have

$$\theta < \theta_c^{(1)}$$

Next, we determine the total time of propagation along the ray $oa_1a_2b_2b_1p$. As is seen from Fig. 1.23a the lengths of segments of this path are

$$o\,a_1 = \frac{h_1}{\cos\theta} = b_1\,p\,, \qquad a_1\,a_2 = \frac{h_2}{\cos\theta_c} = b_2\,b_1 \tag{1.164}$$

and

$$a_2\,b_2 = a_2'\,b_2' = x - 2\,h_1\tan\theta - 2\,h_2\tan\theta_c$$

Thus, the time of arrival of the head wave is equal to

$$t = \frac{2\,h_1}{c_1\cos\theta} + \frac{2\,h_2}{c_2\cos\theta_c} + \frac{x - 2\,h_1\tan\theta - 2\,h_2\tan\theta_c}{c_3}$$

or

$$t = \frac{2\,h_1}{c_1\cos\theta}\left(1 - \frac{c_1}{c_3}\sin\theta\right) + \frac{2\,h_2}{c_2\cos\theta_c}\left(1 - \frac{c_2}{c_3}\sin\theta_c\right) + \frac{x}{c_3} \tag{1.165}$$

From eqs. 1.162–1.163 it follows that

$$t = \frac{2\,h_1}{c_1}\cos\theta + \frac{2\,h_2}{c_2}\cos\theta_c + \frac{x}{c_3} \tag{1.166}$$

or

$$t = \frac{x}{c_3} + \frac{2\,h_1}{c_1\,c_3}\sqrt{c_3^2 - c_1^2} + \frac{2\,h_2}{c_2\,c_3}\sqrt{c_3^2 - c_2^2} \tag{1.167}$$

As we would expect, the traveltime curve is the straight line with the slope equal to $1/c_3$, Fig. 1.23b, and it intersects the time axis at the point t_{i2}:

$$t_{i2} = \frac{2\,h_1}{c_1\,c_3}\sqrt{c_3^2 - c_1^2} + \frac{2\,h_2}{c_2\,c_3}\sqrt{c_3^2 - c_2^2} \tag{1.168}$$

For comparison the traveltime curves of the direct and the head waves, corresponding to the boundary S_1, are also shown in this figure.

Suppose that the velocities c_1 and c_2, as well as the thickness of the upper layer, h_1, are known, for example, from the study of the direct and head waves, arising at the interface S_1. Then, the velocity of an underlying medium, c_3, and the thickness of the second layer, h_2, are easily defined from the slope of the traveltime curve of the head wave and from the intersection time, t_{i2}, eqs. 1.167–1.168. Here it may be proper to make the following comments:

(1) If the second layer has a relatively small thickness, $h_2 \ll h_1$, and velocity c_2 is close to c_1, then the first arrival of the head wave from the S_2 interface may be earlier than from the S_1 interface. Then the head wave from the S_1 interface doesn't appear as first arrival.

(2) If the condition, given by eq. 1.161, is invalid, for instance

$$c_1 > c_2 < c_3 \qquad \text{but} \qquad c_3 > c_1$$

then the head wave arises only on the second interface S_2, and, correspondingly, the boundary S_1 does not manifest itself on the traveltime curve.

(3) In an n-layer medium divided by horizontal interfaces with velocities $c_i < c_n$, $(i = 1, 2, ..., n-1)$, and thicknesses h_i, the traveltime of the head wave, moving along the boundary S_n, is

$$t_{S_n} = t_{0n} + \frac{x}{c_n},$$

where the intercept time t_{0n} is

$$t_{0n} = \sum_{i=1}^{n-1} \frac{2h_i \cos\theta_i}{c_i} \qquad \text{and} \qquad \sin\theta_i = \frac{c_{i-1}}{c_n}$$

Case three: Dipping interface As before, we assume that the source is located at the origin of the Cartesian system of coordinates and the z-axis is directed downward. Since the x-axis has an arbitrary orientation, it forms with the interface an apparent dip, φ_a, Fig. 1.23c. Our goal is to express the time of arrival of the head wave at the observation point p in terms of the distance x from the source, the apparent dipping, φ_a, as well as the slant depth beneath the source h and the critical angle θ_c, because the latter is related to velocities of propagation

$$\theta_c = \sin^{-1}\frac{c_1}{c_2} \tag{1.169}$$

At the beginning, it is useful to determine the slant depth, h_1, beneath point p. As is seen from Fig. 1.24a, we have

$$\angle c\,p_2\,p = \varphi_a\,,$$

since its sides are perpendicular to the x-axis and line l, respectively. Therefore, from triangles: $o_1 c\,p_2$ and $c\,p_2 p_1$ it follows that

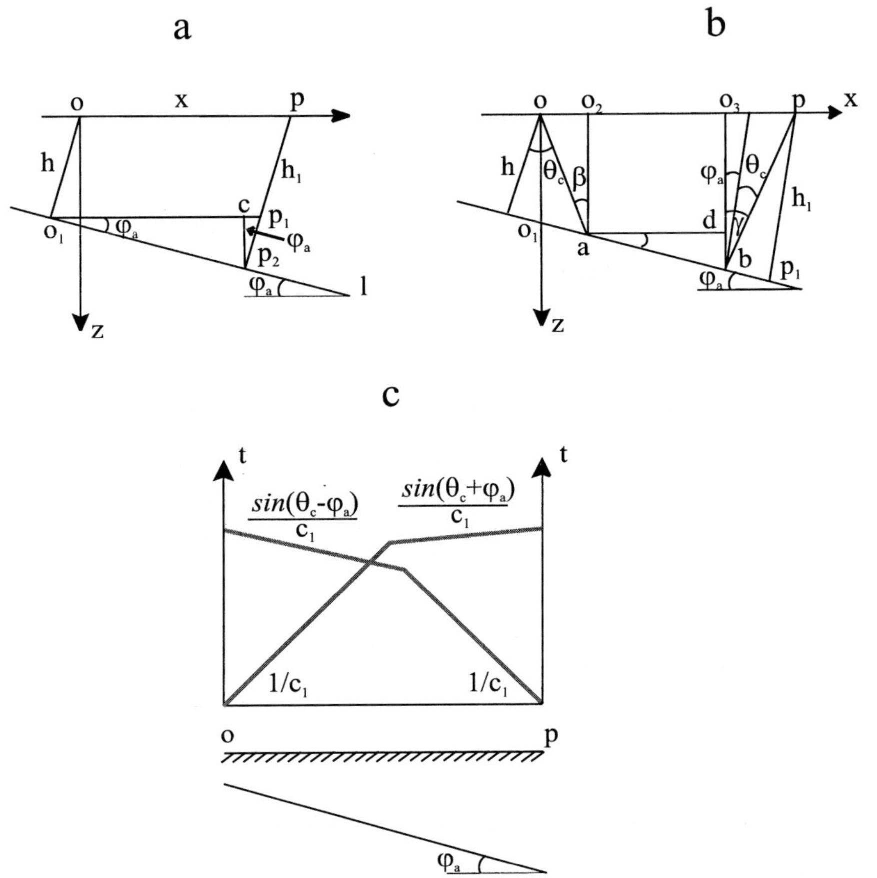

Figure 1.24: Head waves along a dipping interface (a) Definition of the angle φ_a. (b) Illustration of eq. 1.173–1.180. (c) Traveltime curves of direct and head waves for two sources at points o and p. Slopes of curves are indicated.

$$o_1 c = c p_2 \coth \varphi_a \qquad \text{and} \qquad c p_1 = c p_2 \tan \varphi_a$$

Whence

$$x = o_1 c + c p_1 = c p_2 (\coth \varphi_a + \tan \varphi_a)$$

and

$$c p_2 = \frac{x \tan \varphi_a}{1 + \tan^2 \varphi_a} = x \sin \varphi_a \cos \varphi_a$$

Thus,

$$p_1 p_2 = c p_2 \frac{1}{\cos \varphi_a} = x \sin \varphi_a \tag{1.170}$$

and the slant depth beneath the observation point is

$$h_1 = p p_2 = p p_1 + p_1 p_2 = h + x \sin \varphi_a \tag{1.171}$$

It is obvious that if point p is located up-dip with respect to the source, then we have to change the sign in front of the second term of eq. 1.171. Therefore, for both cases we can write

$$h_1 = h + x \sin \varphi_a \tag{1.172}$$

provided that the positive direction of the x-axis corresponds to the down-dip.

Now we are prepared to find the relationship between the time of the first arrival and the distance along the x-axis. From Fig. 1.24b we have

$$o a = \frac{h}{\cos \theta_c} \tag{1.173}$$

Since the angle formed by sides da and ab is equal to the apparent dip, φ_a, we have

$$\left(\frac{\pi}{2} - \theta_c\right) + \beta + \left(\frac{\pi}{2} + \varphi_a\right) = \pi \qquad \text{or} \qquad \beta = \theta_c - \varphi_a \tag{1.174}$$

Then

$$o o_2 = o a \sin \beta = \frac{h}{\cos \theta_c} \sin (\theta_c - \varphi_a) \tag{1.175}$$

The ray segment bp is

$$b p = \frac{h_1}{\cos \theta_c} \tag{1.176}$$

The angle

$$\gamma = \angle o_3\, b\, p$$

is a sum

$$\gamma = \varphi_a + \theta_c \tag{1.177}$$

Hence

$$o_3\, p = b\, p \sin\, \gamma = \frac{h_1}{\cos\, \theta_c} \sin\, (\varphi_a + \theta_c) \tag{1.178}$$

and

$$o_2\, o_3 = a\, d = x - o\, o_2 - o_3\, p$$

or

$$a\, d = x - \frac{h}{\cos\, \theta_c} \sin\, (\theta_c - \varphi_a) - \frac{h_1}{\cos\, \theta_c} \sin\, (\varphi_a + \theta_c) \tag{1.179}$$

As follows from the triangle *adb*

$$a\, b = \frac{a\, d}{\cos\, \varphi_a} \tag{1.180}$$

Thus, the time of arrival of the head wave at point p is

$$t = \frac{o\, a}{c_1} + \frac{a\, b}{c_2} + \frac{b\, p}{c_1} \tag{1.181}$$

where *oa*, *ab*, and *bp* are given by eqs. 1.173, 1.179–1.180, and 1.176, respectively. Taking into account eqs. 1.172, 1.179, the segment *ab* can be represented as

$$\begin{aligned}
a\, b &= \frac{1}{\cos \varphi_a} \left[x - \frac{h}{\cos \theta_c} \sin\, (\theta_c - \varphi_a) - \frac{(h + x\, \sin \varphi_a)}{\cos \theta_c} \sin\, (\theta_c + \varphi_a) \right] \\
&= \frac{1}{\cos \varphi_a} \left[x \left\{ 1 - \frac{\sin \varphi_a}{\cos \theta_c} \sin\, (\theta_c + \varphi_a) \right\} \right. \\
&\qquad \left. - \frac{h}{\cos \theta_c} \left\{ \sin(\theta_c - \varphi_a) + \sin(\theta_c + \varphi_a) \right\} \right] \\
&= \frac{1}{\cos \varphi_a} \left[x\, \{1 - \sin \varphi_a\, (\tan \theta_c\, \cos \varphi_a + \sin \varphi_a)\} - 2\, h\, \tan \theta_c\, \cos \varphi_a \right] \\
&= x\, \cos \varphi_a - x\, \sin \varphi_a\, \tan \theta_c - 2\, h\, \tan \theta_c
\end{aligned} \tag{1.182}$$

Correspondingly, eq. 1.181 becomes

$$t = \frac{h}{c_1 \cos\theta_c} + \frac{x \cos\varphi_a - x \sin\varphi_a \tan\theta_c - 2h \tan\theta_c}{c_2} + \frac{h + x \sin\varphi_a}{c_1 \cos\theta_c} \tag{1.183}$$

By definition, eq. 1.169

$$c_2 = \frac{c_1}{\sin\theta_c}$$

and substituting this into eq. 1.183, we obtain

$$\begin{aligned} t &= \frac{h}{c_1}\left(\frac{1}{\cos\theta_c} - \frac{2\sin^2\theta_c}{\cos\theta_c} + \frac{1}{\cos\theta_c}\right) \\ &+ \frac{x}{c_1}\left(\cos\varphi_a \sin\theta_c - \frac{\sin\varphi_a \sin^2\theta_c}{\cos\theta_c} + \frac{\sin\varphi_a}{\cos\theta_c}\right) \\ &= \frac{2h}{c_1}\cos\theta_c + \frac{x}{c_1}(\cos\varphi_a \sin\theta_c + \sin\varphi_a \cos\theta_c) \end{aligned}$$

or

$$t = \frac{2h\cos\theta_c}{c_1} + \frac{x}{c_1}\sin(\theta_c + \varphi_a) \tag{1.184}$$

Therefore, in the case of the down-dip, the traveltime curve is a straight line with the slope equal to

$$\frac{\sin(\theta_c + \varphi_a)}{c_1} \quad \text{or} \quad \frac{\sin(\theta_c + \varphi_a)}{c_2 \sin\theta_c}$$

Equation 1.184 also describes the case of the up-dip and in order to represent it in a similar manner let us make use of eq. 1.171. Then we have

$$\begin{aligned} t &= \frac{2(h_1 - x\sin\varphi_a)}{c_1}\cos\theta_c + \frac{x}{c_1}\sin(\theta_c + \varphi_a) \\ &= \frac{2h_1}{c_1}\cos\theta_c + \frac{x}{c_1}\sin(\theta_c - \varphi_a) \end{aligned} \tag{1.185}$$

and, correspondingly, in this case the slope of the traveltime line is

$$\frac{\sin(\theta_c - \varphi_a)}{c_1} \qquad \text{or} \qquad \frac{\sin(\theta_c - \varphi_a)}{c_2 \sin\theta_c}$$

Of course, the mutual replacement of the source and an observation point does not change the time of arrival of the head wave, Fig. 1.24c.

Prior to describing the relationship between the traveltime curves and the medium parameters in this case, it is useful to show a different approach allowing us to derive eq. 1.184. As we see from Fig. 1.25a in the direction of the down-dip and up-dip the head wave appears, first of all, at points p_1 and p_2, respectively. By definition, at these points rays of the reflected and head waves coincide and it is obvious that

$$o\,p_1 > o\,p_2$$

To determine a position of point p_1 consider triangles $o^*d\; p_1$ and $o^*d\; o$, where o^* is the mirror reflection of the source with respect to the boundary S. Since the angle

$$\angle\, d\, o^*\, o$$

is formed by sides that are perpendicular to the x-axis and the line l, respectively, we have

$$\angle\, d\, o^*\, o = \varphi_a$$

Thus

$$\angle\, d\, o^*\, p_1 = \theta_c + \varphi_a \qquad \text{and} \qquad \angle\, d\, o^*\, p_2 = \theta_c - \varphi_a\,, \tag{1.186}$$

which gives

$$\begin{aligned} d\;o &= o^*\, o\, \sin\varphi_a = 2\,h\, \sin\varphi_a \\ d\,p_1 &= o^*\, d\, \tan(\theta_c + \varphi_a) = 2\,h\, \cos\varphi_a\, \tan(\theta_c + \varphi_a) \\ d\,p_2 &= o^*\, d\, \tan(\theta_c - \varphi_a) \end{aligned}$$

Whence

$$\begin{aligned} o\,p_1 &= d\,p_1 - d\,o = 2\,h\, \cos\varphi_a\, \tan(\theta_c + \varphi_a) - 2\,h\, \sin\varphi_a \\ &= \frac{2\,h}{\cos(\theta_c + \varphi_a)}\, [\cos\varphi_a\, \sin(\theta_c + \varphi_a) - \sin\varphi_a\, \cos(\theta_c + \varphi_a)] \\ &= \frac{2\,h\, \sin\theta_c}{\cos(\theta_c + \varphi_a)} \end{aligned}$$

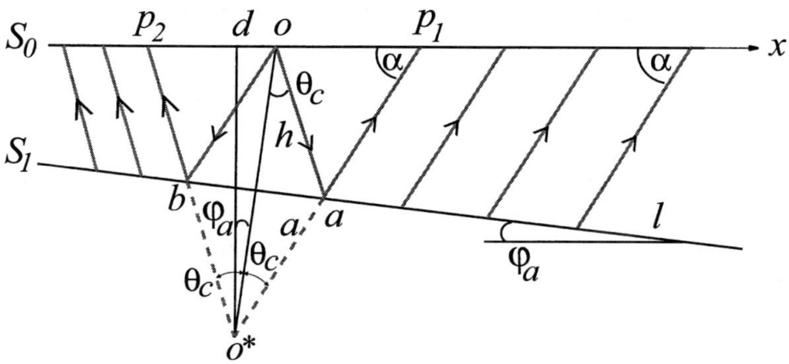

Figure 1.25: Rays of head wave, eq. 1.184.

Thus

$$x\,(p_1) = \frac{2\,h\,\sin\,\theta_c}{\cos\,(\theta_c + \varphi_a)} \tag{1.187}$$

and,

$$\begin{aligned} x\,(p_2) &= o\,d + d\,p_2 = 2\,h\,\sin\,\varphi_a + 2\,h\,\cos\,\varphi_a\,\tan\,(\theta_c - \varphi_a) \\ &= 2\,h\,[\sin\,\varphi_a\,\cos\,(\theta_c - \varphi_a) + \cos\,\varphi_a\,\sin\,(\theta_c - \varphi_a)]\,\frac{1}{\cos\,(\theta_c - \varphi_a)} \\ &= \frac{2\,h\,\sin\,\theta_c}{\cos\,(\theta_c - \varphi_a)} \end{aligned} \tag{1.188}$$

As follows from the triangle $o^* d p_1$, Fig. 1.25, the ray of the head wave at point p_1 forms with the x-axis the angle α, equal to

$$\alpha = \pi - \frac{\pi}{2} - (\theta_c + \varphi_a) = \frac{\pi}{2} - (\theta_c + \varphi_a) \tag{1.189}$$

and it remains constant for all rays if $x \geq x(p_1)$. Therefore, the apparent velocity of the head wave is

$$c_a = \frac{c_1}{\cos\,\alpha} = \frac{c_1}{\sin\,(\theta_c + \varphi_a)} \tag{1.190}$$

Since the traveltime curve, $t = t(x)$, is a straight line, we can write

$$c_a = \frac{x - x_0}{t - t_0},$$

where x_0 is the coordinate of some chosen point along the x-axis, and t_0 is the time of arrival of the wave at this point. In particular, we can choose

$$x_0 = x(p_1) \qquad \text{and} \qquad t_0 = t(p_1) \tag{1.191}$$

Then, we have

$$c_a = \frac{x - x(p_1)}{t - t(p_1)}$$

or

$$t = t(p_1) - \frac{x(p_1)}{c_a} + \frac{x}{c_a} \tag{1.192}$$

Given that the times of arrival of the reflected and head waves coincide at point p_1, we obtain

$$\begin{aligned} t(p_1) &= \frac{o\,a + a\,p_1}{c_1} = \frac{o^*\,a + a\,p_1}{c_1} = \frac{o^*\,p_1}{c_1} \\ &= \frac{o^*\,d}{c_1 \cos(\theta_c + \varphi_a)} = \frac{2\,h \cos\varphi_a}{c_1 \cos(\theta_c + \varphi_a)} \end{aligned} \tag{1.193}$$

Substitution of eqs. 1.187, 1.190 and 1.193 into 1.192 gives

$$t = \frac{2\,h \cos\varphi_a}{c_1 \cos(\theta_c + \varphi_a)} - \frac{2\,h \sin(\theta_c + \varphi_a) \sin\theta_c}{c_1 \cos(\theta_c + \varphi_a)} + \frac{x \sin(\theta_c + \varphi_a)}{c_1}$$

or

$$t = \frac{2\,h\,[\cos\varphi_a - \sin\theta_c \sin(\theta_c + \varphi_a)]}{c_1 \cos(\theta_c + \varphi_a)} + x\,\frac{\sin(\theta_c + \varphi_a)}{c_1} \tag{1.194}$$

The expression in brackets can be simplified

$$\begin{aligned} &\cos\varphi_a - \sin\theta_c \sin(\theta_c + \varphi_a) = \\ &= \cos\varphi_a - \sin^2\theta_c \cos\varphi_a - \sin\theta_c \cos\theta_c \sin\varphi_a \\ &= \cos^2\theta_c \cos\varphi_a - \sin\theta_c \cos\theta_c \sin\varphi_a \\ &= \cos\theta_c\,[\cos\theta_c \cos\varphi_a - \sin\theta_c \sin\varphi_a] \\ &= \cos\theta_c \cos(\theta_c + \varphi_a) \end{aligned}$$

Thus, in place of eq. 1.194 we obtain

$$t = \frac{2\,h\,\cos\,\theta_c}{c_1} + x\,\frac{\sin\,(\theta_c + \varphi_a)}{c_1}\,,$$

and it coincides with eq. 1.184.

Returning to eqs. 1.184–1.185 we have

$$t = \frac{2\,h\,\cos\,\theta_c}{c_1} + \frac{x}{c_1}\,\sin\,(\theta_c + \varphi_a) \qquad \text{if} \quad x > x(p_1)$$

and

$$t = \frac{2\,h\,\cos\,\theta_c}{c_1} + \frac{x}{c_1}\,\sin\,(\theta_c - \varphi_a) \qquad \text{if} \quad x \leq x(p_2) \tag{1.195}$$

As in the case of a horizontal interface S, the head wave is absent within the interval

$$x(p_2) < x < x(p_1)$$

Note that in accordance with eq. 1.187, rays of the head wave do not cross the x-axis, if $x > 0$, provided that

$$\theta_c + \varphi_a \geq \frac{\pi}{2}$$

that is, the "dead" zone increases unlimitedly . The behavior of the traveltime curves is illustrated in Fig. 1.24c.

Now, assuming that the traveltime curves for the down-dip and up-dip are known, eq. 1.195, we can determine velocities c_1 and c_2, the apparent dip, φ_a, as well as the depth to the interface S. It is convenient to introduce notations for the known slopes of these curves

$$k_1 = \frac{1}{c_1}\,\sin\,(\theta_c + \varphi_a) \qquad \text{and} \qquad k_2 = \frac{1}{c_1}\,\sin\,(\theta_c - \varphi_a) \tag{1.196}$$

Thus,

$$\theta_c + \varphi_a = \sin^{-1}\,k_1\,c_1\,, \qquad \theta_c - \varphi_a = \sin^{-1}\,k_2\,c_1$$

Whence

$$\theta_c = \frac{1}{2}\left(\sin^{-1}\,k_1\,c_1 + \sin^{-1}\,k_2\,c_1\right) \tag{1.197}$$

and

$$\varphi_a = \frac{1}{2}\left(\sin^{-1}\,k_1\,c_1 - \sin^{-1}\,k_2\,c_1\right) \tag{1.198}$$

From the set 1.196 the velocity of the upper layer is defined as

$$c_1 = \frac{\sin\,(\theta_c + \varphi_a)}{k_1} \qquad \text{or} \qquad c_1 = \frac{\sin\,(\theta_c - \varphi_a)}{k_2} \tag{1.199}$$

By definition, the velocity of the underlying medium is

$$c_2 = \frac{c_1}{\sin\,\theta_c} \tag{1.200}$$

As follows from equations 1.195 the intersection time t_i is equal to

$$t_i = \frac{2\,h\,\cos\,\theta_c}{c_1}$$

Hence

$$h = \frac{t_i\,c_1}{2\,\cos\,\theta_c} \tag{1.201}$$

In the previous section we found the relationship between the apparent and real dips, (eq. 1.122)

$$\sin\varphi_a = \sin\,\varphi\,\cos\,\beta$$

where β is the angle between the dip line and that of observations. Suppose that apparent dips $\varphi_a^{(1)}$ and $\varphi_a^{(2)}$ are known for two profiles, perpendicular to each other, that is

$$\beta_2 - \beta_1 = \frac{\pi}{2}$$

Thus, we have

$$\sin\varphi_a^{(1)} = \sin\,\varphi\,\cos\,\beta_1,$$

and

$$\sin\varphi_a^{(2)} = \sin\,\varphi\,\cos\,\beta_2 = \sin\,\varphi\,\sin\,\beta_1$$

Taking the ratio we obtain the direction of the dip line:

$$\tan\,\beta_1 = \frac{\sin\varphi_a^{(2)}}{\sin\varphi_a^{(1)}} \tag{1.202}$$

and the real dip can be evaluated as

$$\sin\,\varphi = \frac{\sin\varphi_a^{(1)}}{\cos\,\beta_1} \tag{1.203}$$

It may be proper to notice that the generation of the head waves happens due to the finite curvature radius of the incident wavefront. As we know, the front of the head wave is conical and its curvature radius in the vertical plane, passing through the source, is infinite. Because of this fact the head wave, reflected at the free surface, cannot generate a new head wave at the boundary. Still, any multiple reflected wave can give rise to a head wave and vice versa.

1.4 Wavefront construction and ray tracing

In the previous sections we have considered relatively simple cases of wave fields, when it was possible to express the eikonal and rays in terms of elementary functions. Now we describe qualitatively the main features of methods that allow us to determine the wave fields in the general case. This implies that the media may contain regions, where the velocity is continuous, but there is a discontinuity at boundaries between them. As was pointed out earlier, the wave fields can be determined by using two approaches. One of them is based on solution of the eikonal equation that permits us to find wave surfaces, while the other proceeds from the ray equations and describes the ray families. Of course, both methods are equivalent to each other, since they give the same information about the wave fields. Here it may be appropriate to make one comment, namely, considering transient waves, we assume that they are formed by sufficiently high frequencies and, respectively, the eikonal or isochrone have a clear meaning of surfaces, on which either the phase or time remain constant.

Wavefront

In this light, it may be useful to describe briefly the concept of a wavefront or isochrone of the first arrival. With this purpose in mind, let us formulate a criterion, which characterizes an appearance of the wavefront. In oder to resolve this task, consider a wave field as a function of time at some observation point. By definition, before the first arrival $(t = t_0)$, any wave field (pressure, displacement, or the particle velocity), as well as all its derivatives with respect to time, are equal to zero

$$U(p,t) = 0 \qquad \text{and} \qquad \frac{\partial^n U}{\partial t^n} = 0 \qquad \text{if} \qquad t < t_0 \tag{1.204}$$

Here U is a scalar potential or an arbitrary component of any wave field. Next, consider a wave at some instant t which slightly exceeds t_0

$$t = t_0 + \Delta t \qquad \text{and} \qquad \frac{\Delta t}{t_0} \ll 1$$

Assuming that the field U is a continuous function of time $t > t_0$, it can be expanded in Taylor series

$$U(t) = U(t_0) + \frac{U'(t_0)}{1!}(t-t_0) + \frac{U''(t_0)}{2!}(t-t_0)^2 \ldots + \frac{U^{(n)}(t_0)}{n!}(t-t_0)^n + \ldots \tag{1.205}$$

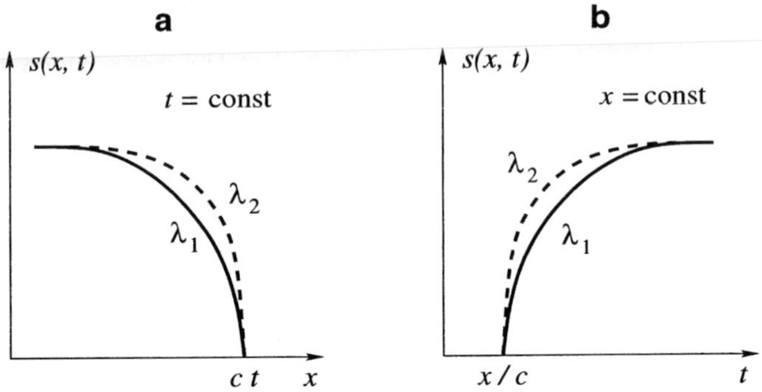

Figure 1.26: Behavior of displacement given by eq. 1.208, $\lambda_2 < \lambda_1$.

It is essential that the wave field and its derivatives in the right-hand side of the equation are taken at the instant t_0 of the wave arrival. Bearing in mind that the wave field generally differs from zero behind the wavefront, we conclude that the series 1.205 contains some nonzero terms. First, suppose that the field U itself has a finite value at the wavefront

$$U(p, t_0) \neq 0 \tag{1.206}$$

Then we can say that the isochrone $t = t_0$ is the surface where the field itself has a discontinuity. Now let us assume that the function U is zero at the wavefront, but its first derivative has a finite value. Correspondingly, the isochrone $t = t_0$ is the surface where the first derivative is discontinuous. Of course, the derivatives of higher order may also have singularities. Generalizing this result, we can say that the wavefront (the isochrone of the first arrival) is a surface where either the field itself or its derivatives or both have singularities. Certainly, this approach defines the wavefronts and even allows us to classify them in accordance with the minimal order of the derivative that is discontinuous. For illustration, consider one simple example.

Plane wave in a homogeneous medium

Suppose that the displacement generated by a plane wave, propagating through a homogeneous medium varies as

$$s(t) = s_0 \begin{cases} 0 & t < 0 \\ 1 - e^{-\lambda t} & t \geq 0 \end{cases} \tag{1.207}$$

Here λ is the parameter that characterizes the rate of change of the displacement and s_0 is its magnitude. As is well known, Part I, the wavelet of the plane wave $s(x, t)$ is the same as that generated by the source. For example, if the wave propagates along the x-axis, the particle displacement is

$$s(x,\, t) = s_0 \begin{cases} 0 & t < x/c \\ 1 - e^{-\lambda\,(t-x/c)} & t \geq x/c \end{cases} \tag{1.208}$$

and its dependence on x and t is shown in Fig. 1.26. All isochrones are planes perpendicular to the x-axis, and the wavefront coincides with the isochrone $t = x/c$. The displacement vanishes at its surface, but the pressure, particle velocity and acceleration differ from zero. In fact, since

$$P_a = -M\,\frac{\partial\, s_x}{\partial x}\,, \qquad v_x = \frac{\partial\, s_x}{\partial t}\,, \qquad a_x = \frac{\partial^2 s_x}{\partial t^2}\,,$$

we have

$$P = \frac{s_0\, M}{c}\,\lambda\, e^{-\lambda\,(t-x/c)}\,, \qquad v_x = s_0\,\lambda\, e^{-\lambda\,(t-x/c)}\,, \qquad a_x = -s_0\,\lambda^2\, e^{-\lambda\,(t-x/c)}\,,$$

where M is the compressibility. In other words, the wavefront is the surface where the wave fields P, v_x and a_x are discontinuous functions. Now let us outline the solution of the boundary value problem.

Wavefront construction

Description of wave fields in terms of wave surfaces can be done in either frequency or time domain because both eikonal and time obey the same equation

$$\text{grad}^2\, L = n^2 \qquad \text{and} \qquad \text{grad}^2\, t = \frac{1}{c^2} \tag{1.209}$$

Because these approaches are equivalent, we will deal with isochrones that can be constructed by either using directly Huygens principle or by applying numerical methods. To visualize application of Huygens principle, consider two situations.

Case one: Surface Q_1 is a wavefront Suppose that the velocity of propagation is a continuous function of spatial coordinates, and the wavefront (the isochrone of the first arrival) coincides with the surface Q_1 at instant t, Fig. 1.27. Let us find the position of the front Q_2 at the instant

$$t_2 = t_1 + \Delta t\,,$$

where Δt is so small that the velocity c along each ray between Q_1 and Q_2 is almost constant. In accordance with the Huygens principle, we treat all the points p_i of the surface Q_1 as elementary sources. Inasmuch as the medium is almost homogeneous in the vicinity of each p_i, the wavefronts are spherical, that is

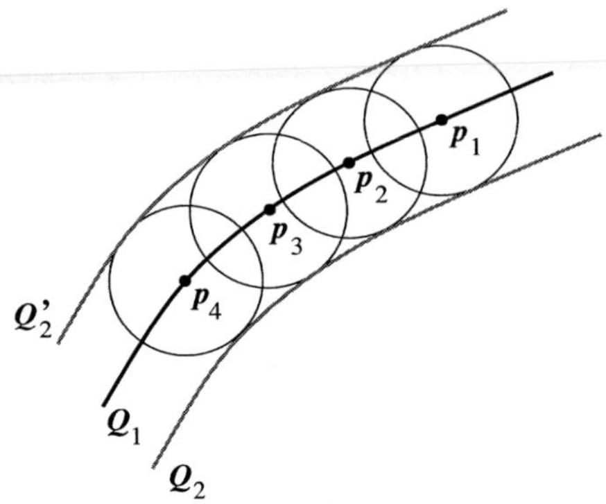

Figure 1.27: The Huygens principle.

$$\Delta R(p_i) = c(p_i)\,\Delta t$$

Here Δt is the same for all the points, but the velocity $c(p_i)$ and, therefore, the radius $\Delta R(p_i)$ may vary.

Making use of the Huygens principle, we state that the envelope of elementary wavefronts defines a new front Q_2. As is seen from Fig. 1.27, however, there are two such surfaces. In order to choose a proper wavefront, it is necessary to know the direction of wave propagation. Then one isochrone, for example, Q_2, is characterized by the time

$$t_2 = t_1 + \Delta t = t_1 + \frac{\Delta R(p_i)}{c(p_i)},$$

whereas the other, Q'_2, has time

$$t'_2 = t_1 - \Delta t = t_1 - \frac{\Delta R(p_i)}{c(p_i)}$$

Considering that the isochrones Q_2 and Q'_2 are given, we can find, in the same manner, the neighboring isochrones and such an approach, which was, in fact, described in Part I, allows us to determine the wave field in a region, where the velocity $c(p)$ is known.

Case two: Surface Q_1 is not a wavefront Let us assume that the surface Q is not an isochrone, but the time of wave arrival is known at each its point. At the beginning, we find a position of the isochrone $t^\star$ of the first arrival, that it, the location of the wavefront at this instant. With this purpose in mind, we take any point p_1 of the surface Q, where the arrival time is t_1, and assume that $t_1 > t^\star$. If some element of the wavefront reaches the point p_1 at the instant t_1, then, at earlier moment, $t^\star$, it was at the distance

$$\Delta R(p_1) = c(p_1)\,(t_1 - t^\star)$$

It is obvious that the spherical surface with this radius and the center at point p_1 is a possible location of the front element at the instant $t^\star$, which reaches the point p_1 at the instant t_1. Next, we take the point p_2, which also belongs to the surface Q and it is close to p_1. In order to establish a possible position of the wave element at the instant $t^\star$, that arrives at the point p_2 when $t = t_2$, we again draw the sphere, but its center is at the point p_2 and the radius is equal to

$$\Delta R(p_2) = c(p_2)\,(t_2 - t^\star)$$

Bearing in mind that the distance between points p_1 and p_2 is small, the front elements, which arrive at these points, are located near the line of intersection of these two spheres. The same procedure, applied to other points p_3, p_4, ..., p_N of the surface Q, gives the system of spherical surfaces, and their envelope is the isochrone $t^\star$. In other words, we define the position of the wavefront at this instant.

As long as one isochrone is found, the others are determined by applying again the Huygens principle. Here it may be appropriate to make several comments:

(a) The time interval $\Delta t_i = t_i - t^\star$ has to be sufficiently small so that the velocity of propagation is almost constant around each point p_i.

(b) As in the first case, Fig. 1.27, there are two envelopes and, correspondingly, it is necessary to know the direction of wave movement to make the correct choice.

(c) It is possible that at instant $t^\star$ the wavefront reached some portion of the surface Q, that is, the latter and the isochrone may partially coincide with each other.

(d) Applying the Huygens principle to the surface Q, we can use small but arbitrary radius $\Delta R(p_i)$. Such an approach gives a system of surfaces Q_n, and the time may change at each of them. In other words, we obtain the time field, $t(p)$, which allows us to derive the isochrones.

(e) Suppose that the medium consists of several regions and there are interfaces where the velocity $c(p)$ is discontinuous. We also assume that the arrival time is known on some surface Q_1. Then, as was shown earlier, we can find the time field in the region around Q_1, including the interface. The latter plays the same role as the surface Q in the second case, which allows us to calculate the field in the next region and so on.

Cauchy initial value problem

As was mentioned earlier, in essence, we describe a solution of the initial boundary value problem for the time field. Such a problem includes the following requirements:

(1) It obeys the equation

$$\text{grad}^2\, t = \frac{1}{c^2}$$

at regular points.

(2) The solution $t(p)$ has to coincide with time $t_0(p)$ at a given surface Q. This condition is usually called the initial one.

(3) The time is a continuous function at the interfaces of media with different velocities.

In accordance with the theorem of existence and uniqueness, time field $t(p)$, satisfying all three conditions, is uniquely defined.

There are two approaches solving this problem. The first one, based on the Huygens principle, has already been described. The second proceeds from equation 1.209 for the time field, which is usually solved in complicated media by the finite difference method. Examples of the isochrones will be given later along with the ray families.

Ray tracing

Now we continue our discussion with the help of rays, and in this light it is convenient to distinguish two problems. The goal of the first one is to define the ray, provided that the initial conditions are given. This procedure is called the ray shooting. The second task is to determine the ray (or rays) passing through two known points, usually, the source and the receiver. This is accomplished by ray bending methods.

System of ray equations of the first order

At the beginning of this chapter we derived three ordinary differential equations of the second order which describe ray behavior, Part I, and they are

$$\frac{d}{ds}\left(n\,\frac{dx}{ds}\right) = \frac{\partial n}{\partial x}\,, \qquad \frac{d}{ds}\left(n\,\frac{dy}{ds}\right) = \frac{\partial n}{\partial y}\,, \qquad \frac{d}{ds}\left(n\,\frac{dz}{ds}\right) = \frac{\partial n}{\partial z} \tag{1.210}$$

Here s is the length of the ray arc, $x(s)$, $y(s)$ and $z(s)$ are the coordinates of the ray points,

$$n = \frac{c_0}{c(x,\, y,\, z)} \tag{1.211}$$

is the refraction index, and c_0 is the velocity at some point.

Let us demonstrate that eqs. 1.210 are equivalent to the system of six ordinary differential equations of the first order. With this purpose in mind, first we introduce the components p_x, p_y and p_z of the slowness vector

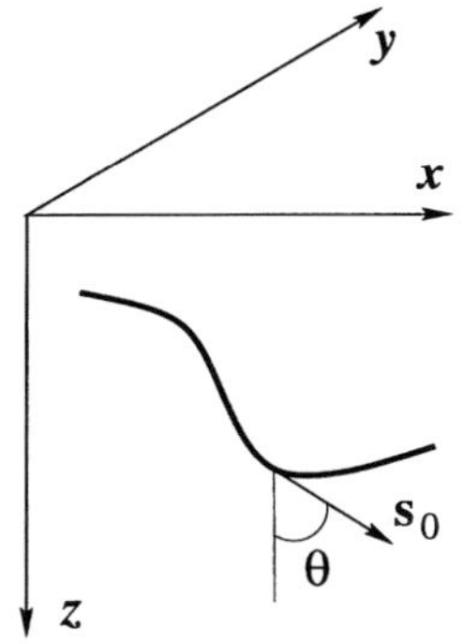

Figure 1.28: Illustration of eq. 1.215.

$$n\,\frac{d\,x}{d\,s} = c_0\,p_x\,, \qquad n\,\frac{d\,y}{d\,s} = c_0\,p_y\,, \qquad n\,\frac{d\,z}{d\,s} = c_0\,p_z \tag{1.212}$$

Multiplication of each equation by the corresponding unit vector, **i**, **j**, **k** and then their summation yields

$$n\,\frac{d\,\mathbf{r}}{d\,s} = c_0\,\mathbf{p} \tag{1.213}$$

Here **p** is the slowness vector and **r** is the radius-vector:

$$\mathbf{p} = p_x\,\mathbf{i} + p_y\,\mathbf{j} + p_z\,\mathbf{k}\,, \qquad \mathbf{r} = x\,\mathbf{i} + y\,\mathbf{j} + z\,\mathbf{k}\,,$$

and the latter defines a ray point. As was shown earlier, the vector $d\,\mathbf{r}/ds$ coincides with the unit vector $\mathbf{s_0}$ tangential to the ray and, therefore, eq. 1.213 becomes

$$\mathbf{p} = \frac{\mathbf{s_0}}{c} \tag{1.214}$$

Thus, the slowness is the vector directed along the ray, and its magnitude is equal to c^{-1}. Except for a few cases, in particular, a homogeneous medium, both the magnitude and the direction of **p** varies along the ray. It is also useful to consider components of this vector, eq. 1.214. Inasmuch as the derivatives dx/ds, dy/ds and dz/ds are directional cosines of the vector $\mathbf{s_0}$, Fig. 1.28, we have

$$p_x = \frac{1}{c_x}\,, \qquad p_y = \frac{1}{c_y}\,, \qquad p_z = \frac{1}{c_z}\,, \tag{1.215}$$

where c_x, c_y and c_z are the apparent velocities of propagation in the direction of the coordinate axes. It may be proper to notice that there are types of inhomogeneous media in which one or two components of the slowness vector remain constant along a ray.

Also, it is convenient to rewrite the set 1.212 in a slightly different form:

$$n\,\frac{d\,x_i}{d\,s} = c_0\,p_i\,, \qquad i = 1,\,2,\,3\,, \tag{1.216}$$

where x_1, x_2, and x_3 are coordinates of the ray point. Assuming that velocity $c(x_1,\,x_2,\,x_3)$ is known, these three differential equations of the first order contain six unknowns:

$$x_1\,,\; x_2\,,\; x_3\,,\; p_1\,,\; p_2\,,\; p_3$$

In order to obtain three additional equations, we substitute eqs. 1.211 and 1.216 into the set 1.210, and it gives

$$\frac{d\,p_i}{d\,s} = -\,\frac{1}{c^2}\,\frac{\partial\,c}{\partial x_i}\,, \qquad i = 1,\,2,\,3 \tag{1.217}$$

The latter is the system of three ordinary differential equations for the three unknowns p_1, p_2 and p_3. After multiplication of these equations by the unit vectors and summation, we have

$$\frac{d\,\mathbf{p}}{d\,s} = -\,\frac{\mathrm{grad}\,c}{c^2} \tag{1.218}$$

To illustrate its meaning, suppose that the velocity c doesn't vary along the x and y axes. Then in accordance with the first equations ($i = 1, 2$) of the set 1.217, we see that the components p_1 and p_2 remain the same at all points of the ray

$$p_1 = \frac{1}{c}\,\frac{d\,x_1}{d\,s} = \mathrm{const}\,, \qquad p_2 = \frac{1}{c}\,\frac{d\,x_2}{d\,s} = \mathrm{const} \tag{1.219}$$

Making use of the incidence angle θ, Fig. 1.28, the first equality can be written as

$$\frac{\sin\theta}{c} = \mathrm{const}\;, \tag{1.220}$$

that is, we arrived at Snell's law in a vertically inhomogeneous media with continuous change of velocity.

Now, combining eqs. 1.213 and 1.218, we obtain the system of six ordinary differential equations of the first order with six unknowns:

$$n\,\frac{d\,\mathbf{r}}{d\,s} = c_0\,\mathbf{p} \qquad \text{and} \qquad \frac{d\,\mathbf{p}}{d\,s} = -\,\frac{\mathrm{grad}\,c}{c^2} \tag{1.221}$$

Let us represent the argument s in terms of time and instead of t use the letter τ

$$ds = c\,d\tau = c\,d\,t \tag{1.222}$$

Here $d\tau$ is the time during which the wave travels the distance ds. Substitution of eq. 1.222 into the set 1.221 gives

$$\frac{d\mathbf{r}}{d\tau} = c^2\,\mathbf{p} \qquad \text{and} \qquad \frac{d\mathbf{p}}{d\tau} = -\frac{\operatorname{grad} c}{c} \tag{1.223}$$

Deriving this system, we assumed that the rays are perpendicular to the eikonal surfaces. They also can be treated as characteristic curves of solution surface of the integral equation, Appendix D.

Thus, we have demonstrated that ray behavior is described by six ordinary differential equations of the first order with six unknowns. They are the coordinates of the ray points and the components of the slowness vector. For example, we have in the Cartesian coordinate system

$$\begin{gathered} \frac{dx}{d\tau} = c^2\,p\,, \qquad \frac{dy}{d\tau} = c^2\,l\,, \qquad \frac{dz}{d\tau} = c^2\,q \\ \text{and} \\ \frac{dp}{d\tau} = -\frac{1}{c}\frac{\partial c}{\partial x}\,, \qquad \frac{dl}{d\tau} = -\frac{1}{c}\frac{\partial c}{\partial y}\,, \qquad \frac{dq}{d\tau} = -\frac{1}{c}\frac{\partial c}{\partial z} \end{gathered} \tag{1.224}$$

Each of the last three equations allows us to find the corresponding component of the slowness with an accuracy of some constant. Then, substituting p, l and q into three other equations, we determine $x(\tau)$, $y(\tau)$ and $z(\tau)$. The latter also contain constants, but all of them can be found from initial conditions

$$\begin{gathered} x(0) = x_0\,, \qquad y(0) = y_0\,, \qquad z(0) = z_0 \\ \text{and} \\ p(0) = p_0\,, \qquad l(0) = l_0\,, \qquad q(0) = q_0 \end{gathered} \tag{1.225}$$

It is interesting to notice that the set 1.224 is similar to that which describes a motion of mass under action of a given force.

As is well known, the system of ray equations and conditions 1.225 form the Cauchy initial value problem and, in accordance with the theorem of uniqueness, they uniquely define the ray family. Suppose that the initial conditions are given on the surface Q, Fig. 1.29. Then, it is obvious that they define a certain pencil of rays, and its cross-section becomes smaller as the area of Q decreases. Correspondingly, the same result is applied to the construction of wavefronts (isochrones).

Runge-Kutta method

In order to obtain a family of rays we usually solve the system 1.224 numerically, and one of the most accurate procedures is the Runge-Kutta method, which has different orders. To illustrate its main idea, consider the simplest initial value problem:

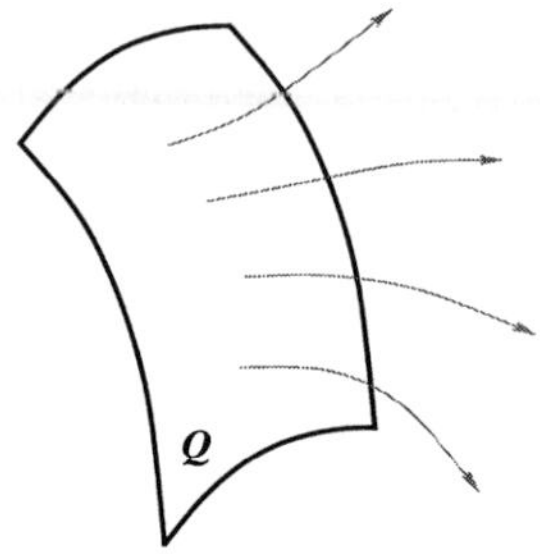

Figure 1.29: Surface with initial conditions.

$$y' = f(x,\, y) \qquad \text{and} \qquad y(x_0) = y_0\,, \tag{1.226}$$

that is, the first derivative $y'(x)$ is given everywhere, as well as the value of the function at the initial point $y(x_0)$. In terms of a ray, this means that we know one point with the coordinates x_0, y_0, through which it passes, and the ray slope at each point. Correspondingly, our goal is to find the function $y(x)$ or the integral curve (the ray) satisfying eqs. 1.226. In other words, we have to determine the ray passing through the initial point. It seems that the most natural "tool" to perform such an extrapolation is the Taylor series

$$y(x+h) \;=\; y(x) + h\,\frac{y'(x)}{1!} + h^2\,\frac{y''(x)}{2!} + h^3\,\frac{y'''(x)}{3!} + R_n(g)\,, \tag{1.227}$$

where h is the distance between two points and

$$R_n(g) = h^{n+1}\,\frac{y^{(n+1)}(g)}{(n+1)!} \tag{1.228}$$

is the remainder of the series and g is some point between x and $x+h$.

First, suppose that the width h of the interval is small, so we can ignore all terms in the right-hand side of eq. 1.227 except the first two. This gives

$$y(x+h) = y(x) + h\,y'(x) = y(x) + h\,f(x,\, y) \tag{1.229}$$

or

$$y(x+h) = y(x) + a\,k_1\,, \tag{1.230}$$

where

$$a = 1 \qquad \text{and} \qquad k_1 = h\,f(x,\, y)$$

In this case, k_1 is the product of the interval width and the value of derivative, $f(x,\, y)$. In accordance with eq. 1.230, calculation of the function at the point $x+h$ requires knowledge only of its first derivative at point x, as well as $y(x)$. In other words, this equation represents the basic Euler formula, (Appendix D). Inasmuch as it contains the multiplier h in the first power, eq. 1.230 is called the first-order Runge-Kutta procedure. As soon as the second point $x+h$, $y(x+h)$, is found, eq. 1.230 is used again to determine other points of the ray. This simple case of direct use of the Taylor series to solve eqs. 1.226 is rather an exception because we need to know only the first derivative $f(x, y)$ which is given. In general, however, when it is desirable to increase accuracy of calculations of $y(x+h)$ and, therefore, preserve more terms, the Taylor series becomes useless, since the derivatives of higher order are unknown. For this reason, Runge and Kutta developed an approach that allows one to calculate the function $y(x+h)$ with the same exact accuracy as the sum of several first terms in the Taylor series. The order of the derivative of the last term of such a sum defines the order of the corresponding Runge-Kutta formula. For instance, eq. 1.230 illustrates the simplest formula of Runge-Kutta method, and its order is unity, because k_1 contains only the first derivative, $f(x, y)$.

Now we derive the second-order formula, which can be represented in the following form

$$y(x+h) = y(x) + a\,k_1 + b\,k_2\,, \tag{1.231}$$

where

$$k_1 = h\,f(x, y) \qquad \text{and} \qquad k_2 = h\,f(x+\alpha\,h,\ y+\beta\,k_1)\,, \tag{1.232}$$

and the coefficients a, b, α and β are unknown. As follows from the main concept of the Runge-Kutta method, they have to be chosen in such a way that eq. 1.231 and the sum of three terms of the Taylor series

$$y(x+h) \;=\; y(x) + h\,f(x, y) + h^2\,\frac{f'_x(x, y)}{2} \tag{1.233}$$

gives the same result. Here

$$y''(x) \;=\; \frac{d\,f(x, y)}{d\,x} \;=\; f'_x(x, y)$$

In order to determine these constants, we expand k_2 in a Taylor series and keep the two first terms. This gives

$$k_2 = h\,\Big[f(x, y) + \alpha\,h\;\; f'_x(x, y) + \beta\,h\,f'_y(x, y)\,f\Big] \tag{1.234}$$

and

$$f'_y(x, y) \;=\; \frac{d\,f(x, y)}{d\,y}$$

Since y is a function of x, we have

$$f'_x(x, y) \;=\; \frac{\partial f}{\partial x} + \frac{\partial f}{\partial y}\,y' \;=\; \frac{\partial f}{\partial x} + \frac{\partial f}{\partial y}\,f \tag{1.235}$$

Then, eqs. 1.231 and 1.233 become

$$y(x+h) = y(x) + a\,h\,f(x,\,y) + b\,h\,\Big[f(x,\,y) + \alpha\,h\,f'_x(x,\,y) + \beta\,h\,f\,f'_y(x,\,y)\Big]$$

and

$$y(x+h) = y(x) + h\,f(x,\,y) + \frac{h^2}{2}\,\Big[f'_x(x,\,y) + f\,f'_y(x,\,y)\Big] \qquad (1.236)$$

Inasmuch as both last equations have to give the same result, we obtain

$$a + b = 1\,, \qquad b\,\alpha = \frac{1}{2}, \qquad \text{and} \qquad b\,\beta = \frac{1}{2} \qquad (1.237)$$

This is the system of three equations and four unknowns, and, therefore, it has infinitely many solutions. For each of them, eq. 1.231 is the second-order Runge-Kutta formula, since it replaces the sum of the first three terms of the Taylor series. In particular, when

$$a = b = \frac{1}{2} \qquad \text{and} \qquad \alpha = \beta = 1\,,$$

we obtain

$$y(x+h) = y(x) + \frac{h}{2}\,f(x,\,y) + \frac{h}{2}\,f(x+h,\,y+k_1)\,, \qquad (1.238)$$

and it corresponds to the improved Euler formula, (Appendix D). Thus, eq. 1.238 permits us to find the function $y(x+h)$ using the initial condition and the values of slope, $f(x,\,y)$, at end points of the interval. It is essential that unlike the Taylor series eq. 1.238 does not require a knowledge of the second derivative. In the same manner we can obtain Runge-Kutta formulas of higher order. For example, in the case of the fourth order it has the form

$$y(x+h) = y(x) + a\,k_1 + b\,k_2 + c\,k_3 + d\,k_4\,, \qquad (1.239)$$

where

$$k_1 = h\,f(x,\,y)\,, \qquad k_2 = h\,f(x+\alpha_1\,h,\,y+\beta_1\,k_1)\,,$$

$$k_3 = h\,f(x+\alpha_2\,h,\,y+\beta_2\,k_1+\beta_3\,k_2)\,, \qquad (1.240)$$

$$k_4 = h\,f(x+\alpha_3\,h,\,y+\beta_4\,k_1+\beta_5\,k_2+\beta_6\,k_3)$$

In order to obtain the same result as the sum of the first four terms of the Taylor series, we must solve a system of 11 equations with 13 unknowns:

$$a\,,\ b\,,\ c\,,\ d\,,\ \alpha_1\,,\ \alpha_2\,,\ \alpha_3\,,\ \beta_1\,,\ \beta_2\,,\ \beta_3\,,\ \beta_4\,,\ \beta_5\,,\ \beta_6$$

As before, there is an infinite number of solutions, and the most commonly used set of values of these constants gives the formula

$$y(x+h) = y(x) + \frac{1}{6}\,(k_1 + 2\,k_2 + 2\,k_3 + k_4)\,, \tag{1.241}$$

where

$$\begin{aligned} k_1 &= h\,f(x,\,y)\,, \qquad k_2 = h\,f\left(x + \frac{1}{2}\,h,\, y + \frac{1}{2}\,k_1\right)\,, \\ k_3 &= h\,f\left(x + \frac{1}{2}\,h,\, y + \frac{1}{2}\,k_2\right)\,, \qquad k_4 = h\,f(x+h,\, y+k_3) \end{aligned} \tag{1.242}$$

Ray behavior in different media

In the next two examples we assume that the velocity of propagation depends on the z-coordinate only, and it is directed downward. Besides, the ray distribution possesses axial symmetry so that it is enough to consider the ray behavior in xz plane. Then, in accordance with eq. 1.224, we have

$$\frac{d\,x}{d\,\tau} = c^2\,p\,, \qquad \frac{d\,z}{d\,\tau} = c^2\,q \tag{1.243}$$

and

$$\frac{d\,p}{d\,\tau} = 0\,, \qquad \frac{d\,q}{d\,\tau} = -\frac{1}{c}\frac{d\,c}{d\,z} \tag{1.244}$$

Here

$$p = p_x \qquad \text{and} \qquad q = p_z$$

Because the function c is independent of x, $(dp/d\tau = 0)$, we conclude that the horizontal component p of slowness is constant along each ray

$$p = \text{const} \tag{1.245}$$

while its vertical component q varies.

As follows from eq. 1.214:

$$q = \sqrt{\frac{1}{c^2} - p^2} \tag{1.246}$$

Making use of eq. 1.243, we obtain

$$\frac{d\,x}{d\,z} = \frac{p}{q} \qquad \text{or} \qquad d\,x = \frac{p}{q}\,d\,z \tag{1.247}$$

and

$$d\,\tau = \frac{1}{c^2\,q}\,d\,z \tag{1.248}$$

Integration of the last two equations yields

$$x = p\int\limits_{z_0}^{z} \frac{c(\zeta)\,d\zeta}{\sqrt{1-p^2\,c^2(\zeta)}} \tag{1.249}$$

and

$$\tau = \int\limits_{z_0}^{z} \frac{d\zeta}{c(\zeta)\,\sqrt{1-p^2\,c^2(\zeta)}} \tag{1.250}$$

Here z_0 is an arbitrary point on the ray. These equations have been derived in Part I, and they demonstrate that in this case, $c = c(z)$, the rays can be described using two alternative approaches. One of them is directly based on the solution of a system of differential equations, while the other proceeds from the integrals 1.249 and 1.250.

Example one Suppose that an elementary source is situated at a point with coordinates $x = 0$, $z = z_0$ and the velocity increases with depth z as

$$c(z) = a\,\sqrt{z}$$

Substitution of the latter into eqs. 1.249 and 1.250 gives

$$x = p\,a\int\limits_{z_0}^{z} \frac{\sqrt{\zeta}\,d\zeta}{\sqrt{1-p^2\,a^2\,\zeta}} \qquad \text{and} \qquad \tau = \frac{1}{a}\int\limits_{z_0}^{z} \frac{d\zeta}{\sqrt{\zeta}\,\sqrt{1-p^2\,a^2\,\zeta}} \tag{1.251}$$

Both integrals are tabular, and we obtain

$$x = \frac{1}{p\,a}\left(\sqrt{z_0}\,\sqrt{1-p^2\,a^2\,z_0} - \sqrt{z}\,\sqrt{1-p^2\,a^2\,z}\right) + \frac{1}{p^2\,a^2}\left(\arcsin p\,a\,\sqrt{z} - \arcsin p\,a\,\sqrt{z_0}\right) \tag{1.252}$$

and

$$\tau = \frac{2}{p\,a^2}\left(\arcsin p\,a\,\sqrt{z} - \arcsin p\,a\,\sqrt{z_0}\right) \tag{1.253}$$

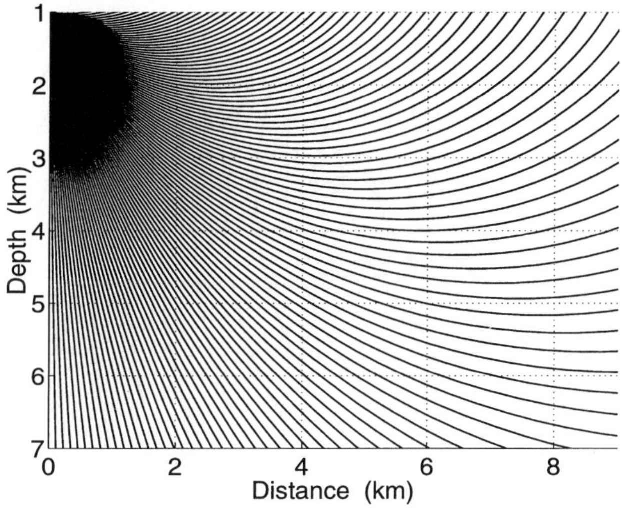

Figure 1.30: Rays radiated from the source at depth $z_0 = 1$ in the medium with velocity $c(z) = a\sqrt{z}$ and $a = 1$.

Results of calculation of the ray family and the system of isochrones, when $z \geq z_0 = 1$ km, are shown in Fig. 1.30 and 1.31, respectively. It is clear that all points are regular, since only one ray passes through each point. Because of the velocity increase with depth z, all rays have turning points where the coordinate z reaches its maximum. Correspondingly, the vertical slowness component vanishes at this point and

$$p = \frac{1}{c(z_{\max})},$$

which allows us to determine $z_{\max}$.

We have considered the ray behavior in the xz-plane. Due to the axial symmetry, the same picture of rays is observed at any vertical plane containing the z-axis. Simplicity of the isochrones is obvious, Fig. 1.31, and the geometry of rays can be visualized easily from their shape.

Example two Unlike the previous case, we now assume that the velocity of propagation decreases with depth:

$$c(z) = \frac{a}{\sqrt{z}}, \tag{1.254}$$

and the source is located at the same point, $x = 0$, $z = z_0$. In accordance with eq. 1.249, the x-coordinate of a ray point

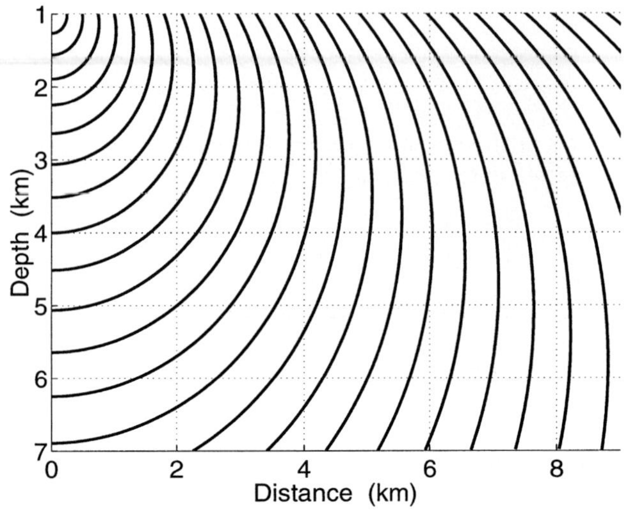

Figure 1.31: Wavefronts plotted with the increment $\Delta\tau = 0.25$.

$$x = p\,a \int\limits_{z_0}^{z} \frac{d\zeta}{\sqrt{\zeta}\sqrt{1 - \dfrac{p^2\,a^2}{\zeta}}} = p\,a \int\limits_{z_0}^{z} \frac{d\zeta}{\sqrt{\zeta - p^2\,a^2}}$$

or

$$x = 2\,p\,a\,\sqrt{z - p^2\,a^2}\,\Big|_{z_0}^{z} \tag{1.255}$$

By analogy, we can obtain an expression for the function $\tau(z)$.

Before we substitute the limits into eq. 1.255, it is important to distinguish two groups of rays. In the first one, rays are directed downward, and they do not have turning points. Correspondingly, the z-coordinate monotonically increases, and eq. 1.255 gives

$$x = 2\,p\,a\left(\sqrt{z - p^2\,a^2} - \sqrt{z_0 - p^2\,a^2}\right) \tag{1.256}$$

Bearing in mind the orientation of rays in this group, we see that the vertical component q of their slowness vectors is positive at all ray points

$$q = \sqrt{\frac{1}{c^2} - p^2} > 0$$

In particular, we have at the source

$$q = \sqrt{\frac{z_0}{a^2} - p^2} > 0$$

The behavior of rays in the second group is different. First, they are directed upward at the source and therefore their component q is negative:

$$q = \sqrt{\frac{z_0}{a^2} - p^2} < 0$$

These rays have turning points, where z reaches its minimum

$$q = 0 \qquad \text{or} \qquad z_{\text{min}} = p^2\, a^2 \tag{1.257}$$

For the ascending part of these rays eq. 1.255 gives

$$x = 2\,p\,a \left(\sqrt{z_0 - p^2\, a^2} - \sqrt{z - p^2\, a^2}\right)$$

For example, the horizontal coordinate of the turning point is

$$x_{\text{min}} = 2\,p\,a \left(\sqrt{z_0 - p^2\, a^2} - \sqrt{z_{\text{min}} - p^2\, a^2}\right)$$

If points are located on the descending branch, eq. 1.255 yields

$$x = x_{\text{min}} + 2\,p\,a \left(\sqrt{z - p^2\, a^2} - \sqrt{z_{\text{min}} - p^2\, a^2}\right)$$

or

$$x = 2\,p\,a \left(\sqrt{z - p^2\, a^2} + \sqrt{z_0 - p^2\, a^2} - 2\sqrt{z_{\text{min}} - p^2\, a^2}\right)$$

Making use of eq. 1.257, we obtain for rays of the second group

$$x = 2\,p\,a \left(\sqrt{z - p^2\, a^2} + \sqrt{z_0 - p^2\, a^2}\right) \qquad \text{if} \qquad x \geq x_{\text{min}} \tag{1.258}$$

As follows from eqs. 1.256 and 1.258 rays are described by parabolas, Fig. 1.32, given by

$$z = p^2\, a^2 + \left(\frac{x}{2\,p\,a} \pm \sqrt{z_0 - p^2\, a^2}\right)^2 \tag{1.259}$$

As we see they are located only within a certain region around the z-axis, and the boundary of this area with the shadow zone is formed by the envelope (caustic) of rays.

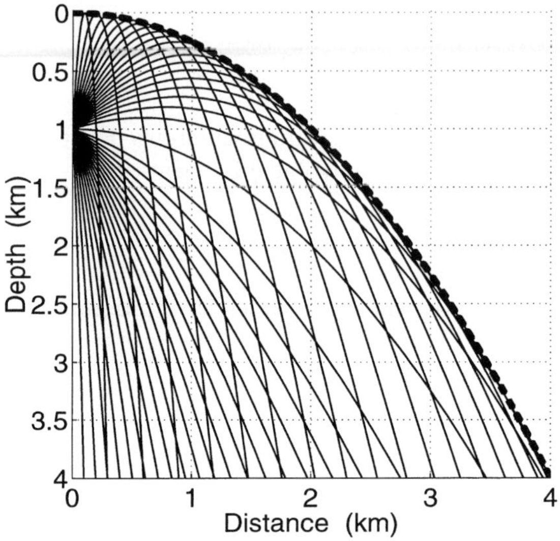

Figure 1.32: Source at depth $z_0 = 1$ km in the medium with velocity $c(z) = a/\sqrt{z}$ and $a = 1\ \mathrm{km}^{1/2}\ \mathrm{s}^{-1}$.

Equation for the caustic

In order to investigate this interesting feature of ray behavior, let us proceed from eqs. 1.256 and 1.258. Combining them, we obtain

$$\frac{x}{2\,p\,a} = \sqrt{z - p^2\,a^2} \pm \sqrt{z_0 - p^2\,a^2}\,, \tag{1.260}$$

where the horizontal component p of the slowness vector plays a role of some parameter. As was demonstrated in Appendix D, the equation for the envelope of a ray family can be obtained by taking the first derivative of the solution of a differential equation with respect to this parameter. Correspondingly differentiation of eq. 1.260 gives

$$-\frac{x}{2\,p^2\,a} = -\frac{p\,a^2}{\sqrt{z - p^2\,a^2}} \mp \frac{p\,a^2}{\sqrt{z_0 - p^2\,a^2}}$$

or

$$\frac{x}{2\,p^3} = \frac{a^3}{\sqrt{z - p^2\,a^2}} \pm \frac{a^3}{\sqrt{z_0 - p^2\,a^2}} \tag{1.261}$$

The left-hand side of this equality is always positive and, eq. 1.261 takes place for all z if we choose the "+" sign between the terms, that is

$$\frac{x}{2\,p^3} = \frac{a^3}{\sqrt{z - p^2\,a^2}} + \frac{a^3}{\sqrt{z_0 - p^2\,a^2}} \tag{1.262}$$

This means that the envelope is related to rays that are directed upwards in the vicinity of the source.

The last equality can be written as

$$\frac{x}{2\,p^3\,a^3} = \frac{\sqrt{z_0 - p^2\,a^2} + \sqrt{z - p^2\,a^2}}{\sqrt{z_0 - p^2\,a^2}\,\sqrt{z - p^2\,a^2}}$$

Making use of eq. 1.258, we have

$$p^2\,a^2 = \sqrt{z_0 - p^2\,a^2}\,\sqrt{z - p^2\,a^2}$$

or

$$p^4\,a^4 = \left(z_0 - p^2\,a^2\right)\left(z - p^2\,a^2\right)$$

Whence

$$p^2\,a^2 = \frac{z\,z_0}{z + z_0} \tag{1.263}$$

Substituting it into the equation 1.258 of rays, we obtain

$$\begin{aligned} x &= 2\sqrt{\frac{z\,z_0}{z+z_0}}\left(\sqrt{z - \frac{z\,z_0}{z+z_0}} + \sqrt{z_0 - \frac{z\,z_0}{z+z_0}}\right) \\ &= 2\sqrt{\frac{z\,z_0}{z+z_0}}\left(\sqrt{\frac{z^2}{z+z_0}} + \sqrt{\frac{z_0^2}{z+z_0}}\right) \end{aligned}$$

Thus, the equation of the caustic is given by the parabola

$$x = 2\sqrt{z\,z_0} \qquad \text{or} \qquad z = \frac{1}{z_0}\left(\frac{x}{2}\right)^2 \tag{1.264}$$

Note that the ray, which is horizontal near the source, divides the ray family into the two groups described above. Inasmuch as $q(z_0) = 0$, we have

$$p^2\,a^2 = z_0$$

Substituting the latter into eq. 1.259, we arrive at the equation of this ray

$$z = z_0 + \frac{1}{z_0}\left(\frac{x}{2}\right)^2, \tag{1.265}$$

which is directed downward and asymptotically approaches the caustic but never touches it.

The behavior of the isochrones in this model is shown in Fig. 1.33. Taking into account that there is a group of rays with turning points, it is not simple to obtain the isochrones directly from Fig. 1.32. Their geometry is rather complicated, in particular,

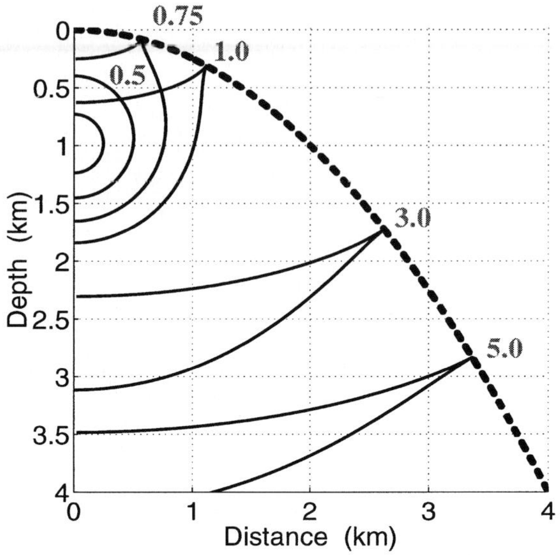

Figure 1.33: Isochrones in the velocity model of eq. 1.254. The bold numbers indicate the values of the time τ.

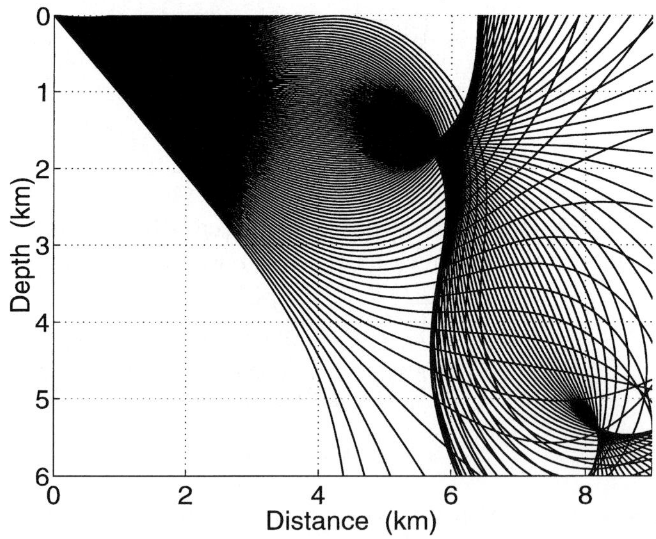

Figure 1.34: Rays radiated from the source located at the coordinate origin in the medium with velocity given by eq. 1.266.

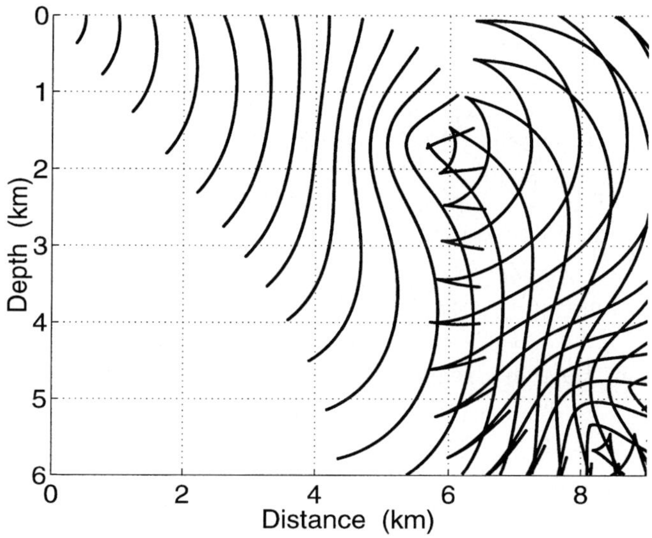

Figure 1.35: System of isochrones corresponding to the rays shown in Fig. 1.34. The isochrones are plotted with the increment $\Delta\tau = 0.25$ s.

some of them intersect each other and form the cusps at points of the caustic.

Example three Suppose that the velocity of propagation varies in both the horizontal and vertical directions as

$$c(x,\, z) = c_0 \begin{cases} 2 & \text{if} \quad z \leq 0\,, \\ 2 + \sin \alpha x \, \sin \beta z & \text{if} \quad z > 0\,, \end{cases} \tag{1.266}$$

where α and β have the dimension km^{-1}, and for this example we choose

$$\alpha = \beta = 1$$

Since the velocity is a function of two spatial coordinates, the system of ray equations becomes

$$\begin{aligned} \frac{dx}{d\tau} &= c^2 p\,, & \frac{dz}{d\tau} &= c^2 q\,, \\ \frac{dp}{d\tau} &= -\frac{1}{c}\frac{\partial c}{\partial x}\,, & \frac{dq}{d\tau} &= -\frac{1}{c}\frac{\partial c}{\partial z} \end{aligned} \tag{1.267}$$

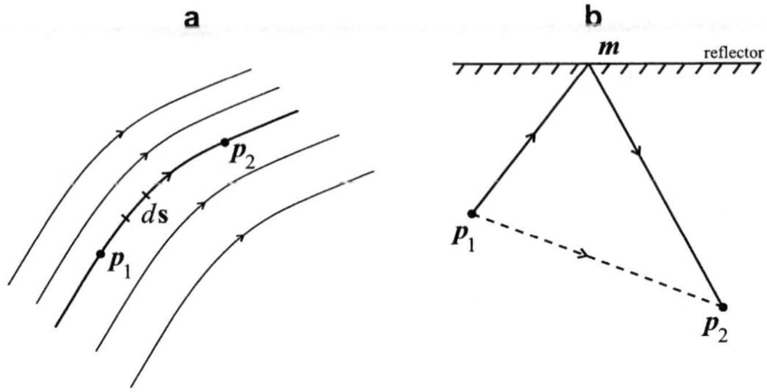

Figure 1.36: Rays in a regular (a) and an irregular (b) neighborhood.

Results of numerical solution of eqs. 1.267 are shown in Fig. 1.34 and 1.35. Because there are zones of high and low velocity, the ray behavior is much more complicated than that in the previous examples. For example, as is seen in Fig. 1.34, several envelopes (caustics) with different shapes are observed.

1.5 Fermat's principle and ray equations

Earlier, proceeding from Newton's laws and the principle of conservation of mass, as well as from the relationship between the density and the pressure, we arrived at Fermat's principle and ray equations. It may seem that those two results characterize completely different features of ray behavior. It turns out, however, that they are related to each other and, for example, ray equations can be derived from the Fermat's principle. Before we demonstrate this important fact, let us formulate this principle which, as was shown in Part I, can be done for two cases. First, suppose that a ray neighborhood is regular. This simply means that only one ray passes through each point, Fig. 1.36a. Then, the acoustic length

$$J = \int_{p_1}^{p_2} n\, ds \tag{1.268}$$

along the actual ray between two arbitrary points p_1 and p_2 is smaller than that of any other curve that connects p_1 and p_2, and if lies in a regular neighborhood of this ray.

Here

$$n = \frac{c_0}{c}$$

is the refraction index, c_0 is the velocity of propagation at some reference point and ds is the displacement along the ray.

It is clear that eq. 1.268 can be written as

$$J = c_0 \int\limits_{p_1}^{p_2} \frac{d\,s}{c} = c_0 \int\limits_{p_1}^{p_2} d\,t = c_0\, t\,, \tag{1.269}$$

and, correspondingly, Fermat's principle is also known as that of least time. In particular, if the velocity c remains constant along the ray, eq. 1.268 becomes

$$J = n \int\limits_{p_1}^{p_2} ds = n\,s\,, \tag{1.270}$$

and the path along the ray is the shortest.

Now it is proper to emphasize that deriving Fermat's principle, Part I, we assumed that the neighborhood of rays is regular. Let us remove this restriction and consider a more general case in which several rays may pass through any point in the ray neighborhood. Such an example was given in Part I and it is again shown in Fig. 1.36b. As we see, there are two rays passing through point p_2, namely, the direct and reflected ones. The same pattern is observed in its vicinity. The acoustic length along the direct ray $p_1 p_2$ is the absolute minimum with respect to other paths connecting these points. The acoustic length of the reflected ray $p_1 m p_2$, however, gives a minimum only relative to other reflected trajectories that have a point at the reflecting boundary. Thus, unlike the previous case of the regular neighborhood, the acoustic length of the reflected ray does not correspond to the absolute minimum, and this happens because there are two families of rays in its vicinity. At the same time, as was shown in Part I, it represents a local minimum, and there are unlimited number of cases, when such a behavior takes place. It turns out that in general, when the condition of regularity is violated, Fermat's principle requires that the acoustic length along the ray be stationary. In other words, the integral 1.268 has an extremum along the ray which can be either a local minimum, or maximum, or a saddle point. This remarkable feature distinguishes rays from any other curves connecting the same points, Fig. 1.36. It may be important to mention that we compare the acoustic lengths along curves located close to each other. In order to analyze Fermat's principle in some detail, it is useful to describe basic elements of variational calculus. First, consider several examples illustrating the problems that can be solved using this branch of mathematics.

Example one: Motion of a particle Suppose that a particle moves through an inhomogeneous medium with a given velocity $v(x,\, y,\, z)$. Let us find the time T of a particle motion along some line l with the end points $(x_0,\, y_0,\, z_0)$ and $(x_1,\, y_1,\, z_1)$. It is obvious that T is defined by the equation

$$T = \int\limits_{l} \frac{d\,s}{v(x,\, y,\, z)}\,, \tag{1.271}$$

and assuming that these points are fixed, the integral, in general, varies when the path l

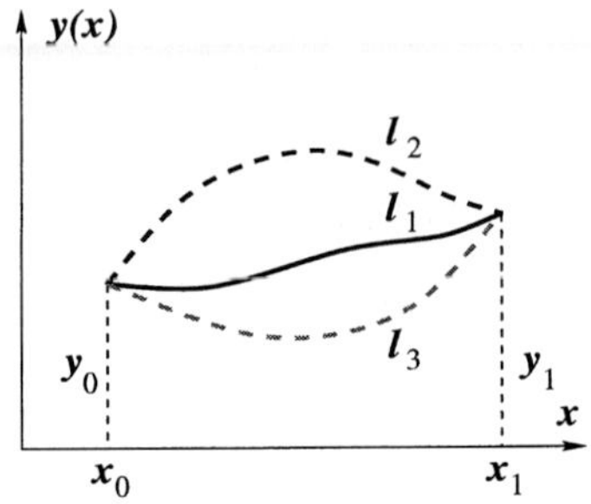

Figure 1.37: Illustration of eq. 1.272.

changes. To emphasize this fact, it is conventional to call this integral functional of l.

If the coordinate x plays a role of a parameter of the line l,

$$y = y(x) \qquad \text{and} \qquad z = z(x)\,,$$

the displacement ds is equal to

$$ds = \sqrt{(dx)^2 + (dy)^2 + (dz)^2} = \sqrt{1 + (y')^2 + (z')^2}\,dx$$

and

$$T = \int\limits_{x_0}^{x_1} \frac{\sqrt{1 + (y')^2 + (z')^2}}{v(x,\,y,\,z)}\,dx \tag{1.272}$$

Here y' and z' are the first derivatives dy/dx and dz/dx. Our goal is to find such functions $y(x)$ and $z(x)$ that the integral $T(l)$ is minimal, provided that these unknowns satisfy the following conditions

$$y(x_0) = y_0\,, \qquad z(x_0) = z_0\,,$$

$$y(x_1) = y_1\,, \qquad z(x_1) = z_1$$

In the case when the line l is located in the plane, Fig. 1.37, the functional T is simplified to

$$T = \int\limits_{x_0}^{x_1} \frac{\sqrt{1 + (y')^2}}{v(x,\,y)}\,dx \tag{1.273}$$

Correspondingly, we have to determine a single function $y(x)$, that is, the line l which obeys the conditions

$$y(x_0) = y_0 \qquad \text{and} \qquad y(x_1) = y_1$$

It is natural that eq. 1.273 also describes one of the problems of geometrical acoustics,

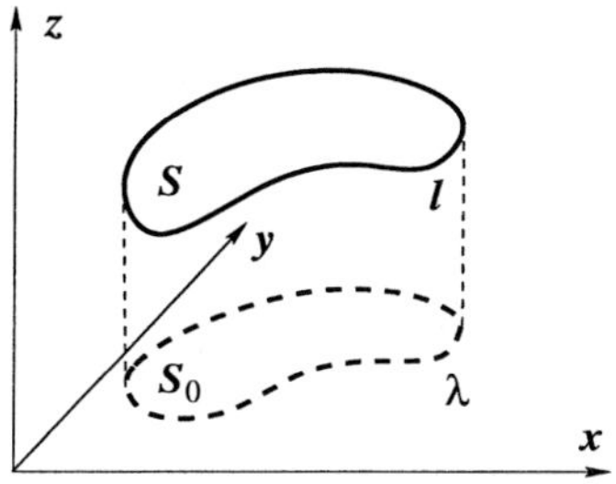

Figure 1.38: Illustration of eq. 1.274

namely, a determination of path l with fixed terminal points such that the functional T has a local minimum. This indicates some common features of mechanics and acoustics.

Example two: Surface integral Another problem is related to surface integration. Suppose that a closed line l is given. Then our task is to find such surface S, bounded by l, Fig. 1.38, which has the smallest area. Let $\mathcal{L}$ be a projection of the line l on the xy-plane; $z(x,\, y)$ is the equation of the unknown surface, while S_0 is the range of integration, bounded by $\mathcal{L}$. Respectively, the area of surface S is described by the integral

$$A = \iint\limits_{S_0} \sqrt{1 + z_x^2 + z_y^2}\, dx\, dy \tag{1.274}$$

Here z_x and z_y are the partial derivatives of $z(x,\, y)$ with respect to x and y. A variation of the surface leads to a change of A, that is, the integral is a functional of S. In our case, the values of function $z(x,\, y)$ at points of the curve $\mathcal{L}$ are the terminal ones, and they give the coordinates z of the closed line l. In general, the main goal of variational calculus is to find the maximal and minimal values of functionals of lines and surfaces, expressed in terms of definite integrals.

Analogy with differential calculus

As is well known, a similar task arises when we want to find the maximal and minimal values of some function that depends on several arguments. The latter is directly related to determination of extrema of a function, that is, such values of independent variables when the function reaches either its maximum or minimum. For illustration, a distribution of extrema of some function $y = y(x)$ is shown in Fig. 1.39. In order to find them, we have to solve the differential equation of the first order

$$y'(x) = 0 \tag{1.275}$$

This means that eq. 1.275 represents a necessary condition for an extremum to exist, and it can be found as a solution of this differential equation. In a similar manner, we

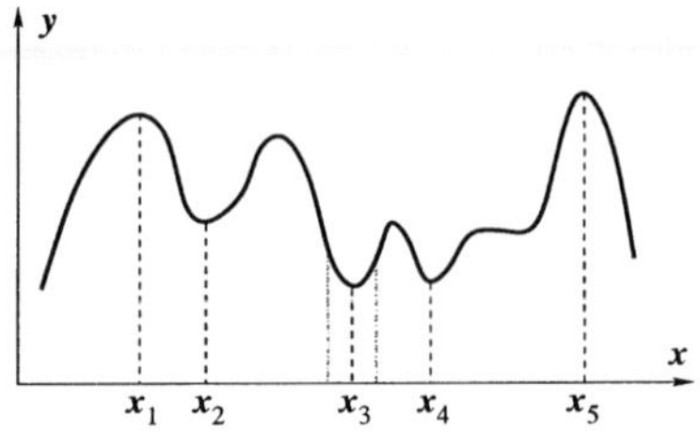

Figure 1.39: Distribution of extrema of $y(x)$.

treat the problem for functionals. For example, in the case described by eq. 1.272, we look for such line l that T has a value which does not exceed those for all other lines, sufficiently close to it. If for some line or surface the functional has a value which is not less (or not more) than that for all other lines or surfaces, situated in their vicinity, then we say that the functional has an extremum at this line or surface.

Now we demonstrate that the line $y(x)$ or surface $z(x, y)$ corresponding to an extremum for some functional has to satisfy certain differential equations, and our task is to find them. The analogy with differential calculus is obvious. Further, we restrict ourselves to functionals for lines only.

Euler's equation for function $\mathbf{F}(\mathbf{x}, \mathbf{y}, \mathbf{y}')$

At the beginning we assume that functionals are considered for lines l which are located in the xy-plane, Fig. 1.37, that is

$$J = \int_{x_0}^{x_1} F(x, y, y')\, dx \,, \tag{1.276}$$

where F is the given function of three arguments: x, $y(x)$, and $y'(x)$.

It is also implied that the integrand F is a continuous function along with its first and second derivatives at all points within the interval of integration. It is clear that if the function $y(x)$, that is, the curve $y = y(x)$ is given, the functional J has a certain value. In addition, we assume that values of the function $y(x)$ are fixed at the terminal points

$$y(x_0) = y_0 \qquad \text{and} \qquad y(x_1) = y_1 \tag{1.277}$$

Besides, we suppose that the unknown function $y(x)$ has a continuous derivative, that is, the slope of the curve l varies gradually.

By definition, the functional J reaches the local extremum for the curve $y(x)$, satisfying the condition 1.277, if the value of the functional for this curve is smaller (or greater) than its value for any other curve, obeying the same condition and located near the line $y(x)$, Fig. 1.40, that is

$$|\, y_1(x) - y(x)\,| < \epsilon \,,$$

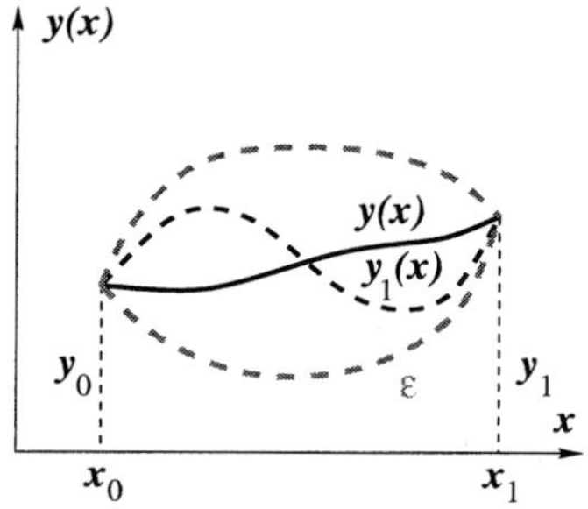

Figure 1.40: Illustration of eq. 1.280.

where ϵ is a small number.

This notation for a local extremum is identical to that for the function. Now we are ready to derive the condition for the function $y(x)$ which provides an extremum of the functional J. First, let us take any function $\eta(x)$ vanishing at the terminal points

$$\eta(x_0) = \eta(x_1) = 0 \tag{1.278}$$

and form a new function

$$y_1(x) = y(x) + \alpha\,\eta(x) \tag{1.279}$$

Here $y(x)$ gives an extremum of functional J and α is a small parameter.

In accordance with eq. 1.278, the function $y_1(x)$ satisfies the same boundary conditions as $y(x)$. Substituting $y_1(x)$ into the integrand of eq. 1.276 and performing integration, we arrive at the function of parameter α

$$J(\alpha) = \int_{x_0}^{x_1} F\left[\,x\,,\, y + \alpha\,\eta(x)\,,\, y' + \alpha\,\eta'(x)\,\right] dx\,, \tag{1.280}$$

For sufficiently small α the difference between the curves $y(x)$ and $y_1(x)$ is small, that is, they are close to each other, and in the limit $\alpha \to 0$ the curves coincide. Changing α, we obtain different values of the functional J corresponding to certain curves $y_1(x)$. Bearing in mind that $y(x)$ gives an extremum of the functional J, we conclude that the function $J(\alpha)$ has an extremum when $\alpha = 0$. Therefore, its first derivative vanishes

$$\frac{\partial\, J(\alpha)}{\partial\, \alpha} = 0 \qquad \text{if} \qquad \alpha = 0 \tag{1.281}$$

Differentiating the integrand, we take into account the fact that a change in α causes variation in two variables y_1, y_1', as shown schematically in Fig. 1.41. This procedure gives the following expression for the derivative of the integrand

$$\frac{\partial\, F}{\partial\, \alpha} = \frac{\partial\, F}{\partial\, y_1}\,\frac{\partial\, y_1}{\partial\, \alpha} + \frac{\partial\, F}{\partial\, y_1'}\,\frac{\partial\, y_1'}{\partial\, \alpha} = \frac{\partial\, F}{\partial\, y}\,\eta(x) + \frac{\partial\, F}{\partial\, y'}\,\eta'(x)$$

Thus,

$$\frac{\partial J(0)}{\partial \alpha} = \int_{x_0}^{x_1} [\, F_y(x,\, y,\, y')\, \eta(x) + F_{y'}(x,\, y,\, y')\, \eta'(x)\,]\; dx \;=\; 0 \qquad \text{if} \quad \alpha = 0 \tag{1.282}$$

Here

$$F_y = \frac{\partial F}{\partial y} \qquad \text{and} \qquad F_{y'} = \frac{\partial F}{\partial y'}$$

The right-hand side of eq. 1.282 can be represented as

$$\int_{x_0}^{x_1} F_y(x,\, y,\, y')\, \eta(x)\, dx + \int_{x_0}^{x_1} F_{y'}(x,\, y,\, y')\, d\eta(x) \tag{1.283}$$

since

$$d\eta(x) \;=\; \eta'(x)\, dx$$

Applying the integration by parts, the second integral becomes

$$\int_{x_0}^{x_1} F_{y'}(x,\, y,\, y')\, d\eta(x) \;=\; F_{y'}(x,\, y,\, y')\, \eta(x)\Big|_{x_0}^{x_1} - \int_{x_0}^{x_1} \frac{d}{dx}\, F_{y'}(x,\, y,\, y')\, \eta(x)\, dx \tag{1.284}$$

Due to the condition 1.278 the latter is simplified, and we obtain

$$\int_{x_0}^{x_1} F_{y'}(x,\, y,\, y')\, \eta'(x)\, dx \;=\; -\int_{x_0}^{x_1} \frac{d}{dx}\, F_{y'}(x,\, y,\, y')\, \eta(x)\, dx \tag{1.285}$$

Substitution of eq. 1.285 into eq. 1.282 yields

$$J'(0) \;=\; \int_{x_0}^{x_1} \left[\, F_y(x,\, y,\, y') - \frac{d}{dx}\, F_{y'}(x,\, y,\, y') \right] \eta(x)\, dx \;=\; 0 \tag{1.286}$$

Inasmuch as the function $\eta(x)$ is arbitrary, we conclude that

$$F_y(x,\, y,\, y') - \frac{d}{dx}\, F_{y'}(x,\, y,\, y') \;=\; 0 \tag{1.287}$$

Thus, the function $y(x)$, describing the curve $y = y(x)$ which provides an extremum of functional J, eq. 1.276, has to satisfy eq. 1.287:

$$F_y - \frac{d}{dx}\, F_{y'} = 0$$

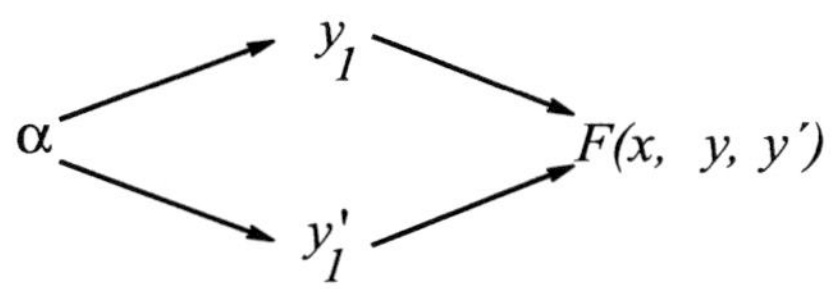

Figure 1.41: Derivation of eq. 1.282.

In order to obtain the derivative $dF_{y'}/dx$, we have to take into account the fact that the function $F_{y'}$ depends on x, y and y'. Correspondingly, a variation in x causes changes in all three arguments of $F_{y'}$. Therefore, we have

$$\frac{d}{dx} F_{y'} = \frac{\partial F_{y'}}{\partial y'} \frac{dy'}{dx} + \frac{\partial F_{y'}}{\partial y} \frac{dy}{dx} + \frac{\partial F_{y'}}{\partial x} = F_{y'y'}\, y'' + F_{y'y}\, y' + F_{y'x}\,,$$

and eq. 1.287 becomes

$$F_{y'y'}\, y'' + F_{y'y}\, y' + F_{y'x} - F_y = 0 \tag{1.288}$$

Here, for example, $F_{y'x}$ is the partial derivative of second order taken with respect to y' and x.

Equation 1.288 was derived by Euler, and it is usually called the Euler's equation. It is the ordinary differential equation of the second order with respect to the unknown function $y(x)$, and its general solution contains two arbitrary constants that are defined by the boundary conditions 1.278.

Euler equation for function F(x, y, y′, z, z′)

Next we consider a more general case in which the functional depends on two functions, for example eq. 1.272 is written as

$$J = \int_{x_0}^{x_1} F(x,\, y,\, y',\, z,\, z')\, dx \tag{1.289}$$

By analogy with the previous case we introduce two functions,

$$y_1(x) = y(x) + \alpha\, \eta(x) \qquad \text{and} \qquad z_1(x) = z(x) + \alpha_1\, \eta_1(x)\,, \tag{1.290}$$

where α and α_1 are small parameters, and $\eta(x)$ and $\eta_1(x)$ are arbitrary functions which are equal to zero at the terminal points

$$\eta(x_0) = \eta(x_1) = \eta_1(x_0) = \eta_1(x_1) = 0 \tag{1.291}$$

Substitution of functions $y_1(x)$ and $z_1(x)$ into the integrand of eq. 1.289 yields

$$J(\alpha,\, \alpha_1) = \int\limits_{x_0}^{x_1} F\left[x,\, y + \alpha\, \eta(x),\, y' + \alpha\, \eta'(x)\,, z + \alpha_1\, \eta_1(x),\, z' + \alpha_1\, \eta_1'(x)\right] dx \qquad (1.292)$$

Since the functions $y(x)$ and $z(x)$ provide an extremum of the functional $J(\alpha,\, \alpha_1)$, eq. 1.289, the partial derivatives $J_\alpha(\alpha,\, \alpha_1)$ and $J_{\alpha_1}(\alpha,\, \alpha_1)$ should vanish when the parameters α and α_1 are equal to zero

$$\frac{\partial\, J(\alpha,\, \alpha_1)}{\partial\, \alpha} = 0 \qquad \text{and} \qquad \frac{\partial\, J(\alpha,\, \alpha_1)}{\partial\, \alpha_1} = 0 \qquad \text{if} \quad \alpha = \alpha_1 = 0 \qquad (1.293)$$

Performing differentiation, we obtain

$$\int\limits_{x_0}^{x_1} \left[\, F_y\, \eta(x) + F_{y'}\, \eta'(x)\, \right] dx = 0$$

and

$$\int\limits_{x_0}^{x_1} \left[\, F_z\, \eta_1(x) + F_{z'}\, \eta_1'(x)\, \right] dx = 0$$

As in the previous case, integration by parts gives

$$F_{y'}\, \eta \bigg|_{x_0}^{x_1} + \int\limits_{x_0}^{x_1} \eta(x) \left(F_y - \frac{d}{dx}\, F_{y'} \right) dx = 0$$

and

$$F_{z'}\, \eta_1 \bigg|_{x_0}^{x_1} + \int\limits_{x_0}^{x_1} \eta_1(x) \left(F_z - \frac{d}{dx}\, F_{z'} \right) dx = 0$$

Making use of eq. 1.291, we arrive at two Euler's equations

$$F_y - \frac{d}{dx}\, F_{y'} = 0 \qquad \text{and} \qquad F_z - \frac{d}{dx}\, F_{z'} = 0 \qquad (1.294)$$

Thus, functions $y(x)$ and $z(x)$ provide an extremum of the functional J, eq. 1.289, if they satisfy the system of two ordinary differential equations of the second order given by eq. 1.294. In other words, the latter represents the necessary conditions for these functions.

Example: Ray as an extremum Now we can demonstrate that the functional J, describing the acoustic length

$$J = \int_{p_1}^{p_2} n\, ds = \int_{p_1}^{p_2} n(x,\, y,\, z)\, \sqrt{1 + (x')^2 + (y')^2}\, dz, \tag{1.295}$$

has an extremum if the integration path coincides with a ray. Here ds is the elementary displacement,

$$ds = \sqrt{(dx)^2 + (dy)^2 + (dz)^2} = \sqrt{1 + (x')^2 + (y')^2}\, dz\,, \tag{1.296}$$

where $x' = dx/dz\,,\ y' = dy/dz\,,\ z$ is the variable of integration, and $n(x,\, y,\, z)$ is the refraction index. Correspondingly, using new notations Euler's equations have the form

$$F_x - \frac{d}{dz}\, F_x' = 0 \qquad \text{and} \qquad F_y - \frac{d}{dz}\, F_y' = 0$$

As follows from eq 1.289, the function F in this case is

$$F = n(x,\, y,\, z)\, \sqrt{1 + (x')^2 + (y')^2} \tag{1.297}$$

First, we transform Euler's equations. It is clear that

$$F_x = \frac{\partial\, n}{\partial\, x}\, \sqrt{1 + (x')^2 + (y')^2} \tag{1.298}$$

and

$$F_y = \frac{\partial\, n}{\partial\, y}\, \sqrt{1 + (x')^2 + (y')^2} \tag{1.299}$$

We also have

$$F_{x'} = n(x,\, y,\, z)\, \frac{x'}{\sqrt{1 + (x')^2 + (y')^2}}$$

or, from eq. 1.296,

$$F_{x'} = n\, \frac{d\, x}{d\, z}\, \frac{d\, z}{d\, s} = n\, \frac{d\, x}{d\, s} \tag{1.300}$$

In the same manner, we obtain

$$F_{y'} = n\, \frac{d\, y}{d\, s} \tag{1.301}$$

Thus, Euler's equations become

$$\frac{\partial\, n}{\partial\, x}\, \sqrt{1 + (x')^2 + (y')^2} - \frac{\partial}{\partial\, z} \left(\frac{n\, x'}{\sqrt{1 + (x')^2 + (y')^2}} \right) = 0\,,$$

$$\frac{\partial n}{\partial y}\sqrt{1+(x')^2+(y')^2}-\frac{\partial}{\partial z}\left(\frac{n\,y'}{\sqrt{1+(x')^2+(y')^2}}\right)=0\,,$$

or

$$\frac{d}{d\,s}\left(n\,\frac{d\,x}{d\,s}\right)=\frac{\partial\,n}{\partial\,x}\,,\qquad \frac{d}{d\,s}\left(n\,\frac{d\,y}{d\,s}\right)=\frac{\partial\,n}{\partial\,y} \tag{1.302}$$

As we know, the latter are the ray equations. In order to derive the third equation of the ray system, we make use of the equality

$$\left(\frac{d\,x}{d\,s}\right)^2+\left(\frac{d\,y}{d\,s}\right)^2+\left(\frac{d\,z}{d\,s}\right)^2=1 \tag{1.303}$$

Its differentiation with respect to s gives

$$\frac{d\,x}{d\,s}\frac{d^2x}{d\,s^2}+\frac{d\,y}{d\,s}\frac{d^2y}{d\,s^2}+\frac{d\,z}{d\,s}\frac{d^2z}{d\,s^2}=0 \tag{1.304}$$

Multiplication of the last two equations by $\partial n/\partial s$ and n, respectively, and their summation gives

$$\frac{\partial\,n}{\partial\,s}\left(\frac{d\,x}{d\,s}\right)^2+n\,\frac{d\,x}{d\,s}\frac{d^2x}{d\,s^2}+\frac{\partial\,n}{\partial\,s}\left(\frac{d\,y}{d\,s}\right)^2+n\,\frac{d\,y}{d\,s}\frac{d^2y}{d\,s^2}$$
$$+\frac{\partial\,n}{\partial\,s}\left(\frac{d\,z}{d\,s}\right)^2+n\,\frac{d\,z}{d\,s}\frac{d^2z}{d\,s^2}=\frac{\partial\,n}{\partial\,s}$$

or

$$\frac{d\,x}{d\,s}\frac{\partial}{\partial s}\left(n\,\frac{d\,x}{d\,s}\right)+\frac{d\,y}{d\,s}\frac{\partial}{\partial s}\left(n\,\frac{d\,y}{d\,s}\right)+\frac{d\,z}{d\,s}\frac{\partial}{\partial s}\left(n\,\frac{d\,z}{d\,s}\right)=\frac{\partial\,n}{\partial\,s} \tag{1.305}$$

Making use of eq. 1.302 in place of eq. 1.305 we have

$$\frac{\partial\,n}{\partial\,x}\frac{d\,x}{d\,s}+\frac{\partial\,n}{\partial\,y}\frac{d\,y}{d\,s}+\frac{d\,z}{d\,s}\frac{\partial}{\partial s}\left(n\,\frac{d\,z}{d\,s}\right)=\frac{\partial\,n}{\partial\,s}$$

On the other hand, by definition, we have

$$\frac{\partial\,n}{\partial\,x}\frac{d\,x}{d\,s}+\frac{\partial\,n}{\partial\,y}\frac{d\,y}{d\,s}+\frac{\partial\,n}{\partial\,z}\frac{d\,z}{d\,s}=\frac{\partial\,n}{\partial\,s}$$

Comparison of the last two equalities yields

$$\frac{d}{d\,s}\left(n\,\frac{d\,z}{d\,s}\right)=\frac{\partial\,n}{\partial\,z}\,, \tag{1.306}$$

which corresponds to the third ray equation.

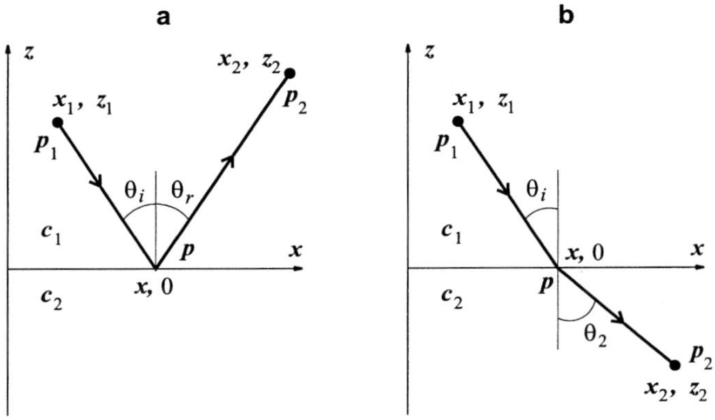

Figure 1.42: Snell's law of reflection (a) and refraction (b).

Snell's Law

Thus, we have shown that Euler's and ray equations coincide with each other, and this means that the acoustic length has an extremum for rays. As was pointed out earlier, ray equations and Fermat's principle were derived independently from the same physical laws. At the same time, as was just demonstrated, the ray equations follow from Fermat's principle. Let us illustrate this approach and, as an example, obtain Snell's law of reflection and refraction. As is seen from Fig. 1.42a, the acoustic length of the path $p_1 \, p \, p_2$ is equal to

$$J = \sqrt{z_1^2 + (x - x_1)^2} + \sqrt{z_2^2 + (x_2 - x)^2}\,, \tag{1.307}$$

since $n = 1$. Here the point $p(x,\, 0)$ is located at the interface of two media.

According to Fermat's principle, the acoustic path is characterized by the variable x for which

$$\frac{\partial J}{\partial x} = \frac{x - x_1}{\sqrt{z_1^2 + (x - x_1)^2}} - \frac{x_2 - x}{\sqrt{z_2^2 + (x_2 - x)^2}} = 0$$

or

$$\sin\theta_i = \sin\theta_r\,, \tag{1.308}$$

where θ_i and θ_r are the incident and reflection angles, respectively. The latter is known as Snell's law of reflection.

In the same manner, we have for the path $p_1 \, p \, p_2$ in Fig. 1.42b

$$J = \frac{\sqrt{z_1^2 + (x - x_1)^2}}{c_1} + \frac{\sqrt{z_2^2 + (x_2 - x)^2}}{c_2}$$

Again, making use of Fermat's principle, an extremum of functional $J(x)$ is defined as

$$\frac{x - x_1}{c_1 \sqrt{z_1^2 + (x - x_1)^2}} - \frac{x_2 - x}{c_2 \sqrt{z_2^2 + (x_2 - x)^2}} = 0$$

or

$$\frac{\sin \theta_i}{c_1} = \frac{\sin \theta_2}{c_2}, \tag{1.309}$$

and we arrived at Snell's law of reflection.

Chapter 2

Dynamics of high-frequency wave fields

At the beginning of the first chapter, we demonstrated that geometrical acoustics is based on the assumption that a wave behaves as the plane one in the vicinity of a regular point. Correspondingly, the complex amplitude of the potential has the form

$$\mathcal{U}(\mathbf{r}) = \mathcal{A}\, e^{i\, k_0\, L}\,, \tag{2.1}$$

where $\mathbf{r}$ is the radius-vector characterizing the position of the observation point. As before we assume that the medium is homogeneous with respect to its density. Otherwise the potential could not be usually introduced.

In general, $\mathcal{A}$ is the complex function

$$\mathcal{A} = |\,\mathcal{A}\,|\, e^{i\varphi}, \tag{2.2}$$

and its amplitude, as well as the argument φ_1, varies from point to point. L is the eikonal, satisfying equation

$$\text{grad}^{\,2} L = n^2 \tag{2.3}$$

and

$$n = \frac{c_0}{c}$$

As follows from eqs. 2.1 and 2.2, the function

$$\psi = k_0\, L + \varphi = k_0 \left(L + \frac{\varphi}{k_0}\right) \tag{2.4}$$

is the phase of the potential. Also, as was shown previously, at the high-frequency spectrum the potential $\mathcal{U}$ can be written in the form of the asymptotic series (Debye expansion)

$$\mathcal{U}(\mathbf{r}) = e^{i\,k_0\,L(\mathbf{r})} \sum_{m=0}^{\infty} \frac{A_m(\mathbf{r})}{(i\,k_0)^m} \tag{2.5}$$

Comparison of eqs. 2.1 and 2.5 suggests that this sum represents the complex amplitude $\mathcal{A}(\mathbf{r})$, while the coefficients in the asymptotic series satisfy the transport equations, eqs. 1.45–1.47.

Now it may be appropriate to make several comments.

1. Suppose that the resultant field consists of several waves, such as the reflected, head waves and others. Then the potential $\mathcal{U}$ can be represented in the form

$$\mathcal{U} = \mathcal{U}_1 + \mathcal{U}_2 + \ldots$$

   and eq. 2.5 describes the asymptotic series for each wave.

2. Different waves have different eikonal functions $L(\mathbf{r})$ but they all obey the eikonal equation

$$\text{grad}^{\,2} L = n^2$$

3. Geometrical acoustics is based on solution of eq. 2.3, as well as the set 1.45–1.47. Thus, the system

$$\text{grad}^{\,2} L = n^2$$

   and

$$2\,(\nabla A_0 \cdot \nabla L) + A_0\,\nabla^2 L = 0$$

   gives the zero-order approximation $(m = 0)$ at the high frequency range.

4. The waves in which the potential exponentially changes with the distance

$$\mathcal{U} \sim e^{-k\,r}$$

   cannot be described by the asymptotic series because each of its terms decays slower.

5. In accordance with eqs. 1.45–1.47, all coefficients A_m are real if the amplitude of zero approximation A_0 is real too.

In the previous chapter the main attention was paid to the ray geometry of waves at the high-frequency spectrum. Now we investigate the behavior of wave amplitudes at regular points as well as at interfaces between media with different physical parameters.

Before we discuss solution of transport equations it is useful to remind the main concepts relating to the the flux of acoustic energy.

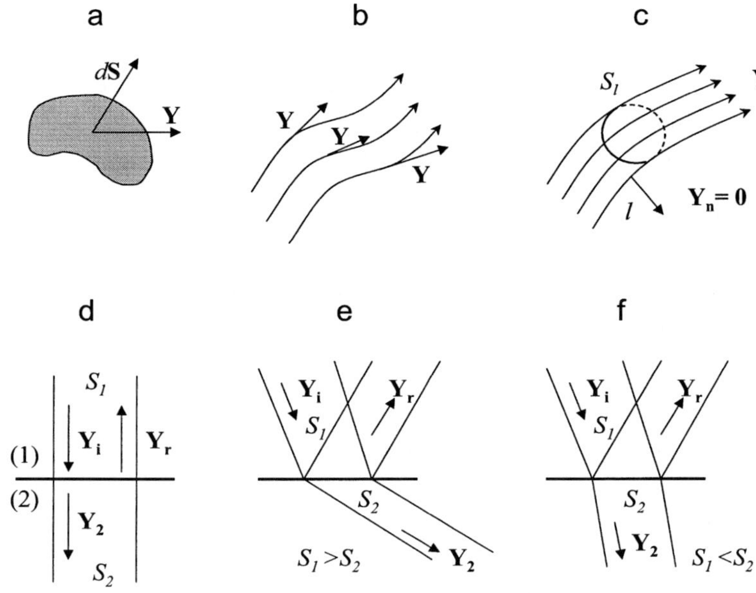

Figure 2.1: Poynting vector and vector tubes.

2.1 Orientation of Poynting vector and rays

As was shown in Part I the amount of energy passing through elementary surface $d\mathbf{S}$ per unit time, Fig. 2.1a, is equal to

$$dF = \mathbf{Y} \cdot d\mathbf{S} \tag{2.6}$$

Here

$$\mathbf{Y} = P\,\mathbf{v} \tag{2.7}$$

is the Poynting vector characterizing the intensity and direction of the energy flux. It is equal to the product of the additional pressure P and the particle velocity $\mathbf{v}$. For instance, if the pressure is positive, than the vectors $\mathbf{Y}$ and $\mathbf{v}$ have the same direction. In contrast, within the extension zone, where P is negative, the velocity and Poynting vector are opposite to each other. It is clear that the former allows us to visualize wave propagation and transmission of energy. By definition, the vector $\mathbf{Y}$ is tangential to its vector lines, Fig. 2.1b, and a family of these lines forms a geometrical model of the field. Let us imagine an arbitrary closed line l, which does not coincide with the vector lines, which are drawn through each of its points Fig. 2.1c. In this way, we obtain the vector tube of the field $\mathbf{Y}$. Inasmuch as its lateral surface S_l is formed by vector lines, the Poynting vectors are tangential to this surface and therefore the normal component $\mathbf{Y}_n$ vanishes on S_l:

$$\mathbf{Y}_n = 0 \tag{2.8}$$

In other words, the flux of energy passing through S_l is equal to zero, and this means that the flux through any cross-section of the vector tube remains the same. In general, the Poynting vector changes along the tube, and it may have different values over its cross-section. At the same time, if the latter is small, the vector **Y** is almost constant at points of this cross-section but may vary along the tube. Such a vector tube is called the elementary one. The Poynting vector is perpendicular to the cross-section and for the energy flux inside it we have

$$Y(p_1)\,S(p_1) = Y(p_2)\,S(p_2) = \ldots = \text{const} \tag{2.9}$$

Here $Y(p)$ and $S(p)$ are the magnitude of the Poynting vector and the area of the cross-section of the elementary tube, respectively. Thus, along the tube the magnitude of **Y** is inversely proportional to the cross-section area

$$Y \sim \frac{1}{S} \tag{2.10}$$

As is well known, at the boundary between media with different parameters the pressure and the component of the displacement normal to the boundary are continuous functions

$$P_1 = P_2 \qquad \text{and} \qquad s_{1n} = s_{2n}\,, \tag{2.11}$$

while the tangential component of the displacement is usually discontinuous.

Performing differentiation of the second equality of the set 2.11 with respect to time, we have

$$v_{1n} = v_{2n} \tag{2.12}$$

Therefore, the normal component of the vector **Y** is a continuous function at the interface but its tangential component, in general, has a discontinuity

$$Y_{1n} = Y_{2n} \qquad \text{and} \qquad Y_{1t} \neq Y_{2t}\,, \tag{2.13}$$

that is, the Poynting vector has different magnitudes and directions at both sides of the interface if $Y_t \neq 0$. For illustration, consider elementary tubes for the incident, reflected and transmitted waves. In accordance with Snell's law the cross-sections of the tubes, corresponding to the first two waves, are the same. However, the Poynting vector magnitude of the reflected wave is smaller than that of the incident wave since a portion of energy generates the transmitted wave. Examples of such tubes are shown in Fig. 2.1d,e,f.

Rays and directions of the energy flux

Now we describe one interesting feature of the flux which takes place when the frequency tends to infinity. By definition, eq. 2.7, components of the Poynting vector in the Cartesian coordinate system are

$$Y_x = P\,v_x\,, \qquad Y_y = P\,v_y\,, \qquad Y_z = P\,v_z \tag{2.14}$$

Assuming the sinusoidal dependence on time

$$U(\mathbf{r}) = \text{Re}\,\left(\mathcal{U}\,e^{-i\,\omega\,t}\right), \tag{2.15}$$

where

$$\mathcal{U} = A\,e^{i\,\varphi}\,e^{i\,k_0\,L}, \tag{2.16}$$

we have

$$U(\mathbf{r}) = A\,\cos(\omega\,t - k_0\,L - \varphi) \qquad \text{or} \qquad U(\mathbf{r}) = A\,\cos(\omega\,t - \psi) \tag{2.17}$$

Here A, L and ψ depend on the observation point and ψ is given by eq. 2.4.

Since an instantaneous value of the Poynting vector does not have any significance, our attention is paid only to its average during a period T. First, consider the x component Y_x^{av}. Bearing in mind that

$$P = -\rho\,\frac{\partial\,\mathcal{U}}{\partial\,t} = \rho\,\omega\,A\,\sin(\omega\,t - \psi)$$

and

$$\mathbf{v} = \text{grad}\;\mathcal{U}\,,$$

that is

$$v_x = \frac{\partial\,U}{\partial\,x} = \frac{\partial\,A}{\partial\,x}\,\cos(\omega\,t - \psi) + A\,\frac{\partial\,\psi}{\partial\,x}\,\sin(\omega\,t - \psi)\,,$$

we have

$$Y_x = \rho\,\omega\,A\,\frac{\partial\,A}{\partial\,x}\,\sin(\omega\,t - \psi)\,\cos(\omega\,t - \psi) + A^2\,\rho\,\omega\,\frac{\partial\,\psi}{\partial\,x}\,\sin^2(\omega\,t - \psi) \tag{2.18}$$

The average value of the x component of the Poynting vector is

$$Y_x^{av} = \frac{1}{T}\int\limits_0^T Y_x(t)\,dt \tag{2.19}$$

Taking into account that

$$\int\limits_0^T \sin(\omega\,t - \psi)\,\cos(\omega\,t - \psi)\,dt = 0$$

and

$$\int_0^T \sin^2(\omega t - \psi)\, dt = \frac{T}{2},$$

eq. 2.19 becomes

$$Y_x^{av} = \frac{\rho\,\omega\, A^2}{2}\frac{\partial \psi}{\partial x} \tag{2.20}$$

By analogy,

$$Y_y^{av} = \frac{\rho\,\omega\, A^2}{2}\frac{\partial \psi}{\partial y} \qquad \text{and} \qquad Y_z^{av} = \frac{\rho\,\omega\, A^2}{2}\frac{\partial \psi}{\partial z} \tag{2.21}$$

Multiplying each of eqs. 2.20–2.21 by the corresponding unit vector and adding them, we obtain

$$\mathbf{Y}^{av} = \frac{\rho\,\omega\, A^2}{2}\,\mathrm{grad}\,\psi$$

or, using eq. 2.4:

$$\mathbf{Y}^{av}(\mathbf{r}) = \frac{\rho\,\omega\, k_0\, A^2}{2}\left(\mathrm{grad}\, L + \frac{1}{k_0}\,\mathrm{grad}\,\varphi\right) \tag{2.22}$$

As we know

$$\mathrm{grad}\, L = n\,\mathbf{s}_0\,,$$

where $\mathbf{s}_0$ being the unit vector tangential to a ray.

Therefore, the Poynting vector, $\mathbf{Y}^{av}$ is, in general, arbitrarily oriented with respect to the ray. However, in the high-frequency limit when $k_0 \to \infty$, the second term in the brackets of eq. 2.22 can be neglected, and it yields

$$\mathbf{Y}^{av} = \frac{\rho\,\omega\, k_0\, A^2}{2}\, n\,\mathbf{s}_0$$

or

$$\mathbf{Y}^{av} = \frac{\rho\,\omega\, k\, A^2}{2}\,\mathbf{s}_0 \qquad \text{if} \qquad \omega \to \infty \tag{2.23}$$

Thus, in this case the Poynting vector $\mathbf{Y}^{av}$ is tangential to rays, that is, the latter are vector lines of the complex field $\mathbf{Y}^{av}$. This behavior is also observed if the argument ψ of the complex function $\mathcal{A}$ is constant and, correspondingly,

$$\mathrm{grad}\,\psi = 0$$

The vector I

Next we derive the same result in a slightly different manner. By definition, we have

$$P = -\rho \frac{\partial U}{\partial t} \qquad \text{or} \qquad P = \omega \rho \operatorname{Re} \left(i\mathcal{U} e^{-i\omega t} \right) \tag{2.24}$$

and

$$\mathbf{v} = \operatorname{Re} \left(\nabla \mathcal{U} e^{-i\omega t} \right) \tag{2.25}$$

Let us introduce the function $\mathcal{U}^*$ which is conjugate to $\mathcal{U}$. It is clear that

$$\operatorname{Re} \left(i\mathcal{U} e^{-i\omega t} \right) = \operatorname{Re} \left[i A e^{-i(\omega t - k_0 L)} \right]$$

$$= [A \sin(\omega t - k_0 L)] = -\operatorname{Im} \left(\mathcal{U} e^{-i\omega t} \right) = \operatorname{Im} \left(\mathcal{U}^* e^{i\omega t} \right)$$

and

$$\operatorname{Re} \left(\mathcal{U} e^{-i\omega t} \right) = \operatorname{Re} \left(\mathcal{U}^* e^{i\omega t} \right)$$

Whence

$$P = \omega \rho \frac{\mathcal{U}^* e^{i\omega t} - \mathcal{U} e^{-i\omega t}}{2i} \tag{2.26}$$

and

$$\mathbf{v} = \frac{\nabla \mathcal{U}^* e^{i\omega t} + \nabla \mathcal{U} e^{-i\omega t}}{2} \tag{2.27}$$

and the instantaneous value of the Poynting vector is

$$\mathbf{Y} = \frac{\omega \rho}{4i} \left(\mathcal{U}^* \nabla \mathcal{U}^* e^{2i\omega t} - \mathcal{U} \nabla \mathcal{U}^* + \mathcal{U}^* \nabla \mathcal{U} - \mathcal{U} \nabla \mathcal{U} e^{-2i\omega t} \right) \tag{2.28}$$

After averaging over a period, the terms containing $\exp[\pm 2 i \omega t]$ vanish, giving us

$$\mathbf{Y} = \frac{\rho \omega^2}{2 c_0} \mathbf{I}, \tag{2.29}$$

where

$$\mathbf{I} = \frac{1}{2 i k_0} \left(\mathcal{U}^* \nabla \mathcal{U} - \mathcal{U} \nabla \mathcal{U}^* \right) \tag{2.30}$$

Bearing in mind that

$$\begin{aligned} \nabla \mathcal{U} &= \nabla A e^{i k_0 L} + A i k_0 \nabla L e^{i k_0 L} \\ \nabla \mathcal{U}^* &= \nabla A e^{-i k_0 L} - A i k_0 \nabla L e^{-i k_0 L}, \end{aligned} \tag{2.31}$$

it is a simple matter to see that eqs. 2.23 and 2.29 coincide. Of course, both vectors, **I** and **Y**, have real magnitudes, which differ from each other by a constant. Let us notice that introducing the vector **I** we assumed that the argument φ, eq. 2.2, is equal to zero.

Consider one important feature of vector **I**, provided that external sources of waves are absent in the vicinity of some point and acoustic energy is not transformed into heat. The latter means that the refraction index n is real. Then, it follows from the Helmholtz equation

$$\nabla^2 \mathcal{U} + n^2\, k_0^2 \mathcal{U} = 0 \tag{2.32}$$

that both the real and imaginary parts of the complex amplitude $\mathcal{U}$ satisfy eq. 2.32. Therefore, $\mathcal{U}^*$ is also a solution of the same equation

$$\nabla^2 \mathcal{U}^* + n^2\, k_0^2 \mathcal{U}^* = 0 \tag{2.33}$$

Taking a divergence of both sides of eq. 2.30 and making use of eqs. 2.32 and 2.33, we obtain

$$\operatorname{div} \mathbf{I} = 0 \tag{2.34}$$

As follows from the Gauss theorem, (Part I),

$$\oint_S \mathbf{I} \cdot d\mathbf{S} = \int_V \operatorname{div} \mathbf{I}\, dV\,, \tag{2.35}$$

the flux of vector **I**, as well as that of the Poynting vector, through a closed surface is equal to zero

$$\oint_S \mathbf{I} \cdot d\mathbf{S} = 0 \tag{2.36}$$

It is obvious that the function $\mathcal{U}^*$ can be written as

$$\mathcal{U}^* = e^{-i\, k_0\, L(\mathbf{r})} \sum_{m=0}^{\infty} \frac{A_m(\mathbf{r})}{(-i\, k_0)^m} \tag{2.37}$$

Then, substitution of both series for $\mathcal{U}$ and $\mathcal{U}^*$ into eq. 2.30 yields

$$\mathbf{I} \approx c_0\, \mathbf{p}\, (A_0)^2 + \frac{1}{k_0^2} \left[c_0\, \mathbf{p}\, (A_1)^2 + A_1\, \nabla A_0 - A_0\, \nabla A_1 \right] \tag{2.38}$$

Here

$$\mathbf{p} = \frac{\mathbf{s}_0}{c} = \frac{\nabla L}{c_0} = \frac{n\, \mathbf{s}_0}{c_0} \tag{2.39}$$

is the slowness vector tangential to the ray. Thus, we arrived again at the same conclusion, namely, the Poynting vector is directed along the ray when $k_0 \to \infty$ and

$$\mathbf{I} \approx c_0 \, \mathbf{p} \, A_0^2 \tag{2.40}$$

As is seen from eq. 2.38, the following terms of expansion of **I** are defined by the components of the Poynting vector directed both along and perpendicular to the ray tube, but they are relatively small.

Suppose that at the high-frequency spectrum of a wave the term with $m = 0$ vanishes, that is, $A_0 = 0$. Then eq. 2.38 becomes

$$\mathbf{I} \approx \frac{1}{k_0^2} \, c_0 \, \mathbf{p} \, A_1^2 \tag{2.41}$$

and the Poynting vector is also directed along the ray in such an approximation. This result can be easily generalized for the case when

$$\mathcal{U} \approx e^{i\,k_0\,L} \, \frac{A_m}{(i\,k_0)^m}$$

and, respectively,

$$\mathcal{U}^* \approx e^{-i\,k_0\,L} \, \frac{A_m}{(-i\,k_0)^m}$$

Then, in accordance with eq. 2.30, we have

$$\mathbf{I} \approx \frac{c_0}{(i\,k_0)^m} \, \mathbf{p} \, A_m^2 \qquad \text{if} \qquad k_0 \to \infty$$

and the leading term of the Poynting vector is directed along the ray. Thus, in approximation of geometrical acoustics, the vector $\mathbf{Y}^{av}$ for any m is tangential to the ray, provided that $k_0 \to \infty$.

2.2 Amplitude A_0 along a ray

In this section we pay the main attention to the amplitude in the zero approximation, A_0, and in accordance with eq. 1.45 it satisfies the transport equation:

$$2\,(\nabla A_0 \cdot \nabla L) + A_0 \nabla^2 L = 0 \tag{2.42}$$

It is proper to notice that an introduction of the scalar potential is possible when $\rho = const.$ In other words, proceeding from eq.2.42 we assume that a medium is homogeneous with respect to density. Later this restriction will be removed. At the same time it turns out that the eikonal $L(\mathbf{r})$ depends on the velocity $c(\mathbf{r})$ only.

Calculation of the eikonal function

From eq. 2.42 it seems that a determination of the amplitude A_0 requires a knowledge of the function $L(\mathbf{r})$ and in order to find it we proceed from the second equation of the set 1.60. Bearing in mind that the magnitude of ∇L coincides with with the maximal derivative of L we have

$$\nabla L = \frac{dL}{ds}\,\mathbf{s}_0$$

Comparison with eq. 2.3 gives

$$\frac{dL}{ds} = n, \tag{2.43}$$

where ds is an elementary displacement along a ray. In further it is natural to assume that a ray distribution is known, that is the coordinates of ray points are defined everywhere. Some methods of ray calculations were described in previous chapter. Integration of 2.43 along a ray gives

$$L(s) \;=\; L(s^0) + \int\limits_{s^0}^{s} n\, ds \tag{2.44}$$

Here $L(s^0)$ is the known value of the eikonal function at the ray point s^0. In particular,

$$L(s) = \int\limits_{0}^{s} n\, ds \tag{2.45}$$

if $s_o = 0$ and $L(0) = 0$. Sometimes it is convenient to use the slowness vector $\mathbf{p}$. As follows from the first ray equation

$$n\frac{d\mathbf{r}}{ds} = c_0\,\mathbf{p} \tag{2.46}$$

Inasmuch as

$$\frac{d\mathbf{r}}{ds} = \mathbf{s}_0 \qquad \text{and} \qquad \nabla L = n\mathbf{s}_0$$

we have

$$\nabla L = c_0\,\mathbf{p} \qquad \text{or} \qquad c_0\, p = \frac{dL}{ds} \tag{2.47}$$

Thus, the gradient of the eikonal and the magnitude of the slowness vector differ from each other by a constant c_0. Correspondingly, in place of equation 2.44 we have

$$L(s) = L(s^0) + c_0 \int\limits_{s^0}^{s} p\, ds \tag{2.48}$$

The latter, as well as eq. 2.44, permits us to calculate the eikonal function at each point of a ray, provided that its position is known. Of course, the same is valid for a time of wave arrival. Now we are ready to derive an expression for the wave amplitude A_0:

$$U(\mathbf{r}) = A_0 e^{ik_0 L(\mathbf{r})},$$

where A_0 is the real number. At the beginning, consider two simple cases when the amplitude A_0 can be easily found.

Example one: Plane wave Suppose that the plane wave moves through a homogeneous medium along the z-axis. Then

$$n = 1 \qquad \text{and} \qquad L(z) = z \tag{2.49}$$

Whence

$$\nabla L = \nabla z = \mathbf{k}$$

Here $\mathbf{k}$ is unit vector, characterizing a direction of the z-axis. Since its orientation does not change

$$\nabla^2 L = 0,$$

and the transport equation of the zero order m = 0, (eq. 2.42), becomes

$$\nabla A_0 \cdot \nabla L = \frac{dA_0}{dz} = 0 \tag{2.50}$$

Its solution is

$$A_0 = \text{const}$$

Certainly, this result is obvious, because the plane wave has a constant amplitude in a homogeneous medium, which can be determined from the initial condition at some plane, $z = \text{const}$.

Example two: Spherical wave We again assume that a medium is homogeneous, but the eikonal function is equal to

$$L(\mathbf{r}) = R \tag{2.51}$$

where R is the spherical coordinate. Inasmuch as in this system of coordinates, (Part I):

$$\nabla L = \frac{\partial L}{\partial R}\,\mathbf{i}_R \qquad \text{and} \qquad \nabla^2 L = \frac{1}{R^2}\frac{d}{dR}\left(R^2\,\frac{dL}{dR}\right),$$

we obtain

$$\nabla L = \mathbf{i}_R \qquad \text{and} \qquad \nabla^2 L = \frac{2}{R} \tag{2.52}$$

Here $\mathbf{i}_R$ is the unit vector in the radial direction. Then, eq. 2.42 has a form:

$$2\frac{dA_0}{dR}+A_0\frac{2}{R}=0 \qquad \text{or} \qquad \frac{dA_0}{dR}=-\frac{A_0}{R} \tag{2.53}$$

Whence

$$A_0=\frac{1}{R}, \tag{2.54}$$

and it describes the amplitude of a wave caused by an elementary source in a homogeneous medium.

Solution of transport equation

Next we consider a more general case, provided that $\rho =$ constant, and derive an expression for the amplitude A_0. As is seen from eq. 2.42, the transport equation contains partial derivatives. However, making use of eq. 2.3, the former can be reduced to the ordinary differential equation of the first order. In fact, since

$$\nabla L = n\mathbf{s}_0\,,$$

we have

$$\nabla A_0 \cdot \nabla L = \frac{dA_0}{ds}\,n \tag{2.55}$$

Then, the transport equation becomes

$$2n\,\frac{dA_0}{ds}+A_0\,\nabla^2 L=0 \tag{2.56}$$

By definition

$$\nabla^2 L = \text{div grad } L$$

or, taking into account again eq. 2.3, we have

$$\nabla^2 L = \text{div}\,(n\mathbf{s}_0) \tag{2.57}$$

Therefore, eq. 2.56 can be rewritten as

$$\frac{dA_0}{ds}+\frac{\text{div}\,(n\mathbf{s}_0)}{2n}A_0=0, \tag{2.58}$$

ant it has the form

$$\frac{dy}{dx}+f(x)y=0$$

or

$$d\ln y = -f(x)\ dx \tag{2.59}$$

Here

$$y = A_0 \qquad \text{and} \qquad f(x) = \frac{\operatorname{div}(n\mathbf{s}_0)}{2n} \tag{2.60}$$

Note that eq. 2.58 does not contain the eikonal function, $L(s)$, and, correspondingly, the amplitude A_0 is independent of L. However, its knowledge is vitally important, since the phase of a wave field or time of the first arrival are defined by this function. Integration of eq. 2.59 yields

$$\ln y = -\int f(x)dx + C$$

or

$$A_0(s) = A_0(s^0)\exp\left(-\int\limits_{s^0}^{s}\frac{\operatorname{div}(n\mathbf{s}_0)}{2n}ds\right), \tag{2.61}$$

where $A_0(s^0)$ is the initial value of the amplitude at the ray point $s = s^0$. Also, it is obvious that limits of integration are functions of x, y, z :

$$s^0 = s(x_0, y_0, z_0)\,, \qquad s = s(x, y, z)$$

Thus, the expression for the amplitude is sufficiently simple, but it requires a calculation of the integral

$$N = \int\limits_{s^0}^{s}\frac{\operatorname{div}(n\mathbf{s}_0)}{2n}ds \tag{2.62}$$

Relationship between $\mathbf{A_0}$ and ray tube cross-section

Now we demonstrate that the latter, eq. 2.62, can be replaced by a different procedure, which is much simpler. With this purpose in mind consider a small part of an elementary ray tube, bounded by two cross sections

$$\sigma(p_1) \qquad \text{and} \qquad \sigma(p_2),$$

as well as the lateral surface S_l, Fig. 2.2a. We assume that the cross-section of this tube, $\sigma(p)$, is so small that the vector $n\mathbf{s}_0$ is the same at all its points. It may be proper to remind that in the zero approximation, $(A = A_0)$, rays are vector lines of the Poynting vector and, respectively, elementary tubes for both vectors $n\mathbf{s}_0$ and $\mathbf{Y}$ coincide.

Inasmuch as the element of the ray tube is very small, it is possible to represent $\operatorname{div}(n\mathbf{s}_0)$ in its the most natural way. By definition, (Part I):

$$\operatorname{div}(n\mathbf{s}_0) = \frac{1}{\Delta V}\oint n\,\mathbf{s}_0\cdot d\mathbf{S} \tag{2.63}$$

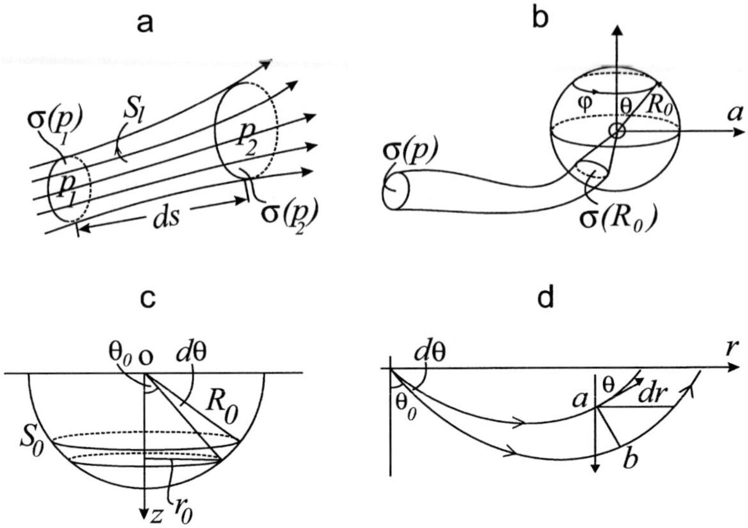

Figure 2.2: Amplitude A_0 along elementary tubes

Here the elementary volume

$$dV = \sigma(p)\, ds \tag{2.64}$$

and $\sigma(p)$ being an area of the cross-section at the middle of the element, while ds is its extension. The surface integral consists of three parts

$$\oint\limits_S n\mathbf{s}_0 \cdot \mathbf{dS} = \int\limits_{S_1} n\mathbf{s}_0 \cdot \mathbf{dS} + \int\limits_{S_2} n\mathbf{s}_0 \cdot \mathbf{dS} + \int\limits_{S_l} n\mathbf{s}_0 \cdot \mathbf{dS} \tag{2.65}$$

where S_1 and S_2 are cross-sections at points p_1 and p_2, respectively, but S_l is the lateral surface. The vector $n\mathbf{s}_0$ is tangential to this surface and, therefore, the last integral vanishes. In accordance with Gauss theorem, (Part I), vectors $\mathbf{dS}_1$ and $\mathbf{dS}_2$ are directed outwards from the volume, while rays are oriented from point p_1 to p_2. For this reason $\mathbf{dS}_1 = -\sigma(p_1)\,\mathbf{s}_0$ and $\mathbf{dS}_2 = \sigma(p_2)\,\mathbf{s}_0$. Since the vector $n\mathbf{s}_0$ remains constant over each cross-section, substitution of the latter into eq. 2.65 gives

$$\oint n\mathbf{s}_0 \cdot \mathbf{dS} \;=\; n(p_2)\,\sigma(p_2) - n(p_1)\,\sigma(p_1) \tag{2.66}$$

Bearing also in mind that distance ds is very small, it is natural to assume the linear change of the flux $n\mathbf{s}_0 \cdot \mathbf{dS}$ between cross-sections $\sigma(p_1)$ and $\sigma(p_2)$. This gives

$$\oint n\mathbf{s}_0 \cdot \mathbf{dS} \;=\; \frac{d(n\sigma)}{ds}\, ds \tag{2.67}$$

Substitution of eq. 2.64 and eq. 2.67 into eq. 2.63 gives

$$\oint \operatorname{div}(n\mathbf{s}_0)\, dS = \frac{1}{\sigma}\frac{d}{ds}(n\sigma) \tag{2.68}$$

Therefore, divergence of the vector $n\mathbf{s}_0$ characterizes a rate of change of its flux along the ray, normalized by an area of the cross-section, $\sigma(p)$. Making use of the latter, the integral N, eq. 2.62, becomes

$$N = \frac{1}{2}\int_{s^0}^{s} \frac{1}{n\sigma}\frac{d}{ds}(n\sigma)\, ds \tag{2.69}$$

This can be easily expressed in terms of an elementary function. Indeed we have

$$\begin{aligned} N &= \frac{1}{2}\int_{s^0}^{s} \frac{d}{ds}\ln(n\sigma)\, ds = \frac{1}{2}\int_{s^0}^{s} d\ln(n\sigma) \\ &= \frac{1}{2}\left[\ln n(s)\,\sigma(s) - \ln n\left(s^0\right)\sigma\left(s^0\right)\right] \\ &= \ln\sqrt{n(s)\,\sigma(s)} - \ln\sqrt{n(s^0)\,\sigma(s^0)} \end{aligned} \tag{2.70}$$

Finally, its substitution into eq. 2.61 gives

$$A_0(s) = A_0\left(s^0\right)\sqrt{\frac{n(s^0)\,\sigma(s^0)}{n(s)\,\sigma(s)}} \tag{2.71}$$

if $\rho = \text{const}$.

The latter describes a relationship between amplitudes along the same ray at two points, located at an arbitrary distance from each other. If the amplitude A_0 and the tube cross-section σ are known at some point $p(s^0)$, then we can calculate the amplitude A_0 at any point of the ray, provided that the area of the cross-section of an elementary tube is known at this point.

Amplitude $\mathbf{A_0}$ and an energy flux

Before we discuss eq. 2.71, as well as a determination of the area $\sigma(p)$, it is useful to derive this equation in a slightly different way, as well as a more general relationship between amplitudes along a ray. First, multiplying both sides of the transport equation by A_0, we have

$$2A_0(\nabla A_0 \cdot \nabla L) + A_0^2\,\nabla^2 L = 0$$

This can be written in a more compact form

$$\operatorname{div}\left(A_0^2\nabla L\right)=0 \qquad \text{or} \qquad \nabla\left(n\,A_0^2\,\mathbf{s}_0\right)=0 \tag{2.72}$$

In accordance with Gauss's theorem we see that the flux of the vector $n\,A_0^2\,\mathbf{s}_0$ through any closed surface is equal to zero

$$\oint n\,A_0^2\,\mathbf{s}_0\cdot\mathbf{dS}=0 \tag{2.73}$$

Considering again a small volume inside an elementary ray tube and discarding the integral over the lateral surface S_l, we obtain

$$A_0^2\,(s)\,n\,(s)\,\sigma\,(s)-A_0^2\left(s^0\right)n\left(s^0\right)\sigma\left(s^0\right)=0$$

or

$$A_0\,(s)=A_0\left(s^0\right)\sqrt{\frac{n\,(s^0)\,\sigma\,(s^0)}{n\,(s)\,\sigma\,(s)}}\,, \tag{2.74}$$

which coincides with eq. 2.71.

As was shown in the previous section the vector $n\,A_0^2\,\mathbf{s}_0$ differs from the Poynting vector by a constant if a medium is homogeneous with respect to its density. This suggests that a behavior of the wave amplitude in zero approximation is defined by a constancy of the flux of the Poynting vector, $\mathbf{Y}^{av}$, through any cross-section of a ray tube. This linkage with the Poynting vector is very important because it allows us to find a relationship between amplitudes when both the density of a medium and the velocity of propagation are functions of point coordinates. In deriving an expression of the Poynting vector as a product of pressure and particle velocity

$$\mathbf{Y}=P\mathbf{v}\,, \tag{2.75}$$

we used the scalar potential, (Part I), which is related to the particle displacement as

$$\mathbf{s}=-\operatorname{grad}U$$

The latter takes place under certain conditions, in particular, in the case of a constant density. Fortunately, eq. 2.75 still remains valid even when ρ is an arbitrary function of a point and there are several ways to prove this. Perhaps, the simplest one is based on physical concepts. Consider an elementary surface dS and suppose that wave propagation causes a force F acting on this element. By definition, we have

$$F=P\,dS$$

The work performed by the force during time interval dt is equal to

$$A = P\,v\,dS\,dt$$

and, correspondingly, it leads to a change of energy at the opposite side of the surface. Thus, the amount of an energy transmitted through the unit area per unit time is $P\,v$ that exactly describes the magnitude of the Poynting vector. Besides both vectors, $\mathbf{v}$ and $\mathbf{Y}$, have the same direction. For this reason, we can use eq. 2.75 in the general case of an isotropic medium. In accordance with the principle of conservation of energy, we have

$$\frac{\partial W}{\partial t} = -\oint_S \mathbf{Y} \cdot \mathbf{dS} \tag{2.76}$$

Here W is the total energy inside the volume V, surrounded by the closed surface S.

As before we study sinusoidal waves and, therefore, it is natural to deal with an average value of the Poynting vector. Then, eq. 2.76 gives

$$\frac{1}{T}\int_{-T/2}^{T/2} \frac{dW}{dt}\,dt = -\oint_S \mathbf{Y}^{av} \cdot \mathbf{dS}$$

The integral at the left-hand side is equal to zero. In fact

$$\int_{-T/2}^{T/2} \frac{dW}{dt}\,dt = \int_{-T/2}^{T/2} dW = W\left(\frac{T}{2}\right) - W\left(\frac{-T}{2}\right) = 0,$$

since the wave is periodic. Respectively, in place of eq. 2.76, we have

$$\oint \mathbf{Y}^{av} \cdot \mathbf{dS} = 0 \tag{2.77}$$

Of course it is assumed that the acoustic energy is not transformed into heat. Bearing in mind that in the vicinity of each point the wave behaves as a plane one (high frequency asymptotic), it is natural to use results, derived in the previous section. In particular, it was found that

$$\mathbf{Y}^{av} = \frac{\rho\,\omega^2 A_0^2}{2c}\,\mathbf{s}_0 \tag{2.78}$$

where A_0 is the amplitude of the potential. Inasmuch is the potential does not have any meaning in a general case, $(\nabla\rho \neq 0)$, we have to express the vector $\mathbf{Y}^{av}$ in terms of the amplitude of either a displacement or a pressure. In order to solve this task suppose that the sinusoidal wave propagates along the z-axis and

$$s_z(k,z) = A_{0S}\sin(\omega t - kz) \tag{2.79}$$

Here

$$k = \frac{\omega}{c}$$

and the amplitude A_{0S} depends on the distance z. Correspondingly, the particle velocity is equal to

$$v_z = A_{0S}\, \omega\, \cos(\omega t - kz)$$

Taking into account that

$$\text{div } \mathbf{s} = -\frac{1}{\rho\, c^2} P\,,$$

we have

$$P = -\rho\, c^2 \frac{\partial s_z}{\partial z} \tag{2.80}$$

Substitution of eq. 2.79 into eq. 2.80 yields

$$P = \rho\, c^2\, k\, A_{0S}\, \cos(\omega t - kz) - \rho\, c^2 \frac{dA_{0S}}{dz} \sin(\omega t - kz) \tag{2.81}$$

Therefore, the magnitude of the Poynting vector, eq. 2.75, is

$$Y = \rho\, c\, \omega^2 A_{0S}^2 \cos^2(\omega t - kz) - \rho\, c^2 \omega A_{0S} \frac{dA_{0S}}{dz} \sin(\omega t - kz) \cos(\omega t - kz) \tag{2.82}$$

and its average value is

$$Y^{av} = \frac{\rho\, c\, \omega^2 A_{0S}^2}{2} \qquad \text{or} \qquad \mathbf{Y}^{av} = \frac{\rho\, c\, \omega^2 A_{0S}^2}{2} \mathbf{s}_0 \tag{2.83}$$

Here $\mathbf{s}_0$ is a direction of the particle velocity, tangential to the ray. Next, applying eq. 2.77 to the surface S, surrounding the volume of an elementary tube, Fig 2.2a, we find

$$\rho(p_1)\, c(p_1)\, A_{0S}^2(p_1)\, \sigma(p_1) = \rho(p_2)\, c(p_2)\, A_{0S}^2(p_2)\, \sigma(p_2) \tag{2.84}$$

where p_1 and p_2 are points characterizing the position of cross-sections $\sigma(p)$.

Introducing the acoustic impedance, (Part I)

$$Z = \rho\, c\,,$$

eq. 2.84 becomes

$$Z(p_1)\, \sigma(p_1)\, A_{0S}^2(p_1) \;=\; Z(p_2)\, \sigma(p_2)\, A_{0S}^2(p_2)$$

Whence

$$A_{0S}(p_2) = \sqrt{\frac{Z(p_1)\sigma(p_1)}{Z(p_2)\sigma(p_2)}}\,A_{0S}(p_1) \tag{2.85}$$

or

$$A_{0S}(p_2) = \frac{1}{\sqrt{F_S}} A_{0S}(p_1)\,, \tag{2.86}$$

where

$$F_S = \frac{Z(p_2)\,\sigma(p_2)}{Z(p_1)\,\sigma(p_1)}\,, \tag{2.87}$$

and it describes a change of the displacement amplitude along the ray. The function F_S is called the spreading factor.

As follows from eq. 2.81 with an increase of the frequency the first term on the right hand side becomes dominant. For this reason, let us assume that

$$P = \rho\, c^2\, k\, A_{0S} \cos(\omega t - kz)$$

and, therefore, the pressure amplitude is

$$A_{0P} = \rho\, c^2\, k\, A_{0S} \qquad \text{or} \qquad A_{0P} = \omega Z\, A_{0S} \tag{2.88}$$

Certainly the second term in eq. 2.81 does not have any influence on $\mathbf{Y}^{av}$.

Substitution of eq. 2.88 into eq. 2.85 gives

$$A_{0P}(p_2) = \sqrt{\frac{Z(p_2)\sigma(p_1)}{Z(p_1)\sigma(p_2)}}\,A_{0P}(p_1) \tag{2.89}$$

or

$$A_{0P}(p_2) = \frac{1}{\sqrt{F_P}} A_{0P}(p_1) \tag{2.90}$$

where

$$F_P = \frac{Z(p_1)\sigma(p_2)}{Z(p_2)\sigma(p_1)} \tag{2.91}$$

is the spreading factor for the pressure. It is essential that, in general functions F_s and F_p differ from each other and their product is equal to the square of the ratio of areas

$$F_S F_P = \frac{\sigma^2(p_2)}{\sigma^2(p_1)} \tag{2.92}$$

For illustration consider several examples.

Case one Suppose that a medium is homogeneous, $c = \text{const}$, and $\rho = \text{const}$. Then eqs. 2.85 and 2.89 become

$$A_{0S}(p_2) = \sqrt{\frac{\sigma(p_1)}{\sigma(p_2)}}\, A_{0S}(p_1) \qquad \text{and} \qquad A_{0P}(p_2) = \sqrt{\frac{\sigma(p_1)}{\sigma(p_2)}}\, A_{0P}(p_1) \tag{2.93}$$

that is a change of both amplitudes is the same, and it is related to either a convergence or divergence of ray tubes. This means that only the geometrical spreading

$$F = F_S = F_P = \frac{\sigma(p_2)}{\sigma(p_1)} \tag{2.94}$$

defines the amplitude behavior along a ray.

Case two Let us imagine that the velocity of propagation depends on one coordinate along which a plane wave propagates. In this case, the area of the cross-section, $\sigma(p)$, remains constant, and it gives

$$A_{0S}(p_2) = \sqrt{\frac{Z(p_1)}{Z(p_2)}}\, A_{0S}(p_1) \qquad \text{and} \qquad A_{0P}(p_2) = \sqrt{\frac{Z(p_2)}{Z(p_1)}}\, A_{0P}(p_1) \tag{2.95}$$

while

$$F_s F_p = 1 \tag{2.96}$$

For instance, eqs. 2.95 show that with a decrease of the density around an observation point p_2 the amplitude A_{0S} increases, and this fact is important in studying possible earthquake hazard.

Case three As follows from eqs. 2.85 and 2.89

$$A_{0S}(p_2) = \sqrt{\frac{c(p_1)\sigma(p_1)}{c(p_2)\sigma(p_2)}}\, A_{0S}(p_1)$$

and

$$A_{0P}(p_2) = \sqrt{\frac{c(p_2)\sigma(p_1)}{c(p_1)\sigma(p_2)}}\, A_{0P}(p_1) \tag{2.97}$$

if

$$\rho(p_1) = \rho(p_2)$$

or a medium is homogeneous with respect to density. At the same time eq. 2.71 gives

$$A_0(p_2) = \sqrt{\frac{c(p_2)\sigma(p_1)}{c(p_1)\sigma(p_2)}}A_0(p_1)$$

Thus, we see the spreading factors for the potential and pressure are the same in a medium with constant density.

Case four Suppose that an elementary source is located at the origin of the spherical system of coordinates, Fig. 2.2b, and the velocity of propagation is an arbitrary function of coordinates. Let us mentally surround a source by a spherical surface of a relatively small radius R_0, and assume that within the range

$$0 < R < R_0$$

the medium is practically homogeneous. Then inside this volume rays are almost radial and for instance, the amplitude A_{0p} at points of the surface S_0 can written as

$$A_{0p}(S_0) = C\,\frac{G(\theta,\varphi)}{R_0} \tag{2.98}$$

Here R_0, θ and φ are coordinates of the point, C characterizes a source strength and $G(\theta,\varphi)$ describes a directional diagram of the source. Consider some elementary ray tube with the cross-section $\sigma(R_0)$ on the surface S_0, Fig. 2.2b. By definition, we have

$$\sigma(R_0) = R_0^2\,d\Omega \tag{2.99}$$

where $d\Omega$ is the solid angle under which this element is seen from the origin.
Since

$$\sigma_0 = R_0^2 \sin\theta\,d\theta\,d\varphi\,, \tag{2.100}$$

we obtain

$$d\Omega = \sin\theta\,d\theta\,d\varphi \tag{2.101}$$

Therefore, eq. 2.89 becomes

$$A_{0p}(R) = C\,G(\theta,\varphi)\sqrt{\frac{Z(R)}{Z(R_0)}\frac{\sin\theta\,d\theta\,d\varphi}{\sigma(R)}}$$

or

$$A_{0p}(R) = C\,G(\theta,\varphi)\sqrt{\frac{Z(R)}{Z(R_0)}\frac{d\Omega}{\sigma(R)}} \tag{2.102}$$

In the same manner we have

$$A_{0S}(R) = C\,G(\theta,\varphi)\sqrt{\frac{Z(R_0)}{Z(R)}\frac{d\Omega}{\sigma(R)}} \tag{2.103}$$

Correspondingly,

$$A_{0S}(R)\,A_{0P}(R) = \frac{d\Omega}{\sigma(R)}\,C^2\,G^2(\theta,\varphi)$$

It may be proper to notice that knowledge of the azimuthal angle θ allows us to avoid a numerical calculation of an area $\sigma(R_0)$, provided that inside S_0 a medium is homogeneous. It is obvious that if the velocity c and density ρ are functions of R only, then ray tubes are radially oriented and eqs. 2.102-2.103 become

$$A_{0P}(R) = \frac{C\,G(\theta,\varphi)}{R}\sqrt{\frac{Z(R)}{Z(R_0)}}$$

and

$$A_{0S}(R) = \frac{C\,G(\theta,\varphi)}{R}\sqrt{\frac{Z(R_0)}{Z(R)}}, \tag{2.104}$$

since

$$\sigma(R) = R^2 d\Omega$$

Case five Next consider a relatively simple but important model of a medium, in which both the velocity and density are functions of the single coordinate z, Fig. 2.2c. The elementary source is located at the origin O. It is convenient to proceed directly from eq. 2.89

$$A_{0P}(p_2) = \sqrt{\frac{Z(p_2)\,\sigma(p_1)}{Z(p_1)\,\sigma(p_2)}}\,A_{0P}(p_1)$$

First let us draw a spherical surface having small radius R_0 and assume as before that a medium inside S_0 is uniform. Then we have

$$A_{0P}(R_0) = \frac{C}{R_0} \qquad \text{if} \qquad G(\theta,\varphi) = 1 \tag{2.105}$$

Inasmuch as a ray distribution possesses by the symmetry with respect to the z-axis it is possible to simplify our task. It is clear that a cross-section area of the elementary tube is the spherical surface inside the cone, bounded by two lateral surfaces: $\theta = \theta_0$ and $\theta = \theta_0 + d\theta$. As is seen from Fig. 2.2c, on surface S_0 this area is equal to

$$\sigma(R_0) = 2\pi\, r_0\, R_0\, d\theta \qquad \text{or} \qquad \sigma(R_0) = 2\pi\, R_0^2\, \sin\theta_0\, d\theta\,, \tag{2.106}$$

since

$$r_0 = R_0 \sin\theta_0$$

In the same manner we define an element of the spherical surface $\sigma(p_2)$, Fig. 2.2d. It is obvious that

$$\sigma(p_2) = 2\pi r\, ab \qquad \text{or} \qquad \sigma(p_2) = 2\pi \cos\theta\ r\, dr \tag{2.107}$$

Bearing in mind that the Poynting vector, $\mathbf{Y}^{av}$, is normal to both surfaces and its flux is the same through both of them, we obtain in place of eq. 2.89

$$A_{0p}(p_2) = \sqrt{\frac{Z(p_2)\sin\theta_0\, d\theta}{Z(p_1)\cos\theta\ r\, dr}}\, C \tag{2.108}$$

By analogy, the amplitude of displacement is

$$A_{0s}(p_2) = \sqrt{\frac{Z(p_1)\sin\theta_0\, d\theta}{Z(p_2)\cos\theta\ r\, dr}}\, C \tag{2.109}$$

Note that the same result can be obtained from the expression for the Poynting vector, which was derived in Part I.

Calculation of an area in the Cartesian system of coordinates

Now let us return to eqs. 2.85 or 2.89. Assuming that the acoustic impedance, $Z(p)$, is known we see that determination of the wave amplitude is reduced to a calculation of the cross-section area, $\sigma(p)$, where p is either the reference or observation point. Certainly, this is a great simplification with respect to eq. 2.61, which requires us to integrate the function

$$\frac{\operatorname{div}(n\,\mathbf{s}_0)}{n}$$

along the ray. In fact, making use of the Cartesian system of coordinates and the well known rules of analytical geometry, it seems that $\sigma(p)$ can be easily found. However, it turns out that this procedure is usually rather cumbersome. To illustrate its main features suppose that an amplitude A_{0p} and an area of the cross-section, $\sigma(R_0)$, are known at some spherical surface S_0, surrounding the primary source, Fig. 2.2b. Correspondingly, for a calculation of an amplitude, we can use eq. 2.102. Inasmuch as the ray tube is elementary, the cross-section $\sigma(p)$ can be treated as planar, and its shape and orientation vary form point to point, Fig. 2.3a. To determine the area $\sigma(p)$ we choose some points

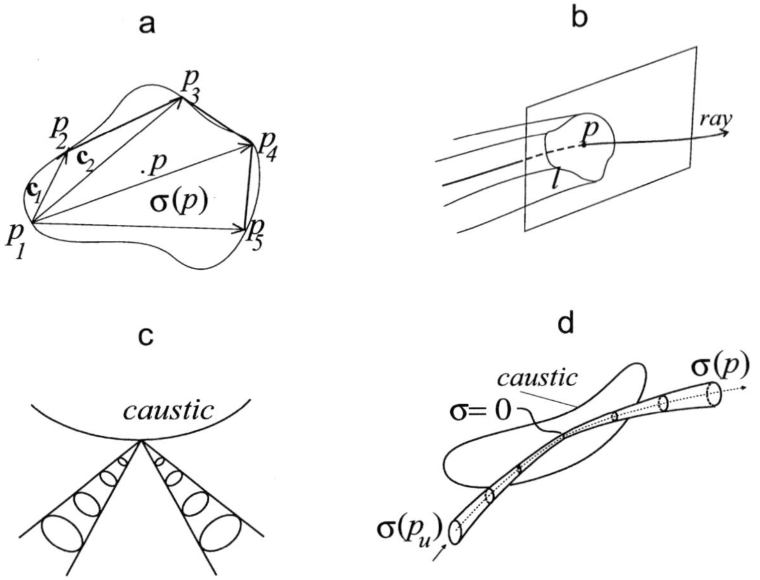

Figure 2.3: Ray behavior: (a) Elementary area as a system of triangles; (b) Cross-section of an elementary tube; (c),(d) Rays near caustics.

p_i on the line l and connect them by straight lines. Then the area $\sigma(p)$ is represented as a sum of triangle areas

$$\sigma(p) = \sum_{i=1} \sigma_i(p) \tag{2.110}$$

Each of them can be easily found and, as an example, consider the triangle with vertices

$$p_1(x_1, y_1, z_1)\,, \qquad p_2(x_2, y_2, z_2)\,, \qquad p_3(x_3, y_3, z_3)$$

As follows from vector algebra, (Part I), we have for vectors $\mathbf{c}_1$ and $\mathbf{c}_2$:

$$\mathbf{c}_1 = (x_2 - x_1)\mathbf{i} + (y_2 - y_1)\mathbf{j} + (z_2 - z_1)\mathbf{k}$$

and

$$\mathbf{c}_2 = (x_3 - x_1)\mathbf{i} + (y_3 - y_1)\mathbf{j} + (z_3 - z_1)\mathbf{k} \tag{2.111}$$

By definition of the cross product the magnitude of $\mathbf{c}_1 \times \mathbf{c}_2$ defines an area of parallelogram, formed by these vectors. Therefore, the area of the triangle $p_1p_2p_3$ is equal to

$$\sigma_1 = \frac{1}{2}\,|\mathbf{c}_1 \times \mathbf{c}_2| \tag{2.112}$$

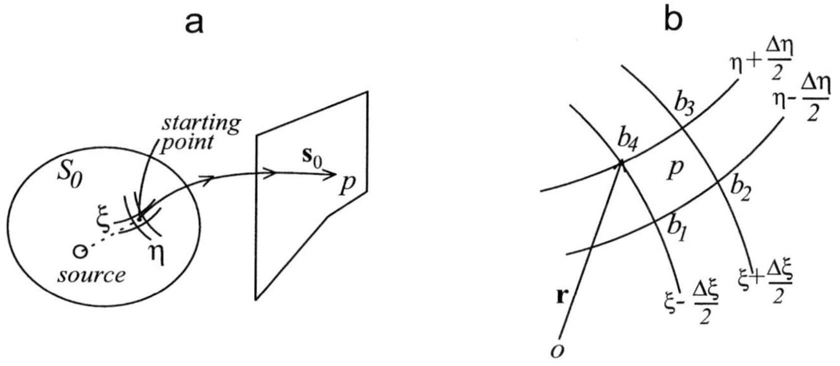

Figure 2.4: (a) Ray coordinates (b) Cross-section of an elementary tube

Repeating these calculations for each triangle, we can find the area $\sigma(p)$ and, correspondingly, the amplitude $A_0(p)$. Of course, with an increase of number of points p_i an accuracy of calculations becomes higher. Thus, a determination of A_{0p} and A_{0s} consists of several steps, namely: (a) solution of ray equations to find the position of several rays; (b) determination of a plane perpendicular to the given ray at point p, and line l, surrounding the cross-section $\sigma(p)$, Fig. 2.3b; (c) calculation of triangle areas and their summation. At the same time, it is necessary to guarantee that a cross-section of a tube is sufficiently small in the vicinity of an observation point. Then the amplitude A_{0p} remains constant over $\sigma(p)$. In order to meet this requirement we often have to form ray tubes with different cross-section areas on the surface S_0, as well as around the point p. It is obvious that if ray tubes are elementary, then eq. 2.102 gives the same value of the amplitude $A_{0p}(p)$. Thus we see that its determination in the Cartesian system of coordinates can be rather cumbersome.

To simplify this procedure ray coordinates are also used, but before we discuss their application, it may be proper to notice the following. As is seen from Fig. 2.3c,d, a mutual position of neighboring rays near a caustic suggests that a cross-section of a ray tube tends to zero. Respectively, the amplitude A_{0p} or A_{0s} becomes infinitely large. Besides, in accordance with transport equations

$$2(\nabla A_m \cdot \nabla L) + A_m \nabla^2 L = -\nabla^2 A_{m-1}$$

amplitudes A_m of m order also unlimitedly increase. This conclusion follows from the recursive character of these equations. Of course, in reality the wave amplitude has everywhere a finite value, and this means that the asymptotic series, (Debye expansion), cannot be applied near a caustic. For this reason in one of next sections we investigate this question in some detail, using a different approach.

Ray coordinates and initial conditions

As discussed previously a solution of the initial value problem has to satisfy the initial conditions at some surface S_0, where the amplitude A_0 and the eikonal L_0 are usually given. In particular, this surface can coincide with the eikonal one. Also we have to know a direction of a ray in the vicinity of S_0. Respectively it is natural to treat a point of an intersection of a ray with this surface as the starting one, though the ray begins around the primary source. To characterize a position of such points on the surface S_0 the system of coordinates ζ and η is introduced, Fig. 2.4a. Its origin can be arbitrarly chosen, including a point where the primary source is situated. In general, the ζ and η coordinate lines are not orthogonal to each other, and, by definition, only one coordinate varies along each line. The numbers ζ and η serve as parameters of each ray, since they are the same for all its points. At the same time, the distance along the ray, s, defines a location of a ray point. It is essential that different rays are characterized by a different set of values of ζ and η. Respectively, three quantities, ζ, η and s are called the ray coordinates. Our main goal is to determine an area of a cross-section of an elementary ray tube, and with this purpose in mind the ζ and η lines are used. It is obvious that the latter are perpendicular to the unit vector $\mathbf{s}_0$, which is tangential to the ray at the observation point. In those cases, when the surface S_0 is not an eikonal one, the vector $\mathbf{s}_0$ can have an arbitrary orientation.

Elementary area. Jacobian.

Now consider an elementary ray tube and suppose that its cross-section is formed by two pairs of coordinate lines, Fig. 2.4b. Coordinates of points of their intersection are

$$\xi - \frac{\Delta\xi}{2},\ \eta - \frac{\Delta\eta}{2},\ \xi + \frac{\Delta\xi}{2},\ \eta - \frac{\Delta\eta}{2},\ \xi + \frac{\Delta\xi}{2},\ \eta + \frac{\Delta\eta}{2},\ \xi - \frac{\Delta\xi}{2},\ \eta + \frac{\Delta\eta}{2}$$

Bearing in mind that the change of coordinates of the parallelogram is very small, we have

$$\mathbf{b}_1\mathbf{b}_2 = \mathbf{r}(\xi + \frac{\Delta\xi}{2}, \eta - \frac{\Delta\eta}{2}) - \mathbf{r}(\xi - \frac{\Delta\xi}{2}, \eta - \frac{\Delta\eta}{2}) = \frac{\partial \mathbf{r}}{\partial \xi} d\xi$$

and (2.113)

$$\mathbf{b}_1\mathbf{b}_4 = \mathbf{r}(\xi - \frac{\Delta\xi}{2}, \eta + \frac{\Delta\eta}{2}) - \mathbf{r}(\xi - \frac{\Delta\xi}{2}, \eta - \frac{\Delta\eta}{2}) = \frac{\partial \mathbf{r}}{\partial \eta} d\eta$$

Here $\mathbf{r}$ is the radius vector of corner points.

As we demonstrated earlier the magnitude of the cross product of $\mathbf{b}_1\mathbf{b}_2$ and $\mathbf{b}_1\mathbf{b}_4$ defines the area $\sigma(p)$, and it is

$$\sigma(p) = |\mathbf{b}_1\mathbf{b}_2 \times \mathbf{b}_1\mathbf{b}_4| = \left|\mathbf{r}_\xi \times \mathbf{r}_\eta\right| d\xi\, d\eta \tag{2.114}$$

where

$$\mathbf{r}_\xi = \frac{\partial \mathbf{r}}{\partial \xi} \qquad \text{and} \qquad \mathbf{r}_\eta = \frac{\partial \mathbf{r}}{\partial \eta}$$

Thus, in order to determine $\sigma(p)$ we have to know three rays. Let us represent eq. 2.114 as

$$\sigma(p) = J\, d\xi\, d\eta \tag{2.115}$$

Here

$$J = \left|\mathbf{r}_\xi \times \mathbf{r}_\eta\right| \tag{2.116}$$

is called Jacobian, and it defines an elementary area $\sigma(p)$. Taking into account that in the Cartesian system of coordinates $\mathbf{r} = x\mathbf{i} + y\mathbf{j} + z\mathbf{k}$ we obtain

$$J = \left\| \begin{array}{ccc} \mathbf{i} & \mathbf{j} & \mathbf{k} \\ \frac{\partial x}{\partial \xi} & \frac{\partial y}{\partial \xi} & \frac{\partial z}{\partial \xi} \\ \frac{\partial x}{\partial \eta} & \frac{\partial y}{\partial \eta} & \frac{\partial z}{\partial \eta} \end{array} \right\| \tag{2.117}$$

As follows from eq. 2.116 in order to find J we have to calculate scalar components of the cross-product and then its magnitude. It is more convenient to represent the Jacobian in a different form. As we know the unit vector $\mathbf{s}_0$, tangential to the ray, can be written as

$$\mathbf{s}_0 = \frac{d\mathbf{r}}{ds} = \mathbf{r}_s,$$

and it is perpendicular to vector $\mathbf{r}_\xi$ and $\mathbf{r}_\eta$, located at the cross-section of the ray tube. This means that

$$\mathbf{r}_\xi \times \mathbf{r}_\eta = |\mathbf{r}_\xi \times \mathbf{r}_\eta|\; \mathbf{s}_0 = J\, \mathbf{r}_s, \tag{2.118}$$

if three vectors: $\mathbf{r}_\xi$, $\mathbf{r}_\eta$, and $\mathbf{r}_s$ obey the right hand rule. Multiplying both sides of eq. 2.118 by unit vector $\mathbf{r}_s$ we have

$$J = (\mathbf{r}_\xi \times \mathbf{r}_\eta) \cdot \mathbf{r}_s \tag{2.119}$$

and it is positive. The right hand side represents the cross-dot product, and, in accordance with rules of vector multiplication, it can be written in the form:

$$J = \left\| \begin{array}{ccc} \frac{\partial x}{\partial \xi} & \frac{\partial y}{\partial \xi} & \frac{\partial z}{\partial \xi} \\ \\ \frac{\partial x}{\partial \eta} & \frac{\partial y}{\partial \eta} & \frac{\partial z}{\partial \eta} \\ \\ \frac{\partial x}{\partial s} & \frac{\partial y}{\partial s} & \frac{\partial z}{\partial s} \end{array} \right\| \tag{2.120}$$

It is interesting to notice the following. The volume of a small element of the ray tube is equal to $dV = \sigma(p)\, ds = J\, d\xi\, d\eta\, ds$, and, therefore, the Jacobian allows us to calculate also a volume. As follows from eq. 2.115 the Jacobian usually has a finite value that differs from zero. However near the caustic the area $\sigma(p)$ vanishes and, correspondingly

$$J(p) = 0 \tag{2.121}$$

Next assume that both $\sigma(p_0)$ and $\sigma(p)$ have non-zero values and they are cross-sections of the same ray tube. Then substitution of eq. 2.115 into eqs. 2.85 and 2.91 gives

$$A_{0S}(p) = \sqrt{\frac{Z(p_0)}{Z(p)} \frac{J(p_0)}{J(p)}}\, A_{0S}(p_0) \qquad \text{and} \qquad A_{0P}(p) = \sqrt{\frac{Z(p)}{Z(p_0)} \frac{J(p_0)}{J(p)}}\, A_{0P}(p_0) \tag{2.122}$$

where p_0 and p are the reference and observation points, respectively. It is proper to notice that eqs. 2.122 do not contain the product $d\xi d\eta$. It is obvious that

$$F_S = \frac{Z(p)}{Z(p_0)} \frac{J(p)}{J(p_0)} \qquad \text{and} \qquad F_P = \frac{Z(p_0)}{Z(p)} \frac{J(p)}{J(p_0)} \tag{2.123}$$

Whence

$$F_P F_S = \frac{J^2(p)}{J^2(p_0)} \tag{2.124}$$

Example For illustration consider again the case in which the source is located at the origin of the spherical system of coordinates and the velocity is a function of R only. Then latter coincides with s, $R = s$, but θ and φ are ray parameters. The known relationship between the Cartesian and spherical (ray) coordinates is

$$x = R \sin\theta \cos\varphi, \qquad y = R \sin\theta \sin\varphi, \qquad z = R\cos\theta$$

Whence

$$\frac{\partial x}{\partial \theta} = R\cos\theta\cos\varphi, \qquad \frac{\partial y}{\partial \theta} = R\cos\theta\sin\varphi, \qquad \frac{\partial z}{\partial \theta} = -R\sin\theta,$$

$$\frac{\partial x}{\partial \varphi} = -R\sin\theta\sin\varphi, \qquad \frac{\partial y}{\partial \varphi} = R\cos\theta\cos\varphi, \qquad \frac{\partial z}{\partial \varphi} = 0,$$

$$\frac{\partial x}{\partial R} = \sin\theta\cos\varphi, \qquad \frac{\partial y}{\partial R} = \sin\theta\sin\varphi, \qquad \frac{\partial z}{\partial R} = \cos\theta$$

and

$$J = \begin{vmatrix} R\cos\theta\cos\varphi, & R\cos\theta\sin\varphi, & -R\sin\theta \\ -R\sin\theta\sin\varphi, & R\cos\theta\cos\varphi, & 0 \\ \sin\theta\cos\varphi, & \sin\theta\sin\varphi, & \cos\theta \end{vmatrix}$$

This yields

$$J = R^2\sin\theta \qquad \text{and} \qquad \sigma(R) = R^2\sin\theta\, d\theta\, d\varphi$$

This is obvious, since in the spherical system of coordinates, the elementary area, formed by the θ and φ lines, is

$$\sigma = h_2 h_3\, d\theta\, d\varphi,$$

where $h_2 = R$ and $h_3 = R\sin\theta$ are metric coefficients, (Part I). Of course, rays are usually bent, and, correspondingly, they are not oriented in the radial direction. This means that the coordinate s differs from R, but we can still use spherical coordinates θ and φ as ray parameters.

As was mentioned earlier an orientation of a cross-section of the ray tube and, therefore, that of the ξ and η lines depends on s in a very complex manner. For this reason, functions $x = x(\xi, \eta, s)$, $y = y(\xi, \eta, s)$, $z = z(\xi, \eta, s)$ cannot usually be expressed in a explicit form.

Determination of $\mathbf{r}_s$, $\mathbf{r}_\xi$, and $\mathbf{r}_\eta$

In this light it is proper to outline the main steps, related to a determination of the area $\sigma(p)$, using ray coordinates. First, from a solution of the ray equation we find the unit vector $\mathbf{s}_0(p)$, Fig. 2.4a

$$\mathbf{s}_0(p) = \frac{d\mathbf{r}}{ds} = \frac{\mathbf{r}\left(s + \dfrac{\Delta s}{2}\right) - \mathbf{r}\left(s - \dfrac{\Delta s}{2}\right)}{\Delta s}$$

This equation defines a position of the plane, perpendicular to the ray at the point p, where a cross-section of an elementary ray tube is located. The starting point of the given ray has coordinates ξ and η. In additional we choose three rays whose starting points have coordinates, Fig. 2.4b

$$\xi-\frac{\Delta\xi}{2},\quad \eta-\frac{\Delta\eta}{2},\quad \xi+\frac{\Delta\xi}{2},\quad \eta-\frac{\Delta\eta}{2}\qquad \text{and}\qquad \xi+\frac{\Delta\xi}{2},\quad \eta+\frac{\Delta\eta}{2}$$

Taking into account that $d\xi$ and $d\eta$ are known, derivatives $\mathbf{r}_\xi$ and $\mathbf{r}_\eta$ are calculated, as well as area $\sigma(p_0)$ at the surface S_0. Then, we find points of an intersection of three additional rays with the plane S, that is the radius vectors: $\mathbf{r}(x_i, y_i, z_i)$.This allows us to calculate $\mathbf{r}_\xi$ and $\mathbf{r}_\eta$ at any point p of the ray, and, correspondingly, the Jacobian and the area $\sigma(p)$.Therefore to determine the amplitude A_{0S} or A_{0P} it is sufficient to know positions of three neighboring rays and coordinates of their points on planes S_0 and S.

Wave fields in zero approximation

Now we are ready to express the pressure or displacement in terms of the spreading factor and its corresponding value at the starting point. First, in accordance with eqs. 2.48, 2.86 and 2.89 we have

$$L(p) = L(p_0) + \int_{p_0}^{p} n\, ds$$

and

$$A_{0P}(p) = \frac{1}{\sqrt{F_p}} A_{0P}(p_0) \qquad\qquad A_{0S}(p) = \frac{1}{\sqrt{F_s}} A_{0S}(p_0)$$

For example it gives

$$P(p) = \frac{P(p_0)}{\sqrt{F_p}} \exp\left(ik_0 \int_{p_0}^{p} n\, ds\right) \tag{2.125}$$

Here $P(p_0)$ is the pressure at the starting point with coordinates ξ and η. The latter clearly shows that the pressure, (displacement), along the ray depends on its value in the vicinity of the starting point only, that is the pressure at other points of the surface S_0 has no effect. As was demonstrated in Part I this means that due to a destructive interference of elementary waves generated in these points, their total effect is negligible in the zero approximation. In contrary, the pressure of the wave, generated around near the starting point, gives the main contribution. In terms of the method of stationary phase (Part I) we can say that this point is stationary, provided that the field is considered along the ray with the same parameters ξ and η. Such behavior reflects the local character of a propagation in the zero approximation. Let us notice that each time, when this locality is not observed, it is conventional to speak about diffraction phenomena .

In those cases when several rays pass through the point, we have

$$P(p) = \sum_m \frac{P(\xi_m, \eta_m)}{\sqrt{F_{pm}}} \exp\left[ik_0 \int_{p_0}^{p} n\, ds + \tau_m\right] \tag{2.126}$$

where F_{pm}, ξ_m and η_m characterize m-ray. In addition, the exponent contains the term τ_m. As will be shown later, it may appear when the ray touches caustic.

Influence of density change

Next we assume that density ρ is a continuous function of the coordinates and proceeding from the system of equations

$$\operatorname{div} \mathbf{s} = -\frac{P}{M}, \qquad \rho \frac{\partial^2 \mathbf{s}}{\partial t^2} = -\operatorname{grad} P \tag{2.127}$$

we obtain the transport equation for the pressure. First of all, as was shown earlier in this case the displacement $\mathbf{s}$ is not a potential field and, correspondingly, the potential cannot be introduced. In fact, as follows from eq. 1.14

$$\operatorname{curl} \mathbf{s} = -\frac{\nabla \rho \times \mathbf{s}}{\rho} \neq 0$$

Because of this we derive the equation, describing the pressure. Making use again of the second Newton's law, we have

$$\nabla \left(\rho \frac{\partial^2 \mathbf{s}}{\partial t^2} \right) = -\nabla^2 P$$

or

$$\frac{\partial^2 \mathbf{s}}{\partial t^2} \cdot \nabla \rho + \rho \frac{\partial^2}{\partial t^2} \nabla \mathbf{s} = -\nabla^2 P$$

or

$$\frac{\nabla P \cdot \nabla \rho}{\rho} + \frac{\rho}{M} \frac{\partial^2 P}{\partial t^2} = \nabla^2 P$$

Whence

$$\nabla^2 P - \frac{1}{c^2} \frac{\partial^2 P}{\partial t^2} - \frac{\nabla P \cdot \nabla \rho}{\rho} = 0$$

Respectively, for sinusoidal waves the latter becomes

$$\nabla^2 \mathcal{P} + k^2 \mathcal{P} - \frac{\nabla \mathcal{P} \cdot \nabla \rho}{\rho} = 0 \tag{2.128}$$

Assuming that the wavelength is relatively small, the complex amplitude of the pressure can be represented in the same form as the potential U when $\rho = \text{const}$:

$$P = A_p \, e^{i k_o L} \tag{2.129}$$

Substituting the latter into eq. 2.128 and taking into account eqs. 1.35-1.36 we have

$$\nabla^2\mathcal{P} + k^2\mathcal{P} - \frac{\nabla\mathcal{P}\cdot\nabla\rho}{\rho} = -k_0^2\,\mathcal{P}\left[D\,(L) - \frac{k^2}{k_0^2}\right]$$

$$+\,2\,ik_0\,\mathcal{P}\;[\nabla L\cdot\nabla\log A_p + \frac{1}{2}\nabla^2 L] - \frac{\nabla A_p\,\nabla\rho}{\rho}$$

$$-\,ik_0\frac{\mathcal{P}\,\nabla L\,\cdot\,\nabla\rho}{\rho} = 0 \qquad (2.130)$$

Neglecting as before the relatively small term (as $k_0 \to \infty$ and $k \to \infty$) $\frac{\nabla A_p \nabla\rho}{\rho}$, we arrive at two equations

$$D\,(L) = \frac{k^2}{k_0^2} \qquad \text{or} \qquad \text{grad}^{\,2}\,L = n^2 \qquad (2.131)$$

and

$$2\,\nabla L\cdot\nabla\log\,A_p - \frac{\nabla L\cdot\nabla\rho}{\rho} = -\,\nabla^2 L \qquad (2.132)$$

Previously in deriving the eikonal equation, we assumed that the density ρ does not vary. However eq. 2.131 shows that it is valid in a general case, when both c and ρ are continuous functions. The left hand side of eq. 2.132 can be rewritten as

$$2\,\nabla L\cdot(\nabla\log\,A_p - \nabla\log\sqrt{\rho}\,) = \nabla L\cdot\nabla\log\frac{A_p}{\sqrt{\rho}}$$

Therefore, the transport equation for the amplitude A_p becomes

$$\nabla L\cdot\nabla\log\frac{A_p}{\sqrt{\rho}} = -\,\nabla^2 L \qquad (2.133)$$

Comparison with eq. 1.40, describing the potential amplitude A, shows that they coincide if we let

$$A = \frac{A_p}{\sqrt{\rho}}$$

Then we obtain the same result for the amplitude in the zero approximation (m = 0), as it follows from a consideration of the energy flux thorough an elementary ray tube.

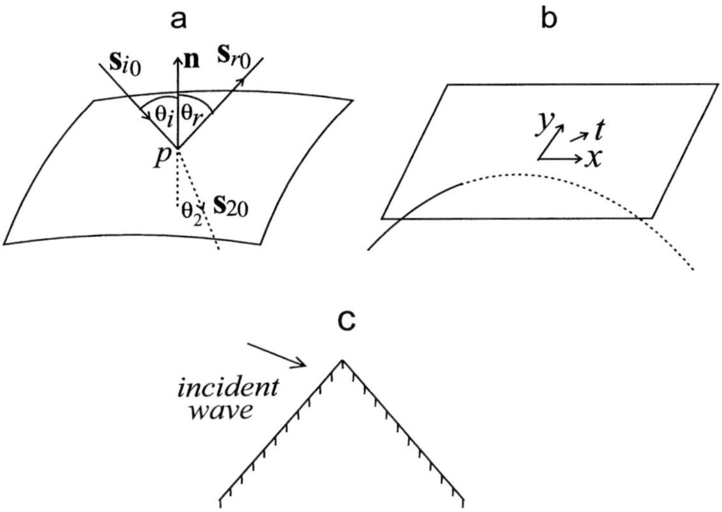

Figure 2.5: (a) Orientation of incident, reflected and transmitted waves (b) Tangential plane and orientation of x, y coordinates (c) Wedge

2.3 Reflected and transmitted waves

We have mainly considered a wave amplitude and eikonal function at regular points where the velocity of propagation and the density of a medium are continuous. Next suppose that two homogeneous media with different physical properties, ρ and c, are separated by the interface S, Fig. 2.5a, which is, in general, a curvilinear one. Inasmuch as at its points either ρ or c or both have a discontinuity, we cannot apply at such places the eikonal and transport equations. In fact, the latter contain derivatives of wave fields with respect to coordinates, as well as parameters of a medium. For this reason, we proceed from boundary conditions and demonstrate that a study of wave fields is greatly simplified in the zero approximation of the Debye expansion because of the principle of locality. Besides, as we know, rays are straight lines in the upper and lower parts of a medium.

Reflection and refraction of waves at the boundary

Let us assume that there is an incident wave and its complex amplitude of the potential is described as

$$\mathcal{U}_i = A_i(r)\, e^{ik_0 L_i(r)} \tag{2.134}$$

Here $A_i(r)$ and $L_i(r)$ are known functions, while $k_0 = \omega/c_0$ is the wave number in some reference point. Under an action of this wave the secondary waves appear at the boundary and in the high-frequency range their potentials have a similar form at each point:

$$\mathcal{U}_r = A_r e^{ik_0 L_r}\,, \qquad \mathcal{U}_2 = A_2 e^{ik_0 L_2} \tag{2.135}$$

where A_r, L_r and A_2, L_2 are unknowns.

By definition, $\mathcal{U}_r$ and $\mathcal{U}_2$ are potentials of the reflected and transmitted waves, (Part I). At the beginning we determine amplitudes at the interface and then, making use of the transport equations, they can be found at any point of a medium.

Snell's law

As was shown in Part I, at both sides of the interface wave fields are related to each other and this linkage can be represented at every point of the surface S as

$$\mathcal{U}_i(p) + \alpha\,\mathcal{U}_r(p) = \beta\,\mathcal{U}_2(p)$$

or

$$A_i e^{ik_0 L_i} + \alpha A_r e^{ik_0 L_r} = \beta A_2 e^{ik_0 L_2} \tag{2.136}$$

Here α and β are some constants. Since

$$k_0 L_i \gg 1\,, \qquad k_0 L_r \gg 1\,, \qquad k_0 L_2 \gg 1$$

all three exponents are rapidly varying functions. Correspondingly, in order to satisfy the boundary condition, eq. 2.136, the equality

$$L_i = L_r = L_2 \tag{2.137}$$

on S has to be met. This means that eikonal functions of all three waves have the same value at each point of the interface. Let us assume that there is a point at which these values are not equal. Then, from eq. 2.136, it follows that amplitudes also change very rapidly. The latter contradicts to one of postulation of the geometrical acoustics, namely: $A(r)$ is a slowly varying function. It may be proper to notice that eq. 2.137 can be treated as the initial condition for eikonal functions of the secondary waves, since they are equal to the known function L_i. Also from this equality we arrive at the simple relationship between amplitudes:

$$A_i + \alpha A_r = \beta A_2$$

Now we determine a mutual position of rays of the incident and secondary waves. By definition,

$$\nabla L_i = n_i \mathbf{s}_{0i}\,, \qquad \nabla L_r = n_i \mathbf{s}_{0r}\,, \qquad \nabla L_2 = n_2 \mathbf{s}_{02} \tag{2.138}$$

Here $\mathbf{s}_{0i}$, $\mathbf{s}_{0r}$ and $\mathbf{s}_{02}$ are unit vectors, oriented along corresponding rays, Fig. 2.5a, and

$$n_1 = \frac{c_0}{c_1}, \qquad n_2 = \frac{c_0}{c_2}$$

At each point of the surface S it is convenient to draw the tangential plane and the unit vector $\mathbf{n}$ is perpendicular to it. The plane, which contains the unit vectors $\mathbf{s}_{0i}$ and $\mathbf{n}$, is called the incident one. The similar planes of secondary waves are formed by pairs: $\mathbf{s}_{0r}$, $\mathbf{n}$ and $\mathbf{s}_{02}$, $\mathbf{n}$, respectively.

Next we demonstrate that all three planes coincide, that is vectors $\mathbf{s}_{0i}$, $\mathbf{s}_{0r}$ and $\mathbf{s}_{02}$ are coplanar. With this purpose in mind let us introduce the Cartesian system of coordinates x and y, as shown in Fig. 2.5b, and consider vector components of ∇L at the tangential plane. By the definition of a gradient, we have

$$(\nabla L)_t = \frac{\partial L}{\partial t}\mathbf{t}_0 \tag{2.139}$$

Here $\mathbf{t}_0$ is the unit vector, tangential to the plane S. The latter can be also rewritten as

$$\begin{aligned} (\nabla L_i)_{t_i} &= \frac{\partial L_i}{\partial x}\mathbf{i} + \frac{\partial L_i}{\partial y}\mathbf{j} \\ (\nabla L_r)_{t_r} &= \frac{\partial L_r}{\partial x}\mathbf{i} + \frac{\partial L_r}{\partial y}\mathbf{j} \\ (\nabla L_2)_{t_2} &= \frac{\partial L_2}{\partial x}\mathbf{i} + \frac{\partial L_2}{\partial y}\mathbf{j} \end{aligned} \tag{2.140}$$

Inasmuch as all three eikonals are equal to each other at every point of the boundary, eq. 2.137, their derivatives coincide in any tangential direction, in particular, along coordinate axes, x and y. Respectively, from eqs. 2.140 it follows that

$$(\nabla L_i)_t = (\nabla L_r)_t = (\nabla L_2)_t \tag{2.141}$$

or making use of eqs. 2.138, we obtain

$$n_1 \mathbf{s}_{it} = n_1 \mathbf{s}_{rt} = n_2 \mathbf{s}_{2t} \tag{2.142}$$

where $\mathbf{s}_{it}$, $\mathbf{s}_{rt}$, $\mathbf{s}_{2t}$ are vector components of unit vectors $\mathbf{s}_{0i}$, $\mathbf{s}_{0r}$,and $\mathbf{s}_{02}$ on the tangential plane. Thus, we see that the three vectors, given by eq. 2.142, have the same magnitude and direction. The latter means that unit vectors along rays $\mathbf{s}_{0i}$, $\mathbf{s}_{0r}$, and $\mathbf{s}_{02}$ are situated in one plane. Certainly, this is the remarkable feature of a ray behavior, which was derived from eq. 2.137. In this light it is proper to notice that this condition, eq. 2.137, does not provide an equality of normal components of the vector ∇L.

As seen from Fig. 2.5a

$$\mathbf{s}_{it} = \sin\theta_i\,\mathbf{t}_0\,, \qquad \mathbf{s}_{rt} = \sin\theta_r\,\mathbf{t}_0\,, \qquad \mathbf{s}_{2t} = \sin\theta_2\,\mathbf{t}_0\,, \tag{2.143}$$

where θ_i, θ_r and θ_2 are the incident, reflection and transmission angles, respectively. Then, in place of eq 2.142 we obtain

$$n_1 \sin\theta_i = n_1 \sin\theta_r = n_2 \sin\theta_2$$

or

$$\frac{\sin\theta_i}{c_1} = \frac{\sin\theta_r}{c_1} = \frac{\sin\theta_2}{c_2} \tag{2.144}$$

and the latter is Snell's law of reflection and refraction for an arbitrary surface. As we already know this result is of a great importance in seismology and eq. 2.144 is usually written in the form of Snell's laws of reflection and refraction

$$\sin\theta_i = \sin\theta_r \qquad \text{or} \qquad \theta_i = \theta_r \tag{2.145}$$

as well as

$$\frac{\sin\theta_i}{c_1} = \frac{\sin\theta_2}{c_2}\,, \tag{2.146}$$

which describe ray behavior of the reflected and transmitted waves. It is clear that the distribution of the rays of these waves is extremely simple in the homogeneous parts of a medium and, correspondingly, knowing initial conditions for the eikonals, eq. 2.137, it is easy to determine its value at any point of a ray. In fact, we have

$$L_r(s) = L_i + n_1 s\,, \qquad L_2(s) = L_2 + n_2 s$$

Here s is the distance along a ray from the boundary.

Derivation of eq. 2.142 and Snell's law may suggest that the latter is valid under certain conditions, and one of them is related to a curvature of an interface. For instance, at points of the wedge corner, the tangential plane, as well as angles θ_i, θ_r and θ_2 are not defined, Fig 2.5c and, respectively, Snell's law is not applied. Of course, this is the limiting case, but it allows us to expect that the surface curvature is an important factor and if its value is sufficiently large, eqs. 2.145-2.146 become invalid. A study of this subject shows that in every case and at certain value of the wavelength λ, there is a minimal value of the curvature radius, R_c, when Snell's law cannot be used.

The second limitation is clearly seen from Snell's law of refraction. In fact, when the velocity c_2 in the lower medium is greater, $c_2 > c_1$, we have for the refraction angle

$$\sin\theta_2 = \frac{c_2}{c_1}\sin\theta_i > 1 \qquad \text{if} \qquad \sin\theta_i > \frac{c_2}{c_1} \tag{2.147}$$

Inasmuch as eq. 2.147 is impossible for real values of θ_2, we have to conclude that in such a case Snell's law does not provide any information about ray direction below the boundary. At the same time, the law of reflection: $\theta_r = \theta_i$ remains valid. The meaning of an inequality 2.147 is very interesting and it will be studied in detail in one of next chapters. It may be proper to notice the following. At the beginning we represented the secondary waves in terms of reflected and transmitted waves. However, the analysis suggests that this assumption is valid until the incident angle does not exceed the critical one, $\theta_i < \theta_c$, where

$$\sin\theta_c = \frac{c_1}{c_2} \tag{2.148}$$

Boundary conditions and wave amplitudes

Before we find amplitudes of secondary waves at the interface it is useful to derive expressions for normal components of the gradient of L, (∇L). It is clear that this component can be written as

$$(\nabla L)_n = \nabla L - n\mathbf{s}_t \qquad \text{or} \qquad (\nabla L)_n = n\mathbf{s}_0 - n\mathbf{s}_t \tag{2.149}$$

Taking into account that

$$(\nabla L)_n = \frac{\partial L}{\partial n}\mathbf{n}$$

and making use of eqs. 2.142-2.143, we have for each wave

$$\frac{\partial L_i}{\partial n} = -\sqrt{n_1^2 - n_1^2\sin^2\theta_i} = -n_1\cos\theta_i$$

and

$$\frac{\partial L_r}{\partial n} = \sqrt{n_1^2 - n_1^2\sin^2\theta_i} = n_1\cos\theta_i \tag{2.150}$$

Thus

$$\frac{\partial L_r}{\partial n} = -\frac{\partial L_i}{\partial n} \tag{2.151}$$

Also

$$\frac{\partial L_2}{\partial n} = -\sqrt{n_2^2 - n_1^2\sin^2\theta_i} \tag{2.152}$$

It is important to emphasize that the negative sign at the right hand side of $\partial L_i/\partial n$ and

$\partial L_2/\partial n$ appears because the unit vectors $\mathbf{s_{01}}$ and $\mathbf{s_{02}}$ form angles exceeding $\pi/2$ with the vector normal to the interface, Fig. 2.6a.

Now we are prepared to derive equations for amplitudes. As is well known, at the boundary the pressure and normal components of particle displacement are continuous functions, and they are related to the scalar potential as

$$P = -\rho \frac{\partial^2 U}{\partial t^2} \qquad \text{and} \qquad \mathbf{s} = \text{grad } U$$

Correspondingly, at the interface we have

$$\rho_1(U_i + U_r) = \rho_2 U_2 \qquad \text{and} \qquad \frac{\partial U_i}{\partial n} + \frac{\partial U_r}{\partial n} = \frac{\partial U_2}{\partial n} \tag{2.153}$$

Since we consider the range of relatively high frequencies each potential can be presented as the asymptotic series

$$\begin{aligned} U_i &= \sum_{m=0}^{\infty} \frac{A_{im}}{(ik_0)^m} e^{ik_0 L_i} \\ U_r &= \sum_{m=0}^{\infty} \frac{A_{rm}}{(ik_0)^m} e^{ik_0 L_r} \\ U_2 &= \sum_{m=0}^{\infty} \frac{A_{2m}}{(ik_0)^m} e^{ik_0 L_2} \end{aligned} \tag{2.154}$$

Substitution of eqs. 2.154 into the first equation of the set 2.153 and equating terms with equal power of m gives

$$\rho_1 A_{im} + \rho_1 A_{rm} = \rho_2 A_{2m} \tag{2.155}$$

Thus, the condition for the pressure is written for wave amplitudes of the same order only. The equality of the normal components of the displacement at the boundary requires differentiation of eikonals and, therefore, the second equation of the set 2.153 yields

$$\begin{aligned} &\sum_{m=0}^{\infty} \frac{\partial A_{im}}{\partial n} \frac{1}{(ik_0)^m} + \sum_{m=0}^{\infty} A_{im} \frac{1}{(ik_0)^{m-1}} \frac{\partial L_i}{\partial n} \\ &+ \sum_{m=0}^{\infty} \frac{\partial A_{rm}}{\partial n} \frac{1}{(ik_0)^m} + \sum_{m=0}^{\infty} A_{rm} \frac{1}{(ik_0)^{m-1}} \frac{\partial L_r}{\partial n} \\ &= \sum_{m=0}^{\infty} \frac{\partial A_{2m}}{\partial n} \frac{1}{(ik_0)^m} + \sum_{m=0}^{\infty} A_{2m} \frac{1}{(ik_0)^{m-1}} \frac{\partial L_2}{\partial n} \end{aligned} \tag{2.156}$$

Collecting terms with the same degree of ik_0 gives

$$A_{i0}\frac{\partial L_i}{\partial n} + A_{r0}\frac{\partial L_r}{\partial n} = A_{20}\frac{\partial L_2}{\partial n} \qquad \text{if} \quad m=0$$

and (2.157)

$$\frac{\partial A_{i,m-1}}{\partial n} + A_{im}\frac{\partial L_i}{\partial n} + \frac{\partial A_{r,m-1}}{\partial n} + A_{rm}\frac{\partial L_r}{\partial n} = \frac{\partial A_{2,m-1}}{\partial n} + A_{2m}\frac{\partial L_2}{\partial n} \qquad \text{if} \quad m \geq 1$$

Thus, the system of equations with respect to unknown amplitudes is

$$\begin{aligned} \rho_1 A_{i0} + \rho_1 A_{r0} &= \rho_2 A_{20} \\ \frac{\partial L_i}{\partial n}A_{i0} + \frac{\partial L_r}{\partial n}A_{r0} &= \frac{\partial L_2}{\partial n}A_{20} \qquad \text{if} \quad m=0\,, \end{aligned} \tag{2.158}$$

and

$$\begin{aligned} \rho_1 A_{im} + \rho_1 A_{rm} &= \rho_2 A_{2m} \\ \frac{\partial A_{i,m-1}}{\partial n} + A_{im}\frac{\partial L_i}{\partial n} + \frac{\partial A_{r,m-1}}{\partial n} + A_{rm}\frac{\partial L_r}{\partial n} &= \frac{\partial A_{2,m-1}}{\partial n} + A_{2m}\frac{\partial L_2}{\partial n} \qquad \text{if} \quad m \geq 1 \end{aligned} \tag{2.159}$$

Taking into account eqs. 2.150-2.152 and solving eqs. 2.158, we arrive at the following expressions for amplitudes in the zero approximation

$$A_{r0} = R_r A_{i0} \qquad \text{and} \qquad A_{20} = R_t A_{i0},$$

where

$$R_r = \frac{Z_2\cos\theta_i - Z_1\cos\theta_2}{Z_2\cos\theta_i + Z_1\cos\theta_2} \tag{2.160}$$

and

$$R_t = \frac{2\rho_1 Z_2\cos\theta_i}{\rho_2(Z_2\cos\theta_i + Z_1\cos\theta_2)} \tag{2.161}$$

are usually termed the reflection and transmission coefficients, respectively.

Each set of eqs. 2.159 contains four unknowns of $m-1$ and m orders, and they allow us to obtain the recursive relationships between them, which have the form

$$A_{rm} = R_r\, A_{im} + \frac{\rho_2}{2\rho_1} R_t\, \delta_{m-1} \qquad \text{if} \quad m \ge 1 \tag{2.162}$$

and

$$A_{2m} = R_t\, A_{im} + \frac{1}{2} R_t\, \delta_{m-1} \tag{2.163}$$

Here

$$\delta_{m-1} = \frac{1}{n_1 \cos\theta_i} \left[\frac{\partial A_{2,m-1}}{\partial n} - \frac{\partial A_{r,m-1}}{\partial n} - \frac{\partial A_{i,m-1}}{\partial n} \right] \qquad \text{if} \quad m \ge 1 \tag{2.164}$$

and

$$n_1 = \frac{c_0}{c_1}\,, \qquad n_2 = \frac{c_0}{c_2}$$

Now it is appropriate to make several comments

1. At the beginning we assumed that the secondary fields were represented by the reflected and transmitted waves only, and that by definition they obey Snell's law. Knowing amplitudes A_{rm} and A_{2m} allows us to improve the accuracy of determination of these waves. However, using eqs. 2.158 - 2.159 we cannot find other waves, if they exist.

2. The expressions for the amplitudes A_{r0} and A_{20}, called Fresnel formulas, play a leading role in studying the reflection and refraction of plane waves in the presence of a planar interface, (Chapters 4 and 5).

3. Knowing amplitudes in the zero approximation at a boundary S and the use of the spreading factor permit us to calculate these amplitudes at any point above and beneath the interface.

4. The function δ_{m-1} is defined by a rate of change of wave amplitudes along the normal.

5. In the case when the incident wave is described by the zero term only,

$$A_{im} = 0 \qquad \text{if} \quad m \ge 1\,,$$

we have

$$A_{rm} = \frac{\rho_2}{2\rho_1} R_t\, \delta_{m-1} \qquad \text{and} \qquad A_{2m} = \frac{1}{2} R_t\, \delta_{m-1} \tag{2.165}$$

that is,

$$A_{2m} = \frac{\rho_1}{\rho_2} A_{rm} \qquad \text{if} \quad m > 0 \tag{2.166}$$

6. As follows from eqs. 2.160–2.161 wave amplitudes in the zero approximation are defined by acoustic impedances, ratio of densities and the incident angle θ_i. Of course, in general, the latter varies along the boundary and the use of the spreading factor emphasizes the principle of locality. As was demonstrated earlier such a behavior occurs because an average value of the Poynting vector is directed along rays.

7. At the same time amplitudes of higher order are subjected to the influence of diffraction phenomena. First of all, it follows from the fact that δ_{m-1} is a function of values of derivatives $\partial A_{m-1}/\partial n$. In other words, the amplitude A_{m-1} at points of the boundary depends on its values above and beneath. Besides, the solution of the transport equation indicates an influence of a wave field at neighboring points. Correspondingly, the concept of locality is not applied if $m > 0$ and part of the energy moves through a lateral surface of ray tubes.

8. Thus, the reflected and transmitted waves, which obey Snell's law are in general subjected to diffraction.

9. We may expect that with an increase of the curvature of a wave front of the incident wave as well as that of the boundary, wave fields will vary more rapidly along the normal, **n**. Correspondingly, a relative contribution of amplitude of higher order increases.

Next consider several examples, that illustrate the behavior of amplitudes of the reflected and transmitted waves in zero approximation.

Case one: Plane incident wave in the presence of plane interface, $(\theta < \theta_c)$ By definition, the incident plane wave

$$\mathcal{U}_i = A_{i0}\, e^{ik_1 L_i}$$

is described by the term of zero order, $m = 0$, of the asymptotical series only and we can use eqs. 2.165. Inasmuch as the incident angle θ_i is constant, amplitudes of secondary waves, A_{r0} and A_{20}, remain the same at points of the boundary. Besides the cross-section of these ray tubes does not change and, therefore, amplitudes A_{r0} and A_{20} do not vary above and beneath the interface. In other words, the secondary waves are also plane waves in the zero approximation. In particular, this means that $\partial A_{r0}/\partial n$ and $\partial A_{20}/\partial n$ are equal to zero. Taking into account that $A_{im} \equiv 0$ if $m > 0$, we conclude that

$$A_{r,m} = A_{2,m} = 0 \tag{2.167}$$

at points of the boundary since $\delta_0 = 0$, eqs. 2.162 - 2.164. Applying the transport equation, we see that these amplitudes are equal to zero everywhere. Thus, we have demonstrated that in the high frequency range the incident plane wave gives rise to secondary plane waves too. As will be shown in Chapter 5 this result remains valid for any frequency. Summarizing we have

$$A_r = \frac{Z_2 \cos\theta_i - Z_1 \cos\theta_2}{Z_2 \cos\theta_i + Z_1 \cos\theta_2} A_i \tag{2.168}$$

and

$$A_2 = \frac{2\rho_1}{\rho_2} \frac{Z_2 \cos\theta_i}{Z_2 \cos\theta_i + Z_1 \cos\theta_2} A_i \tag{2.169}$$

It may be proper to notice that the same equation can be derived if we assume at the beginning that secondary waves are planar.

Case two: Linear source above a plane boundary, $(\theta < \theta_c)$ Suppose that a linear source of the incident wave is located in the upper medium and is parallel to the boundary, Fig. 2.6a. As was shown in Part I the complex amplitude of potential is described by

$$\mathcal{U}_i = A\, H_0^{(1)}(k, r) \tag{2.170}$$

Here $H_0^{(1)}$ is the Hankel function of zero order and first type.

If the argument is sufficiently large, $k_1 r \gg 1$, the latter can be represented in the form of the asymptotic series with respect to powers: $(i\, k_0)^{-m}$ and, correspondingly, all coefficients A_{im} of the Debye expansion are known. In our case, the incident wave is described by the zero term, $m = 0$, that is

$$\mathcal{U}_i = \frac{C}{\sqrt{r}}\, e^{ik_1 r} \tag{2.171}$$

where r is the distance between the source and an observation point.

Now let us find amplitudes of secondary waves in the zero approximation. First of all in accordance with eqs. 2.160–2.161 at points of the interface we have

$$A_{r0} = \frac{Z_2 \cos\theta_i - Z_1 \cos\theta_2}{Z_2 \cos\theta_i + Z_1 \cos\theta_2} \frac{C}{\sqrt{r}}$$

and (2.172)

$$A_{20} = \frac{2\rho_1}{\rho_2} \frac{Z_2 \cos\theta_i}{Z_2 \cos\theta_i + Z_1 \cos\theta_2} \frac{C}{\sqrt{r}}$$

where θ_i and θ_2 obey Snell's law:

$$\frac{\sin\theta_2}{c_2} = \frac{\sin\theta_i}{c_1}$$

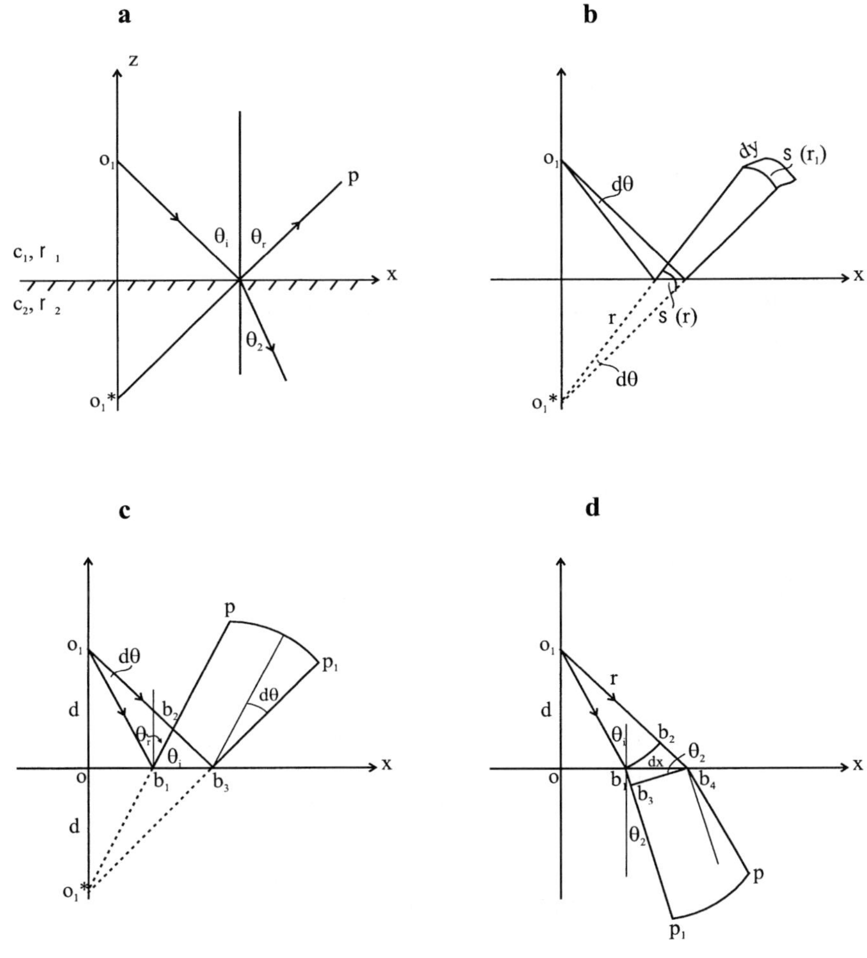

Figure 2.6: Spreading factor: (a) linear source above plane boundary; (b),(c) elementary tube of reflected wave and spreading factor; (d) elementary tube of transmitted wave and spreading factor.

Because of a change of the incident angle and the distance from the source r both amplitudes vary at the interface. Taking into account the fact that secondary waves arise in the vicinity of the boundary, we can say that eq. 2.172 describes, to some extent, a distribution of linear sources of these waves in the zero approximation.

In order to determine amplitudes at any point of a medium we make use of the concept of spreading factor . First, consider the reflected wave. As follows from Fig. 2.6a

$$o\,o_1 = o\,o_1^*$$

since $\theta_i = \theta_r$, and, correspondingly, all rays of this wave intersect at the same point o_1^*, which is a mirror reflection of the point o. As we know, this is a very important fact and it greatly simplifies a study of the reflected wave. For instance, the cross-section of an elementary tube can be easily calculated, Fig. 2.6b. Indeed, we have

$$\sigma(r) = r\,d\theta\,dy \qquad \text{and} \qquad \sigma(r_1) = r_1\,d\theta\,dy \tag{2.173}$$

where dy is an elementary extension of the ray tube along the y axis. Thus, the spreading factor is equal to

$$F_r = \frac{r_1}{r} \tag{2.174}$$

and the amplitude of the reflected wave is

$$A_{r0} = \frac{1}{\sqrt{F_r}}\,A_{r0}(r) \tag{2.175}$$

Here r_1 is the distance from an observation point to o_1^*. Substitution of eq. 2.172 and 2.174 into eq. 2.175 gives

$$A_{r0}(r_1) = \frac{Z_2\cos\theta_i - Z_1\cos\theta_2}{Z_2\cos\theta_i + Z_1\cos\theta_2}\,\frac{C}{\sqrt{r_1}}, \tag{2.176}$$

while the eikonal function is

$$L_r = \frac{k_1}{k_0}\,r_1 = r_1$$

since $k_0 = k_1$. This means that in the zero approximation the reflected wave behaves as the cylindrical one, and it is caused by a fictitious source, located at the point o_1^*, which has the same intensity as real one. This conclusion remains valid for all terms of the asymptotic series, because they are characterized by the same eikonal and, correspondingly, the total reflected wave is cylindrical too.

It is interesting to derive the spreading factor, eq. 2.174, in a slightly different way. Inasmuch as the cross-section of elementary tubes is very small, Fig. 2.6c, we can treat arcs b_1b_2 and b_2b_3 as sides of the triangle $b_1b_2b_3$, where

$$b_1b_2 = b_2b_3 = r\,d\theta \tag{2.177}$$

As is seen from this figure

$$pp_1 = b_2b_3 + p_1b_3\, d\theta = (r + p_1b_3)\, d\theta$$

Whence

$$F = \frac{r + p_1b_3}{r} = \frac{r_1}{r}$$

since

$$r + p_1b_3 = o_1^* p = r_1$$

From eq. 2.176 it follows that the complex amplitude of the potential of the reflected wave is

$$\mathcal{U}_r = \frac{Z_2 \cos\theta_i - Z_1 \cos\theta_2}{Z_2 \cos\theta_i + Z_1 \cos\theta_2} \frac{C}{\sqrt{r_1}} e^{ik_1r_1}\,, \tag{2.178}$$

and this shows that the wave field varies on the eikonal surface, $r_1 = \text{const}$. Certainly such a behavior indicates the presence of a tangential component of the Poynting vector, however its average value, $\mathbf{Y}^{av}$, is directed along the normal to this surface.

Next we derive an amplitude of the transmitted wave in the lower medium. Again consider elementary tubes of this and incident waves, Fig. 2.6d, and find a relationship between their cross-sections near the boundary. It is obvious that

$$b_1b_2 = r\, d\theta_i = dx\, \cos\theta_i$$

and

$$b_3b_4 = dx\, \cos\theta_2$$

Whence

$$b_3b_4 = \frac{b_1b_2}{\cos\theta_i} \cos\theta_2 = \frac{r\cos\theta_2}{\cos\theta_i}\, d\theta_i\,, \tag{2.179}$$

which characterizes an area of the tube cross-section in the vicinity of an interface.

As seen in Fig. 2.6d, near the observation point p we have

$$p_1p = b_3b_4 + pb_4\, d\theta_2$$

or

$$p_1p = \frac{r\cos\theta_2}{\cos\theta_i}\, d\theta_i + pb_4\;\, d\theta_2 \tag{2.180}$$

In order to relate small changes of angles we make use Snell's law

$$\sin\theta_2 = \frac{c_2}{c_1}\, \sin\theta_1$$

Its differentiation gives

$$\cos\theta_2\, d\theta_2 = \frac{c_2}{c_1}\cos\theta_i\, d\theta_i$$

and

$$d\theta_2 = \frac{c_2}{c_1}\frac{\cos\theta_i}{\cos\theta_2}\, d\theta_i$$

Whence

$$p\,p_1 = \left(\frac{r\cos\theta_2}{\cos\theta_i} + \frac{c_2}{c_1}\frac{\cos\theta_i}{\cos\theta_2}\, pb_4\right)\, d\theta_i \tag{2.181}$$

Correspondingly, the spreading factor of the transmitted wave is

$$F_t = \frac{pp_1}{b_3b_4} = 1 + \frac{c_2}{c_1}\frac{\cos^2\theta_i}{\cos^2\theta_2}\frac{pb_4}{r} \tag{2.182}$$

Here $b_4\,p$ is the distance along the ray from the boundary to an observation point p. Also it is convenient to represent eq. 2.182 as

$$F_t = \frac{pp_1}{b_3b_4} = 1 + \frac{c_2}{c_1}\frac{z}{d}\frac{\cos^3\theta_i}{\cos^3\theta_2} \tag{2.183}$$

since

$$z = pb_4\cos\theta_2 \qquad \text{and} \qquad d = r\cos\theta_i\,,$$

which characterize the position of the observation point and the source, respectively. The behavior of the transmitted wave differs essentially from that of a reflected wave. First of all, it is not a cylindrical wave, and the eikonal surface, Fig. 1.21b, is defined from equality

$$k_1 r + k_2\, pb_4 = \text{const}$$

or

$$\frac{d}{\cos\theta_i} + \frac{c_1}{c_2}\frac{z}{\cos\theta_2} = \text{const} \tag{2.184}$$

Moreover, the dependence of the amplitude A_{20} on the distance from the source is more complicated than in the case of the reflected wave. It happens because the spreading factor is also a function of the incident angle and the ratio of velocities, c_2/c_1.

Note that if θ_i is small, we have

$$F_t \approx 1 + \frac{c_2}{c_1}\frac{z}{d}$$

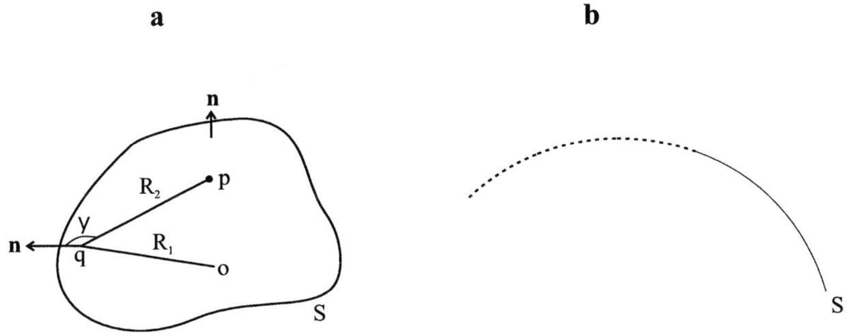

Figure 2.7: (a) Illustration of eq. 2.194 (b) Model of screen

and it varies within the range

$$\frac{c_2}{c_1}\frac{z}{d} < F_t < 1 \qquad \text{if} \qquad \theta_i < 1$$

By definition, the potential of the transmitted wave is

$$\mathcal{U}_2 = \frac{2\rho_1}{\rho_2}\,\frac{Z_2\cos\theta_i}{Z_2\cos\theta_i + Z_1\cos\theta_2}\,\frac{C}{\sqrt{F_t}}\,\frac{e^{i(k_1 r + k_2 p b_4)}}{\sqrt{r}} \tag{2.185}$$

As before, at each instant there is a flux of energy along the eikonal surface, but its average value is equal to zero. Also the amplitude A_{20} is defined at each point of the lower medium by corresponding values at the boundary, (the principle of locality). In this light it may be proper to notice that in deriving eq. 2.183 we assumed that amplitudes at points of elements b_1b_4 and b_3b_4 are the same. This takes place because these elements are very close to each other.

In accordance with eq. 2.185 the amplitude A_{20} is inversely proportional to $\sqrt{p\,b_4}$, if an observation point is far from the boundary and $p\,b_4/r \gg 1$.

Case three: Elementary spherical source above a plane interface ($\theta_i < \theta_c$) As we know this source gives rise to the incident spherical wave

$$\mathcal{U}_i = C\,\frac{e^{ik_1R}}{R}, \tag{2.186}$$

where R is the distance from the source. Correspondingly

$$A_{im} = 0 \qquad \text{if} \qquad m > 0$$

Because of the axial symmetry consider a wave field in the plane xoz only. In accordance with eqs. 2.158 at points of the boundary

$$A_{r0} = R_r \, \frac{e^{ik_1 R}}{R} \, C \tag{2.187}$$

$$A_{20} = R_t \, \frac{e^{ik_1 R}}{R} \, C \tag{2.188}$$

Here

$$R = \sqrt{x^2 + d^2},$$

but R_r and R_t are reflection and transmission coefficients, given by eqs. 2.160 - 2.161.

As is seen from Fig. 2.6c rays of the reflected wave intersect each other at the same point o_1^*, which is a mirror reflection of o_1. Therefore, as in the case of the cylindrical wave, we can imagine that a fictitious source of the wave is situated at the point o_1^* and

$$\mathcal{U}_{r0} = \frac{Z_2 \cos\theta_i - Z_1 \cos\theta_2}{Z_2 \cos\theta_i + Z_1 \cos\theta_2} \, \frac{e^{ik_1 R_1}}{R_1} \, C \tag{2.189}$$

where

$$R_1 = \sqrt{d^2 + x^2}$$

Again we observe the remarkable result, namely, at the high frequency spectrum the eikonal surfaces of this wave are spherical and their center coincides with the point o_1^*. At the same time the wave amplitude varies at these surfaces. It may be proper to notice the following. Inasmuch as the acoustic potential depends on the distance R_1 and the incident angle θ, the particle displacement usually has both components

$$s_{R_1} = \frac{\partial U}{\partial R_1} \qquad \text{and} \qquad s_\theta = \frac{\partial U}{\partial \theta},$$

and they are shifted in phase. This results in the elliptical polarization, that is, during each period, the particle moves along an ellipse, (Part I). Also it is important to emphasize that due to a simplicity of the wave field we need not calculate the spreading factor.

Next, by analogy with the case of the linear source, we derive an expression for the amplitude of a transmitted wave. Let us again use Fig. 2.6d, bearing in mind that unlike two dimensional case, the cross-section of an elementary tube also changes in the direction perpendicular to the plane xoz. For instance, near points b_1 and a we have

$$dl(b_1) = ob_1 \, d\varphi\,, \qquad dl(p) = x \, d\varphi \tag{2.190}$$

Here x is the coordinate of an observation point and $d\varphi$ is the angle between two rays located in the plane, perpendicular to xoz. Therefore, areas of cross-sections in the vicinity of points b and p are

$$\sigma(b_1) = b_3 b_4 \, ob_4 \, d\varphi\,, \qquad \sigma(p) = p_1 p \, x \, d\varphi,$$

and the spreading factor is

$$F_t = \frac{p_1 p\, x}{b_3 b_4\; ob_4} \tag{2.191}$$

As follows from eq. 2.183 and Fig. 2.6d

$$F_t = \frac{p_1 p}{b_3 b_4}\,\frac{x}{d\tan\theta_i} = \frac{x}{d\tan\theta_i}\left[1 + \frac{c_2}{c_1}\,\frac{z}{d}\,\frac{\cos^3\theta_i}{\cos^3\theta_2}\right], \tag{2.192}$$

and the potential of the transmitted wave in the zero approximation, $(m = 0)$, is

$$\mathcal{U}_{20} = \frac{2\rho_1}{\rho_2}\,\frac{Z_2\cos\theta_i}{(Z_2\cos\theta_i + Z_1\cos\theta_2)\sqrt{F_t}}\,\frac{e^{ik(k_1R + k_2b_3p)}}{R} \tag{2.193}$$

Helmholtz formula and secondary waves

Until now our study of waves in zero approximation was based on calculation of their amplitudes at the interface and then the use of spreading factor. Also it is useful to derive the same result in a completely different way, namely, proceeding from the Huygens - Fresnel principle, (Part I). The latter states that a vicinity of any point of the wave field can be treated as a volume, where sources of elementary spherical waves are situated. Moreover, a summation of these waves allows us to determine a total field at an observation point. Developing this approach Helmholtz found a relationship between the complex amplitude of the potential $\mathcal{U}(p)$ inside the volume V and values of $\mathcal{U}$, as well as $\partial\mathcal{U}/\partial n$, on the surface S, surrounding it, Fig. 2.7a. As was described in Part I, this expression is

$$\mathcal{U}(p) = \frac{1}{4\pi}\int\limits_S \left[\frac{e^{ikR}}{R}\,\frac{\partial\mathcal{U}(q)}{\partial n} - \mathcal{U}(q)\,\frac{\partial}{\partial n}\,\frac{e^{ikR}}{R}\right] dS \tag{2.194}$$

Here the unit vector $\mathbf{n}$ is directed outwards and R is the distance between an observation point p and any point q of the surface S.

In accordance with the Helmholtz formula, eq. 2.194, the complex amplitude $\mathcal{U}(q)$ inside the volume is the result of a summation of the terms

$$F_1 = \frac{\partial\mathcal{U}(q)}{\partial n}\,\frac{e^{ikR}}{R} \qquad \text{and} \qquad F_2 = \mathcal{U}(q)\,\frac{\partial}{\partial n}\,\frac{e^{ikR}}{R} \tag{2.195}$$

This suggests that in the vicinity of each point of the surface S there are two types of sources with strength proportional to $\partial\mathcal{U}/\partial n$ and $\mathcal{U}$, respectively. At the same time, a wave at any point of the volume V is a superposition of elementary waves, caused by these sources.

The function F_2 can be represented as

$$F_2 = \mathcal{U}(q)\,\frac{\partial}{\partial R}\left(\frac{e^{ikR}}{R}\right)\cos\psi \tag{2.196}$$

where

$$\cos\psi = \frac{\partial R}{\partial n} \tag{2.197}$$

and ψ is the angle between the vector $\mathbf{R}$ and the unit vector $\mathbf{n}$, Fig. 2.7a.

As follows from the Helmholtz formula, each element of the surface generates two spherical waves into volume V. One of them is described by the function F_1. The amplitude of this spherical wave has a constant value on the eikonal surface with the center at the point q. However, the second wave, given by eq. 2.196, is different. Its phase surface is still spherical, but the amplitude varies as a function of the angle ψ. Because as both waves are spherical and have a common origin at the point q, the resultant wave is spherical too. In fact, performing a differentiation in eq. 2.196, the Helmholtz formula can be written as

$$\mathcal{U}(q) = \frac{1}{4\pi}\oint\limits_S F(q)\,\frac{e^{ikR}}{R}\,dS \tag{2.198}$$

where

$$F(q) = \frac{\partial\mathcal{U}(q)}{\partial n} - (ikR-1)\,\frac{\mathcal{U}(q)}{R^2}\cos\psi \tag{2.199}$$

Kirchhoff diffraction theory

This equation found numerous applications in studying interference of electromagnetic and acoustic waves. One of them is Fraunhofer and Fresnel diffraction in the presence of nontransparent screens. Then the surface of integration, eq. 2.198, consists of two parts, Fig. 2.7b. In accordance with Kirchhoff assumptions at the internal side of the rigid screen the wave field is absent, as are its normal derivatives

$$\mathcal{U}(q) = 0, \qquad \frac{\partial\mathcal{U}(q)}{\partial n} = 0$$

At the same time, at points of the second part, S_0 (dotted line), the wave field coincides with the incident one, that is

$$\mathcal{U}(q) = \mathcal{U}(q_i) \qquad \text{and} \qquad \frac{\partial\mathcal{U}(q)}{\partial n} = \frac{\partial\mathcal{U}_i(q)}{\partial n} \tag{2.200}$$

This means that the integrand in eq. 2.198 becomes known and it is equal to

$$F(q) = \frac{\partial\mathcal{U}_i(q)}{\partial n} - (ikR-1)\,\frac{\mathcal{U}_i(q)}{R^2}\cos\psi \tag{2.201}$$

Assuming that the wave length is relatively small, eqs. 2.198 and 2.201 allowed one to study an influence of screens with a different geometry. As was demonstrated in Part I, in spite of an approximative character of Kirchhoff method, it permits us to observe such known patterns of diffraction as an exponential decay of the wave in the shadow zone and a rapid oscillations of its intensity.

Secondary waves and Kirchhoff diffraction theory

Taking into account a great success of this theory to study an interference of waves in the presence of screens it is natural to apply the same approach, eq. 2.198, to investigate reflection and transmission. With this purpose in mind suppose that the surface S coincides with the boundary between two media with different physical parameters. Let us emphasize that in the case of screens points of the surface S_0 (opening) are regular, since in their neighborhood a medium is homogeneous. For this reason conditions 2.200 are used. However, at the interface either the velocity c or density ρ or both of them are discontinuous and the reflected and transmitted waves arise. Because of this in the vicinity of the boundary the wave field differs from the incident one and, since the high-frequency range is considered, all waves behave as plane waves. Further we assume that the secondary waves at points of S are described by the zero approximation, $m = 0$. Thus, we have at both sides of surface S

$$\mathcal{U}_1 = \mathcal{U}_i + \mathcal{U}_{r0} \qquad \text{and} \qquad \mathcal{U}_{20}\,, \tag{2.202}$$

and their substitution into eq. 2.194 gives for secondary waves

$$\mathcal{U}_r(p) = \frac{1}{4\pi}\int\limits_S \left[\frac{e^{ik_1R_2}}{R_2}\frac{\partial\,\mathcal{U}_{r0}(q)}{\partial\,n} - \mathcal{U}_{r0}(q)\,\frac{\partial}{\partial n}\left(\frac{e^{ik_1R_2}}{R_2}\right)\right] dS \tag{2.203}$$

and

$$\mathcal{U}_{20}(p) = \frac{1}{4\pi}\int\limits_S \left[\frac{e^{ik_2R_2}}{R_2}\frac{\partial\,\mathcal{U}_{20}(q)}{\partial\,n_1} - \mathcal{U}_{20}(q)\,\frac{\partial}{\partial n_1}\left(\frac{e^{ik_2R_2}}{R_2}\right)\right] dS \tag{2.204}$$

Here $\mathbf{n}_1 = -\mathbf{n}$

Assuming that the incident wave is caused by an elementary spherical source, we have

$$\mathcal{U}_i(q) = C\,\frac{e^{ik_1R_1}}{R_1} \tag{2.205}$$

where R_1 is the distance between the source and the point q, while R_2 is the distance between q and an observation point p. Also

$$\mathcal{U}_{r0}(q) = R_r \frac{e^{ik_1R_1}}{R_1} C \qquad \text{and} \qquad \mathcal{U}_{20}(q) = R_t \frac{e^{ik_1R_1}}{R_1} C, \tag{2.206}$$

where R_r and R_t are the reflection and transmission coefficients, given by eqs. 2.160 - 2.161 and they depend on impedances, as well as on the incident angle.

Both equations, 2.203 - 2.204, contain derivatives along the normal. For instance, taking into account eqs. 2.206, and performing a differentiation we have:

$$\begin{aligned}\frac{\partial \mathcal{U}_{r0}}{\partial n} &= \frac{\partial R_r}{\partial n} C \frac{e^{ik_1R_1}}{R_1} + C\,R_r \frac{\partial}{\partial n}\left(\frac{e^{ik_1R_1}}{R_1}\right) = \\ &= \frac{\partial R_r}{\partial n} C \frac{e^{ik_1R_1}}{R_1} + C\,R_r \left(\frac{ik_1}{R_1} - \frac{1}{R_1^2}\right) e^{ik_1R_1} \cos\psi_1\end{aligned}$$

and

$$\frac{\partial}{\partial n}\left(\frac{e^{ik_1R_2}}{R_2}\right) = \left(\frac{ik_1}{R_2} - \frac{1}{R_2^2}\right) e^{ik_1R_2} \cos\psi_2$$

Here ψ_2 is the angle between the normal $\mathbf{n}$ and $\mathbf{r}_2$.

Inasmuch as we are interested mainly in the high-frequency spectrum, terms proportional to k_1 play a dominant role and, thus, eq. 2.203 becomes

$$\mathcal{U}_r(p) = \frac{ik_1C}{4\pi} \oint\limits_S R_r(q)\,(\cos\psi_1 - \cos\psi_2) \frac{e^{ik_1(R_1+R_2)}}{R_1R_2}\, dS \tag{2.207}$$

In the same manner we obtain

$$\frac{\partial \mathcal{U}_2(q)}{\partial n_1} = -\frac{\partial \mathcal{U}_2}{\partial n} = -\frac{ik_1}{R_1} R_t(q)\, C\, e^{ik_1R_1} \cos\psi_1$$

and

$$\frac{\partial}{\partial n_1}\left(\frac{e^{ik_1R_2}}{R_2}\right) = -\frac{ik_2}{R_2} e^{ik_2R_2} \cos\psi_2$$

Whence

$$\mathcal{U}_2(p) = -\frac{ik_1C}{4\pi} \int\limits_C \frac{R_t(q)\left(\cos\psi_1 - \frac{c_1}{c_2}\cos\psi_2\right)}{R_1R_2} e^{i(k_1R_1 + k_2R_2)}\, dS \tag{2.208}$$

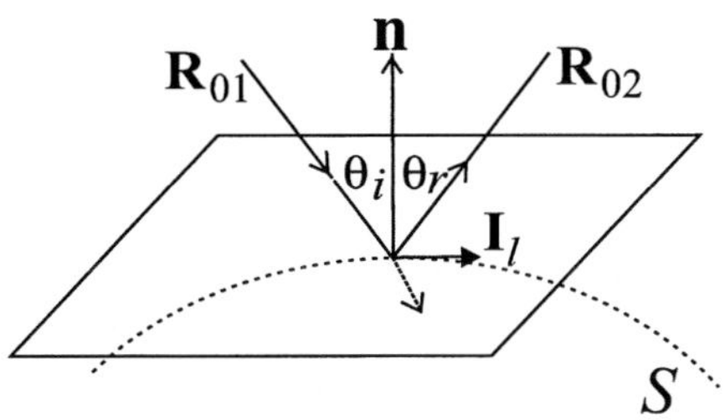

Figure 2.8: Snell's law and stationary point

In essence, eqs. 2.207 - 2.208 are the well known Fresnel-Kirchhoff diffraction formulas. They show that at each point of a medium the wave is a superposition of elementary spherical waves, arising at the boundary. It is clear that this process can have either the constructive or destructive character and so it is natural to use the concept of the stationary point and repeat some derivations performed previously. First, consider the reflected wave, eq. 2.207, where the integral at the right hand side can be written as

$$\mathcal{U}_r(p) = \frac{ik_1 C}{4\pi} \int_S \Phi_r(q)\, e^{i\varphi_r(q)}\, dS \tag{2.209}$$

Here q is the point of the surface and $\Phi_r(q)$ is a relatively slowly varying function, while the term $exp\,[k_1(R_1+R_2)]$ rapidly changes from point to point because $k_1(R_1+R_2) \gg 1$. In such a case, the integral is defined by the behavior of the integrand in the vicinity of the stationary point, where

$$\frac{\partial \varphi_r}{\partial l} = 0 \qquad \text{or} \qquad \frac{\partial}{\partial l}(R_1 + R_2) = 0 \tag{2.210}$$

and dl is the displacement along line l, which is an intersection of two planes. One of them is formed by rays of the incident, reflected, and transmitted waves, as well as the normal $\mathbf{n}$, Fig. 2.8. The secondary plane is tangential to the boundary at point q.

The left side of eq. 2.210 is a sum of directional derivatives and, by definition, can be represented as

$$\mathbf{i}_l \nabla R_1 + \mathbf{i}_l \nabla R_2 = 0 \qquad \text{or} \qquad \mathbf{i}_l \mathbf{R}_{01} - \mathbf{i}_l \mathbf{R}_{02} = 0\,, \tag{2.211}$$

since $\nabla R_1 = \mathbf{R}_{01}$ and $\nabla R_2 = -\mathbf{R}_{02}$.

Here $\mathbf{R}_{01}$ and $\mathbf{R}_{02}$ are unit vectors, characterizing a direction of the incident and reflected waves, and $\mathbf{i}_l$ is the unit vector along the line l. Thus we have

$$\cos(\mathbf{i}_l, \mathbf{R}_{01}) = \cos(\mathbf{i}_l, \mathbf{R}_{02}) \tag{2.212}$$

As seen from Fig. 2.8 we again arrive at Snell's law

$$\sin\theta_i = \sin\theta_r \tag{2.213}$$

This means that for a given source-receiver geometry, the stationary point is a point of the boundary, where the orientation of the incident and reflected rays obeys eq. 2.213. This result can be interpreted in the following way. In principle, at any observation point p the wave is a sum of elementary waves arising at different points of the interface. However, only the superposition of waves, which start in the vicinity of the stationary point, has a constructive character, and, therefore, they give the main contribution. It happens because these waves have almost equal phases at the point p. At the same time, waves generated by sources at other points of the boundary, have essentially different phases at point p, and their interference does not produce a noticeable effect. Applying the method of stationary phase, it is not difficult to derive the known expression for the amplitude of the reflected wave at an observation point, eq. 2.189. Certainly we can obtain the same results by using the spreading factor derived earlier.

It may be proper to notice that the method of stationary phase vividly illustrates the principle of locality for the zero term of the asymptotical series, $m = 0$, since the wave at all points of the ray, ($\theta_r = \text{const}$), arrives from the stationary point.Thus it becomes clear that an absence of the flux of the Poynting vector through a lateral surface of the ray tube means a weak influence of elementary waves, arising at other places of the boundary.

By analogy with the previous case, the potential of the transmitted wave is written in the form

$$\mathcal{U}_2(p) = \frac{ik_1 C}{4\pi} \int\limits_S \Phi_2(q)\, e^{i\varphi_2(q)}\, dS \tag{2.214}$$

where $\varphi_2(q) = k_1 R_1 + k_2 R_2$. Correspondingly, the stationary point is determined from the equality

$$\frac{\partial}{\partial l}(k_1 R_1 + k_2 R_2) = 0 \qquad \text{or} \qquad \frac{\partial R_1}{\partial l} + \frac{c_1}{c_2}\frac{\partial R_2}{\partial l} = 0\,,$$

that is

$$\mathbf{R}_{01}\,\mathbf{i}_l - \frac{c_1}{c_2}\mathbf{R}_{02}\,\mathbf{i}_l = 0 \tag{2.215}$$

As is seen from Fig. 2.8b we again arrive at Snell's law of refraction

$$\frac{\sin\theta_i}{c_i} = \frac{\sin\theta_2}{c_2} \tag{2.216}$$

This means that if the incident angle does not exceed the critical one, ($\theta_i < \theta_c$), at all points of the ray the transmitted wave is caused by sources in the vicinity of the stationary point. In other words, each point of the boundary is stationary for both waves if observation points are located on rays whose orientation obeys Snell's law. However, when $\theta_i > \theta_c$, the point of an intersection of the incident ray with the boundary is not stationary one, eq. 2.215.

Area of application of the zero approximation

As was described earlier, Snell's law does not define the angle θ_2 if the incident angle exceeds the critical one. In other words, the zero approximation of the transmitted wave is not able to describe a wave field at points where $\theta_i > \theta_c$. The other limitation is related to the reflected wave. In accordance with eq. 2.189 we have

$$\mathcal{U}_{r0} = \frac{Z_2 \cos\theta_i - Z_1\sqrt{1-\sin^2\theta_i\,/n^2}}{Z_2 \cos\theta_i + Z_1\sqrt{1-\sin^2\theta_i\,/n^2}}\, C\, \frac{e^{ik_1R_1}}{R_1},$$

Here $n = c_1/c_2$.

We determine the rate of change of this function with respect to the angle θ_i. Differentiation by θ_i gives

$$\frac{\partial \mathcal{U}_{r0}}{\partial \theta_i} = \frac{2Z_1Z_2\,(1/n^2-1)\,\sin\theta_i}{(Z_2\,\cos\theta_i + Z_1\,\cos\theta_2)^2\sqrt{1-\dfrac{\sin^2\theta_i}{n^2}}}\,\frac{e^{ik_1R}}{R_1}$$

This shows that in the vicinity of the critical angle $(\sin\theta_i = n)$ the wave amplitude changes very quickly, and in the limit $\theta_i \to \theta_c$ it tends to infinity. At the same time, one of assumptions of geometrical acoustics is small variation of the wave amplitude. Thus, the reflected wave, described by eq. 2.189, does not satisfy this condition, when the incident angle is close to the critical one. Similar behavior takes place for the transmitted wave. Now it may be proper to make one comment. In principle, equations 2.162 - 2.163 allow us to determine the next terms of the asymptotic expansion, A_{rm} and A_{2m} for the reflected and transmitted waves. However, these recursive relations contain derivatives of amplitudes in the direction normal to the boundary. For this reason their calculation becomes rather cumbersome for both the plane and curvilinear surfaces. In the first case an information about these terms is usually obtained indirectly, when we solve a boundary value problem.

Influence of a weak boundary

These equations also permit us to study the effect of a so called weak boundary. At such a surface both the density and velocity of propagation are continuous functions

$$\rho_1 = \rho_2 \qquad \text{and} \qquad c_1 = c_2, \tag{2.217}$$

but some of its derivatives along the normal to the boundary have a discontinuity. Respectively it is natural to classify these boundaries in accordance with the minimal order of the derivative which is discontinuous. For instance, if

$$\frac{\partial^m c_1}{\partial n^m} = \frac{\partial^m c_2}{\partial n^m} \qquad \text{if} \qquad m < M \tag{2.218}$$

and

$$\frac{\partial^M c_1}{\partial n^M} \neq \frac{\partial^M c_2}{\partial n^M}$$

we deal with a weak interface of M-order.

In particular, the ordinary boundary, where $c_1 \neq c_2$ or $\rho_1 \neq \rho_2$, corresponds to a discontinuity of the zero order. As follows from eqs. 2.160 - 2.161 at a weak interface the reflection and transmission coefficients of the zero approximation are equal to zero and unity, respectively. Therefore, recursive formulas have the form

$$A_{rm} = \frac{1}{2}\,\delta_{m-1}\,, \qquad A_{2m} = A_{im} + \frac{1}{2}\,\delta_{m-1}\,, \tag{2.219}$$

where

$$\delta_{m-1} = \frac{1}{n_1 \cos\theta_i}\,\frac{\partial}{\partial n}(A_{2,m-1} - A_{i,m-1} - A_{r,m-1}) \tag{2.220}$$

If $m < M$, all values of δ_{m-1} are equal to zero. Correspondingly we have $A_{rm} = 0$, and $A_{2m} = A_{im}$.

Moreover, δ_{M-1} differs from zero, and it gives

$$A_{rM} = \frac{1}{2n_1 \cos\theta_i}\,\frac{\partial}{\partial n}(A_{2,M-1} - A_{i,M-1}) \tag{2.221}$$

Thus, as follows from the asymptotical series the reflected wave is

$$\mathcal{U}_r \approx \frac{A_{rM}}{(ik_0)^M}\, e^{ik_0 L} \tag{2.222}$$

that is in the presence of a weak boundary the wave, $\mathcal{U}_r$, is described mainly by the series term of M-order. This result is rather obvious and it can be expected. In fact, the weaker the interface is the smaller the reflected wave is. In the limiting case, when parameters of a medium and its derivatives of any order are continuous functions, in place of reflection we can observe scattering, which is exponentially small, and, respectively, it cannot be taken into account by Debye expansion.

2.4 Wave field near caustic

Earlier we introduced the concept of a caustic as an envelope of ray's family where they are tangential to this surface and intersect each other in its vicinity, Fig. 2.3. As a result,

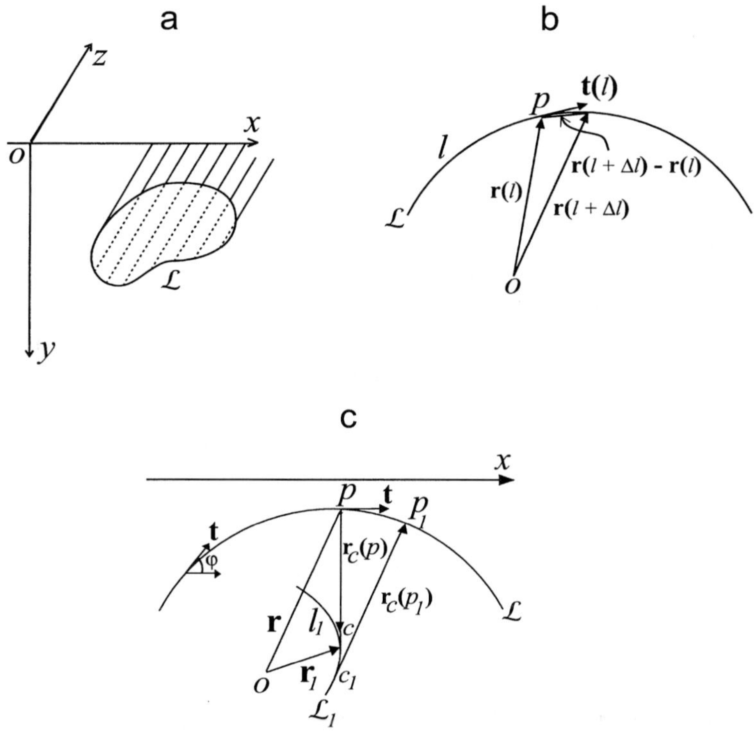

Figure 2.9: (a) Phase surface of cylindrical wave (b) Radius vector $\mathbf{r}(l)$ (c) Relationship between curve $\mathcal{L}$ and its evolute $\mathcal{L}_\infty$

a cross-section of ray tubes tends to zero in such places and the wave amplitude seems to become infinitely large. Of course, such behavior is impossible and indicates that the amplitude representation in the form of the asymptotical series is not able to describe a wave near a caustic. For this reason, we investigate this subject in a different way and we start from the simplest two dimensional case.

Cylindrical wave in a homogeneous medium

Suppose that some part of a medium is homogeneous and an arbitrary system of linear sources, oriented along the z-axis, produces a cylindrical wave. It is clear that an intersection of its phase surface with the XY plane gives the line $\mathcal{L}$, Fig. 2.9a, and since we deal with two dimensional case it is sufficient to study a wave field in this plane. Also it is proper to notice that our main attention is paid to a region, where the medium is homogeneous and, correspondingly, rays are rectilinear ones. First we remind some basic concepts, characterizing a plane curve, which in this case represents the wave surface.

Suppose that a position of the point p on the curve is defined by the parameter l, which is the arc length, counted from a reference point located on this curve. Respectively, the radius vector $\mathbf{r}(l)$, drawn from some point o, defines the position of the curve, Fig. 2.9b. Forming the difference

$$\mathbf{r}(l+\Delta l)-\mathbf{r}(l),$$

we see that the derivative $d\mathbf{r}/dl$ is equal to the unit vector $\mathbf{t}$, tangential to the curve, and its direction coincides with that of an increase of l,

$$\frac{d\mathbf{r}}{dl}=\mathbf{t} \tag{2.223}$$

Now we introduce the curvature vector $\mathbf{N}$ as

$$\mathbf{N}=\frac{d\mathbf{t}}{dl} \tag{2.224}$$

It is obvious that the magnitude of this vector characterizes a rate of change of the direction of the tangential vector $\mathbf{t}$, called the curvature of the curve. It is a simple matter to prove that vectors $\mathbf{t}$ and $\mathbf{N}$ are perpendicular to each other. In fact, a differentiation of the equality

$$\mathbf{t}\cdot\mathbf{t}=1$$

with respect to l yields

$$\mathbf{t}\cdot\frac{d\mathbf{t}}{dl}+\frac{d\mathbf{t}}{dl}\cdot\mathbf{t}=0 \qquad \text{or} \qquad \mathbf{t}\cdot\frac{d\mathbf{t}}{dl}=0 \tag{2.225}$$

Thus, the dot product of $\mathbf{N}$ and $\mathbf{t}$ is equal to zero, that is the curvature vector $\mathbf{N}$ is perpendicular to the tangent and, correspondingly, it is directed along the normal to the curve. From a definition it follows that $\mathbf{N}$ is oriented toward a concavity of the curve, since the vector

$$\mathbf{t}(l+\Delta l)-\mathbf{t}(l),$$

has the same direction, provided that $\Delta l>0$.

It is convenient to represent the magnitude of the curvature vector as

$$|\mathbf{N}|=\frac{1}{r_c} \tag{2.226}$$

Here r_c is the radius of curvature and it is inversely proportional to $|\mathbf{N}|$. Therefore, we can write

$$\mathbf{N}=\frac{1}{r_c}\mathbf{n}, \tag{2.227}$$

where $\mathbf{n}$ is the unit vector along the normal to the curve. As follows from eqs. 2.223-2.224

$$\mathbf{N}=\frac{d^2\mathbf{r}}{dl^2} \qquad \text{or} \qquad \mathbf{N}=\frac{d^2x}{dl^2}\mathbf{i}+\frac{d^2y}{dl^2}\mathbf{j} \tag{2.228}$$

where $\mathbf{i}$ and $\mathbf{j}$ are unit vectors along coordinate axes, and the curvature vector, as well as r_c, is expressed in terms of the second derivative of the function $\mathbf{r}(l)$.

Evolute of a curve (caustic of a cylindrical wave surface)

Let us take an arbitrary point p on the curve $\mathcal{L}$ and draw along the normal in the direction of concavity the segment pc, equal to the radius of curvature r_c at this point, Fig. 2.9c. The point c is called the center of curvature of curve $\mathcal{L}$ at point p. If p moves along $\mathcal{L}$ then a position of the point c also moves forming the curve $\mathcal{L}_1$, which is called the evolute of the curve $\mathcal{L}$. Thus, the former is the locus of curvature centers of the curve $\mathcal{L}$. Bearing in mind that we consider the two-dimensional case, it is natural to treat $\mathcal{L}_1$ is a result of an intersection of some cylindrical surface S with the XY plane. In this light it is proper to notice that lines pc, directed along the vector $\mathbf{n}$, represent rays, since they are normal to the wave surface.

In order to study some important features of the evolute $\mathcal{L}_1$, it is necessary to determine the derivative $\partial\mathbf{n}/\partial l$. Inasmuch as $\mathbf{n}$ is unit vector, a differentiation of the product $\mathbf{n}\cdot\mathbf{n} = 1$ gives

$$\frac{\partial \mathbf{n}}{\partial l}\cdot\mathbf{n} = 0 \tag{2.229}$$

and therefore the vector $\partial\mathbf{n}/\partial l$ is parallel to the tangent.

Differentiating the equality $\mathbf{t}\cdot\mathbf{n} = 0$ with respect l, we have

$$\mathbf{N}\cdot\mathbf{n} + \mathbf{t}\cdot\frac{d\mathbf{n}}{dl} = 0 \qquad \text{or} \qquad \mathbf{t}\cdot\frac{d\mathbf{n}}{dl} = -\frac{1}{r_c}, \tag{2.230}$$

since

$$\mathbf{N}\cdot\mathbf{n} = \frac{1}{r_c}$$

Taking into a account that $|\mathbf{t}| = 1$ and both vectors: $\mathbf{t}$ and $d\mathbf{n}/dl$ are tangential to curve $\mathcal{L}$, we conclude that they have opposite directions. Therefore

$$\frac{d\mathbf{n}}{dl} = -\frac{\mathbf{t}}{r_c} \tag{2.231}$$

Consider two pairs: $\mathbf{r}$, l and $\mathbf{r}_1$, l_1, which are the radius-vector and the arc length for curves $\mathcal{L}$ and $\mathcal{L}_1$, respectively. As is seen from Fig. 2.9c

$$\mathbf{r}_1 = \mathbf{r} + r_c\,\mathbf{n} \tag{2.232}$$

Its differentiation with respect to l yields

$$\frac{d\mathbf{r}_1}{dl} = \frac{d\mathbf{r}}{dl} + \frac{dr_c}{dl}\mathbf{n} + r_c\frac{d\mathbf{n}}{dl}$$

or

$$\frac{d\mathbf{r}_1}{dl} = \mathbf{t} + \frac{dr_c}{dl}\mathbf{n} + r_c\frac{d\mathbf{n}}{dl} \tag{2.233}$$

Making use of eq. 2.231 we obtain

$$\frac{d\mathbf{r}_1}{dl} = \mathbf{t} + \frac{d\mathbf{r}_c}{dl}\mathbf{n} - \mathbf{t}$$

that is

$$\frac{d\mathbf{r}_1}{dl} = \frac{d\mathbf{r}_c}{dl}\mathbf{n} \tag{2.234}$$

Considering points of the evolute it is natural to deal with arc length l_1. The use of the known rule of differentiation of the function $\mathbf{r}_1[\mathbf{r}(l)]$ gives

$$\frac{d\mathbf{r}_1}{dl} = \frac{d\mathbf{r}_1}{dl_1}\frac{dl_1}{dl} = \frac{dl_1}{dl}\mathbf{t}_1 \tag{2.235}$$

where $\mathbf{t}_1$ is the unit vector tangential to the evolute. Substitution of eq. 2.235 into eq. 2.234 yields

$$\frac{dl_1}{dl}\mathbf{t}_1 = \frac{dr_c}{dl}\mathbf{n} \tag{2.236}$$

The right and left hand sides of this equality represent vectors, directed along the normal to the curve $\mathcal{L}$ and tangential to the evolute $\mathcal{L}_1$, respectively. Therefore, the normal of the curve $\mathcal{L}$ is parallel to the evolute tangential. Since both of them pass through the common point c, Fig. 2.9c, we discover the first feature of the evolute: the normal to the curve $\mathcal{L}$ touches the evolute at the corresponding point. This means that the evolute is an envelope of normals to the curve $\mathcal{L}$. Taking into account that

$$|\mathbf{t}_1| = |\mathbf{n}| = 1,$$

eq. 2.236 gives

$$\left|\frac{dl_1}{dl}\right| = \left|\frac{dr_c}{dl}\right| \qquad \text{or} \qquad |dl_1| = |dr_c| \tag{2.237}$$

For simplicity, assuming that within a certain interval of the curve and an evolute both l_1 and r_c increase, we can write

$$dl_1 = dr_c$$

Its integration shows that an increment of the evolute arc coincides with that of the curvature radius of the curve $\mathcal{L}$. Thus we obtain one more feature of the evolute, namely, a change of the curvature radius of the curve $\mathcal{L}$ is equal to the arc length of the evolute, provided that r_C varies monotonically. For instance, as is seen from Fig. 2.9c

$$r_c(p_1) - r_c(p) = c\,c_1 \tag{2.238}$$

Note that this equality will be used in studying the wave amplitude near caustic. Suppose that the angle between the x-axis and the tangent t is equal to φ, Fig. 2.9c. Then we have

$$\mathbf{t} = \cos\varphi\,\mathbf{i} + \sin\varphi\,\mathbf{j}$$

Differenting by l gives

$$\mathbf{N} = -\sin\varphi\,\frac{d\varphi}{dl}\,\mathbf{i} + \cos\varphi\,\frac{d\varphi}{dl}\,\mathbf{j}$$

Correspondingly, the square of the magnitude of $\mathbf{N}$ is

$$\frac{1}{r_c^2} = \left(-\sin\varphi\,\frac{d\varphi}{dl}\right)^2 + \left(\cos\varphi\,\frac{d\varphi}{dl}\right)^2$$

or

$$\frac{1}{r_c} = \left|\frac{d\varphi}{dl}\right|$$

and this expression for the curvature was derived in Part I.

Equation of the evolute

Earlier we obtain equations of rays and eikonal surfaces. Now let us find an equation describing the evolute. Suppose that the curve $\mathcal{L}$ is given in the explicit form

$$y = f(x) \tag{2.239}$$

and X, Y are coordinates of any point of the normal. Also we assume that x and y characterize a position of the point p, where the normal intersects the curve $\mathcal{L}$. Then, it is obvious that the vector

$$\mathbf{M} = (X - x)\,\mathbf{i} + (Y - y)\,\mathbf{j}$$

is oriented along the normal. At the same time elementary vector $\mathbf{dl}$, which is tangential to the curve $\mathcal{L}$, can be written as

$$\mathbf{dl} = dx\,\mathbf{i} + dy\,\mathbf{j}$$

Inasmuch as vectors $\mathbf{M}$ and $\mathbf{dl}$ are perpendicular to each other, we have

$$\mathbf{M} \cdot \mathbf{dl} = 0$$

or

$$(X - x)\,dx + (Y - y)\,dy = 0$$

or

$$(X - x) + y'(x)(Y - y) = 0 \tag{2.240}$$

The latter represents the equation of a family of normals, where x plays the role of the parameter, since y is the function of x. In order to find an equation of the envelope of this family, we have to use two equations (Appendix D). Eq.2.240 is one of them. The second is obtained by a differentiation of the former with respect to the parameter x. This yields

$$\begin{cases} (X - x) + y'(Y - y) = 0 \\ -1 + y''(Y - y) - (y')^2 = 0 \end{cases} \tag{2.241}$$

Eliminating x, we arrive at the relationship between X and Y that represents an equation of the envelope of normals, that is the evolute. Also we can obtain a different form of the evolute equation by solving the system 2.241 with respect to X and Y:

$$X = x - \frac{y'(1 + y'^2)}{y''}, \qquad Y = y + \frac{1 + y'^2}{y''} \tag{2.242}$$

Thus, knowing coordinates of the point p, as well as the first and second derivatives of the function $y(x)$, we can determine coordinates of point c, located on the evolute. This is the parametric form of the evolute equation. Finally suppose that in place of eq. 2.239, the curve $\mathcal{L}$ is given as

$$x = \varphi_1(t) \qquad \text{and} \qquad y = \varphi_2(t) \tag{2.243}$$

where t is some parameter, for instance, the arc length along $\mathcal{L}$. To find coordinates X and Y of the evolute points it is necessary to express derivatives of $y(x)$ in terms of differentials. For example,

$$y' = \frac{dy}{dx}, \qquad y'' = \frac{d}{dx}\left(\frac{dy}{dx}\right) = \frac{d^2y\,dx - d^2x\,dy}{dx^2}$$

Substitution of the latter into eq. 2.242 yields

$$X = x - \frac{dy(dx^2 + dy^2)}{d^2y\,dx - d^2x\,dy}, \qquad Y = y + \frac{dx(dx^2 + dy^2)}{d^2y\,dx - d^2x\,dy} \tag{2.244}$$

Now consider three examples, that illustrate the behavior of evolutes.

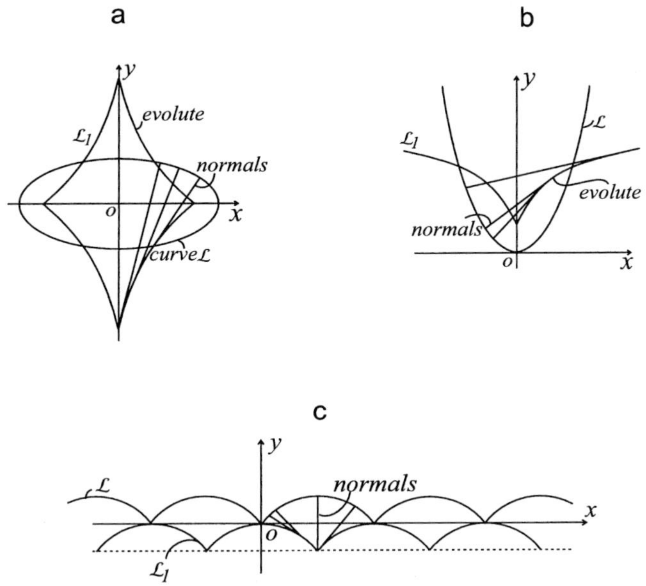

Figure 2.10: (a) Ellipse, its evolute and normals (b) Parabola, its evolute and normals (c) Cycloid, its evolute and normals

Example one: Evolute of an ellipse In this case the curve $\mathcal{L}$ is an ellipse,

$$\frac{x^2}{a^2} + \frac{y^2}{b^2} = 1 \qquad \text{and} \qquad a > b$$

The parametric form of this equation is

$$x = a \cos t\,, \qquad y = b \sin t$$

The latter allows us to determine differentials of x and y of the first and second orders. Substituting them in eqs. 2.244 gives

$$X = \frac{a^2 - b^2}{a} \cos^3 t\,, \qquad Y = -\frac{a^2 - b^2}{b} \sin^3 t \tag{2.245}$$

Multiplying the first and second equations by a and b, respectively, then taking the power 2/3 and adding them, we obtain the equation of the evolute

$$a^{2/3}\, X^{2/3} + b^{2/3}\, Y^{2/3} = (a^2 - b^2)^{2/3},$$

which is shown in Fig. 2.10a. It is interesting to notice that at the apogees and perigees of the ellipse its radius of curvature has maximal and minimal values, and at points located on ellipse axis, the evolute has singularities, namely points of return.

Example two: Evolute of a parabola Now we find the evolute of the parabola

$$y = a\,x^2$$

Since

$$y' = 2\,a\,x\,, \qquad y'' = 2\,a \qquad \text{and} \qquad (y')^2 = 4\,a^2\,x^2,$$

eq. 2.242 gives

$$X = x - \frac{2ax\,(1 + 4a^2x^2)}{2a} = -4a^2\,x^3$$

and

$$Y = ax^2 + \frac{(1 + 4a^2x^2)}{2a} = \frac{1}{2a} + 3ax^2$$

Eliminating the parameter x, we obtain the explicit form of the evolute equation

$$Y = \frac{1}{2a} + \frac{3}{2\sqrt{2a}}\,X^{2/3},$$

and it is shown in Fig. 2.10b.

Example three: Evolute of a cycloid Consider the cycloid, Fig. 2.10c, given by equations

$$x = a(t - \sin t)\,, \qquad y = a(1 - \cos t)$$

Making use of eq. 2.244, we arrive at the evolute equation

$$X = a(t + \sin t)\,, \qquad Y = -a(1 - \cos t)$$

This curve is also a cycloid, but it is located differently with respect to the coordinate axes.

Main features of caustic in two dimensional case

As the curve $\mathcal{L}$ describes an intersection of a wave surface with the XY plane, the curve $\mathcal{L}_1$, represents the caustic, Fig. 2.9c. Now we will outline its main features. First of all at each its point the ray is tangential to this surface and, correspondingly, it is an envelope of a ray family. Second, the distance along the ray between the wave surface and the caustic is equal to the curvature radius of the former. In other words, the caustic consists of curvature centers of the eikonal surface. In this light consider an elementary displacement dl of the line $\mathcal{L}$, Fig. 2.11a. Inasmuch as it can be treated as an element of

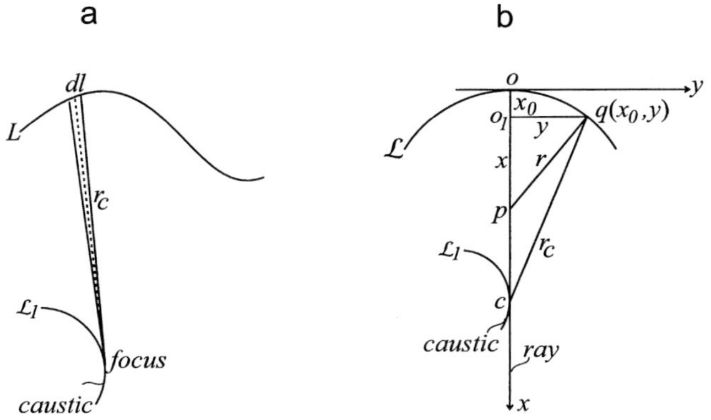

Figure 2.11: (a) Caustic is locus of foci (b) Illustration of eq. 2.253

the circle arc with the radius $r_c(p)$, all rays, passing through dl, merge at the same point c of the caustic. Thus, each its point is crossed by an infinite number of rays. Such a point is called focus. Correspondingly, we may say that the caustic is formed by foci.

If we assume that the wave field is described by the zero approximation of Debye series, then, as is well known, the flux of the Poynting vector along any ray tube does not change. In particular, for the elementary tube with cross-section area, S, we have

$$YS = \text{const}\,,$$

From Fig. 2.11a it is clearly seen that in the vicinity of the caustic, the latter, S, tends to zero, and therefore the Poynting vector becomes infinitely large, as well as the wave amplitudes. Earlier we pointed out that this result shows only that the ray theory, based on the asymptotical expansion of the wave field, is invalid in the vicinity of caustic.

Often the caustic is called the evolute of the wave surface, but the latter is an evolvent of the caustic.

Behavior of the phase, (eikonal), near the caustic

Because as the asymptotic expansion, (Debye series), cannot describe a wave near the caustic, we make use of the Kirchhoff formula. Unlike the series

$$\mathcal{U} = e^{ik_0 L} \sum_{m=0}^{\infty} \frac{A_m}{(ik_0)^m},$$

this representation is valid at any wave length. In particular, at the high frequency spectrum and for two dimensional case we have

$$\mathcal{U}(p) = \int_{\mathcal{L}} \Phi(k,q)\, \frac{e^{ikr}}{\sqrt{r}}\, dl, \tag{2.246}$$

where $\mathcal{L}$ is the line of the wave surface and r is the distance between the point q and an observation point p,and k is the wave number. Since we assume that $kr \gg 1$, the exponential term, $exp(ikr)$, varies rapidly as a function of the point q. At the same time, the product

$$\frac{\Phi(k,q)}{\sqrt{r}}$$

changes relatively slowly. From a physical point of view, the integrand describes the potential of the wave, caused by an elementary linear source, passing through the point q. Correspondingly, the total wave, $\mathcal{U}(q)$, is the result of a superposition of all elementary waves. Next, making use of eq. 2.246, consider the behavior of a phase along any ray touching the caustic. With this purpose in mind let us choose the Cartesian system of coordinates, shown in Fig. 2.11b. Then an observation point p has coordinates: x, 0, while the point $q:\ x_0,\ y$. Whence

$$r = \sqrt{(x-x_0)^2 + y^2} \tag{2.247}$$

It is almost obvious that with an increase of the distance from the origin, the phase difference between elementary waves arising at neighboring points, becomes greater. Respectively, a superposition of these waves has a rather destructive character. In other words, the contribution of waves generated in such parts of the line $\mathcal{L}$ is relatively small. For this reason we assume that the potential $\mathcal{U}(p)$ is defined mainly by those elementary waves, which appear at points q, located sufficiently close to the x-axis. This condition may be written as

$$x_0 \ll r_c \qquad \text{and} \qquad y \ll r_c \tag{2.248}$$

The latter implies that these points are practically situated on the circle with the radius r_c. Since we study an eikonal in the vicinity of caustic, that is near the center of curvature, it is natural to relate coordinates x_0 and y of a point q to the radius r_c. As is seen from the triangle $0\,q\,c$, Fig. 2.11b, we have

$$(r_c - x_0)^2 + y^2 = r_c^2$$

Neglecting by the term x_0^2 we obtain

$$x_0 \simeq \frac{y^2}{2r_c} \tag{2.249}$$

This allows us to represent the distance r as

$$\begin{aligned} r &= [(x-x_0)^2 + y^2]^{1/2} \approx (x^2 - 2\,x\,x_0 + y^2)^{1/2} \\ &= x\left(1 - \frac{2x_0}{x} + \frac{y^2}{x^2}\right)^{1/2} \approx x + \frac{y^2}{2x} - x_0 \end{aligned}$$

Substitution of eq. 2.249 gives

$$r \approx x + \frac{y^2}{2}\left(\frac{1}{x} - \frac{1}{r_c}\right) \tag{2.250}$$

Of course, if the point q is located at the origin we have $r = x$. It is interesting to notice that the derivative $\partial r / \partial y$ is equal to zero when $y = 0$. This means that the method of stationary phase also suggests that elementary waves, arising near the origin, $y = 0$, play the dominant role. Bearing in mind that the function $\Phi(k, q) \,/\, \sqrt{r}$ is the almost a constant within a small interval of integration around this point, we can write eq. 2.246 in the form

$$\mathcal{U}(p) \approx C \int_{-\infty}^{\infty} e^{ikr} \, dy, \tag{2.251}$$

since $dl \approx dy$. Then, taking into account eq. 2.250, we have

$$\mathcal{U}(p) \approx C \, e^{ikx} \, I \,, \tag{2.252}$$

where

$$I = \int_{-\infty}^{\infty} exp\left[ik\left(\frac{1}{x} - \frac{1}{r_c}\right)\frac{y^2}{2}\right] dy \tag{2.253}$$

Before we evaluate this integral it may be useful to make several comments.

a. In essence, an integration has to be performed within a small range of variable y around the origin. Beyond that the contribution is negligible. For this reason, we can use infinite limits, allowing us to find a simple expression for the integral I.

b. The first derivative of the function

$$\varphi(y) = \left(\frac{1}{x} - \frac{1}{r_c}\right)\frac{y^2}{2}$$

is

$$\varphi'(y) = \left(\frac{1}{x} - \frac{1}{r_c}\right) y \tag{2.254}$$

and it vanishes at the point $y = 0$, where the ray intersects the wave front. This clearly shows again that to evaluate the integral I we can apply the method of stationary phase, provided that

$$k\left(\frac{1}{x} - \frac{1}{r_c}\right) \gg 1 \tag{2.255}$$

In particular, we cannot use eq. 2.253 at the center of curvature, where the ray touches caustic. Thus in order to find the amplitude and phase of the wave along the ray it is necessary to perform a numerical integration, eq. 2.251, where the integrand is a rapidly oscillating function.

c. In passing the point c, Fig. 2.11b, the function $\varphi(y)$ changes a sign

$$\varphi(y) > 0 \quad \text{if} \quad x < r_c \qquad \text{and} \qquad \varphi(y) < 0 \quad \text{if} \quad x > r_c \tag{2.256}$$

Respectively we have to find two integrals

$$I_1 = \int_{-\infty}^{\infty} e^{ik\alpha y^2}\, dy \qquad \text{and} \qquad I_2 = \int_{-\infty}^{\infty} e^{-ik\alpha y^2}\, dy\,, \tag{2.257}$$

where

$$\alpha = \frac{1}{2}\left|\frac{1}{x} - \frac{1}{r_c}\right| \tag{2.258}$$

d. As follows from eq. 2.250 the distance between point q and observation point p exceeds its coordinate x $(r > x)$, provided that $x < r_c$. At the point where the ray is tangent to the caustic they are equal $(r = x)$, and if $x > r_c$, the distance r is smaller than the coordinate x, $(r < x)$. Such a relationship between r and x defines the behavior of the phase near the caustic.

Evaluation of integrals I_1 and I_2

First consider the integral

$$I_1 = \int_{-\infty}^{\infty} e^{ik\alpha y^2}\, dy$$

Introducing a new variable

$$u = \sqrt{k\,\alpha}\, y\,,$$

we have

$$du = \sqrt{k\,\alpha}\, dy$$

and therefore

$$I_1 = \frac{1}{\sqrt{k\,\alpha}} \int_{-\infty}^{\infty} e^{iu^2}\, du$$

or

$$I_1 = \frac{1}{\sqrt{k\,\alpha}} \left(\int_{-\infty}^{\infty} \cos u^2\, du + i \int_{-\infty}^{\infty} \sin u^2\, du \right)$$

In a similar manner, the second integral of eqs. 2.257 is equal to

$$I_2 = \frac{1}{\sqrt{k\,\alpha}} \left(\int_{-\infty}^{\infty} \cos u^2\, du - i \int_{-\infty}^{\infty} \sin u^2\, du \right)$$

Since

$$\int_{-\infty}^{\infty} \cos u^2 \, du = \int_{-\infty}^{\infty} \sin u^2 \, du = \sqrt{\frac{\pi}{2}}$$

we have

$$I_1 = \sqrt{\frac{\pi}{2\,k\,\alpha}}\,(1+i)\,, \qquad I_2 = \sqrt{\frac{\pi}{2\,k\,\alpha}}\,(1-i) \tag{2.259}$$

Substitution of the latter into eq. 2.252 yields

$$\mathcal{U}(p) = C\,\sqrt{\frac{\pi}{k\,\alpha}}\,e^{i\,(k\,x+\pi/4)} \qquad \text{if} \quad x < r_c\,,\ k\alpha \gg 1 \tag{2.260}$$

and

$$\mathcal{U}(p) = C\,\sqrt{\frac{\pi}{k\,\alpha}}\,e^{i\,(k\,x-\pi/4)} \qquad \text{if} \quad x > r_c\,,\ k\alpha \gg 1 \tag{2.261}$$

Thus, in the neighborhood of the caustic the phase has two values

$$i\left(k\,r_c + \frac{\pi}{4}\right) \qquad \text{and} \qquad i\left(k\,r_c - \frac{\pi}{4}\right),$$

and, correspondingly, in passing the point c, where $x = r_c$, Fig. 2.11b, it decreases by $\pi/2$.

Wave amplitude near a caustic

As was noted previously, eqs. 2.260 - 2.261 allow us to find the wave amplitude along the ray. However, in the vicinity of a caustic it unlimitedly increases and become infinitely large ($\alpha \to 0$), where the ray touches the caustic. For this reason an evaluation of the amplitude can be done making use of eq. 2.246. Now we consider its behavior in the direction perpendicular to the ray, Fig. 2.12a. As before we have

$$\mathcal{U} = C \int_{\mathcal{L}} e^{ikr}\, d\ell, \tag{2.262}$$

where $d\ell$ is the elementary displacement along line $\mathcal{L}$, representing the wave surface; r is the distance between any point q of the curve $\mathcal{L}$ and an observation point with coordinates x and y, Fig. 2.12a. In order to evaluate the integral at the right hand side of eq. 2.262 we first express the distance q_1c in terms of the coordinate x and the angle θ. As follows from eq. 2.238, the difference of the curvature radii of the line $\mathcal{L}$ at points q and q_1 is equal to the arc length cc_1 of the evolute, that is

$$x - q_1c_1 = c_1c \qquad \text{or} \qquad q_1c_1 = x - r_0\,\theta \tag{2.263}$$

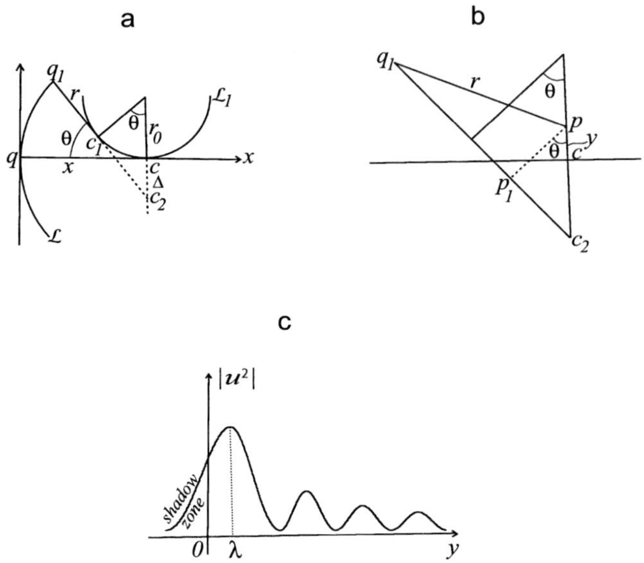

Figure 2.12: (a) Illustration of eq. 2.263 (b) Illustration of eq. 2.265 (c) Wave intensity near caustic

Here we use the fact that the distance $c\,c_1$ is small, and it can be represented as

$$c\,c_1 = r_0\,\theta,$$

where r_0 is the curvature radius of the evolute at the point c.
As is seen from Fig. 2.12a, the distance $q_1\,c_2$ is equal to

$$q_1 c_2 \;=\; q_1 c_1 + c_1 c_2 \;=\; x - r_0\,\theta + (r_0 + \Delta)\sin\theta \tag{2.264}$$

Here

$$\Delta = c\,c_2$$

Bearing in mind that $r \gg (y + \Delta)$, Fig. 2.12b, we have

$$q_1\,p = q_1\,p_1$$

and, therefore

$$r = q_1 c_2 - p_1 c_2 \qquad \text{or} \qquad r = q_1 c_2 - (y + \Delta)\sin\theta \tag{2.265}$$

Substitution of eq. 2.264 into eq. 2.265 gives

$$r = x - r_0\,\theta + r_0 \sin\theta + \Delta\sin\theta - y\,\sin\theta - \Delta\,\sin\theta$$

Then

$$r = x - r_0\,\theta + r_0 \sin\theta - y\,\sin\theta \tag{2.266}$$

Thus, we found a relationship between the distance r and the coordinates of the observation point, as well as the parameter of the evolute r_0 and the angle θ. Inasmuch as the potential $\mathcal{U}$ is mainly defined by sources near the point q, Fig. 2.12a, the angle θ is rather small. Then expanding $\sin\theta$ in the power series

$$\sin\theta \approx \theta - \frac{\theta^3}{6},$$

we obtain

$$\mathcal{U} \approx C\, e^{ikx} x \int\limits_{-\infty}^{\infty} e^{-ik\left(y\,\theta + \frac{\theta^3}{6} r_0\right)} d\theta \tag{2.267}$$

since

$$d\ell = x\, d\theta$$

The imaginary part of the integrand is an odd function of θ and, correspondingly, eq. 2.267 becomes

$$\mathcal{U}(y) = 2\,C\,e^{ikx}\,x\,T\,, \tag{2.268}$$

where

$$T = \int\limits_0^{\infty} \cos\left(k\,y\,\theta + \frac{k\,r_0}{6}\,\theta^3\right)\, d\theta \tag{2.269}$$

The latter is related to the Airy function $\Phi(w)$

$$\Phi(w) = \frac{1}{\sqrt{\pi}} \int\limits_0^{\infty} \cos\left(\xi\,w + \frac{\xi^3}{3}\right)\, d\xi \tag{2.270}$$

In fact, introducing a new variable

$$\xi = \left(\frac{k\,r_0}{2}\right)^{1/3} \theta\,,$$

the integral T becomes

$$T = \left(\frac{2}{k\,r_0}\right)^{1/3} \int\limits_0^{\infty} \cos\left(w\,\xi + \frac{\xi^3}{3}\right)\, d\xi \tag{2.271}$$

or

$$T = \left(\frac{2\,\pi^{3/2}}{k\,r_0}\right)^{1/3} \Phi(w)\,, \tag{2.272}$$

and the parameter w is

$$w = y\,\left(\frac{2\,k^2}{r_0}\right)^{1/3} \tag{2.273}$$

Airy function is well studied and, in particular, for large and positive values of w, $(y > 0)$:

$$\Phi(w) \simeq \frac{1}{2\,w^{1/4}}\,e^{-2/3\,w^{3/2}}\,,$$

that is it decays exponentially. Certainly, this corresponds to the shadow zone, where the wave field rapidly decreases towards the center of the evolute curvature. If the parameter w is negative, $(y < 0)$, but is still large, then

$$\Phi(w) \approx \frac{1}{(-w)^{1/4}}\,\sin\left[\frac{2}{3}(-w)^{3/2} + \pi/4\right]\,,$$

and it oscillates with the magnitude proportional to $(-w)^{1/4}$. This shows that at the opposite side of the caustic there is a diffraction zone, where constructive and destructive interference of elementary waves take place. Of course, the wave field has everywhere a finite value. For instant the Airy function attains the largest value, equal 0.949, when $w = 1.02$. In particular, with an increase of the frequency, this maximum approaches the caustic. Results of calculation of the function $|\mathcal{U}|^2$, eq. 2.268, which characterizes the wave intensity, are given in Fig. 2.12c. As we already know from the asymptotic expressions for the Airy function, wave behavior at both sides of the caustic is completely different. In one part the field rapidly decays, while the other vividly demonstrates a diffraction phenomenon. A width of this zone is approximately equal to the distance Λ, between the caustic and the first interference maximum. Because of an influence of diffraction near the caustic a flux of the energy through a lateral surface of ray tubes plays an important role. At the same time this factor is neglected in the zero approximation of the geometrical acoustic and, as result, the wave amplitude becomes infinitely large.

Three-dimensional case

Now we assume again that some part of a medium is homogeneous and, correspondingly, rays are rectilinear, but the phase surface S may have an arbitrary shape, Fig. 2.13a. Consider a plane passing through some ray, which crosses S at point p. An intersection of the plane with S gives the curve C. By definition, the ray is normal to the wave surface and, respectively, the center of curvature is located at the ray. Suppose that the plane rotates around the ray by 180°. Then the curve C changes, as well as its curvature, and it passes through maximal and minimal values. From analytical geometry it is well

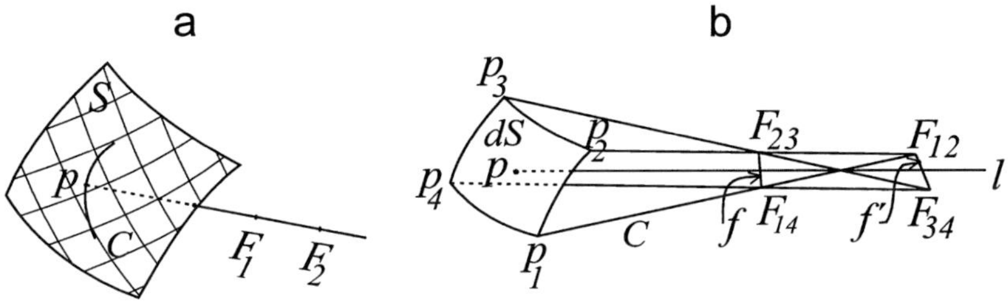

Figure 2.13: (a) Lines of curvature (b) Focal lines

known that two planes, which contain the largest and smallest radius of curvature, are perpendicular to each other, and they are called the principal planes. Correspondingly, these radii are named the principal radii of curvature of the surface at the point p. Curves on this surface S, that are parallel to the principal planes, form two mutually orthogonal families, called lines of curvature, Fig. 2.13a. In general, two rays, starting at two adjacent points of the wave surface, do not intersect each other. However, if these points are situated at the same line of curvature, the rays intersect, and the corresponding point is called the focus. Thus, at each rectilinear ray, there can be two foci, and they are principal centers of curvature. This means that a pencil of rays has two branches of the caustic surface. It is obvious, that in the limiting case of the cylindrical wave one principal radius of curvature becomes infinitely large, and, therefore, there is only a single caustic surface. Such model of a wave field was studied earlier. For illustration consider a concave element of the wave surface, which serves as the cross-section of a ray tube, Fig. 2.13b. It is convenient to suppose that this element is bounded by two pairs of curvature lines, which can be treated as circle arcs. As is clearly seen from Fig. 2.13b all rays, passing through the arc $p_1 p_2$, intersect at focus F_{12}. In the same manner, rays, starting at the arc $p_3 p_4$, intersect at the focus F_{34}. The line f', which connects F_{12} and F_{34}, is called the focal line f', and in this case it is horizontal. Also it is obvious that rays, passing through any vertical curvature line on the surface dS, meet at some point of the segment f'. By analogy, rays from horizontal lines of curvature form the vertical focal line f. Certainly, the focal lines are places where the cross-section of a ray tube becomes equal to zero. In other words, these lines are elements of the caustic surface.

In conclusion we note the following.

(a) As we know from the two-dimensional case, passage through a point, where the ray is tangential to a caustic, results in a decrease of the phase by $\pi/2$. Inasmuch as the rectilinear ray touches, in general, two caustics, the total change of the phase is equal to $-\pi$.

(b) It is possible to demonstrate that the number of caustic surfaces increases, if rays are curvilinear.

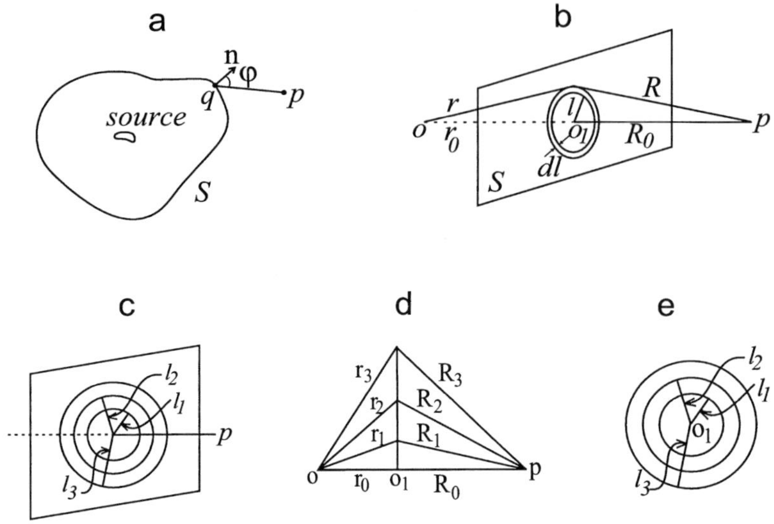

Figure 2.14: (a) Model of a medium (b) Plane surface S (c),(d),(e) Fresnel zones

2.5 Fresnel volume and physical meaning of ray

Earlier we considered a wave field, described by a zero approximation of the asymptotic series, and found out that the Poynting vector is tangential to the ray. This means that the energy flux moves through an elementary tube. This may suggest that with the decrease of this cross-section a wave always propagates along a single ray. In particular, it may seem that if a medium is homogeneous, then a wave travels from the source to an observation point along a rectilinear line, connecting them. However, such point of view contradicts to basic concepts of wave phenomena. In order to resolve this discrepancy we can proceed from either the Helmholtz-Kirchhoff representation or the Huygens - Fresnel principle. Since both of them give a similar result, it is more convenient for our purposes to use the last approach.

Huygens - Fresnel principle

Before we remind main features of this principle, (Part I), let us assume that an arbitrary surface S surrounds the source of the primary wave, Fig. 2.14a. In particular, this surface may coincide with the phase surface. Then, as was shown in Part I, this principle is based on the following assumptions

1. Each element dS of this surface generates outside the spherical wave.
2. The magnitude of this elementary wave at any point p is directly proportional to
   (a) the small area dS in the vicinity of some point q of the surface S.

(b) the wave field at this point q. For instance, it can be the potential U of the wave, caused by the primary source in a homogeneous medium.

(c) some factor $K(\varphi)$, that depends on the angle φ between the normal, $\mathbf{n}$, to the element dS and the line, connecting points q and p, Fig. 2.14a. It becomes smaller with an increase of φ.

3. At an observation point p the wave is a result of superposition of spherical waves, which arise at all elements of a surface S.

For example, if outside S a medium is homogeneous, the complex amplitude, $d\mathcal{U}(p)$, of the elementary wave generated at the point q is

$$d\mathcal{U}(p) = K(\varphi)\,\mathcal{U}(q)\,\frac{e^{ikR}}{R}\,dS \tag{2.274}$$

Here R is the distance between points q and p, while k is a wavenumber.

Finally, taking into account the fact that the resultant wave is a superposition of elementary waves, we obtain for the complex amplitude $\mathcal{U}(p)$ of this wave

$$\mathcal{U}(p) = \oint_S K(\varphi)\,\mathcal{U}(q)\,\frac{e^{ikR}}{R}\,dS \tag{2.275}$$

Here it may be appropriate to make several comments.

(a) We assume that a medium between the surface S and an observation point is homogeneous.

(b) The parameter K is usually called the inclination factor. It reaches the maximal value in the direction of the normal to dS, $\varphi = 0$, but becomes smaller with an increase of φ. Finally, K is equal to zero when the line qp is tangential to the element $dS(q)$.

(c) Unlike the Huygens - Fresnel principle the approach, based on the Helmholtz - Kirchhoff representation, allows us to have the exact relationship between the potential at the observation point and values of both the potential and its normal derivative at points of the surface S.

(d) If the distance R greatly exceeds the wave length λ, $R \gg \lambda$, then the two approaches give the same result. Besides the Helmholtz - Kirchhoff formula permits us to find the expression for the inclination factor K.

The most important conclusion, which follows from eq. 2.275, is that the field at any point p is a superposition of secondary waves arriving from different points of any surface S. In other words, the wave is not only transmitted from the source to an observation point along the ray, connecting them, but also it arrives at point p from different places of the surface S.

Fresnel zones and Fresnel volume

In this light it is useful to evaluate a dimension of a part of the surface S, where secondary sources of elementary waves make the main contribution to the total field at the point p. To investigate this question, we consider several examples, starting with the simplest one, an elementary source, located in a homogeneous medium. As we know in this case rays are rectilinear and radially oriented, and the potential $\mathcal{U}$ is defined by the equality, (Part I)

$$\mathcal{U}(p) = C\,\frac{e^{ikb}}{b}, \tag{2.276}$$

where C is some constant and b is the distance from the source o to an observation point p.

Case one First suppose that the surface S, eq. 2.275, is a plane perpendicular to the ray op and crosses it somewhere between these points, Fig. 2.14b. Then, in accordance with the Huygens principle, we can treat the field $\mathcal{U}(p)$ as a superposition of elementary waves, caused by sources that are located at this plane. Certainly, it is not a phase surface, since the latter is spherical, where $kr = const$. Because at points of the plane, Fig. 2.14b

$$\mathcal{U}(q) = C\,\frac{e^{ikr}}{r},$$

eq. 2.275 is rewritten as

$$\mathcal{U}(p) = C\int\limits_S K(p,q)\,\frac{e^{ik(r+R)}}{rR}\,dS\,, \tag{2.277}$$

where $K(p,q)$ is unknown inclination factor.

However, this is not important because the total field is given, eq. 2.276, and our purpose is only to compare a relative contribution of different parts of the plane. To perform this task, Fresnel took into account the axial symmetry with respect to the line op and represented the plane as a system of concentric rings, which are called Fresnel zones. Examples of the first three zones are shown in Fig. 2.14c, d, e. It is essential that the width of every zone is such that a phase difference at the point p between elementary waves, which arise near its external and internal boundaries, is equal to π, that is

$$k\,(r_n + R_n) - k\,(r_{n-1} + R_{n-1}) = \pi$$

or

$$(r_n + R_n) - (r_{n-1} + R_{n-1}) = \frac{\lambda}{2}, \tag{2.278}$$

where n is the number of the zone. This condition can be also written in the form

$$(r_1 + R_1) - b = \frac{\lambda}{2}, \qquad (r_2 + R_2) - b = 2\,\frac{\lambda}{2} \qquad \text{and} \qquad (r_n + R_n) - b = n\,\frac{\lambda}{2} \tag{2.279}$$

where $b = r_0 + R_0$. These allow us to construct zones, since $r_n + R_n$ is easily defined.

Thus, the resultant wave is represented as a sum of waves, $\mathcal{U}_n$, caused by sources inside each zone,

$$\mathcal{U}(p) = \sum_{n=1}^{\infty} \mathcal{U}_n(p) \tag{2.280}$$

For instance,

$$\mathcal{U}_1 = C \int_{S_1} K(p,q) \frac{e^{ik(r+R)}}{rR} dS$$

$$\mathcal{U}_2 = C \int_{S_2} K(p,q) \frac{e^{ik(r+R)}}{rR} dS$$

where S_1 and S_2 are areas of the first and second zones, respectively.

Furthermore we make an important assumption, namely, the wave length λ is much smaller than distances r_n and R_n.

$$r_n \gg \lambda \qquad \text{and} \qquad R_n \gg \lambda \tag{2.281}$$

As follows from eqs. 2.278 and Fig. 2.14d, the last inequality means that the width of zones,

$$\Delta \ell_n = \ell_n - \ell_{n-1}$$

is relatively small.

$$\Delta \ell_n \ll r_n \qquad \text{and} \qquad \Delta \ell_n \ll R_n$$

By definition, Fig. 2.14d, we have

$$r_1 = \sqrt{r_0^2 + \ell_1^2} \qquad R_1 = \sqrt{R_0^2 + \ell_1^2}$$

$$r_2 = \sqrt{r_0^2 + \ell_2^2} \qquad R_2 = \sqrt{R_0^2 + \ell_2^2}$$

and

$$r_n = \sqrt{r_0^2 + \ell_n^2} \qquad R_n = \sqrt{R_0^2 + \ell_n^2} \tag{2.282}$$

Considering only the first Fresnel zones, when

$$\ell_n \ll r_0 \qquad \text{and} \qquad \ell_n \ll R_0 \tag{2.283}$$

it is not difficult to find simple expressions for their radii and areas. Using eqs. 2.282 - 2.283, we obtain

$$r_n \simeq r_0 + \frac{1}{2}\frac{\ell_n^2}{r_0}, \qquad R_n \simeq R_0 + \frac{1}{2}\frac{\ell_n^2}{R_0}$$

or

$$r_n + R_n = r_0 + R_0 + \frac{1}{2}\frac{R_0 + r_0}{r_0\,R_0}\,\ell_n^2 \tag{2.284}$$

As follows from eqs. 2.279

$$n\,\frac{\lambda}{2} = \frac{1}{2}\frac{R_0 + r_0}{r_0\,R_0}\,\ell_n^2$$

Then

$$\ell_n = \sqrt{n\,\lambda\,\frac{r_0\,R_0}{r_0 + R_0}}, \tag{2.285}$$

that is in this approximation the radius of the Fresnel zone increases proportional to the square root of its number. From this equation we conclude that all zones have the same area, and it is their second important feature. Indeed, the area of the zone is

$$S_n = \pi\,\ell_n^2 - \pi\,\ell_{(n-1)}^2$$

and substitution of eq. 2.285 gives

$$S_n = S = \frac{\pi\,\lambda\,r_0\,R_0}{r_0 + R_0} \tag{2.286}$$

It depends on distances from the source and an observation point to the plane, as well as the wave length. Assuming that the distance b is given, it is convenient to rewrite eq. 2.286 as

$$S = \frac{\pi\,\lambda}{b}\,r_0\,R_0 = \frac{\pi\,\lambda}{b}\,r_0\,(b - r_0) \tag{2.287}$$

It is obvious that in approaching either the source or an observation point, the area become smaller, but it reaches the maximal value

$$S_{max} = \frac{\pi\,\lambda\,b}{4} \tag{2.288}$$

when the plane is located at the middle, $r_0 = b/2$.

Evaluation of function $\mathcal{U}_n$

Now we are prepared to estimate integrals in eq. 2.280. Consider the potential at point p, caused by sources of the nth zone:

$$\mathcal{U}_n(p) = C \int\limits_{S_n} K(p,q)\, \frac{e^{ik(r+R)}}{rR}\, dS \tag{2.289}$$

Because the width of the zone is small with respect to r and R, the product $r\,R$ remains constant. Moreover, the angles between the normal to the plane and lines R and r change only slightly, that is the inclination factor $K(p,q)$ is also almost constant. Respectively, in place of eq. 2.289 we can write

$$\mathcal{U}_n(p) \simeq C \left(\frac{K}{R\,r}\right)_n I_n\,, \tag{2.290}$$

where

$$I_n = \int\limits_{S_n} e^{ik(r+R)}\, dS \tag{2.291}$$

and the value of the term

$$\left(\frac{K}{R\,r}\right)_n$$

corresponds to points somewhere at the middle of the n-th zone.

Bearing in mind that distances R and r are much greater than the wave length λ, eq. 2.281, even very small variation of R and r can result in a large change of the exponent

$$\exp\,[\,i\,k(r+R)\,]\,, \tag{2.292}$$

and for this reason it remains under the integral.

In spite of these simplifications, the integral I_n, eq. 2.291, can be evaluated only numerically. However, the special form of the integrand, eq. 2.292, allows us to compare without calculations the contributions of different Fresnel zones. With this purpose in mind, consider at the beginning the first zone and represent it as a system of N thin concentric rings, Fig. 2.14b. Then the integral I_1 is approximately equal to

$$I_1 \simeq \Delta s \sum_{m=1}^{N} e^{ik\,(r_m + R_m)} \tag{2.293}$$

and therefore the complex amplitude of the potential is

$$\mathcal{U}_1(p) = C \left(\frac{K}{r\,R_1}\right)_1 \Delta s \sum_{m=1}^{N} e^{ik\,(r_m + R_m)} \tag{2.294}$$

Here we assume that thin rings are chosen in such way that their areas Δs are the same.

It is clear that the phase of an elementary wave at the observation point p, caused by sources in each ring, is equal to

$$k\,(r_m + R_m) \tag{2.295}$$

Certainly a transition to the neighboring ring slightly changes the phase, but the magnitude of the wave remains the same, since

$$|\exp ik(r+R)| = 1$$

For example, the potential $\mathcal{U}_{11}$, due to sources of a small disk around point o_1, Fig. 2.14a, is

$$\mathcal{U}_{11} = C\left(\frac{K}{r\,R}\right)_1 \Delta s\, e^{ik(r_0 + R_0)} \tag{2.296}$$

To visualize a process of superposition of waves, caused by the first zone, it is convenient to recall that a summation of complex numbers can be done in the same manner as for vectors, (Part I). Applying this approach, we replace the complex amplitude

$$\mathcal{U}_{1m} = C\left(\frac{K}{r\,R}\right)_1 \Delta s\, e^{ik\,(r_m + R_m)}$$

of the mth ring, by an elementary vector $\mathcal{U}_{1m}$, which has the same magnitude and a phase (argument), as $\mathcal{U}_{1m}$. Therefore, instead of sum

$$\mathcal{U}_1 = \sum_{m=1}^{N} \mathcal{U}_{1m}$$

we obtain

$$\mathcal{U}_{1m} = \sum_{m=1}^{N} \mathcal{U}_{1m}, \tag{2.297}$$

and an addition of elementary waves due to each ring gives a system of small vectors $\mathcal{U}_{1m}$ of the same length, Fig. 2.15a. Their direction gradually alters corresponding to a monotonic change of phase of these waves.

As follows from eq. 2.297 the slope of the vector $\mathcal{U}_{11}$ is equal to

$$k\,(r_0 + R_0)\,,$$

but for convenience it is directed along the horizontal axis, Fig. 2.15a. Then, in accordance with eq. 2.278, the last vector $\mathcal{U}_{1m}$, characterizing the effect of the thin ring with radius ℓ_1, has the phase

$$k(r_0 + R_0) + \pi\,,$$

that is, $\mathcal{U}_{11}$ and $\mathcal{U}_{1N}$ are opposite to each other. In fact, the angle between these vectors is equal to the phase shift of corresponding elementary waves. At the limit, when the

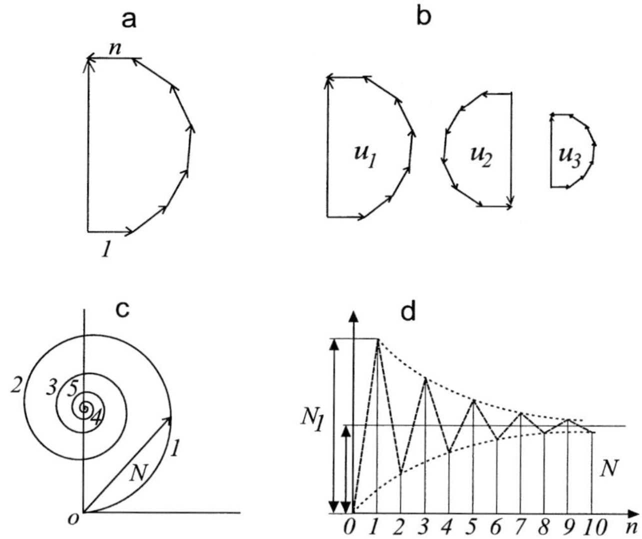

Figure 2.15: (a) Superposition of elementary waves of the first zone (b) Superposition of these waves, corresponding to the first, second, and third zones (c) Cornu spiral (d) Summation of series 2.301

ring thickness tends to zero, the system of elementary vectors forms a half circle and the magnitude of the resultant vector, $|\mathcal{U}_1|$, becomes equal to its diameter, Fig. 2.15a. Also we see that the vector $\mathcal{U}_1$ is perpendicular to the first one, $\mathcal{U}_{11}$. In other words, the phase shift between the resultant wave due to sources of the zone and the first elementary wave is $\pi/2$. Making use of eq. 2.297, we write

$$\mathcal{U}_1 = Q_1 \exp\left[i\left(k(r_0 + R_0) + \frac{\pi}{2}\right)\right] \tag{2.298}$$

where Q_1 is the amplitude of the wave, caused by the first Fresnel zone.

Next consider a superposition of elementary waves, arising in the second zone. As before, it is represented as a system of thin concentric rings and, therefore

$$\mathcal{U}_2 = \sum_{m=1}^{N} \mathcal{U}_{2m}$$

We note the following. First, two thin rings, adjacent to each other and located beneath and above the boundary $\ell = \ell_1$, are at the same distance from an observation point. This means that elementary waves caused by them have equal phases. Second, with an increase of the ring radius ℓ, distances R and r become greater and the phase of corresponding elementary waves gradually increases too. For this reason in order to determine the resultant wave, caused by sources of the second zone, we have to form a system of vectors, as is shown in Fig. 2.15b. As before, the phase difference between elementary waves due to the first and last rings is equal to π, eq. 2.278, and, therefore, vectors $\mathcal{U}_1$ and $\mathcal{U}_2$ have opposite directions. Correspondingly, complex amplitudes $\mathcal{U}_1$ and $\mathcal{U}_2$ differ by sign, and from eq. 2.298 we have

$$\mathcal{U}_2 = -Q_2 \exp i \left[k(r_0 + R_0) + \frac{\pi}{2}\right], \tag{2.299}$$

where Q_2 is the magnitude of the wave, caused by the second Fresnel zone.

It is obvious that

$$Q_2 < Q_1$$

since

$$\left(\frac{K}{r\,R}\right)_1 > \left(\frac{K}{r\,R}\right)_2$$

Bearing in mind that the zones have a relatively small width, $\Delta\ell$, values of Q_1 and Q_2 are close to each other. Now we apply the same approach to other zones, and the resultant wave, generated by all sources on the plane, is described by the alternating series

$$\mathcal{U}(p) = (Q_1 - Q_2 + Q_3 - Q_4 + ...) \exp\, i \left[k(r_0 + R_0) + \frac{\pi}{2}\right], \tag{2.300}$$

where

$$Q_n < Q_{n-1}\,,$$

and they have close values. Before we evaluate this series it may be appropriate to make two comments:

1. Within each zone a maximal change of the phase of elementary waves is rather large and it is equal to π. At the same time a superposition of these waves has a constructive character. This fact may be useful for evaluation of the interval of integration when the method of stationary phase is applied (Part I).

2. If we perform a summation of elementary vectors for different Fresnel zones, then we arrive at the known Cornu spiral, (Part I). The latter is a combination of half circles, shown in Fig. 2.15c, and it is obvious that the vector **N** characterizes the resultant wave, caused by some system of elementary waves. In particular, the distance between terminal points of the spiral defines the wave due to all sources of the plane. In order to determine the latter consider the series

   $$Q = Q_1 - Q_2 + Q_3 - Q_4 + ...\,,$$

   which can be rewritten as

   $$Q = \frac{Q_1}{2} + \left(\frac{Q_1}{2} - Q_2 + \frac{Q_3}{2}\right) + \left(\frac{Q_3}{2} - Q_4 + \frac{Q_5}{2}\right) + ... \tag{2.301}$$

First, suppose that

$$Q_n \geq \frac{Q_{n-1} + Q_{n+1}}{2}$$

Then brackets in eq. 2.301 are negative and correspondingly

$$Q \leq \frac{Q_1}{2} \tag{2.302}$$

Also the same series can be represented in the form

$$Q = Q_1 - \frac{Q_2}{2} - \left(\frac{Q_2}{2} - Q_3 + \frac{Q_4}{2}\right) - \left(\frac{Q_4}{2} - Q_5 + \frac{Q_6}{2}\right) - \ldots \tag{2.303}$$

Again all brackets are negative, and therefore

$$Q \geq Q_1 - \frac{Q_2}{2} \tag{2.304}$$

Bearing in mind that values of Q_1 and Q_2 are almost equal, the latter yields

$$Q \geq \frac{Q_1}{2} \tag{2.305}$$

Comparison of eqs. 2.302 and 2.305 allows us to conclude that

$$Q = \frac{Q_1}{2} \tag{2.306}$$

Applying the same approach, we obtain again this equality, when

$$Q_n \leq \frac{Q_{n-1} + Q_{n+1}}{2}$$

This process of summation is shown in Fig. 2.15d.

Taking into account eq. 2.300, the complex amplitude of the potential at the point p is

$$\mathcal{U}(p) = \frac{Q_1}{2} \exp i \left[k_0(r_0 + R_0) + \frac{\pi}{2}\right] \tag{2.307}$$

or

$$\mathcal{U}(p) = \frac{1}{2} \mathcal{U}_1(p) \tag{2.308}$$

Thus, the wave, caused by the sources in the first Fresnel zone, is twice greater than that generated by all sources on the plane. This suggests that approximately half of the zone creates a wave with magnitude equal to $|\mathcal{U}|$.

From eq. 2.286 it follows that the area of this part is

$$\frac{S_n}{2} = \frac{S_1}{2} = \frac{\pi\,\lambda\,r_0\,R_0}{2(r_0 + R_0)} \tag{2.309}$$

The latter allows us to determine the radius ℓ of a disk, whose sources generate a wave with magnitude $|\mathcal{U}|$ at point p. We see that

$$\ell = \sqrt{\frac{\lambda}{2}\,\frac{r_0\,R_0}{b}} \tag{2.310}$$

As we already know, ℓ has a maximum when $r_0 = R_0 = b/2$, and

$$\ell_{max} = \frac{1}{2}\sqrt{\lambda\,R_0} \tag{2.311}$$

Let us mentally change the position of the plane S between the source and an observation point. Then the area of the disk with radius ℓ also varies and it forms cross-sections of the ellipsoid of rotation with foci at points o and p. To some extent the latter represents the volume of a homogeneous medium, which is involved mainly in a transmission of the wave from the source to an observation point and it is called the Fresnel volume, Fig. 2.16a. The dimension of this part of a medium depends on distances r_0 and R_0, as well as the wave length. As follows from eq. 2.311 the small semi-axis of the ellipsoid is

$$\ell_{max} = \frac{1}{2\sqrt{2}}\,\sqrt{\lambda\,b} \tag{2.312}$$

For example, suppose that $b = 1000$m, $\lambda = 50$m, then $\ell_{max} \approx 80$m.

Case one Next consider the limiting case when the source is located at infinity, $r_0 \to \infty$. This means that we deal with a plane wave and the plane S coincides with its phase surface. Correspondingly, eq. 2.310 gives

$$\ell = \sqrt{\frac{\lambda}{2}\,R_0} \tag{2.313}$$

and the Fresnel volume has the shape of the paraboloid of rotation. In particular, with an increase of the distance between the plane and point p the radius of the cross-section tends to infinity. This last equation allows us to estimate a part of an interface, which gives the main contribution to the reflected wave, provided of the normal incidence, Fig. 2.16b. For illustration, assume that $R_0 = 1000$m and $\lambda = 50$m. Then eq. 2.313 gives $\ell \approx 160$m.

We see that even the area of half Fresnel zone is rather large.

Case two Suppose that the source of the primary waves is located at the point o and the incident and reflected waves satisfy Snell's law at the point q, Fig. 2.16a. In such

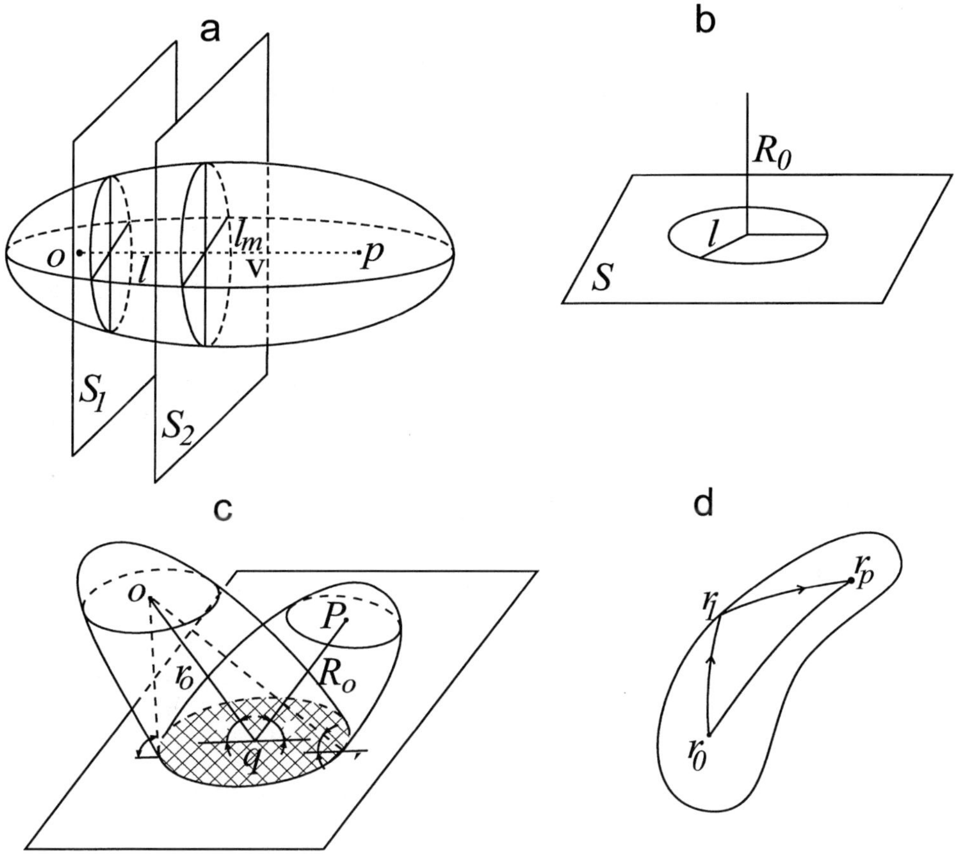

Figure 2.16: (a) Fresnel volume (b) Fresnel zone and plane wave (c) Reflection and Fresnel volumes (d) Fresnel volume in inhomogeneous medium.

the case Fresnel volume consists of two ellipsoids of rotation, and their intersection with the interface forms an ellipse. Its dimensions are functions of the wave length, as well as positions of points o and p. As is clearly seen from Fig. 2.16c the reflected wave arrives at the point p along rays having different reflection angles θ_r:

$$\theta_1 < \theta_r < \theta_2 ,$$

where θ_r follows from Snell's law.

Certainly, with a decrease of the wave length the ellipse area at the boundary becomes smaller and, correspondingly, the range of a change of θ_r narrows. This means that the reflected wave arrives at the point p along an elementary ray tube only, and its amplitude is described by the zero approximation of the asymptotic series.

Fresnel volume in an inhomogeneous medium

Suppose that both the source and an observation point are located in an inhomogeneous medium where the velocity of propagation and the density vary relatively slowly, Fig. 2.16d.

As we know at the high-frequency spectrum the potential $\mathcal{U}(\mathbf{r})$ is described by

$$\mathcal{U}(\mathbf{r}) = A\, e^{i\, k_0\, L(\mathbf{r})} \tag{2.314}$$

Consider the acoustic length along the ray between the source o and observation point p. Then, by definition, we have

$$L(\mathbf{r}_0,\, \mathbf{r}_p) = \int_0^p n\, ds = L_0 \tag{2.315}$$

here

$$n = \frac{c_0}{c(\mathbf{r})}$$

Now we take the point $q(\mathbf{r}_1)$, located near our basic ray, and draw a ray from o to q. Assuming that this point is the source of an elementary spherical wave, we draw the ray from q to p. The broken ray $\mathbf{r}_0 \to \mathbf{r}_1 \to \mathbf{r}_p$ is usually called a virtual ray and its acoustic length is

$$L = L(\mathbf{r}_0,\, \mathbf{r}_1) + L(\mathbf{r}_1,\, \mathbf{r}_p) \tag{2.316}$$

Let us construct in the vicinity of the basic ray a surface, which is described by the equation

$$|L(\mathbf{r}_0,\, \mathbf{r}_1,\, \mathbf{r}_p) - L(\mathbf{r}_0,\, \mathbf{r}_1)| = \frac{\lambda_0}{2} \tag{2.317}$$

Points of this surface, $\mathbf{r}_1$, are such that the eikonal $L(\mathbf{r}_0,\, \mathbf{r}_1,\, \mathbf{r}_p)$ along any virtual ray differs from L_0 by $\lambda_0/2$. This surface represents the envelope of the first Fresnel zones, provided that the wave length is equal to λ_0, $(k_0 = 2\pi/\lambda_0)$, and, correspondingly, it surrounds the Fresnel volume. Of course, this approach allows us only approximately evaluate dimensions of a medium through which the main signal arrives at the observation point, since the wave length is not equal to λ_0, but in general varies.

From the mathematical point of view the ray is an infinitely thin line, which satisfies ray equations. At the same time we found out that the resulting wave arrives at an observation point through a volume, surrounding the ray, that is Fresnel volume. This happens because of the destructive interference of waves, caused by sources in other zones. It is obvious that with a decrease of the wave length the cross-section of Fresnel volume becomes smaller and in the limit, $\lambda \to 0$, it vanishes. Then, the Fresnel volume is transformed into the ray.

Chapter 3

Basics of Kirchhoff migration

First in this chapter we demonstrate how concepts related to time fields allow us to determine, without the use of wave amplitude, a position of the reflector at which secondary waves arise. This kinematical approach was successfully applied before the computer revolution when it became possible to bring information about amplitudes of waves arriving at all observation points. Our description of the main features of this approach is based on the classic paper by J. Hagedorn, where he also formulated some principles of the modern interpretation of the reflection method. To simplify our discussion, it is usually assumed that the medium is two-dimensional.

3.1 Time fields and reconstruction of reflector

Source and receiver at the same point (zero offset)

In this case the ray of the reflected wave arriving at the receiver, Fig. 3.1a, coincides with that of the incident wave, and the traveltime of both waves is

$$t = 2\,t_i = 2\,t_r \tag{3.1}$$

where, t_i is the time of propagation of the incident wave between the source and the point of reflection Q.

It is clear that due to the geometry of the medium a study of the time field can be restricted by a vertical plane containing the line of observations. At the beginning we assume that the surface S (reflector) has a relatively small curvature and each its element, Δs, with dimensions, exceeding the wave length, can be treated as a straight segment. Then, in accordance with Snell's law, the ray OQ or QO is normal to the plane, which is tangential to the boundary S at the point Q. To determine the position of the reflector we imply that the velocity of propagation is known in the upper layer and introduce the concept of a surface of equal reflection times.

Let us start from the simplest case, when the velocity c is constant, and plot a family of a semi-spherical surface with radii

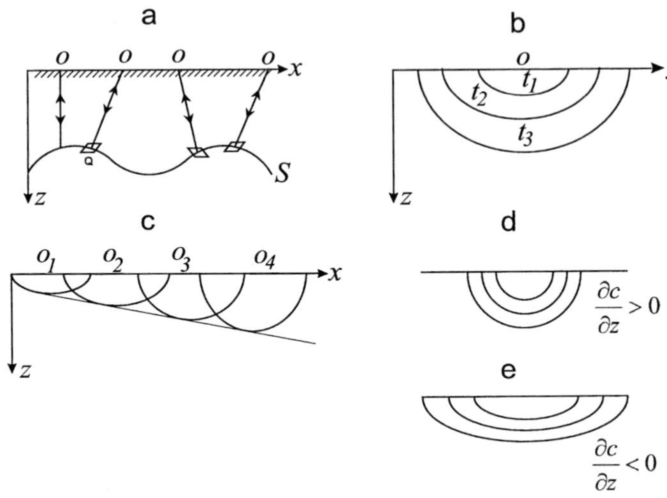

Figure 3.1: (a) Source and receiver at the same point (b) System of lines of equal reflection time , $c = \text{const}$ (c) Reflector as an envelope (d),(e) Lines of equal reflection time when c varies.

$$R = ct_i = ct_r \tag{3.2}$$

and the common origin at the point O. A cross section of them is a system of semicircles, shown in Fig. 3.1b, and taking into account the fact that we deal with a two-dimensional model of a medium, it is sufficient to consider only these lines. Suppose that the wave arrives at the point of observation at the instant t. Then each point of a semicircle with the radius

$$r = \frac{ct}{2} \tag{3.3}$$

can be considered as a place where the reflected wave may arise. In fact, by definition the total time of propagation of the wave from the source to any point of this line and back to the receiver O is equal to t. Bearing in mind that the traveltime of the reflected wave remains constant at such points ($t_i = t_r = \text{const}$), these semicircles are called the lines of equal reflection times. It is clear that in a three-dimensional case we deal with surfaces of equal reflection time.

Assuming that the time of the first arrival of the reflected wave, t, is known along the line of observations, we plot a system of semicircles of equal reflection times for each position of the source (receiver), Fig. 3.1c. Inasmuch as each element of the reflector S is tangential to the corresponding circle of equal reflection times, the envelope of them describes the reflector. The accuracy of its determination depends on several factors, in particular, the distance between observation points and the shape of the boundary S. The influence of the last factor is discussed later. Also it is clear that errors in calculating the velocity c have a strong effect on the estimation of the reflector position.

We note the following.

(a) If the velocity of propagation is a function of the z-coordinate a shape of lines or surfaces of equal reflection times, in general, differs from that of a circle or sphere. However, the axial symmetry with respect to the vertical line passing through the source, is still preserved, Fig. 3.1d,e.

(b) When the distance between the source and the receiver is equal to zero, lines of equal reflection times and those of equal arrival time of the incident wave have the same shape.

(c) In essence, we described the construction of a reflector, (tangent method), introduced almost 80 years ago.

(d) It is also interesting to notice that such an approach can be interpreted as if a fictitious source is situated at the point O and gives rise to an elementary wave with a spherical phase surface. An envelope of this waves gives the reflector. A similar concept of sources is used in the Kirchhoff formula, and its application is described in the next section.

(e) It may be worth to mention that points, where lines of equal reflection times touch their envelope, are stationary ones. In other words, in the high-frequency spectrum the reflected wave, arriving at the receiver, is defined mainly by surface sources around a corresponding stationary point.

Source and receiver are located at different points (non-zero offset)

To determine a position of lines (surfaces) of equal reflection times in such a case, we plot two families of isochrones with the origin at the source and receiver, respectively, Fig. 3.2a. For instance, if the velocity c is constant, then these isochrones are semicircles and their radii

$$r_1 = c\,t_i \qquad \text{and} \qquad r_2 = c\,t_r \tag{3.4}$$

Here t_i and t_r are times of propagation between points of corresponding isochrones and their centers.

The intersection of two isochrones with indexes t_i and t_r is a point, which can be characterized by a sum:

$$t = t_i + t_r \tag{3.5}$$

The latter can be treated as the time of the arrival of the reflected wave at the receiver, provided that the point of intersection belongs to the reflector.

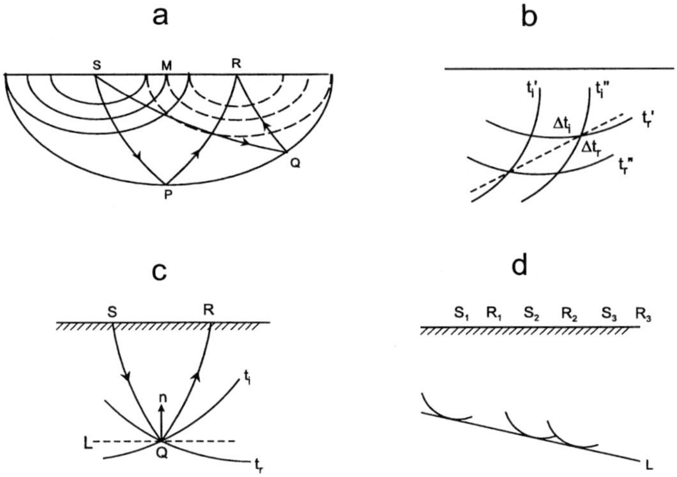

Figure 3.2: (a) Two systems of isochrones (b) Construction of a line of equal reflection time (c) Ray behavior near the line of equal reflection time (d) Envelope of lines of equal reflection times.

Changing values of t_i and t_r so that their sum, t, remains constant, we obtain a line, and its middle point M is situated on the z-axis, Fig. 3.2a. In the simplest case of the constant velocity c the line $t = \text{const}$ coincides with the ellipse, since the sum of distances from the source and receiver to each point of the line is the same

$$ct = r_1 + r_2 = c\,(t_i + t_r) = \text{const} \tag{3.6}$$

In general, when the velocity of propagation is variable the shape of this line becomes more complicated. It is clear that a variation of time t produces a system of such lines and, as before, they are called lines of equal reflection times. However, there is a difference between these cases. In the first one, when the source and receiver are in the same place, t characterizes the time of traveling from a source to the line, eq. 3.3. In the second case, it is a time of wave propagation from the source to the line and then to the receiver, that is t is the time of the arrival of the reflected wave.

It is essential that in both cases any point of these lines may belong to the reflector, and with a decrease of the distance between the source and receiver lines with parameters t_r and t_i, $(t = t_i + t_r)$, coincide. To visualize better a construction of lines of equal reflection times, $(t = \text{const}\,)$, note that, as is seen in Fig. 3.2b, an intersection of neighboring isochrones of both families forms a curvilinear parallelogram. Points, corresponding to opposite vertexes, are located at the same line of equal reflection times and, therefore, if one point of the line is known, it is a simple matter to determine the other.

Suppose that the time of the arrival of the reflected wave is equal to t. Then, by definition, at least one point of the line of equal reflection times with the index equal t, for example Q, belongs to the reflector, Fig. 3.2a. In the vicinity of this point the element of the boundary is tangential to the line of equal reflection times, since we assume that Snell's law is satisfied, Fig. 3.2c. As in the case of the zero offset the envelope of lines of equal reflection times forms the reflector. Thus, a process of reconstruction of the reflector consists of two main steps, namely,

1. calculation of a line of equal reflection times for each position of the array;

2. determination of the envelope of these lines.

Record horizon (two-dimensional case)

Describing this first stage of an interpretation of the reflection method, which was before the use of computers, let us introduce the procedure of "the vertical plotting" and the concept of a record horizon. Further, it is assumed that observations are performed with a constant offset between the source and receiver, in particular, the zero offset.

To determine the position of a line of equal reflection times, t, corresponding to the wave arrival, we find first a location of its point P. As is seen from Fig. 3.2a, it is situated on the vertical, drawn at the middle of the interval between the source and the receiver. If the velocity of propagation c is constant, this point can be easily found. In fact, an intersection of two semicircles with the radius

$$r = \frac{ct}{2}$$

and centers at points S and R gives the point P. In a more general case, for instance when the velocity depends on the z-coordinate, this point is derived from ray equations. The latter is greatly simplified for zero offset, since in such a case, the point P is located on the vertical ray. This procedure of obtaining points P for each position of the array represents "the vertical plotting" and a set of them is called the record horizon. Of course, in the three-dimensional case we arrive at the record surface. Suppose that the boundary between two media is planar and horizontal. Then, in accordance with Snell's law the point P is situated on the actual reflector and therefore the record horizon gives correct information about the interface. However, in general, they have different positions. In other words, the point P and that of the reflector Q are usually located at different places of the same line of equal reflection times, Fig. 3.2a. It happens because of the curvature of the boundary, as well as an inclination if it is a planar one. Let us notice that the point P is the deepest point of the line of equal reflection times, and, correspondingly, the record horizon is the locus of these points. The procedure of finding the reflector from the record horizon is sometimes called the migration. As was demonstrated above, the reflector is defined as the envelope to lines of equal reflection times. Thus, a transition from such lines to their envelope is equivalent to that from the record horizon to the reflector. To illustrate this process of migration consider several models of a medium.

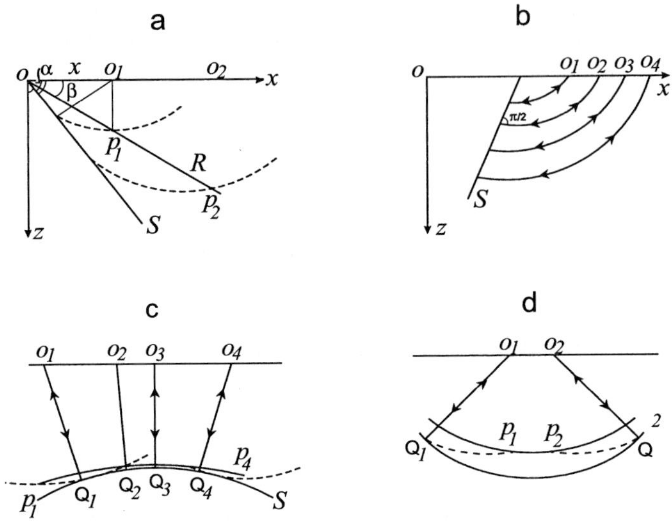

Figure 3.3: (a) Illustration of eq. (3.7) (b) Dipping angle exceeds $\pi/2$ (c) Reflector is convex (d) Reflector is a cylindrical trough.

Example one Suppose that the offset is equal to zero and the velocity at the upper layer is constant. The reflector is a planar dipping interface S. As can be seen from Fig. 3.3a, the envelope of lines of equal reflection times allows us to easily construct the dipping boundary. Now we determine its position analytically. Inasmuch as the radius of these semicircles linearly increases with the x-coordinate, the record horizon R is also a straight line, and it intersects the x-axis at the same point as the reflector. Then we have

$$O_1Q_1 = x \sin\alpha \qquad \text{and} \qquad O_1P_1 = x \tan\beta$$

Since

$$O_1Q_1 = O_1P_1$$

we obtain

$$\sin\alpha = \tan\beta \tag{3.7}$$

where α and β are dip angles of the reflector S and record horizon R respectively.

Thus, knowing the angle, β, formed by the record horizon and the x-axis, the dip of the boundary is defined as

$$\alpha = \sin^{-1}(\tan\beta)$$

It is also interesting to consider the case of the vertical contact when $\alpha = \pi/2$. Then the ray of the reflected wave coincides with the x-axis, and the line of arrivals is equal to

$$t = \frac{2x}{c}$$

Whence

$$O_1P_1 = \frac{ct}{2} = x \qquad \text{and} \qquad \tan\beta = 1$$

Correspondingly, the dip angle of the record horizon is $\pi/4$. If velocity c is constant and angle α exceeds $\pi/2$, the reflected waves do not reach the line of observations and, hence, the reflector cannot be detected. However, when velocity c increases with depth, its position can be defined, provided that rays form with the interface the angle which is equal to $\pi/2$, Fig. 3.3b.

Example two Consider a medium, in which a reflector is a convex line. It is clear that the record horizon also has a shape of an anticline, which is wider than the actual one and closer to the line of observations, Fig. 3.3c. Of course, drawing lines of equal reflection times and their envelope, we define again a position of the reflector.

Example three Suppose that the reflector is a cylindrical trough and its cross section is shown in Fig. 3.3d. If the curvature of the boundary is relatively small, then rays of reflected waves intersect each other somewhere above the line of observations and only a single reflected wave arrives at the receiver. Correspondingly, the record horizon also has the shape of a syncline, but it is narrower than the actual trough. As before, the envelope of lines of equal reflection times coincides with the reflector. At the same time with an increase of the reflector curvature, ray behavior becomes more complicated, as we discuss next.

Example four Unlike the previous case the curvature of the reflector can be sufficiently large. First of all, as is seen from Fig. 3.4a, some rays intersect each other beneath the line of observations, and, therefore, several reflected waves may be observed at the same receiver. In general we can distinguish four sets of rays arising at the interface, Fig. 3.4a. The first group of rays is caused by a reflection on the convex part of the boundary, $1a$, where the x-coordinate varies as $0 \le x \le a$. Here a is the point of a change of the reflector curvature.

Inasmuch as this part of the line is convex, reflected rays diverge with respect to each other. The corresponding element of the record horizon, $1a$, is shown in Fig. 3.4b. In particular, the ray reflected from the point a arrives at the receiver, located at the point $x = x_1$. The second set of rays results from reflection in the interval ac, where c is the deepest point of the syncline. Because of the concavity of this part of the boundary, the rays converge. Therefore, in approaching the point c the reflected rays are observed at

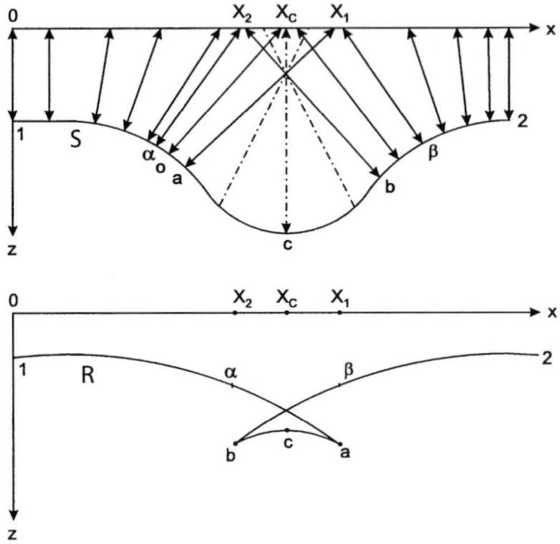

Figure 3.4: Formation of a triplet.

smaller values of x. For instance, when the source, (receiver), is situated at the point, x_c, the reflection occurs at the syncline bottom and the ray is vertical. The record horizon caused by this element of the reflector is described by the curve ac. As we see from Fig. 3.4b, the behavior of the record horizon, due to the rest part of the boundary, is similar. For instance, elements cb and $b2$ of the reflector give rise to the third and fourth sets of reflected rays.

This analysis shows that only one reflected wave arrives at the receiver if its coordinate x is relatively small, $x < x_2$, and the corresponding element of the record horizon is 1α.

Within the interval $x_2 < x < x_1$ three groups of waves arrive from the reflector elements: αa, ab and $b\beta$, respectively. And finally the single wave is observed at the receiver if $x > x_1$. It is essential to emphasize that reflected waves, that arise at the central part of the syncline, appear after the other waves.

Thus, at different receivers we can observe one, two, or three waves. This consideration shows that if only the time of the earliest arrival is used then the envelope of lines of equal reflection times may not reproduce the reflector. Such an example is given in Fig. 3.5. The dash-dot line represents the record horizon and its parts $A'P'$ and $B'Q'$ are caused by a reflection from two parts of the reflector: AP and BQ, respectively. Certainly, the use of an envelope of lines of equal reflection times allows us to find a position of these elements of the boundary. However, as we have already seen, reflected waves, that arise at points of the interval PQ, may arrive at receivers later than other waves. For simplicity, suppose that lines l_p and l_q of equal reflection times, passing through points P and Q, coincide. At the same time this line differs from the reflector within its central parts, Fig. 3.5. In other words, for the reflector to be found its curvature cannot exceed

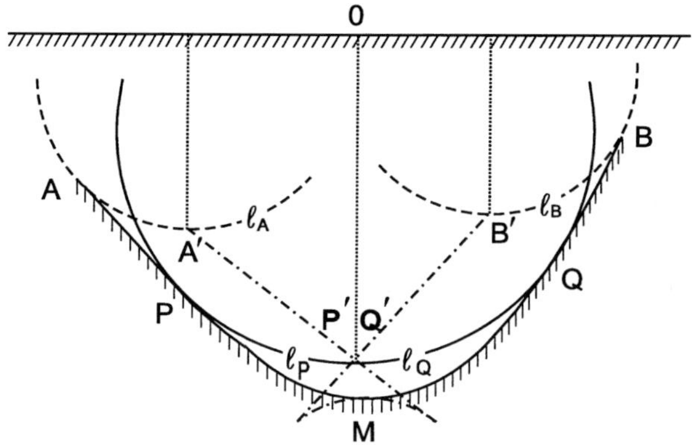

Figure 3.5: Envelope of lines of equal reflection time and reflector curvature.

that of the corresponding line of equal reflection times.

The limitation of this approach, based on the use of the earliest arrival time, vanishes, when all waves are considered at each point of observation. This implies that waves, reflected from different parts of the reflector, arrive at receivers at completely different moments, that is, we can neglect by their superposition. In such a favorable case the section of the record horizon 1α permits us to reconstruct the left part of the syncline, Fig. 3.4. The next element of the record horizon, αa, gives the next interval of the boundary, oa. And finally, the element of the record horizon, ac, based on an observation of the third arrival, permits us to draw lines of equal reflection times, and their envelope coincides with the deepest element of the boundary. The other side of the reflector can be found in a similar manner.

Before we consider the next example of migration it is useful to discuss briefly an influence of the distance between the line of observations and the boundary. It is obvious that with a decrease of the distance between the observation line and the reflector, a triplet of the central part of the record horizon becomes smaller and finally disappears. Such improvement, performed with the system of parallel lines of observations, is shown in Fig. 3.6. This clearly illustrates one of the main ideas of the modern migration.

Example five In this case, the boundary is formed by two straight segments with different dips, Fig. 3.7a. They intersect at point P. We assume that, except in the vicinity of P, rays, reflected from AP or PB, obey Snell's law. This means that the reflected ray, which returns back to the source, is normal to the reflector. For instance, rays from the side AP are observed within the interval of the observation line

$$x_1 < x < x_2 \,,$$

and, as is seen from Fig. 3.7a, the straight element $A'P_1'$ describes the corresponding portion of the record horizon, if $c = \text{const}$.

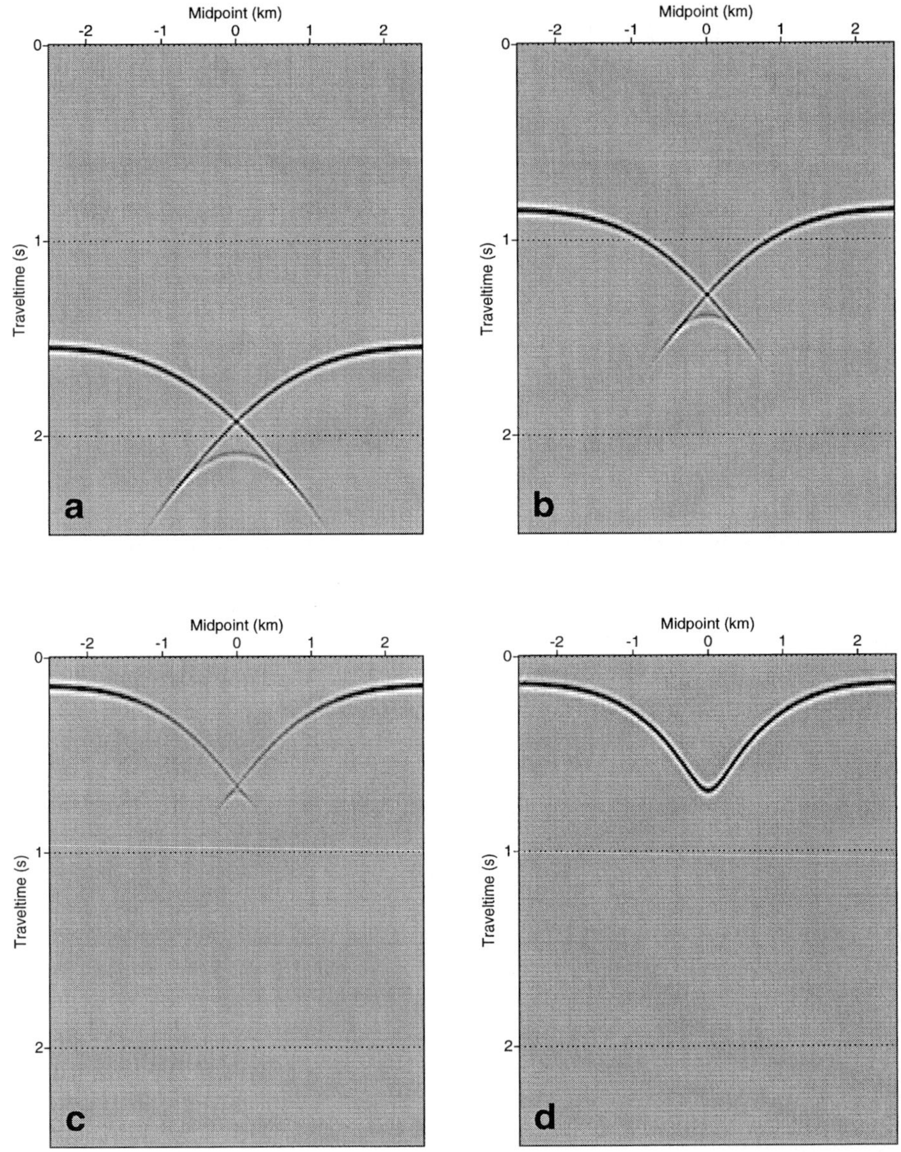

Figure 3.6: Influence of the distance between an observation line and reflector.

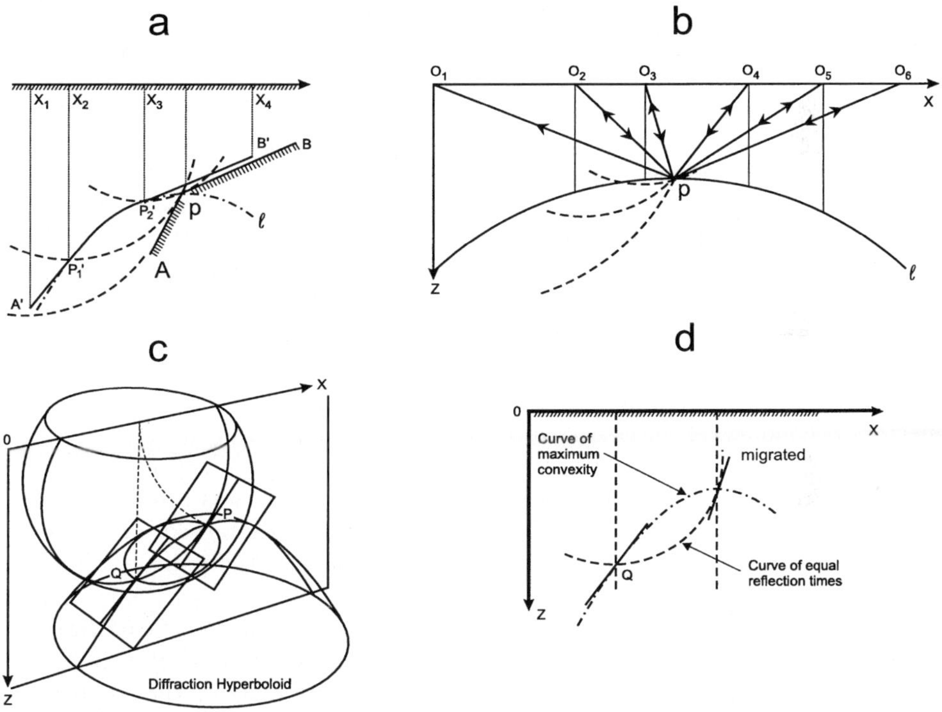

Figure 3.7: (a) Intersection of lines of equal reflection times (b) Diffraction point (c) Diffraction hyperboloid (d) Curves of maximal convexity and equal reflection times.

It is clear that rays, arising at points of the other side, BP, appear at receivers inside the interval

$$x_3 < x < x_4,$$

and their record horizon is $P_2'B'$.

Between two segments, $A_1'P_1'$ and $P_2'B'$ there is a gap and it is related to a new phenomenon, namely, diffracted waves. Inasmuch as near the edge point P the boundary curvature tends to infinity, Snell's law is not valid and, respectively, waves, arising around the point P, propagate in all directions. These diffracted waves arrive after reflected ones within intervals:

$$x_1 < x < x_2 \qquad \text{and} \qquad x_3 < x < x_4\,,$$

but they exist only if $x_2 < x < x_3$.

The part of the record horizon, caused by them, is represented by curve $P_1'P_2'$, Fig. 3.7a. There is of course, some transition zone between intervals. It is obvious that each pair of points P and P_2', P and P_1' is located on the corresponding line of equal reflection times. Therefore, these lines intersect each other at point P. The same is true for any point of the curve $P_1'P_2'$, that is, all lines of equal reflection times, crossing this part of the record horizon, have the common point P. This means that unlike the case of reflected waves, these lines do not have an envelope. Thus, migration in such case is based on two important features. First of all, the envelope of lines of equal reflection times, describing reflected waves, allows us to find the position of two sides of the reflector. Secondary, an intersection of lines of equal reflection times, characterizing diffracted waves, defines a location of the point P. However, accuracy of determination of the reflector position near P may not be sufficient, since as was already mentioned, a transition from the reflected waves, obeying Snell's law, to diffracted ones takes place gradually. In this light it is useful to introduce a new concept, the point of diffraction.

Example six Suppose that reflection occurs at a relatively small element of the interface around point P, so that reflected waves propagate in all directions, Fig. 3.7b. Then, performing observations with zero offset along the x-axis and plotting the record horizon, we arrive at the line of the second order, which is a hyperbola. Its apex is located at the point P, ($c = $ const).

As was mentioned earlier, all points of the record horizon are the deepest points of the corresponding lines of equal reflection times. Because of this, the line l is called by Hagedorn the curve of maximal convexity or the diffraction curve, Fig. 3.7a. In this simplest case a migration is a transformation of all points of the record horizon into one point. Thus, the point of diffraction can be represented by either the single diffraction curve or a system of lines of equal reflection times. Their envelope is absent, but instead of it they have the common point P.

Example seven Next consider a three-dimensional medium and assume that a reflector is an element of the planar surface around P, while its record horizon is in the vicinity of point Q, Fig. 3.7c. It is obvious that in place of the diffraction curve we have a hyperboloid with the apex at the point P if $c =$ const, (zero offset). Let us treat the reflector as a system of diffraction points. Then, the element of the record horizon around a point Q is an envelope of hyperboloids corresponding to each such point, that is it is tangential to the diffraction surface. For instance, in the vertical plane xoz the linear segment of the record horizon is tangential to the hyperbola with the apex at P, Fig. 3.7d. At the same time we know that the actual reflector is tangential to the line of equal reflection times at point P. Taking into account the orientation of the boundary element and that of the record horizon with respect to lines of equal reflection times for the reflected and diffracted waves, a migration in two dimensional case can be performed in the following way. First, drawing a convex curve, tangential to the element of the record line near point Q, we find the position of the apex of this curve, that is, the point P, that belongs to the reflector. Then the line of equal reflection times, passing through points Q and P, is plotted, Fig. 3.7d. The linear element, tangential to this curve at the point P, defines a position of the reflector at this place. This important subject will be discussed again but in different way in the next section.

Now we can say that in the case, shown in Fig. 3.7a, the portion of the record horizon between points with coordinates x_2 and x_3 is described by the convex hyperbola. The latter has an apex P and it is tangential to the element $A'P'$ of the record horizon.

3.2 Principles of Kirchhoff migration

We continue to describe some features of migration, which allow one to determine a reflector position, provided that the velocity of propagation is known. Unlike the approach based on the use of lines of equal reflection times and diffraction curves (section 3.1), the main attention is paid here to an evaluation of wave magnitude at points beneath the observation surface and, specially, at the reflector. In general, such information permits us to locate a reflector position with much higher accuracy. As was shown earlier, if an observation surface approaches the reflector, the shape of the record horizon is close to that of the interface. Because of this, it is not accidental that all conventional methods of migration calculate wave fields everywhere between the observation surface and the reflector. As an example, consider the approach based on the use of the Kirchhoff integral. Its description mainly follows the pioneering paper of W. Schneider. At the beginning we assume that sinusoidal waves propagate through a medium, that is homogeneous with respect to the density ρ, and the velocity of propagation, $c(x, y, z)$, is a continuous function of the position. Also it is implied that the wave field is known at all points of the observation surface S_0. From a mathematical point of view, the problem of a wave calculation inside a medium is formulated in the following way. The complex amplitude of the potential $\mathcal{U}(p)$ satisfies at any point of a medium, the Helmholtz equation,

$$\nabla^2 \mathcal{U} + k^2 \mathcal{U} = 0\,, \tag{3.8}$$

and $k(p) = \omega/c(p)$ is the wave number.

Besides the function $\mathcal{U}(p,\omega)$ and its normal derivative are given at the surface S, surrounding the volume V, Fig. 3.8a. Later this condition will be modified. Solution of this task can be done by different methods, and one of them is discussed here.

Helmholtz formula

First we outline the main steps in deriving the Helmholtz formula. Making use of the Gauss formula:

$$\int_V \operatorname{div} \mathbf{M}\, dV = \oint_S \mathbf{M} \cdot \mathbf{dS}, \tag{3.9}$$

the second Green formula was deduced (Part I),

$$\int_V \left(\mathcal{U}\,\nabla^2 G - G\,\nabla^2 \mathcal{U}\right) dV = \oint_S \left(\mathcal{U}\,\frac{\partial G}{\partial n} - G\,\frac{\partial \mathcal{U}}{\partial n}\right) dS \tag{3.10}$$

where $\mathcal{U}$ is the complex amplitude of the acoustic potential, related to pressure and a particle displacement as

$$P = \rho\omega^2 \mathcal{U}, \qquad \mathbf{s} = \operatorname{grad} \mathcal{U}, \tag{3.11}$$

while G is some function and $\mathbf{n}$ is the unit vector, directed outward, $\mathbf{dS} = \mathbf{n}\, dS$.

It is important to emphasize that the second Green formula is valid if both functions, $\mathcal{U}$ and G, have continuous first and second derivatives at each point of the volume V. Otherwise operations $\nabla^2\mathcal{U}$ and $\nabla^2 G$, as well as $\partial\mathcal{U}/\partial n$ and $\partial G/\partial n$, are meaningless. Certainly, there is an infinite number of such functions G and some of them are called Green functions. Also, as follows from eq. 3.11, a function $\mathcal{U}$ can be replaced by the complex amplitude of a pressure P.

The second Green formula is already a relationship between function $\mathcal{U}(p)$ in the volume and values of $\mathcal{U}$ and its normal derivative, $\partial\mathcal{U}/\partial n$, on surface S. However, the unknown function $\mathcal{U}$ is not represented in the explicit form, instead, it is a term of the integrand in eq. 3.10. To overcome this problem, we impose two conditions on a function $G(q,p)$, where q is an arbitrary point of the volume V and surface S, but p is the point in the volume V, where the wave field, for instance, the pressure is observed. The first condition requires that in approaching p the function G behaves as

$$G(q,p) \to \frac{1}{R} \qquad \text{if} \qquad q \to p \tag{3.12}$$

Here R is the distance between points q and p, Fig. 3.8a. Because of the singularity at the point p we cannot apply eq. 3.10 to the volume V. To avoid this obstacle point p is surrounded by the surface of "safety". Then the second Green formula is applied to the volume V_*, enclosed by surfaces S and S_s, Fig. 3.8a. Therefore, eq. 3.10 becomes

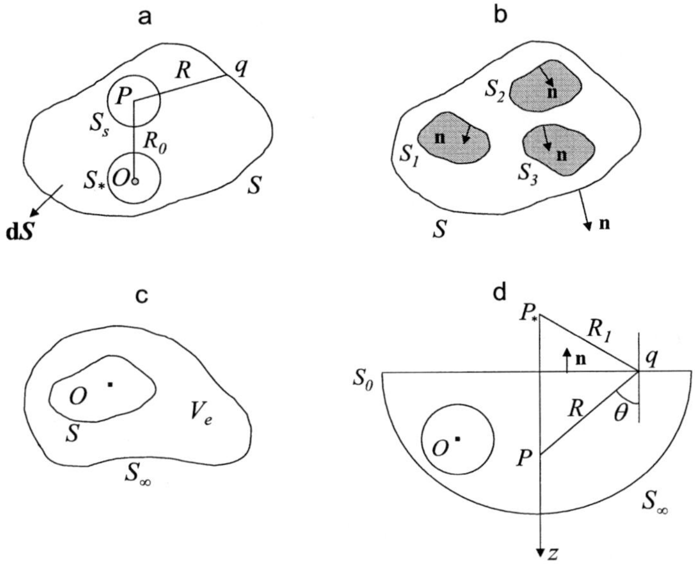

Figure 3.8: (a) Illustration of eq. 3.13 (b) Volume surrounded by several surfaces (c) Illustration of eq. 3.30 (d) Model of a homogeneous half-space.

$$\int\limits_{V_*} \left(\mathcal{U}\,\nabla^2 G - G\,\nabla^2 \mathcal{U}\right)\, dV = \oint\limits_{S} \left(\mathcal{U}\,\frac{\partial G}{\partial n} - G\,\frac{\partial \mathcal{U}}{\partial n}\right)\, dS + \oint\limits_{S_s} \left(\mathcal{U}\,\frac{\partial G}{\partial n} - G\,\frac{\partial \mathcal{U}}{\partial n}\right)\, dS \quad (3.13)$$

Assuming that S_s is a spherical surface with a origin at the point p and its radius R tends to zero, that is $V_* \rightarrow V$, consider the second integral at the right hand side of eq. 3.13. Also, we suppose that the function $\mathcal{U}$ has finite values everywhere within the volume, including, of course, points of the surface S_s.

In contrary, as follows from eq. 3.12, both functions, $\mathcal{G}$ and $\partial G/\partial n$, increase unlimitedly when the radius R goes to zero, and we have

$$G \rightarrow \frac{1}{R} \qquad \text{and} \qquad \frac{\partial G}{\partial n} = -\frac{\partial G}{\partial R} \rightarrow \frac{1}{R^2}\,, \quad (3.14)$$

The last equality takes place because the radius vector $\mathbf{R}$ and the normal $\mathbf{n}$ have opposite directions. Respectively, the integral over the surface S_s becomes

$$\oint\limits_{S_s} \left(\mathcal{U}\,\frac{\partial G}{\partial n} - G\,\frac{\partial \mathcal{U}}{\partial n}\right)\, dS = \oint\limits_{S_s} \mathcal{U}\,\frac{dS}{R^2} - \oint\limits_{S_s} \frac{\partial \mathcal{U}}{\partial n}\,\frac{dS}{R} \quad (3.15)$$

The function $\mathcal{U}$ is continuous and, therefore, when $R \to 0$ its values on the surface S_s coincide with the corresponding value at point p:

$$\mathcal{U}(S_s) \to \mathcal{U}(p) \qquad \text{if} \qquad R \to 0$$

This gives

$$\oint_{S_s} \mathcal{U} \frac{dS}{R^2} = \frac{\mathcal{U}(p)}{R^2} \oint_{S_s} dS = 4\pi \mathcal{U}(p) \tag{3.16}$$

Also the use again of the mean value theorem yields

$$\oint_{S_s} \frac{\partial \mathcal{U}}{\partial n} \frac{dS}{R} = \left(\frac{\partial \mathcal{U}}{\partial n}\right)_m \frac{1}{R} \oint_{S_s} dS = 4\pi \left(\frac{\partial \mathcal{U}}{\partial n}\right)_m R \to 0$$

Substitution of the latter into eq. 3.13 yields

$$\mathcal{U}(p) = \frac{1}{4\pi} \int_V \left(\mathcal{U} \nabla^2 G - G \nabla^2 \mathcal{U}\right) dV + \frac{1}{4\pi} \oint_S \left(G \frac{\partial \mathcal{U}}{\partial n} - \mathcal{U} \frac{\partial G}{\partial n}\right) dS \tag{3.17}$$

It is clear that this equation does not yet solve our problem, because it contains the volume integral, depending on the function $\mathcal{U}$ and $\nabla^2 \mathcal{U}$. Now we demonstrate how to remove the last difficulty. The second condition for the function $G(q,p)$ takes into account the fact that the complex amplitude of the potential obeys eq. 3.8. Then from among functions G we choose ones that also satisfy the same Helmholtz equation as the potential $\mathcal{U}$, except at point p, that is

$$\nabla^2 G + k^2 G = 0 \,, \tag{3.18}$$

and wave number k generally varies in the volume V.

Let us notice that functions $G(q,p)$ satisfying both conditions, eqs. 3.12 and 3.18, are called Green functions. As follows from eqs. 3.8 and 3.18, the volume integral in eq. 3.17 vanishes and it gives

$$\mathcal{U}(p) = \frac{1}{4\pi} \oint_S \left(G \frac{\partial \mathcal{U}}{\partial n} - \mathcal{U} \frac{\partial G}{\partial n}\right) dS \,, \tag{3.19}$$

where on the right-hand side the potential $\mathcal{U}$ and its normal derivative, $\partial \mathcal{U}/\partial n$, are taken at points q of the surface S, and the Green function depends on points q and p. Thus, we have solved our problem and established the relationship between the wave field at any point of the volume and its values on the surface S. The latter, eq. 3.19, is the well known Helmholtz formula and it permits us to determine the wave spectrum inside the volume, if it is given on the surface S (Part I). In accordance with eqs. 3.11 this implies that, in general, both the pressure and normal component of the displacement are known at all points of S. Moreover, there is another strong requirement, namely, the Green function should satisfy the Helmholtz equation for a certain distribution of velocity $c(q)$ in the volume V. Except some special cases, a determination of such Green function is usually a very difficult mathematical problem. Note that if the volume V is surrounded by several surfaces, Fig. 3.8b, then the wave fields have to be known at all their points and an integration is performed over these boundaries.

Primary source in volume V

From the beginning it was assumed that the wave field is finite everywhere inside the volume V. In other words, sources of the primary wave are located outside the surface S. In fact, replacing the real source by the point source, the field becomes infinitely large at its vicinity, that is the function $\mathcal{U}(p)$ has a singularity. Now we generalize eq. 3.19 and suppose that such a source is located at some point O of the volume V, Fig. 3.8a and the potential at its vicinity behaves as

$$\mathcal{U}_o(q) = A\frac{e^{ikR_o}}{R_o} \qquad \text{if} \qquad q \rightarrow 0 \tag{3.20}$$

Here k is the wave number at the point O and R_o is the distance between the source and an observation point. Then in order to apply the second Green formula, eq. 3.10, it is necessary to surround the point O by a small spherical surface S_* and, correspondingly, in place of eq. 3.19 we obtain

$$\mathcal{U}(p) = \frac{1}{4\pi}\oint\limits_S \left(G\frac{\partial \mathcal{U}}{\partial n} - \mathcal{U}\frac{\partial G}{\partial n}\right) dS + \frac{1}{4\pi}\oint\limits_{S_*} \left(G\frac{\partial \mathcal{U}}{\partial n} - \mathcal{U}\frac{\partial G}{\partial n}\right) dS, \tag{3.21}$$

where $\mathcal{U}(p)$ is the potential, caused by all sources of the wave, located outside of and within the volume V. Unlike the observation point p, Green function $G(q,p)$ has a finite value at the point O, but the potential $\mathcal{U}$ and its derivative tend to an infinity as

$$\mathcal{U} \rightarrow \frac{A}{R_o}, \qquad \frac{\partial \mathcal{U}}{\partial n} = -\frac{\partial \mathcal{U}}{\partial R} \longrightarrow \frac{A}{R_o^2} \qquad \text{if} \qquad R_o \rightarrow 0 \tag{3.22}$$

Respectively, the integral over the surface S_* becomes

$$\frac{1}{4\pi}\oint\limits_{S_*} \left(G\frac{\partial \mathcal{U}}{\partial n} - \mathcal{U}\frac{\partial G}{\partial n}\right) dS \longrightarrow A\,G(O,p) \tag{3.23}$$

The Green function depends on two points, O and p, and it does not change, when they replace each other. Such procedure means that the singularity of the function $G(O,p)$

occurs at the point O. Hence, the right-hand side of eq. 3.23 describes the potential $\mathcal{U}_o(p)$, caused by the primary source at the point O. For this reason, eq. 3.21 is written as

$$\mathcal{U}(p) = \mathcal{U}_o(p) + \frac{1}{4\pi} \oint_S \left[G(q,p) \frac{\partial \mathcal{U}(q)}{\partial n} - \mathcal{U}(q) \frac{\partial G(q,p)}{\partial n} \right] dS \tag{3.24}$$

The latter is easily generalized in the case in which there is a system of such primary sources inside the volume V. In accordance with eq. 3.24, function $\mathcal{U}(p)$ represents the total potential

$$\mathcal{U}(p) = \mathcal{U}_o(p) + \mathcal{U}_s(p) \tag{3.25}$$

Its substitution into eq. 3.24 gives

$$\mathcal{U}_s(p) = \frac{1}{4\pi} \oint_S \left(G \frac{\partial \mathcal{U}_o}{\partial n} - \mathcal{U}_o \frac{\partial G}{\partial n} \right) dS + \frac{1}{4\pi} \oint_S \left(G \frac{\partial \mathcal{U}_s}{\partial n} - \mathcal{U}_s \frac{\partial G}{\partial n} \right) dS \tag{3.26}$$

The first integral on the right hand side is independent of $\mathcal{U}_s$. In particular, letting in eq. 3.26 $\mathcal{U}_s = 0$, we obtain

$$\oint_S \left(G \frac{\partial \mathcal{U}_o}{\partial n} - \mathcal{U}_o \frac{\partial G}{\partial n} \right) dS = 0 \,,$$

and therefore eq. 3.26 has the form

$$\mathcal{U}_s(p) = \frac{1}{4\pi} \oint_S \left(G \frac{\partial \mathcal{U}_s}{\partial n} - \mathcal{U}_s \frac{\partial G}{\partial n} \right) dS \tag{3.27}$$

Thus, the complex amplitude of the potential, associated with the secondary wave inside the volume, is defined by its values on the surface S. Let us notice that eqs. 3.27 and 3.19 coincide and it is not accidental, since we assume that the secondary field $\mathcal{U}_s$ does not have singularities in the volume V.

By definition potential $\mathcal{U}_o$ describes the wave field in a medium, if the chosen Green function everywhere satisfies the Helmholtz equation, except the point O. At the same time $\mathcal{U}_s$ characterizes the secondary wave, which may appear if, outside the volume V either there are primary sources or the Green function does not obey the Helmholtz equation.

Inasmuch as the main formulas of Kirchhoff migration are derived from eq. 3.24 or 3.27 let us consider several examples illustrating the meaning of functions $\mathcal{U}_o$, $\mathcal{U}_s$, and G.

Further it is assumed that primary sources are absent at infinity and, correspondingly, the field vanishes

$$\mathcal{U} \to 0 \qquad \text{and} \qquad \frac{\partial \mathcal{U}}{\partial n} \to 0 \tag{3.28}$$

Also, implying that Sommerfeld conditions are met, (Part I), the integral in eq. 3.27 over the spherical surface with an infinitely large radius disappears

$$\oint_S \left(G\frac{\partial \mathcal{U}}{\partial n} - \mathcal{U}\frac{\partial G}{\partial n} \right) dS \to 0 \qquad \text{if} \qquad R \to \infty \tag{3.29}$$

Case one Suppose that Green function $G(q,p)$ satisfies the Helmholtz equation at all points of the medium. Then, applying the second Green formula, eq. 3.10, to the volume V_e, bounded by the surface S and S_∞, Fig. 3.8c, we see that the volume integral is equal to zero:

$$\oint_{V_e} \left(\mathcal{U}\nabla^2 G - G\nabla^2 \mathcal{U} \right) dV = 0 \tag{3.30}$$

Then

$$\oint_S \left(G\frac{\partial \mathcal{U}_S}{\partial n} - \mathcal{U}_s\frac{\partial G}{\partial n} \right) dS = 0\,,$$

that is,

$$\mathcal{U}_s = 0 \tag{3.31}$$

Thus, the acoustic potential is defined by the primary potential only

$$\mathcal{U}(p) = \mathcal{U}_o(p) = A\,G(O,p)\,, \tag{3.32}$$

and the function $G(O,p)$ is called the Green function of the whole space.

In deriving eq. 3.31 it was assumed that the primary sources are absent in the volume V_e. Otherwise they have to be taken into account, and, in general we have

$$\mathcal{U}(p) = \sum_i A_i\,G(O_i,p)\,,$$

where O_i characterizes a position of the $i$primary source.

Case two Now we assume that Green function $G(q,p)$ obeys the Helmholtz equation in the volume V only. Then outside V the left hand side of eq. 3.10 differs from zero, and it happens because functions $\mathcal{U}(q)$ and $G(q,p)$ satisfy the Helmholtz equation with different wave numbers. Correspondingly, the integral over the surface S, eq. 3.27, has a non-zero value, and it defines the wave field, caused by secondary sources outside S. The same result is observed when primary sources are located in the volume V_e.

Case three: Homogeneous half-space Because the wavenumber k is constant, it is easy to find a solution of eq. 3.18, which has a necessary singularity at the observation point p. For instance, both functions

$$\frac{e^{ikR}}{R} \qquad \text{and} \qquad \frac{e^{-ikR}}{R} \tag{3.33}$$

satisfy these conditions. Respectively, eq. 3.27 is written as

$$\mathcal{U}_s(p) = \frac{1}{4\pi} \oint\limits_S \left(\frac{\partial \mathcal{U}_s}{\partial n} \frac{e^{ikR}}{R} - \mathcal{U}_s \frac{\partial}{\partial n} \frac{e^{ikR}}{R} \right) dS \tag{3.34}$$

The surface S consists of two parts, Fig. 3.8d. One is observation plane S_o, and the other is a semispherical surface S_∞ with an infinitely large radius. As was demonstrated in Part I, the integral over S_∞ in eq. 3.34 is equal to zero. This also follows from a physical point of view, eq. 3.29. Thus,

$$\mathcal{U}_s(p) = \frac{1}{4\pi} \oint\limits_{S_o} \left[\mathcal{U}_s(q) \frac{\partial}{\partial z_q} \left(\frac{e^{ikR}}{R} \right) - \frac{e^{ikR}}{R} \frac{\partial \mathcal{U}_s(q)}{\partial z_q} \right] dS \tag{3.35}$$

Here it is assumed that the z-axis is directed downward, that is $\partial/\partial z_q = -\partial/\partial n$ and z_q is a coordinate of some point in the vicinity of S_o.

Bearing in mind that the normal component of the displacement, $\partial \mathcal{U}/\partial n$, is usually unknown, we introduce a new Green function, which is equal to zero at all points of the horizontal plane S_o. Then the term $G\,\partial \mathcal{U}/\partial n$ in eq. 3.27 vanishes. In the case of the plane surface S_o such function $G(q,p)$ is well known and has the form

$$G_1 = \frac{e^{ikR}}{R} - \frac{e^{ikR_1}}{R_1} \qquad \text{or} \qquad G_1 = \frac{e^{-ikR}}{R} - \frac{e^{-ikR_1}}{R_1}, \tag{3.36}$$

where

$$R = \left[(x_q - x_p)^2 + (y_q - y_p)^2 + (z_q - z_p)^2 \right]^{1/2}$$

and

$$R_1 = \left[(x_q - x_p)^2 + (y_q - y_p)^2 + (z_q + z_p)^2 \right]^{1/2}, \tag{3.37}$$

while x_q, y_q, z_q and x_p, y_p, z_p are coordinates of points q and p, respectively.

It is essential that the new Green function, G_1, has the same singularity at the point p, since $R \to 0$. As is seen from Fig. 3.8d, R_1 is the distance between the point q and p_*, which is a mirror reflection of p with respect to the plane S_o. It is obvious that at all its points, $z_q = 0$, the function $G_1(q,p)$ vanishes and the Helmholtz formula, eq. 3.27 becomes

$$\mathcal{U}_s(p,\omega) = \frac{1}{4\pi} \oint\limits_{S_o} \mathcal{U}_s(q,\omega) \frac{\partial G_1}{\partial z_q} dS \tag{3.38}$$

Here, the derivative is taken at points of the surface S_o and $G_1(q,p)$ is defined by eq. 3.36.

Now we perform some elementary transformations, that simplifies the right-hand side of the eq. 3.38. It is clear that

$$\frac{\partial}{\partial z_q} \frac{e^{ikR}}{R} = \frac{\partial}{\partial R} \left(\frac{e^{ikR}}{R} \right) \frac{\partial R}{\partial z_q} \qquad \text{and} \qquad \frac{\partial}{\partial z_q} \frac{e^{ikR_1}}{R_1} = \frac{\partial}{\partial R_1} \left(\frac{e^{ikR_1}}{R_1} \right) \frac{\partial R_1}{\partial z_q}, \tag{3.39}$$

while

$$\frac{\partial R}{\partial z_q} = \frac{z_q - z_p}{R}, \qquad \frac{\partial R_1}{\partial z_q} = \frac{z_q + z_p}{R_1}, \tag{3.40}$$

At the observation plane we have

$$\frac{\partial}{\partial z_q}\frac{e^{ikR}}{R} = -\frac{\partial}{\partial z_q}\frac{e^{ikR_1}}{R_1}$$

and, therefore, eq. 3.38 can be written as

$$\mathcal{U}_s(p,\omega) = \frac{1}{2\pi}\oint\limits_{S_o} \mathcal{U}_s(q,\omega)\frac{\partial}{\partial z_q}\frac{e^{ikR}}{R}\,dS \tag{3.41}$$

Also, taking into account that

$$\frac{\partial}{\partial z_q}\frac{e^{ikR}}{R} = -\frac{\partial}{\partial z_p}\frac{e^{ikR}}{R}$$

eq. 3.41 becomes

$$\mathcal{U}_s(p,\omega) = -\frac{1}{2\pi}\int\limits_{S} \mathcal{U}_s(q,\omega)\frac{\partial}{\partial z_p}\frac{e^{ikR}}{R}\,dS \tag{3.42}$$

It is useful to represent the latter in a different form. As follows from eq. 3.40

$$\frac{dR}{dz_q} = -\cos\theta\,, \tag{3.43}$$

where θ is the angle formed by the z-axis and the line connecting point q with p, Fig. 3.8d. Since

$$\frac{\partial}{\partial z_q}\left(\frac{e^{ikR}}{R}\right) = \left(\frac{ik}{R} - \frac{1}{R^2}\right)e^{ikR}\cos\theta,$$

we have

$$\mathcal{U}_s(p,\omega) = -\frac{1}{2\pi}\int\limits_{S_o} \mathcal{U}_s(q,\omega)\frac{e^{ikR}}{R^2}(ikR-1)\cos\theta\,dS \tag{3.44}$$

In particular, in the wave zone

$$\mathcal{U}_s(p,\omega) = -\frac{ik}{2\pi}\int\limits_{S_o} \mathcal{U}_s(q,\omega)\frac{e^{ikR}}{R}\cos\theta\,dS \tag{3.45}$$

Thus, in accordance with eqs. 3.44–3.45, we can calculate the potential at any point of a homogeneous medium if this function is known at the observation plane S_o. In other words, the Helmholtz formula allows us to solve the forward problem in the frequency domain, that is it performs the downward continuation of the wave-field.

Now let us make several comments:

a. Because the complex amplitudes of the potential and pressure differ by a constant, eqs. 3.41, we can replace $\mathcal{U}$ by $\mathcal{P}_s$ in eqs. 3.44 - 3.45. For instance in the wave zone:

$$\mathcal{P}_s(p,\omega) = -\frac{ik}{2\pi} \int\limits_{S_o} P_s(q,\omega) \frac{e^{ikR}}{R} \cos\theta \, dS \tag{3.46}$$

b. The Helmholtz formula is valid regardless of the wave length and, in particular, at the high frequency spectrum it gives the same wave distribution, as it follows from the ray theory.

c. The surface S_o is a boundary with a free space and, hence, a superposition of waves is such that at points of S_o the pressure vanishes

$$\mathcal{P}_s(x,y,0,\omega) \equiv 0 \tag{3.47}$$

Then, in accordance with eq.3.27 the pressure $\mathcal{P}_s$ is equal to zero inside a homogeneous half-space. At the same time we apply eq. 3.46 to the up-going wave only, and a separation of the up-going and down-going waves is accomplished out by measurements of the pressure slightly below S_o.

d. It is evident that the pressure on the plane S_o, caused by the up-going reflected wave depends on the position and an intensity of the primary source, as well as parameters of a medium.

e. As is well known, the presence of the interface is the reason why the reflected waves appear. This means that a medium beneath the surface S_o is inhomogeneous and therefore the Green function, given by eq. 3.36, does not satisfy everywhere the Helmholtz equation with a constant value of the wave number. Correspondingly, we have to change an area of integration in the Helmholtz formula, eq. 3.34.

Homogeneous layer

Suppose that a medium between plane S_o and a reflector is homogeneous, and at the interface, S_r, the velocity of propagation and density are discontinuous, Fig. 3.9a.

As follows from a continuity of the pressure, as well as the Helmholtz equation:

$$\nabla^2 \mathcal{U} = -k^2 \mathcal{U},$$

the potential is a discontinuous function at points on the interface S_r. This means, that we cannot apply the second Green formula to the whole space, surrounded by the observation plane S_o and S_∞. Because of this consider the volume V, enclosed by three surfaces, namely S_o, S_r and the cylindrical one, S_ℓ, located at an infinity. Neglecting by the integral over the last surface, eq. 3.19 can be written as

$$\mathcal{U}_S(p) = \frac{1}{4\pi}\int\limits_{S_o}\left[G(q,p)\frac{\partial \mathcal{U}_s(q)}{\partial n} - \mathcal{U}_S(q)\frac{\partial G(q,p)}{\partial n}\right] dS$$

$$+\frac{1}{4\pi}\int\limits_{S_r}\left[G(q,p)\frac{\partial \mathcal{U}_S(q)}{\partial n} - \mathcal{U}_S(q)\frac{\partial G(q,p)}{\partial n}\right] dS \tag{3.48}$$

Inasmuch as the potential $\mathcal{U}(q)$ and its normal derivative, $\partial\mathcal{U}_s(q)/\partial n$, are unknowns in the vicinity of the reflector, it seems that the Helmholtz formula does not allow us to perform the downward continuation and, therefore, determine the wave field at any point of the upper layer. At the same time, it is possible to show that under certain conditions the integral over the reflector surface can be disregarded.

Suppose that the primary source is situated at point O of the layer (Fig. 3.9b) and, thus, the potential of the incident wave is

$$\mathcal{U}_i(q) = A\,\frac{e^{ik_1 R_o}}{R_o}, \tag{3.49}$$

where

$$R_o = \left[(x_q - x_o)^2 + (y_q - y_o)^2 + (z_q - z_o)^2\right]^{1/2}$$

Since the reflected wave appears at the boundary S_r and propagates in the direction opposite to the z-axis, we choose the following Green function

$$G(q,p) = \frac{e^{-ik_1 R_p}}{R_p}, \tag{3.50}$$

where

$$R_p = \left[(x_q - x_p)^2 + (y_q - y_p)^2 + (z_q - z_p)^2\right]^{1/2}$$

and k_1 is the wavenumber of the upper layer.

Further, we assume that the field is considered in the high-frequency spectrum, and therefore it is possible to make use of results, that follow from the ray theory. For instance, as was shown earlier, the rays of the incident and reflected waves obey Snell's law, and the reflected wave behaves as the spherical one:

$$\mathcal{U}_s(p) = \alpha\, A\,\frac{e^{ik_1 R_1}}{R_1} \tag{3.51}$$

where α is the Fresnel coefficient (Chapter 2) and

$$R_1 = \left[(x_p - x_{o*})^2 + (y_p - y_{o*})^2 + (z_p - z_{o*})^2\right]^{1/2}$$

The position of the fictitious source, o_*, satisfies the equality, (Fig. 3.9b)

$$R_o = oq = o_* q \tag{3.52}$$

In accordance with eqs. 3.49 and 3.51 at points of the surface S_r the potential or pressure of the incident and reflected waves are proportional to each other

$$\mathcal{U}_s(q,\omega) = \alpha\,\mathcal{U}_i(q,\omega) \qquad \text{on} \qquad S_r \tag{3.53}$$

In other words, they change synchronously.

This fact is of a great importance for migration and it will be discussed later. Let us notice that this equality represents the first assumption of the Kirchhoff theory of diffraction. To evaluate the integrand in eq. 3.48, corresponding to the second integral, we have to find the normal derivatives of functions $G(q,p)$ and $\mathcal{U}(q)$ at points of the reflector. Inasmuch as the wave field is studied at the high-frequency spectrum, in calculating derivatives we preserve only terms proportional to the wave number.

By definition, this gives

$$\frac{\partial G}{\partial n} = -ik_1 \frac{e^{-ik_1 R_p}}{R_p} \nabla R_p \cdot \mathbf{n} \qquad \text{and} \qquad \frac{\partial \mathcal{U}_s}{\partial n} = Aik_1\alpha \frac{e^{ik_1 R_1}}{R_1} \nabla R_1 \cdot \mathbf{n} \tag{3.54}$$

Substitution of eqs. 3.49 - 3.54 into the second integral of eq. 3.48 yields:

$$\frac{A}{4\pi}\, i\, k_1 \alpha \int\limits_{S_r} \frac{e^{ik_1(R_1 - R_p)}}{R_1 R_p} \left(\nabla R_p \cdot \mathbf{n} + \nabla R_1 \cdot \mathbf{n}\right) dS \tag{3.55}$$

As is well known, $\nabla R_p = \mathbf{r}_p$ and $\nabla R_1 = \mathbf{r}_1$ are unit vectors, shown in Fig. 3.9b.

By definition these vectors have opposite directions, and the integral over the reflector vanishes (Snell's law). Therefore, the wave field at points of the upper layer is defined by values of $\mathcal{U}_s(q)$ and $\partial \mathcal{U}_s(q)/\partial n$ on the observation surface S_o only, eq. 3.48. Certainly, this result is of a great importance.

It may be proper to notice that the use of the Green function

$$\frac{e^{-ik_1 R_p}}{R_p}$$

does not eliminate the integral over the observation surface S_o. Also it is interesting to find the relationship between derivatives

$$\frac{\partial \mathcal{U}_i(q)}{\partial n} \qquad \text{and} \qquad \frac{\partial \mathcal{U}_s(q)}{\partial n}$$

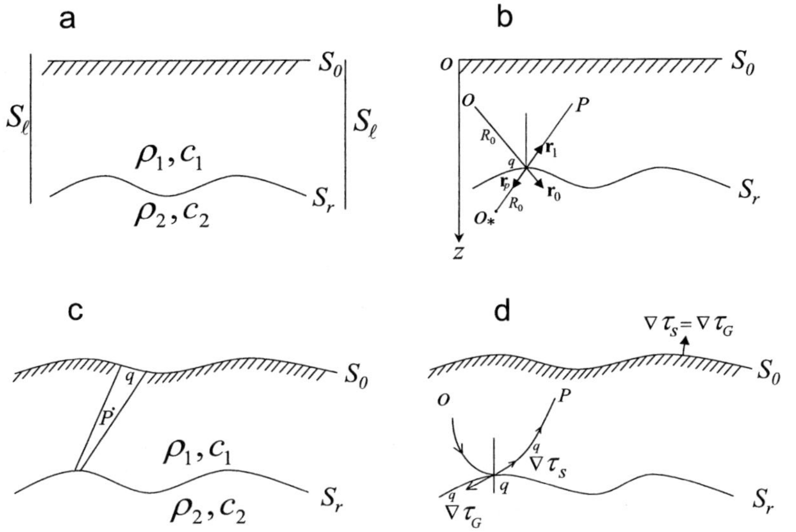

Figure 3.9: (a) Medium with one interface (b) Illustration of eqs. 3.56 and 3.59 (c) Snell's law and stationary point (d)Upper layer is inhomogeneous.

at points of S_r.

As is seen from Fig. 3.9b,

$$\frac{\partial \mathcal{U}_i\,(q)}{\partial n} = ik_1 A\,\frac{e^{ik_1 R_o}}{R_o}\,\overset{q}{\nabla}\,R_o\cdot\mathbf{n} = ik_1 A\,\frac{e^{ik_1 R_o}}{R_o}\,\mathbf{r_o}\cdot\mathbf{n}$$

Taking into account Snell's law we conclude that

$$\frac{\partial \mathcal{U}_s\,(q)}{\partial n} = -\alpha\,\frac{\partial \mathcal{U}_i\,(q)}{\partial n}, \qquad \text{on} \qquad S_r \tag{3.56}$$

which represents the second boundary condition of the Kirchhoff theory of diffraction.

Next, consider the first integral in eq. 3.48, which contains the unknown function $\partial\mathcal{U}_s\,(q)/\partial n$, but this difficulty can be easily overcome. As follows from eq. 3.54

$$\frac{\partial \mathcal{U}_s\,(q)}{\partial n} = ik_1\,\mathcal{U}_s\,(q)\,(\nabla R_1\cdot\mathbf{n}) \tag{3.57}$$

By analogy,

$$\frac{\partial\, G(q,p)}{\partial n} = -ik_1\,G(q,p)\,(\nabla R_p\cdot\mathbf{n}) \tag{3.58}$$

Substituting into eq. 3.48 gives

$$\mathcal{U}_s(p) = \frac{ik_1}{4\pi} \int\limits_{S_o} G(q,p)\,\mathcal{U}_s(q)\,(\mathbf{r_1} + \mathbf{r_p}) \cdot \mathbf{n}\, dS \tag{3.59}$$

As follows from the ray theory the wave arrives at the point q of the surface S_o along the elementary tube. Respectively, the potential at the point p is mainly defined by its values on the elementary surface around q, which is situated at the ray of the reflected wave, passing point p. This means that

$$\mathbf{r_1} \cdot \mathbf{n} = \mathbf{r_p} \cdot \mathbf{n},$$

and it yields

$$\mathcal{U}_s(p) = \frac{ik_1}{2\pi} \int\limits_{S_o} \mathcal{U}_s(q)\, G(q,p)\,(\mathbf{r_p} \cdot \mathbf{n})\, dS \tag{3.60}$$

Thus, we were able to express the function $\mathcal{U}_s(p)$ in terms of its values on the surface S_o only.

It is essential that this result is also applied to an arbitrary surface S_o as soon as the normal **n** is defined at each its point, and therefore the incidence and reflection angles obey Snell's law.

Previously we arrived at Snell's law making use of the method of the stationary phase. This shows that the main contribution to the reflected wave at some point p comes from the vicinity of the corresponding stationary point, and both of them belong to the same ray tube, Fig. 3.9c. This approach also allows us to demonstrate that the integral over the reflector can be neglected. This conclusion is valid if the specular waves are considered, which, by definition, obey Snell's law. In other words, an influence of the head, diffraction and evanescent waves is not taken into account. Besides it is assumed that the radius of the reflector curvature greatly exceeds the wave length. Let us represent eq. 3.60 in terms of the pressure. Bearing in mind eqs. 3.11 we have

$$\mathcal{P}_s(p) = \frac{ik_1}{2\pi} \int\limits_{S_o} P(q)\, \frac{e^{-ik_1 R_p}}{R_p}\,(\mathbf{r_p} \cdot \mathbf{n})\, dS$$

Inasmuch as

$$\frac{\partial}{\partial n}\frac{e^{-ik_1 R_p}}{R_p} = -i\,k_1\,\frac{e^{-ik_1 R_p}}{R_p}\,(\overset{q}{\nabla}\, R_p \cdot \mathbf{n}),$$

we obtain

$$\mathcal{P}_s(p) = -\frac{1}{2\pi} \int\limits_{S_o} P(q)\, \frac{\partial}{\partial n}\frac{e^{-ik_1 R_p}}{R_p}\, dS \tag{3.61}$$

and in the case of the plane surface S_o eqs. 3.42 and 3.61 differ only slightly from each other.

Inhomogeneous upper layer

Consider a more general case when the wavenumber of the layer is a continuous function, Fig. 3.9d. Then, in accordance with ray theory, we can represent the reflected wave and the Green function in the form

$$\mathcal{U}_s\,(q) \;=\; \mathcal{A}_S(q)\,e^{i\,\omega\,\tau_S} \qquad \text{and} \qquad G(q,p) \;=\; \mathcal{A}_G(q,p)\,e^{-i\,\omega\,\tau_G} \tag{3.62}$$

Here τ_S and τ_G are functions, that characterize the time of wave traveling along corresponding rays, and $\mathcal{A}_S$ and $\mathcal{A}_G$ are amplitudes.

Preserving terms, proportional to the frequency, we have

$$\frac{\partial\,\mathcal{U}_s\,(q)}{\partial\,n} \;=\; i\omega\,\mathcal{A}_S(q)\,e^{i\,\omega\,\tau_S}\left(\overset{q}{\nabla}\,\tau_S\cdot\mathbf{n}\right)$$

and (3.63)

$$\frac{\partial\,G(q,p)}{\partial\,n} \;=\; -i\,\omega\,\mathcal{A}_G(q)\,e^{-i\,\omega\,\tau_G}\left(\overset{q}{\nabla}\,\tau_G\cdot\mathbf{n}\right)$$

Substitution of eqs. 3.62 - 3.64 into the second integral of eq. 3.48 gives

$$\mathcal{U}_s\,(p) \;=\; \frac{i\omega}{4\pi}\int\limits_{S_r}\mathcal{A}_S(q)\,\mathcal{A}_G(q,p)\,e^{i\omega\,(\tau_S-\tau_G)}\left(\overset{q}{\nabla}\,\tau_S+\overset{q}{\nabla}\,\tau_G\right)\cdot\mathbf{n}\,dS$$

Again applying the concept of a ray tube, we see that

$$\overset{q}{\nabla}\,\tau_S = -\,\overset{q}{\nabla}\,\tau_G$$

and therefore the integral over the reflector vanishes. In a similar manner it is easy to demonstrate that the integral over the observation surface gives

$$\mathcal{U}_s\,(p) \;=\; \frac{i\omega}{2\pi}\int\limits_{S_o}\mathcal{U}_s(q)\,G(q,p)\;\overset{q}{\nabla}\,\tau_G\cdot\mathbf{n}\,dS \tag{3.64}$$

Next consider transient waves.

Kirchhoff formula and downward continuation

In order to find a relationship for non-stationary waves, similar to the Helmholtz formula, we make use of Fourier's integral, (Part I):

$$P(q,t) \;=\; \frac{1}{2\pi}\int\limits_{-\infty}^{\infty} P(q,\omega)\,e^{-i\omega t}\,d\omega \tag{3.65}$$

After differentiation with respect to time the latter gives

$$\frac{\partial P(q,t)}{\partial t} \;=\; -\frac{i}{2\pi}\int\limits_{-\infty}^{\infty}\omega P(q,\omega)\,e^{-i\omega t}\,d\omega\,, \tag{3.66}$$

where q is an arbitrary point of a medium, including observation point p. Now taking Fourier's integral from both sides of eq. 3.44, we obtain

$$\begin{aligned} P(p,t) &= \frac{1}{2\pi} \int\limits_{S_o} \left[\frac{\cos\theta}{2\pi R^2} \int\limits_{-\infty}^{\infty} P(q,\omega)\, e^{-i\omega\left(t-\frac{R}{c_1}\right)}\, d\omega \right. \\ &\quad \left. - \frac{i\cos\theta}{2\pi R c_1} \int\limits_{-\infty}^{\infty} \omega\, P(q,\omega)\, e^{-i\omega\left(t-\frac{R}{c_1}\right)}\, d\omega \right] dS \end{aligned}$$

As follows from eqs. 3.65 - 3.66 the last equality can be rewritten in the form

$$P(p,t) = \frac{1}{2\pi} \int\limits_{S_o} \left[\frac{P(q,t_1)}{R^2} + \frac{\dot{P}(q,t_1)}{c_1\, R} \right] \cos\theta\, dS \tag{3.67}$$

Here

$$\dot{P}(q,t_1) = \frac{\partial P}{\partial t} = \frac{\partial P}{\partial t_1} \qquad \text{and} \qquad t_1 = t - \frac{R}{c_1} \tag{3.68}$$

This is an example of the Kirchhoff formula for a homogeneous layer, which allows us to calculate the pressure at any point p of the volume V, provided that it is known for earlier times t_1 on the observation plane S_o.

Bearing in mind that

$$\frac{\partial}{\partial z} \left[\frac{P(q,t_1)}{R} \right] = - \left[\frac{P(q,t_1)}{R^2} + \frac{\dot{P}(q,t_1)}{c_1\, R} \right] \cos\,\theta,$$

we have from eq. 3.67

$$P(p,t) = -\frac{1}{2\pi} \frac{\partial}{\partial z} \int\limits_{S_o} P(q,t_1)\, \frac{dS}{R} \tag{3.69}$$

Since z is a coordinate of an observation point, the order of an integration and differentiation can be changed.

Representation, given by eq. 3.69, has a very clear physical meaning because the ratio R/c_1 shows the time of wave propagation from the point q to p. This form is convenient when we consider a down-going wave, arising, for instance, at the surface S_o. However, in studying the up-going reflected wave, which appears at the reflector, it is natural to change slightly eq. 3.69. With this purpose in mind, we make use of the second Green function $G_1(q,p)$, given by eq. 3.36. The latter gives the relationship in the frequency domain

$$P(p,\omega) = \frac{1}{2\pi} \int\limits_{S_o} P(q,\omega)\, \frac{\partial}{\partial z_q} \frac{e^{-ik_1 R}}{R}\, dS\,,$$

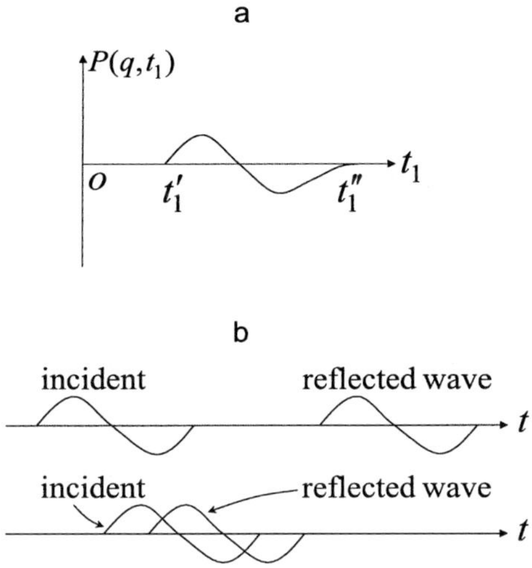

Figure 3.10: (a) Pressure at the receiver (b) Relationship between the incident and reflected waves.

which almost coincides with eq. 3.41.

Applying Fourier's transform, we obtain

$$P(p,t) = -\frac{1}{2\pi}\frac{\partial}{\partial z}\int\limits_{S_o}\frac{P(q,t_1)}{R}\,dS \tag{3.70}$$

where

$$t_1 = t + \frac{R}{c_1} \tag{3.71}$$

Of course, the same result follows from eq. 3.61. Equation 3.70 is more preferable for our purpose. Indeed, the reflected wave propagates upward and reaches a receiver on the surface S_o later than points inside the volume V. For illustration, consider an arbitrary point p of the medium and suppose that the pressure at the receiver varies as is shown in Fig. 3.10a.

Then in order to calculate its contribution to the pressure $P(p,t)$ at point p and at instant t, we have to define this function at the receiver when

$$t_1 = t + \frac{R(q,p)}{c_1}$$

For example, inequalities

$$t_1 < t_1' \qquad \text{and} \qquad t_1 > t_1''$$

correspond to cases when either the wave has not yet arrived at point p or has already passed it. Therefore,

$$P(q, t_1) = 0\,,$$

and the integrand at such instances is equal to zero.

By definition, eq. 3.70 allows us to perform a downward continuation, that is we can determine a transient wave as a function of time at different points of a homogeneous medium between the observation plane and the reflector.

Introduction to migration

Now we demonstrate how knowledge of the pressure inside the layer permits us to find a position of the reflector and, moreover, evaluate to some extent acoustic parameters of a medium. Let us assume that the pressure of the incident and reflected waves is known at each point of the upper layer and describe some features of these waves, that help to determine the reflector position. First, taking into account the fact that the reflected wave arises at the interface it is natural to treat the latter as a system of secondary sources (scatterers). Each of them gives rise to an elementary wave with an infinitesimal amplitude. At the same time, a superposition of this unlimited number of waves produces a wave with a finite magnitude, (Huygens principle).

In particular, the pressure at any receiver is, in principle, a sum (integral) of pressures, caused by all scatterers. This presentation does not contradict the fact that at the high-frequency spectrum the wave, arriving at the receiver, is mainly due to sources, situated around the stationary point of the reflector. As we already know, the influence of other scatterers becomes negligible because of the destructive interference.

Next, proceeding either from the Kirchhoff theory of diffraction or the ray theory, let us pay attention to the wave behavior in the vicinity of a reflector. At the beginning the first approach found its application when wave diffraction was studied behind screens, provided that dimensions of the opening greatly exceeded the wavelength, (Part I). In such the case, Kirchhoff assumed that a strength of the secondary source, (scatterer), is directly proportional to the incident wave at the same point and that both of them change synchronously. In the frequency domain this means that the phase shift between them is equal either to zero or π.

Of course, in general, the magnitude of the secondary wave at some point of the reflector is defined by all elementary sources. However, with an increase of the frequency the main contribution comes from the nearest scatterers. In other words, the reflected and incident waves vary in the same manner at the interface, and, as was pointed out earlier, this feature is of great importance. A similar conclusion follows from the principle of locality (Chapter 2) in accordance to which the wave moves along the ray tubes and at points of the reflector (eq. 3.53)

$$P_s(p,\omega) \;=\; \alpha\, P_i(p,\omega), \tag{3.72}$$

where $P_s(p,\omega)$ and $P_i(p,\omega)$ are the pressure of the incident and reflected waves.

As was already mentioned, α is the reflection (Fresnel) coefficient, and it is a function of the incident angle and impedances of a medium. It is essential that $\alpha(q)$ is independent of the frequency. Certainly, such a relationship between both waves is an approximate one, and it becomes more accurate with a decrease of the wavelength. Respectively, at the high-frequency spectrum, eq. 3.72 can be treated as the criterion, that allows us to determine a reflector position. First assume that the pressure, $P(q,\omega)$, is known at receivers. Then, performing a down continuation of the reflected wave, as well as a calculation of the incident wave, we can evaluate at each point the phase difference between them and the ratio $P_r(q,\omega)/P_i(q,\omega)$. This information defines a location of the reflector as a locus, where the phase shift is absent. Besides there may exist other surfaces with null phase, that is fictitious reflectors. However, unlike the real boundary, their location varies with a frequency, and therefore it is possible to distinguish the actual reflector. Taking into account this fact, the quantity

$$\int \frac{P_s(p,\omega)}{P_i(p,\omega)}\, d\omega \tag{3.73}$$

can be applied to determine a reflector position. Indeed, if an observation point p is located above the reflector, there is in general a phase shift between these waves and, correspondingly, the integral is small. On the other hand, at the boundary this quantity, eq. 3.73, is relatively large, since regardless of the frequency $P_s(p,\omega)$ and $P_i(p,\omega)$ change synchronously. In reality, in place of eq. 3.73, the so-called migrated image $m(p)$ is used, and it is defined as

$$m(p) = \frac{1}{2\pi}\int F(\omega)\,\frac{P_s(p,\omega)}{P_i(p,\omega)}\, d\omega \tag{3.74}$$

Here $F(\omega)$ is a function, (filter), that allows us to consider waves within a limited band of frequencies. Note that a theoretical foundation for the parameter $m(p)$ is the well known imaging principle, which states that reflectors exist in such places where the down-going incident wave is coincident with the up-going wave. We see that it corresponds exactly to the Kirchhoff approximation and also follows from the principle of locality, (Chapter 2).

As was indicated earlier, eq. 3.72, the ratio of amplitudes at points of the boundary defines the reflection coefficient of the plane wave, and it is the same for any frequency, as soon as the ray theory is applied. Then, calculating the incident angle, we can evaluate the impedance of the medium beneath the reflector. In essence, it is already an important step for solving the inverse problem.

In the same manner we can determine the position of the reflector when waves are transient. For illustration consider some point p of the medium and, first, suppose that it does not belong to the interface S_r. It is clear that in such a case the incident wave arrives earlier at this point than the reflected wave and two possible situations are shown in Fig. 3.10b. There is always the time interval, when the reflected wave is absent and, therefore, the ratio $P_s(p,t)/P_i(p,t)$ is equal to zero. Within the interval, where both waves are present, this ratio varies with time. In contrary at points of the reflector the ratio $P_s(p,t)/P_i(p,t)$ is independent of time and, as we already know, it defines the

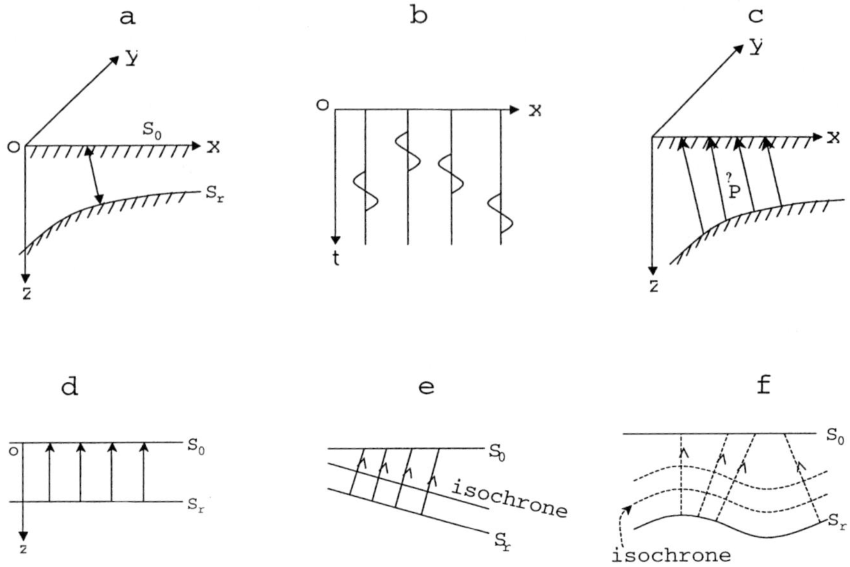

Figure 3.11: (a) Zero offset (b) Wavelets (c) Rays of reflected wave (d), (e), (f) Rays and isochrones of reflected wave

reflection coefficient. This analysis shows that the integral

$$\int P_s(p,t)P_i(p,t)\,dt \tag{3.75}$$

may serve as the integrated image for transient waves. Of course, performing Fourier's integral we can again use the parameter $m(p)$. Certainly, a transition from synchronous to nonsynchronous behavior occurs gradually, and this factor makes an influence on an accuracy with which the reflector position is defined.

Until now the main attention was paid to a relationship between the reflected and incident waves and its role for a determination of the reflector. Next we outline some basic principles of Kirchhoff migration. We start from the case of the simplest array and assume that the upper layer is homogeneous and isotropic.

Source and receiver at the same point

First suppose that at some point of the plane S_o the pressure, $P(t)$, is measured with the single combined array, (zero offset), when the primary source and receiver are situated at the same point, Fig. 3.11a. As we know, the incident wave, caused by the source, reaches the interface S_r, where the reflected wave appears. In accordance with Snell's law, the latter returns to the receiver along the same path, that is rays of the incident and reflected waves coincide with each other, and they are normal to reflector S_r. Performing these measurements at different points on the plane S_o, we obtain the pressure distribution due to the up-going wave, and it is represented in the form of wave traces, Fig. 3.11b. Here t is the time of observation, counted from the instant, when the corresponding primary source was turned on. New wave field, $P(q,t)$, is the result of a combination of measurements with single arrays, and it is difficult to reproduce such field. For instance, if we imagine that all primary sources simultaneously generate incident waves, then each receiver would measure reflected waves arriving along different rays.

Because of such a system of arrays the wave field, $P(q,t)$, has several remarkable features. First of all, at any point of the upper layer the wave arrives only along the line, (ray), perpendicular to the reflector and passing through a corresponding receiver, Fig. 3.11a. Second, for each single array a time of wave traveling from the primary source to the reflector is equal to that in the opposite direction. In other words, we can say that pressure $P(q,t)$, measured by the receiver at the moment t, is defined by the scatterer of the reflector at the instant, $t/2$. Therefore, the distance between the reflector and the primary source along the ray is equal to

$$\ell = \frac{ct}{2} \tag{3.76}$$

It may be proper to notice that secondary sources, (scatterers), arise at different times, if we assume that all primary sources are turned on simultaneously, Fig. 3.11c.

Inasmuch as rays of the wave field are perpendicular to the reflector, the latter should coincide with the wavefront or isochrone. However, there is a serious contradiction. In fact, at each point of the interface, S_r the reflected wave arises at the same time as its source. Because scatterers appear at different moments, $t/2$, the reflector cannot be treated as an isochrone, even if rays are normal to S_r. This emphasizes the artificial nature of the wave field $P(q,t)$. In order to overcome this discrepancy, we will do the following. In reality, the reflected wave arises at the instant $t(q)/2$ and, propagating with the velocity c, arrives at the receiver at the moment t. Instead of this let us assume that each scatterer appears at the instant $t = 0$, but its reflected wave moves at half the velocity, $c/2$. Correspondingly, as before, its time of arrival at the receiver is equal to t. Such a wave travels twice longer along the path ℓ, given by eq. 3.76. This change has not effect on a ray geometry, nor on the time of observation at the plane S_o, but there is now a consistency between a ray direction and the geometry of isochrones. In particular, the reflector can be treated as the isochrone when $t = 0$. In other words, this transformation allows us to assume that the reflected wave arises simultaneously at all points of the boundary S_r at the instant $t = 0$.

For illustration, consider several examples. First suppose that both surfaces S_o and S_r are parallel to each other, Fig. 3.11d. In this simplest case the observation plane S_o coincides with the isochrone and rays are directed along the z-axis. Then, the layer thickness is

$$h = \frac{c\,t}{2}$$

Of course, the reflector, as well as other horizontal planes, is an isochrone.

In the next case, the reflector is the dipping plane, Fig. 3.11e, but unlike S_r, the observation plane does not coincide with the isochrone. It is obvious that all of them are parallel to the reflector. A more general case is given in Fig. 3.11f. In the vicinity of the reflector isochrones have almost the same shape as S_r, but with an increase of distance it changes.

Now, knowing the main features of the wave field, measured by a system of combined arrays, we outline several procedures of migration, that allow us to determine the reflector location. Some of them are described only to illustrate a behavior of the pressure above the interface S_r. First of all we can proceed from eq. 3.70 and determine the pressure $P_s(q,t)$ at different points of the upper layer. Then comparison with the incident wave, eq. 3.75, defines a surface S_r, where the migrated image has a maximal value.

The same task can also be solved using the concept of scatterers, arising at the surface S_r simultaneously at instant $t = 0$. This approach is based on the fact that rays are normal to the reflector. Letting in eq. 3.70 the velocity of propagation, equal $c/2$ we obtain

$$P(p,t) = -\frac{1}{2\pi}\frac{\partial}{\partial z}\int\limits_{S_o} P(q,t_1)\frac{dS}{R} \tag{3.77}$$

Here

$$t_1 = t + \frac{2R}{c}$$

By definition, the function $P(p,t)$ describes the pressure of the up-going reflected wave in the upper layer. Correspondingly, the use of eq. 3.77 gives us information about this field at any instant of time, and, in particular, the position of a wavefront. Then, we can draw from each receiver a line, (ray), that is normal to the front and plot a segment with the length, equal to $c\,t_1/2$. It is obvious that low ends of these elements form the reflector surface. However, there is another procedure of migration, which is certainly simpler. Assuming that $t = 0$ eq. 3.77 becomes

$$P(p,0) = -\frac{1}{2\pi}\frac{\partial}{\partial z}\int\limits_{S_o} P\left(q,\frac{2R}{c}\right)\frac{dS}{R}, \tag{3.78}$$

and it defines the pressure beneath the observation plane at the first moment only. Bearing in mind that the scatterers appear exactly at this instant, we conclude that the pressure $P(p,0)$ is equal to zero everywhere except at points of the reflector. For this reason a determination of its position is a straightforward task, and it is illustrated in Fig. 3.12a. As we see all points of the line $z = z_1$ and between x_1 and x_2 are situated at a finite distance from the boundary S_r and, respectively, the pressure $P(x,y,z,o)$ is equal to zero. The same behavior is observed along the profile $z = z_2$, but the line

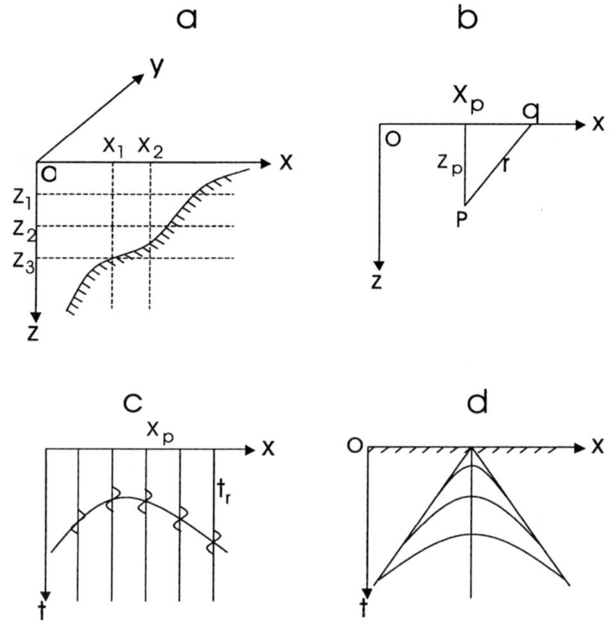

Figure 3.12: (a) Illustration of eq. 3.78 (b) Scatterer at the point p (c) Diffraction curve (d) System of hyperbolas.

$z = z_3$ intersects the surface S_r at the point with coordinates, x_1, y_1 and z_3, where the pressure has a nonzero value. Repeating these calculations along lines: $z = \text{const}$, located at different distances from S_o, we are able to find secondary sources, (scatterers) and, hence, the reflector position. It may be proper to emphasize that eq. 3.77 allows one to perform Kirchhoff migration for the combined array only, because in this case the wave field behaves as if its sources start to act simultaneously. For instance, such an approach cannot be used when the array consists of one common source and a system of receivers. This is related to the fact that the time of traveling of the incident wave from the primary source to the reflector and the corresponding time for the reflected wave are, in general, not equal to each other. As will be demonstrated below, there are at least two more approaches, based on the use of eq. 3.70, that also allow us to find a reflector position. Unlike the last one, they do not require that secondary sources arise at the interface S_r at the same time $t = 0$.

For illustration, consider a procedure of migration in the frequency domain, which is similar to that for transient waves. Making use of the Helmholtz formula we can calculate the complex amplitude of the pressure at any point of the upper layer and, in particular, its phase. Earlier, it was pointed out that the wave field, $P(q, \omega)$, is rather artificial, and it is not surprising that rays are not usually perpendicular to phase surfaces. Suppose that the phase shift between oscillations in the primary source and receiver, located at

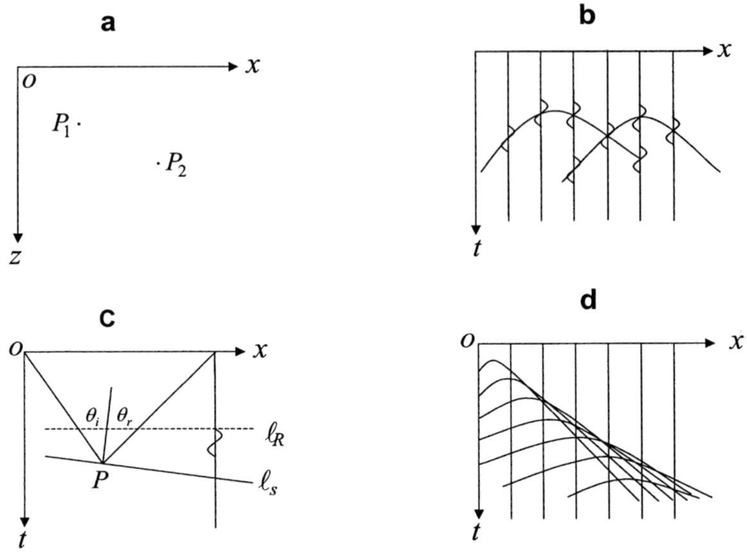

Figure 3.13: (a) Two scatterers (b) Superposition of two sets of wavelets (c) Model of the reflector, l_s, and record horizon l_r (d) System of diffraction curves.

the same point, is equal to φ. It appears due to a wave motion from the source to the reflector and back to the receiver. Then we have

$$P_s = P_o \sin(\omega t - \varphi) = P_o \sin \omega \left(t - \frac{2\ell}{c}\right) \tag{3.79}$$

As before, ℓ is the ray length, and therefore the phase of reflected wave at the boundary S_r is a function of a point, and it is equal to

$$\frac{\varphi}{2} = \frac{\ell(q)}{c}\omega \tag{3.80}$$

By analogy with the case of transient waves, let us assume that the velocity of propagation is two times smaller, but at each point of the reflector the phase of the secondary wave is zero. It is obvious that such a transformation does not change a phase shift between the primary source and the reflected wave in the receiver, but the phase surfaces become normal to rays. Then, the Helmholtz formula eq. 3.70 becomes

$$P(p,\omega) = -\frac{1}{2\pi}\frac{\partial}{\partial z}\int\limits_{S_o} P(q,\omega)\frac{e^{-2ik_1R}}{R}\,dS \tag{3.81}$$

If one of these surfaces is known, it is a simple matter to find a reflector position. For instance, let us draw a line, which is perpendicular to the phase surface and plot the segment, equal to

$$\ell = \frac{\varphi c}{2\omega} \tag{3.82}$$

Its low end defines the point of the reflector. The same procedure for other receivers defines the reflector position.

In principle, we can solve our task differently. Calculating the pressure in the vicinity of some point one can find derivatives of the phase in the horizontal and vertical direction. The latter gives the gradient of the phase, that is a ray direction. Making use of eq. 3.80 we find a reflector position.

Scatterers and diffraction curves

Now we describe some features of migration, that follow from the concepts of the diffraction point and diffraction surface. This approach is based on the principle of superposition and an assumption that the reflector can be treated as a place where sources (scatterers) of the reflected wave are located. For simplicity, we restrict ourselves to the two-dimensional case and consider several examples.

One source

First, suppose that there is a single scatterer, situated at the point p with coordinates $x_p, 0, z_p$, Fig. 3.12b. It starts to generate the reflected wave at instant t_{op}, when the incident wave arrives at point p. Then the time of arrival of the reflected wave at receivers, located along the x-axis, is defined from the equation

$$t(x) = t_{op} + \frac{r_{pq}}{c} \qquad \text{or} \qquad t(x) = t_{op} + \frac{\sqrt{(x-x_p)^2 + z_p^2}}{c} \tag{3.83}$$

As is well known, the latter describes the hyperbola, and it is called the diffraction curve, Fig. 3.12c. Its apex has the same coordinate x_p as the scatterer. Also, wavelets are shown in the vicinity of points where the hyperbola is intersected by the vertical traces. Inasmuch as receivers are situated at different distances, r, from the source, wavelets are shifted in time with respect to each other. With an increase of distances r, the maximal magnitude becomes smaller, and it is related to the influence of the spreading factor. Nevertheless, it is assumed that a shape of waves, including a duration, remains the same.

Now let us mentally translate all wavelets measured along the observation line at the apex of the hyperbola and add them together. Since this procedure has a constructive character we obtain the wavelet with a rather large magnitude. In fact, the pressure, as a function of time, can be written in the form

$$P(p,q,t) = P_o(q)\, f\{a\,[t - (t_{op} + t_{pq})]\} \tag{3.84}$$

Here t is a time of observation and t_{pq} is a time interval during which the wave travels from the scatterer to the receiver.

By definition, the function $P(p, q)$ is equal to zero when

$$t < t_{op} + t_{pq}$$

In this light, note that "translation" means a decrease of the observation time, t, by the value equal to t_{pq} or r_{pq}/c. Correspondingly, the argument of the pressure becomes equal to

$$a(t - t_{op})$$

regardless of the receiver position. This is why the constructive interference takes place, and it is follows because the incident wave and all wavelets change synchronously. For instance, all secondary waves appear at instant $t = t_{op}$, when the incident wave arrives at the point p. In contrary, if this summation is performed for any other point of a medium, then the destructive interference is observed. Such behavior can be expected because in this case a translation of wavelets is not governed by the separation between the receiver and the scatterer, r_{qp}, and, respectively, they are shifted with respect to each other.

It is clear that the same process of summation is carried out in the Kirchhoff formula, describing a migration. For instance, in both cases terms of such sums are situated at a hyperbola. However, there is also a strong difference, since, unlike this equation, the influence of the ratio $\Delta S/2\pi R$ is not taken into account and differentiation with respect to z is absent. In this light, we note the following. With an increase of the distance from the scatterer, the reflected wave becomes smaller. However, a multiplication by the factor $1/R$, eq. 3.70, may produce an opposite impression, namely, in approaching the source, that is point p, the wave field decreases. This confusion is easily removed, since calculation of the pressure requires integration, and the main contribution is defined by the area, which grows more rapidly than R.

By definition, in a homogeneous medium the diffraction curve (hyperbola) is observed for the zero offset and an array with a common source, and, in accordance with eq. 3.83, its shape essentially depends on the position of the scatterer along the z-axis. For instance, if it is located in the vicinity of the observation surface, $(z_p \to 0)$, the hyperbola is transformed into the straight lines

$$t(x) = t_{op} + \frac{x - x_p}{c} \qquad \text{if} \qquad x > x_p$$

and (3.85)

$$t(x) = \frac{x_p - x}{c} + t_{op} \qquad \text{if} \qquad x < x_p$$

With an increase of the depth, z_p, the hyperbola shape changes, and it becomes wider, as shown in Fig. 3.12d. At the same time, the straight lines, given by eq. 3.86, serve as asymptotics of this family of hyperbolas. As follows from eq. 3.83 the derivative of the function $t(x)$ is

$$\frac{dt(x)}{dx} = \frac{x - x_p}{c\sqrt{(x - x_p)^2 + z_p^2}}, \tag{3.86}$$

and it varies within the range

$$0 \leq \frac{dt(x)}{dx} < \frac{1}{c} \tag{3.87}$$

Thus, at the beginning, $(x \sim x_p)$, the derivative is very small and then, with an increase of the distance $(x-x_p)$, it asymptotically approaches to $1/c$. Such behavior is not difficult to explain. In fact, the wave arrives almost simultaneously at receivers, located close to point $x = x_p$, and it corresponds to the flat part of the hyperbola. In contrast, at large distances, $x - x_p > z_p$, the shift in time between two receivers becomes more noticeable and tends to the value

$$\Delta t = \frac{x_2 - x_1}{c}, \tag{3.88}$$

where x_1 and x_2 are coordinates of observation points.

Further we will see that these features of the diffraction curve are useful for understanding the constructive and destructive interference of elementary waves, arising at the reflector.

Two sources

Next suppose that there are two scatterers, located at different distances from the primary source, Fig. 3.13a. Then, diffraction curves for each of them are

$$t(x) = t_{op_1} + \frac{\sqrt{(x - x_{op_1})^2 + z_{p_1}^2}}{c}$$

and (3.89)

$$t(x) = t_{op_2} + \frac{\sqrt{(x - x_{op_2})^2 + z_{p_2}^2}}{c}$$

Assuming, as before, that wavelets, caused by scatterers, are known, we perform their superposition and obtain the resultant wave at each receiver. In other words, the forward problem is solved in this very simple case. If the scatterers are located at sufficiently great distances wavelets generated by them are almost separated in time, Fig. 3.13b, that is they remain the same as in the case of the single scatterer. An exception is the wavelet in the receiver, located at the line, where hyperbolas are intersected. As in the first example, a translation of wavelets to apexes of hyperbolas and their summation produce unlike other points, a wavelet with a rather large magnitude. Of course, these points are directly visualized from a comparison of waves traces. With a decrease of the separation, p_1p_2, Fig. 3.13a, we begin to observe an overlapping of wavelets. This becomes especially noticeable when a duration of wavelets greatly exceeds a time of traveling between p_1 and p_2, that is the dominant wavelength is greater than the separation of scatterers. Certainly, these two factors influence on the accuracy with which a position of these sources is determined.

Reflector as a system of scatterers

Next consider a general case when the reflected wave arises at the interface S_r. In accordance with Huygens principle each point can be treated as a source, (scatterer), of the secondary elementary wave, traveling toward the observation surface S_o. Thus, strictly speaking, any receiver measures all these waves, which usually arrive at different times and may form an almost unlimitedly long trace. However, as we already know, a completely different picture is observed in the high-frequency spectrum, and it happens due to remarkable features of an interference of elementary waves. Let us briefly remind them. First of all, it turns out that the reflected wave at each receiver is practically caused by only scatterers around some point p of the reflector, where Snell's law is valid, Fig. 3.13c. Moreover, this wave behaves as a spherical one, and it arrives at the observation point q at the instant

$$t(q) = t_{op} + t_{pq} \tag{3.90}$$

Therefore, an interference of elementary waves, generated by sources in the vicinity of the point p, (the first Fresnel zone), has a constructive character, and, correspondingly, the wave energy travels inside the ray tube from the point p to q, Fig. 3.9c. In contrast, a superposition of waves due to all other scatterers is destructive and their influence is neglected. As result of both types of an interference each wave trace, (wavelet), starts at the instant $t(q)$, eq. 3.90, and its duration is defined by that of the primary source, located at the point O. Of course, it may happen, that the reflector has several points, where Snell's law is valid for the same receiver. Then a wave trace consists of a group of wavelets. Thus, proceeding from assumptions of the geometrical acoustics we were able to solve the forward problem, that is, to determine the wave field, (Chapter 2).

The same result can be obtained differently, and with this purpose in mind we make use again diffraction curves and, as usual, consider the zero offset and a homogeneous medium. It is convenient to treat the reflector as a system of scatterers, located sufficiently close to each other. For example we assume that the distance between them $\Delta\ell$ satisfies the condition

$$\frac{\Delta\ell}{c} \ll T, \tag{3.91}$$

where T is the wavelet duration.

Of course, with an increase in the number of scatterers this representation becomes more accurate. Let us draw in the x, t plane hyperbolas for each scatterer, and imagine the corresponding wavelets, Fig. 3.13d. Then, performing a superposition of these waves we obtain the resultant wave at each receiver as a function of time. Thus, if elementary waves are known, then using the principle of superposition we are able again to solve the forward problem. After multiplication of the time by $c/2$, the system of curves, shown in Fig. 3.13d, has several important features. First of all, the locus of the hyperbola apexes represents an image of the reflector. Also we see that the family of these hyperbolas has an envelope and, in general, apexes do not belong to this line. An exception is the case

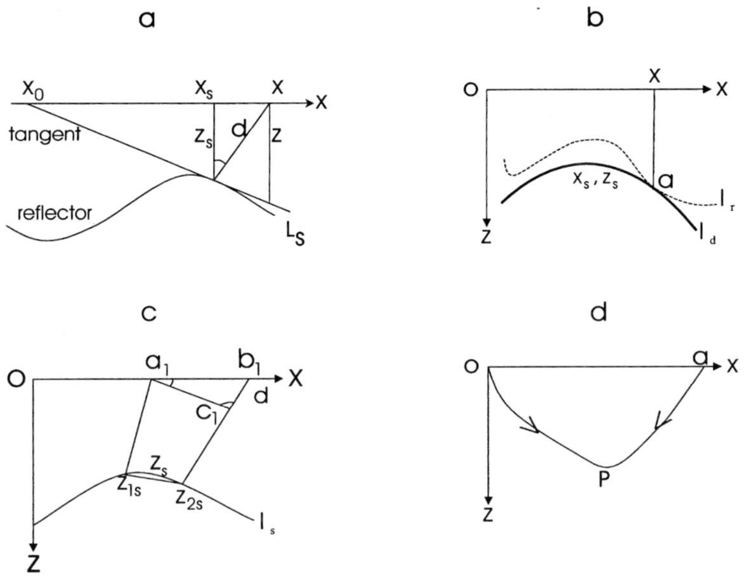

Figure 3.14: (a) Illustration of eqs. 3.93 – 3.96 (b) Record line and diffraction curve (c) Slope of the record line (d) Ray tracing

when the reflector is the horizontal plane and observations are performed with the zero offset.

Finally, it turns out that this envelope is the place of the constructive interference of elementary waves. In other words, it coincides with the record line, ℓ_r, that is at points, where a hyperbola touches this line, both have the same tangent. To prove this fact, suppose that the primary source and receiver are situated at the same point and this array moves along the x-axis, Fig. 3.14a. It is convenient to introduce the following notations: x_s, z_s are coordinates of the scatterer and

$$z_s = f(x_s) \tag{3.92}$$

describes the reflector line, ℓ_s. Also x is the coordinate of the receiver, that measures the reflected wave, arriving from the vicinity of the point x_s, z_s. Such specular wave arises due to the constructive interference of elementary waves, and their sources are located at the reflector around the point with coordinates x_s, z_s. In accordance with Snell's law, the ray of this wave is normal to the reflector, while its tangent intersects the x-line at the point x_0. First, we derive the equation of the record line, ℓ_r, where the interference is constructive.

The relationship between time t_r and the coordinate x directly follows from Fig. 3.14a.

In fact, the time of arrival of the reflected wave is

$$t_r = \frac{2d}{c} = \frac{2z\cos\theta}{c}, \tag{3.93}$$

where θ is the angle formed by the tangent and the x-axis, and as is seen from Fig. 3.14a:

$$z = (x - x_o)\,\tan\theta \tag{3.94}$$

It is obvious that with a change of the receiver position both the coordinate z and the angle θ vary.

In order to find a slope of the record line, ℓ_r, Fig. 3.14b, let us choose two points of the reflector, z_{1S} and z_{2S}, located close to each other, as shown in Fig. 3.14c, and draw rays from these points. Their intersection with the observation line ox gives points a_1 and b_1. Consider the triangle $a_1b_1c_1$, where the position of point c_1 satisfies the condition

$$z_{2s}\,c_1 \;=\; z_{1s}\,a_1 \tag{3.95}$$

As follows from this triangle

$$\frac{\Delta x}{\sin\alpha} = \frac{\Delta d}{\sin\theta} \qquad \text{or} \qquad \frac{\Delta d}{\Delta x} = \frac{\sin\theta}{\sin\alpha},$$

and, as before, θ characterizes the direction of the tangent to the reflector at some intermediate point z_s, $\Delta d = c_1b_1$.

By definition, we have for the distance difference along two rays (zero offset)

$$\Delta d = \frac{c\,\Delta t_r}{2}$$

Hence the previous equality becomes

$$\frac{\Delta t_r}{\Delta x} = \frac{2\sin\theta}{c\sin\alpha}$$

In the limit, when $\Delta x \to 0$, we have $\alpha \to \pi/2$ and, correspondingly

$$\frac{\partial t_r}{\partial x} = \frac{2\sin\theta}{c} \tag{3.96}$$

Next consider the diffraction curve, ℓ_d, related to the scatterer with coordinates x_s, z_s, Fig. 3.14b. As is well known the time of arrival of the diffraction wave is

$$t_d(x, x_s) = \frac{2\,r}{c} = \frac{2\sqrt{(x - x_s)^2 + z_s^2}}{c} \tag{3.97}$$

Correspondingly, its slope is defined as

$$\frac{\partial t_d}{\partial x} = \frac{2(x - x_s)}{\sqrt{(x - x_s)^2 + z_s^2}} = \frac{2\sin\theta}{c} \tag{3.98}$$

Therefore, from comparison of eqs. 3.96 and 3.97 it follows that at the point where the record line, ℓ_r, and the hyperbola, $(t_r = t_d)$, touch they have the same slope. In other words, we demonstrated that the record line is the envelope of diffraction curves. For instance, changing a position of the scatterer, that is its coordinates, x_s, z_s, we obtain a new hyperbola, which touches the envelope at a different point. Of course, there is unlimited number of diffraction curves, which intersect the trace $x = \text{const}$, but only one touches the record line. It is essential that the apex of this hyperbola corresponds to the certain scatterer. It is located within the part of the reflector, that defines the reflected wave, observed by the receiver with the coordinate x. This fact may serve as a foundation of the graphical method of migration, described in the previous section. As was shown in the previous section, drawing the hyperbola, which is tangential to the record line at any point a, we determine a position of its apex. Repetition of this procedure for other points of the reflecting line gives the image of the reflector in x, t plane. Since the apex and the corresponding scatterer have the same coordinate, x_s, the use of eq. 3.97 allows us to find the location of the reflector, ℓ_s, itself. It may be proper to notice that in the three-dimensional case we deal with the record surface and the diffraction hyperboloid, but the procedure of migration in principle remains the same.

Preserving the main concept, this graphical approach can be replaced by a much more efficient one, which in essence, was described for cases of the one and two scatterers. As before, let us choose an arbitrary point in the upper layer and assume that it belongs to the reflector, that is there is a scatterer at this place. Then its wavelets, corresponding to different traces, are located around the same hyperbola. Inasmuch as coordinates of the possible source are known, it is a simple matter to draw this diffraction curve in the x, t plane. At the points, where the hyperbola intersects traces, we take values of the pressure and perform their summation. If the scatterer is really located at the chosen point, then the sum is rather large, while in the other case, because of the destructive interference, it is much smaller. Repetition of this procedure for different points allows us to determine a position of the reflector. Certainly, the use of this approach gives a strip, and its upper boundary coincides with the reflector, but whose width is defined by a wavelet duration.

Ray tracing

Until now we supposed that the upper layer is homogeneous and diffraction curves are hyperbolas. Next assume that the velocity of propagation is an arbitrary function of coordinates. In such a complicated case, proceeding again from the concept of scatterers, we describe some features of ray tracing, which permits us to find the reflector location. Consider the array with the common source at the point O and a system of wave traces for different receivers, Fig. 3.14d. By analogy with the case of a homogeneous layer, let us take some point p beneath the observation surface. From the primary source O we draw a ray, passing through the point p. Inasmuch the medium is, in general, inhomogeneous, the ray may have a rather complicated shape. Also the time of wave traveling between these points, t_{op}, is evaluated. The second step is a ray plotting from the point p to q, where receivers are situated, as well as calculation of time t_{pq}, during which the wave travels along these paths. Thus, we know the total time

$$t = t_{op} + t_{pq} ,$$

and this allows us to find corresponding values of the pressure at each trace.

Now, performing a summation of these pressures we obtain a relatively large number, (constructive interference), provided that the point p belongs to the reflector. In the case when the point p is above the reflector the sum is small. Certainly, this approach is a generalization of the method of diffraction surfaces. Bearing in mind that it is necessary to repeat the ray tracing for many points, it is clear that this approach requires an extremely long time. However, with an increase of a speed of operations and a memory of computers it becomes more and more available. Besides, a knowledge of ray tubes of the down and up-going waves allows us to find amplitudes of these waves at points of the reflector and therefore to estimate the reflection coefficient. In this light, it is important to locate places where rays may touch caustics, and a diffraction phenomenon is observed.

Method of ellipses and the Kirchhoff formula for forward modeling

Earlier we have demonstrated that Kirchhoff's integrals permit us to find the pressure, $P(p,t)$, at any point of the upper layer, provided that it is known either at the reflector or the observation surface. In other words, we can solve both the forward and inverse problems. For this reason, it is natural to expect that these formulas may serve as a theoretical foundation of the method of ellipses described in the previous section. In order to establish a relationship between them, let us review briefly this graphical approach of migration. Suppose that the upper layer is homogeneous and receivers are located at different distances from the primary source of the incident wave, Fig. 3.15a. Also, examples of wave traces for several receivers are shown in Fig. 3.15b, and, as is well known, they arise due to the constructive interference.

As before, we assume that a shape of wavelets everywhere remains the same. In the section 3.1 it was shown that a surface of equal reflection time, corresponding to the wave, arriving at some receiver, is an ellipsoid. In particular, its intersection by the vertical plane, passing through an array, gives an ellipse, so that the primary source and the receiver are situated at its foci, Fig. 3.15c. The surface of equal reflection time can be treated as a possible reflector for the given source and receiver, and it has several important features, related to each other.

By definition, the time of wave traveling along straight lines, connecting primary source with any point of such surface and the latter with the receiver, is constant. In accordance with Fermat's principle, (Chapter 1), this means that all these lines can be rays along which the incident wave moves toward the ellipsoid and then, after a reflection, the wave arrives at the receiver. From this fact follows the second remarkable feature of the surface of equal reflection time, namely, at each point the incident and reflection angles are equal to each other (Snell's law). It may be proper to notice that even in a general case, when the velocity of propagation varies and this surface becomes much more complicated, Snell's law still remains valid.

From the physical point of view it is clear that at each point of such surface the incident and possible reflected wave change synchronously. In other words, the imaging principle of migration takes place, and this is the third important feature of ellipsoids. Finally, at least one point of the surface belongs to the real reflector, because the total time of traveling both the incident and reflected waves is the parameter of the surface of equal reflection time.

Now, following the method of ellipses, we plot these surfaces for each pair of the source-receiver pair and determine a position of the reflector, as an envelope of such surfaces. The same approach is applied to the case in which the primary source and receiver move along the observation lines with the constant separation, Fig. 3.15d. Again we obtain a system of ellipsoids, as well as their envelope. For example, in the case of zero offset if the upper layer is homogeneous, the surfaces of equal reflection time are spherical. Intuitively it is clear that the envelope of surfaces of equal reflection time coincides with the reflector, but it still requires a prove.

Case of the zero offset In order to accomplish this task consider, for simplicity, the plane xoz, where the line of equal reflection time is the circle, and the array is located at its center. As we know, at all points of the circle possible rays are normal to this line, (Snell's law), and the imaging principle holds. Of course, the same behavior occurs at points of the reflector. In other words, these two features do not allow us to select the point of the line of equal reflection time, which also belongs to the real reflector

$$z = f(x) \tag{3.99}$$

Fortunately, there is another factor, the slope of the circle, $(\tan\theta)$, and it changes along the line as

$$-\infty < \tan\theta < \infty$$

Here θ is the angle formed by the tangent of the circle and the x-axis.

Then, it seems obvious that there is some point of the line where its slope coincides with that of the reflector. Correspondingly, it suggests that the envelope of circles represents the reflector. Let us discuss this question in some detail. First consider the reflected wave, that arrives at some observation point x_o at the instant t. As is well known, this wave appears, due to the constructive interference, at certain element of the reflector, and in order to determine its position we make use of the Kirchhoff integral. Certainly, the point which characterizes such an element, also belongs to the circle with the center at the point x_o, because in both cases the traveltime of the wave is the same, and it is equal to t.

Earlier we demonstrated that the complex amplitude of the pressure $\mathcal{P}(x_o, \omega)$ can be represented in the form

$$\mathcal{P}(x_o, \omega) = \int F(x, z, \omega)\, e^{ikr}\, dx\,, \qquad (3.100)$$

where x and z are coordinates of a reflector point and

$$r = \sqrt{(x - x_o)^2 + z^2(x)} \qquad (3.101)$$

Since the reflector element, $d\ell$, shown in Fig. 3.16a, is characterized by the stationary point of the integrand in eq. 3.100, we have

$$\frac{\partial r}{\partial x} = 0 \qquad \text{or} \qquad \frac{(x - x_o) + z\dfrac{dz}{dx}}{\sqrt{(x - x_o)^2 + z^2}} = 0\,,$$

that is

$$(x - x_o)\, dx + z\, dz = 0 \qquad (3.102)$$

Here dx and dz are scalar components of the vector

$$\mathbf{d\ell} = dx\,\mathbf{i} + dz\,\mathbf{k}\,,$$

which defines an orientation of the reflector at some point with coordinates x and z.

As is seen from Fig. 3.16a the radius vector $\mathbf{r}$ is

$$\mathbf{r} = (x - x_o)\,\mathbf{i} + z\,\mathbf{k}\,, \qquad (3.103)$$

and, by definition, it is normal to the tangent at each point of the circle.

Hence, eq. 3.102 can be rewritten in the form

$$\mathbf{r} \cdot \mathbf{d\ell} = 0\,, \qquad (3.104)$$

that is, the reflector element coincides with the tangent of the circle. This allows us to conclude that the reflector is an envelope of the family of curves of equal reflection time. Of course, the same result remains valid for three dimensional case and an arbitrary type of array.

It is also instructive to write eq. 3.102 as

$$x - x_o + z\,\frac{dz}{dx} = 0 \qquad \text{or} \qquad x = x_o + z\tan\theta \qquad (3.105)$$

From Fig. 3.16a it follows that

$$z\tan\theta = z\,\frac{\sin\theta}{\cos\theta} = \frac{c\,t}{2}\sin\theta$$

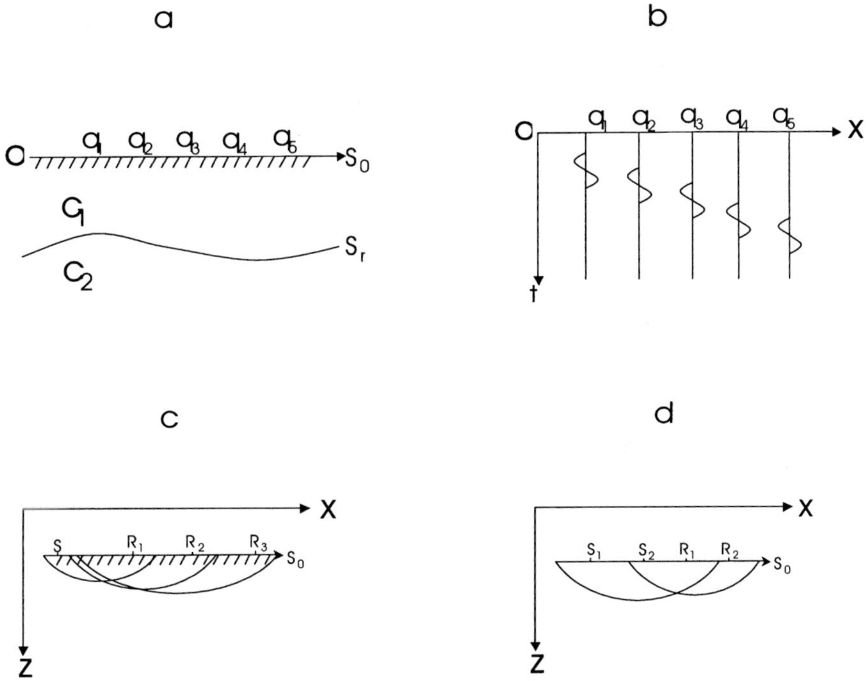

Figure 3.15: (a) Model of a medium (b) Wave traces (c) Array with one source (d) Array with a constant separation.

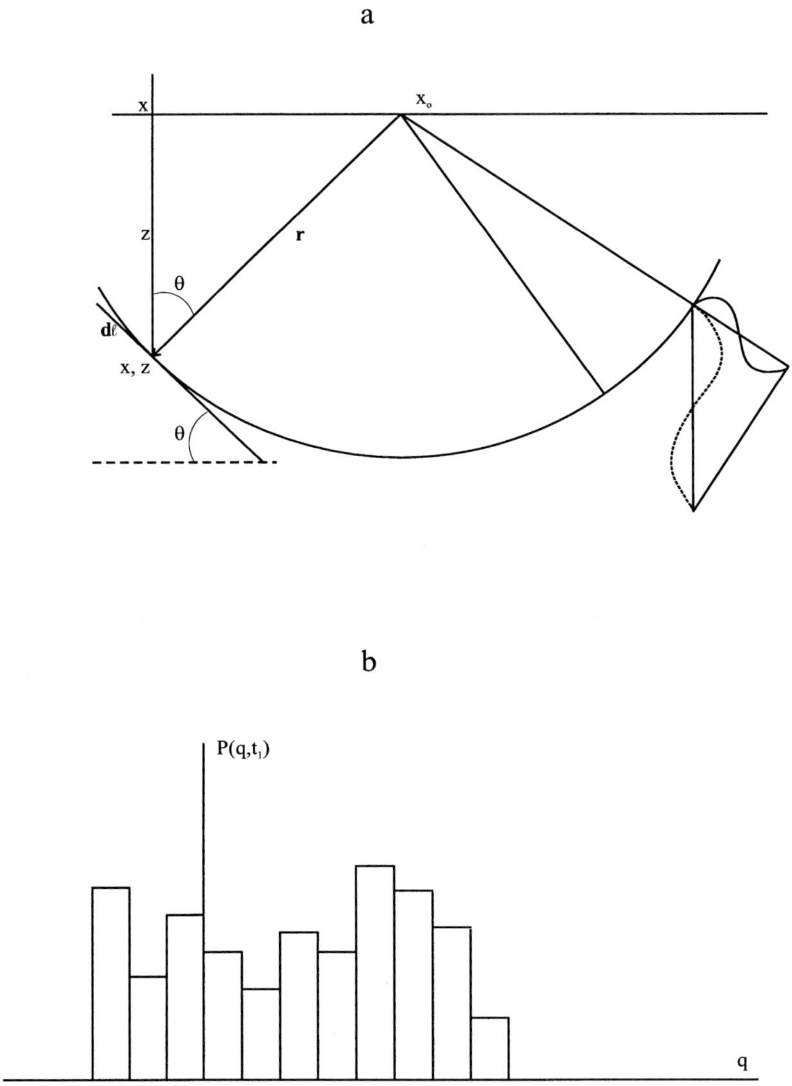

Figure 3.16: (a) Illustration of eq. 3.106 (b) Distribution of pressure.

Thus

$$x = x_o + \frac{ct}{2}\sin\theta \tag{3.106}$$

The right hand side of this equality is known. In fact, the position of the receiver, x_o, and the time of wave arrival, t, are given. As concerns $\sin\theta$, it can be calculated from the well-known expression for the apparent velocity:

$$c_a = \frac{dx_o}{dt} = \frac{c}{\sin\theta},$$

since c_a can be found from observation data. Thus, eq. 3.106 illustrates the procedure of migration, because it defines the x-coordinate of the reflector element, tangential to the circle. In other words, knowing of the apparent velocity permits us to determine a ray direction, and its intersection with the line of equal reflection time gives a location of the reflector element. We see that each circle defines the position and orientation of such elements, and their combination forms the envelope of the family of circles, that is the reflector.

Pressure at points of the probable reflector

Next we demonstrate that the concept of surfaces of equal reflection times offers a slightly different approach for a numerical integration of Kirchhoff's integral over the observation surface. With this purpose in mind it is necessary to study a distribution of the pressure at the probable reflector. We proceed from eq. 3.70

$$P(p,t) = -\frac{1}{2\pi}\frac{\partial}{\partial z}\int\limits_{S_o}\frac{P(q,t_1)}{R}\,dS_o$$

Here

$$t_1 = t + \frac{R}{c}$$

is the time of arrival of the reflected wave at the receiver, and R is its distance to any point of the ellipsoid, where $t_1 = \text{const}$. Also

$$t = \frac{R_s}{c} \tag{3.107}$$

is the time during which the incident wave propagates from the primary source to this surface of equal reflection time. By definition, at all its points

$$R_s + R = \text{const}$$

Rays of the wave, that appears at such probable reflector, are focused at one point of the observation surface. Respectively, the pressure, $P(q,t)$ differs from zero only at this point. In this light let us mentally represent the measured pressure, $P(q,t_1)$, as a system of sufficiently narrow impulses, Fig. 3.16b. Within an elementary area ΔS_o the pressure associated with the given ellipsoid remains the same, but outside it disappears. Therefore, the Kirchhoff formula, eq. 3.70, is greatly simplified for this special case and we obtain

$$P(p,t) = -\frac{1}{2\pi}\frac{\partial}{\partial z}\frac{P(q,t_1)}{R}\Delta S_o \tag{3.108}$$

or

$$P(p,t) = \frac{\Delta S_o}{2\pi}\left[\frac{P(q,t_1)}{R^2} - \frac{1}{cR}\frac{\partial P(q,t_1)}{\partial t_1}\right]\cos\theta\,, \tag{3.109}$$

where

$$\cos\theta = \frac{z}{R}$$

and θ is the angle between the z-axis and the line connecting points q and p, Fig. 3.16a.

This equation defines the pressure at points of the probable reflector, that is at the surface of equal reflection time. In other words, it characterizes the pressure of the reflected wave at the instant t when the incident wave arrives at some point of the ellipsoid. By definition, the pressure, measured by the receiver at moment t_1 is due to elementary waves, arising at different times at points of such reflector. As follows from eq. 3.109, the function $P(p,t)$ depends on point p, because of the directional cosine, $\cos\theta$, and a change of distances R and R_s.

To illustrate the concept of the probable reflector, assume, as before, the simplest case of the zero offset and constant velocity of propagation in the upper layer. Then $R = R_s$, and in place of the ellipsoid we have a spherical surface where the reflected wave arises simultaneously at all its points. Considering the plane xoz, we obtain for the pressure at points of the semicircle

$$P(p,t) = \frac{\Delta S_o}{2\pi R^2}\left[P(q,t_1) - \frac{R}{c}\frac{\partial P(q,t_1)}{\partial t_1}\right]\cos\theta, \tag{3.110}$$

Here

$$R = \text{const} \qquad \text{and} \qquad t_1 = \frac{2R}{c} \tag{3.111}$$

If the distance R is sufficiently large, $R/c \gg 1$, eq. 3.110 can be approximately replaced by

$$P(p,t) = -\frac{\Delta S_o}{2\pi Rc}\frac{\partial P(q,t_1)}{\partial t_1}\cos\theta\,, \tag{3.112}$$

Thus, the pressure along the semicircle is defined by the derivative of this function at the receiver. It is also valid for any array, as soon as separations R and R_s exceed the wave length, corresponding to the dominant frequency of the spectrum. It is clear that the pressure of the reflected wave varies synchronously at points of the semicircle, but its magnitude changes as $\cos\theta$. For instance, beneath the receiver ($\theta = 0$) it has a maximal value and then gradually decreases in approaching the observation line.

Inasmuch as rays characterize a direction of the wave propagation, it is natural that wavelets are functions of the semicircle radius R, Fig. 3.16a. However, wave traces are plotted along the z-axis. Performing a translation of wavelets in this direction, they become distorted, and, in particular, the sinusoidal wave changes its period. As is seen from Fig. 3.16a, the apparent period along the vertical line is equal to:

$$T_a = \frac{T}{\cos\theta} \qquad \text{or} \qquad \omega_a = \omega\,\cos\theta \tag{3.113}$$

For instance, at the point, located beneath a receiver, $\cos\theta = 0$, the frequency does not change. At the same time, with an increase of θ a distortion of the function $P(p,t)$ becomes more profound. In the limit, when $\theta \to \pi/2$, that is in the vicinity of an observation line, the apparent frequency tends to zero. This analysis shows that in the case of the zero offset and constant velocity c, the pressure varies along the probable reflector for two reasons. One of them is an influence of the directional cosine, but the second is due to a wave presentation along the z-axis. It is obvious that if a pressure is an arbitrary function of time, then the last factor leads to an apparent increase of the low frequency spectrum in approaching the line of observation, $\theta \to \pi/2$. In a more general case, when the primary source and receiver are located at some distance from each other the similar behavior is also observed.

Let us notice that, by definition, at least one point of the line of equal reflection time belongs to the envelope of this family. Correspondingly, eq. 3.109 characterizes the pressure around some point of the real reflector.

Summation of pressure for different ellipses

To understand the method of ellipses, suppose there is an array with the common source located at point s, but measurements are performed on the surface S_0 at points q. We consider the case when the medium is homogeneous above the reflector, and the velocity of propagation is c_1. First, we evaluate the Kirchhoff integral at the instant when the direct wave arrives at the observational point p, that is $t = R_1/c_1$. Here R_1 is a distance between the primary source and point p. Then, in accordance with eq. 3.70, the pressure of the reflected wave is defined as

$$P\left(p, \frac{R_1}{c_1}\right) = -\frac{1}{2\pi}\frac{\partial}{\partial z}\oint\limits_{S_0} P\left(q, \frac{R+R_1}{c_1}\right)\frac{1}{R}\,dS, \tag{3.114}$$

where R is the distance between points p and q. It is obvious that if point p is situated above the reflector, then the direct wave does not generate the reflected wave at this point and therefore

$$P\left(p, \frac{R_1}{c_1}\right) = 0. \tag{3.115}$$

Thus, using this simple physical fact, we can prove that the integral in eq. 3.114 is equal to zero. In contrast, if point p belongs to the reflector, the incident wave gives rise to the reflected wave, and

$$P\left(p, \frac{R_1}{c_1}\right) \neq 0. \tag{3.116}$$

We see that calculation of the pressure caused by the reflected wave allows us to determine a position of the reflector provided that the observation time corresponds to the arrival of the incident (direct) wave. Note that integration in eq. 3.114 performs a downward continuation and simultaneously uses the imaging principle. Particularly, in the case of zero offset we have $R = R_1$ and

$$P\left(p, \frac{R_1}{c_1}\right) = -\frac{1}{2\pi}\frac{\partial}{\partial z}\oint\limits_{S_0} P\left(q, \frac{2R}{c_1}\right)\frac{1}{R}\,dS \tag{3.117}$$

We note the following:

1. Eq. 3.114 is valid at the high-frequency range where the ray theory is applicable. Because of this, we were able to discard the surface integral over the reflector in the Kirchhoff formula. Moreover, we applied the concept of rays in deriving eq. 3.116. We used the fact that the total distance from point s to any point of the reflector and then to the observation point p is greater than R_1.
2. As usual, it is assumed that only up-going waves are studied.
3. Previously we obtained eq. 3.117 using the model of the explosive reflector.

Now we are ready to discuss the same method of ellipses, using the concept of the constructive and destructive interference. First, suppose that observations are performed with the zero offset. In accordance with eq. 3.110 or 3.112 we define the pressure at any point of the semicircle, including that of the real reflector, at the instant $t = R_1/c_1$, when the incident wave arrives.

Consider a wavelet at some point p of the probable reflector. This wavelet is intersected by a group of semicircles, and each of them is characterized by own wavelet at this point. In general, they are shifted in phase. Performing the summation of these wavelets, we obtain the pressure of the reflected wave at point p, and this procedure is equivalent to numerical integration in eq. 3.114. The value of the function P is defined practically by a limited number of participating wavelets. As follows from eqs. 3.115 or 3.116, the result of this integration is already known. We see that summation of wavelets has a destructive character if point p is located above the reflector, but at the points belonging to the reflector constructive interference occurs. Because of this, we define the reflector as an envelope of probable reflectors. The same occurs when the source and receivers are separated and the lines of equal reflection times are ellipses.

Chapter 4

Plane waves in layered media (normal incidence)

4.1 Waves in media with a single planar interface

First, consider a medium with one horizontal interface shown in Fig. 4.1. The density and velocity of propagation in the upper and lower parts of the medium are ρ_1, c_1, and ρ_2, c_2, respectively. We choose the Cartesian coordinate system x, y, z with the origin O located at the interface $z = 0$, so that the z-axis is directed downward. It is natural to begin our study with the simplest case when the primary (incident) wave propagates along the z-axis. Correspondingly, its phase surface N_i is parallel to the boundary, Fig. 4.1. In order to determine wave fields, we formulate the boundary value problem in terms of the potential U, which is related to the displacement $\mathbf{s}$ as

$$\mathbf{s} = \text{grad } U \tag{4.1}$$

At the beginning, let us assume that the wave is a sinusoidal function of both time t and distance z. As was shown in Chapter 1, the complex amplitude of the potential $\mathcal{U}$

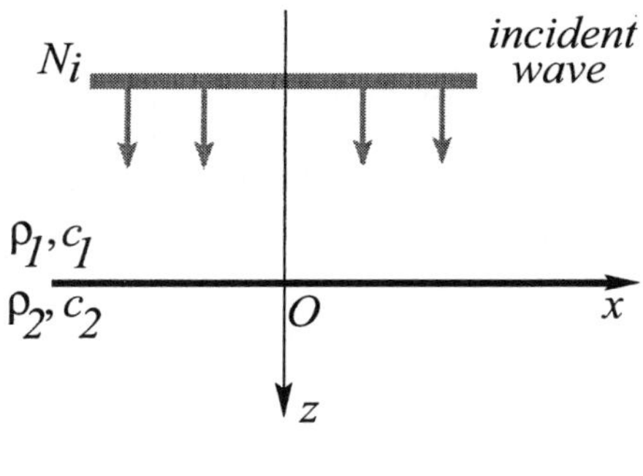

Figure 4.1: Normal incidence of plane wave.

satisfies the Helmholtz equation:

$$\nabla^2 \mathcal{U}_1 + k_1^2 \mathcal{U}_1 = 0 \qquad \text{if} \quad z < 0 \tag{4.2}$$

and

$$\nabla^2 \mathcal{U}_2 + k_2^2 \mathcal{U}_2 = 0\,, \qquad \text{if} \quad z > 0, \tag{4.3}$$

where

$$k_1 = \frac{\omega}{c_1} \qquad \text{and} \qquad k_2 = \frac{\omega}{c_2} \tag{4.4}$$

are the wavenumbers.

We also know that the complex amplitudes of the displacement and pressure can be represented as

$$\mathcal{S} = \text{grad}\, \mathcal{U} \qquad \text{and} \qquad \mathcal{P} = \omega^2 \rho\, \mathcal{U} \tag{4.5}$$

Here ω is the angular frequency and ρ is the density.

As follows from the continuity of pressure and normal component of the displacement

$$\rho_1 \mathcal{U}_1 = \rho_2 \mathcal{U}_2 \qquad \text{and} \qquad \frac{\partial \mathcal{U}_1}{\partial z} = \frac{\partial \mathcal{U}_2}{\partial z} \tag{4.6}$$

at the boundary $z = 0$.

Equations 4.2, 4.3, and 4.6 constitute the boundary value problem and our goal is to find the function $\mathcal{U}$ that satisfies these conditions. By definition, when waves have sinusoidal dependence on time t, the function $\mathcal{U}$ relates to the acoustic potential U as

$$U = \text{Re}\; \mathcal{U}\, e^{-i\,\omega\, t} \tag{4.7}$$

and even though $\mathcal{U}$ and U are different, we will still call the former the acoustic potential. The same convention is used for the displacement, velocity and pressure.

Solution of the boundary value problem

Since phase surfaces of the incident wave are parallel to the boundary, we assume that the wave fields depend on a single coordinate z. Then, the Helmholtz equation is simplified to

$$\frac{d^2 \mathcal{U}}{d\, z^2} + k^2 \mathcal{U} = 0 \tag{4.8}$$

As is well known, its solution is

$$\mathcal{U}(k,\, z) = C\, e^{i\, k\, z} + D\, e^{-i\, k\, z} \tag{4.9}$$

where C and D are some coefficients.

Taking into account the fact that the incident wave propagates in the upper medium along the z-axis and time dependence has the form $exp\,(-i\,\omega\,t)$, the potential of this wave is

$$\mathcal{U}_i(k_1,\, z) = \frac{s_0}{i\,k_1}\,\exp(i\,k_1\,z)\,, \tag{4.10}$$

where s_0 is the coefficient that describes the amplitude of displacement which, in general, can be a function of frequency.

As follows from eqs. 4.5, the fields of the incident wave are

$$\mathcal{S}_i = s_0\,\exp(i\,k_1\,z)\,, \qquad \mathcal{V}_i = -i\,\omega\,s_0\,\exp(i\,k_1\,z)\,,$$
$$\mathcal{P}_i = -i\,\omega\,s_0\,Z_1\,\exp(i\,k_1\,z)\,, \tag{4.11}$$

where

$$Z_1 = c_1\,\rho_1 \tag{4.12}$$

is the acoustic impedance of the upper medium, $\mathcal{S}_i$ and $\mathcal{V}_i$ are the complex amplitudes of vertical component of the particle displacement and its velocity. $\mathcal{P}_i$ is the complex amplitude of the pressure. Since a plane wave propagates along the vertical axis, horizontal components of the displacement are absent:

$$\mathcal{S}_x = \mathcal{S}_y = \mathcal{V}_x = \mathcal{V}_y = 0 \tag{4.13}$$

We see that the velocity and pressure are related as

$$\frac{\mathcal{V}_i}{\mathcal{P}_i} = \frac{1}{Z_1} \tag{4.14}$$

They are both shifted in phase by $-\pi/2$ with respect to the displacement. In accordance with eqs. 4.11, the wave fields depend on the product

$$k_1\,z = \frac{\omega}{c_1}\,z = \frac{2\,\pi}{\lambda_1}\,z \tag{4.15}$$

Here $\lambda_1 = c_1\,T$ is the wavelength in the upper part of the medium and T is the period.

When the product $k_1\,z$ becomes small, the exponential term in eqs. 4.11 is approximately equal to unity. This means that the medium particles move almost synchronously, and the phenomenon of wave propagation is negligible. Such behavior occurs at a greater volume for lower frequency or higher propagation velocity. In contrast, with a decrease of the wavelength the change in phase with distance becomes more profound.

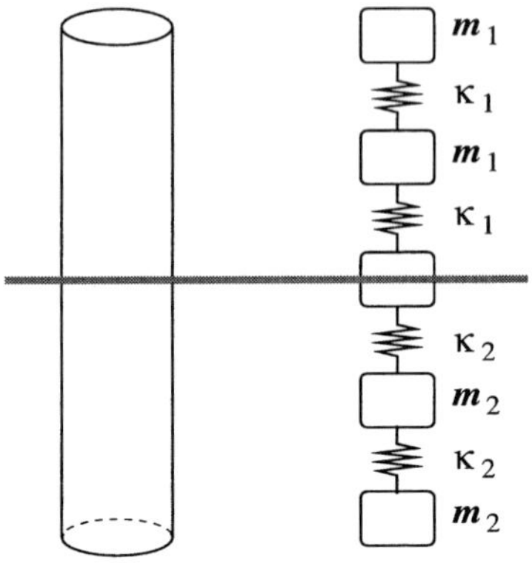

Figure 4.2: Representation of a medium as a system of masses and springs.

In order to predict the results of solving the boundary value problem, it may be useful to imagine an infinitely long cylinder of the medium directed perpendicular to the interface, Fig. 4.2. Each element of the cylinder, characterized by some density and compressibility of the corresponding part of the medium, can be mentally replaced by an elementary mass and spring. In such a way, we model the cylinder by an equivalent system of masses and springs, and their parameters change at the boundary. As was demonstrated in Part I, the wave, propagating along this system and reaching its interface, gives rise to the reflected and transmitted waves. Bearing in mind this fact, it is natural to assume that the plane boundary, Fig. 4.1, also produces those waves. Correspondingly, we represent the potentials in the upper and lower parts of the medium as

$$\mathcal{U}_1 = \frac{1}{i\,k_1}\,[s_0 \exp(i\,k_1\,z) + B \exp(-i\,k_1\,z)] \qquad \text{if} \qquad z < 0 \tag{4.16}$$

and

$$\mathcal{U}_2 = \frac{1}{i\,k_2}\,C \exp(i\,k_2\,z) \qquad \text{if} \qquad z > 0 \tag{4.17}$$

where B and C are unknown coefficients.

It is obvious that functions $\mathcal{U}_1$ and $\mathcal{U}_2$ satisfy the Helmholtz equation in the same manner as $\mathcal{U}_i$, eq. 4.10. The "$-$" sign of the argument $-i\,k_1\,z$ in the second term of eq. 4.16 indicates a wave traveling away from the boundary. To determine unknown coefficients, we use eqs. 4.6. Substitution of eqs. 4.16 and 4.17 into this set yields

$$\rho_1\,c_1\,(s_0 + B) = \rho_2\,c_2\,C \qquad \text{and} \qquad s_0 - B = C \tag{4.18}$$

The solution of this system is

$$B = \frac{\rho_2\, c_2 - \rho_1\, c_1}{\rho_2\, c_2 + \rho_1\, c_1}\, s_0 \qquad \text{and} \qquad C = \frac{2\,\rho_1\, c_1}{\rho_2\, c_2 + \rho_1\, c_1}\, s_0 \tag{4.19}$$

or

$$B = \frac{Z_2 - Z_1}{Z_2 + Z_1}\, s_0 \qquad \text{and} \qquad C = \frac{2\, Z_1}{Z_2 + Z_1}\, s_0\,, \tag{4.20}$$

where $Z_1 = \rho_1\, c_1$ and $Z_2 = \rho_2\, c_2$ are the acoustic impedances.

Thus, we solved the boundary value problem and proved that the incident wave produces two more waves at the interface. One of them, propagating through the upper medium away from the boundary, is called the reflected wave. The other, which moves in the lower medium along the z-axis, is the transmitted wave. Since the study of these waves is of great importance in seismology, next we will discuss their main features in details.

Reflected and transmitted waves

As follows from eq. 4.5,

$$\mathcal{S} = \mathcal{S}_z = \frac{\partial\, \mathcal{U}}{\partial\, z}\,, \qquad \mathcal{V} = \mathcal{V}_z = -i\,\omega\, \mathcal{S}_z\,, \qquad \mathcal{P} = \omega^2\, \rho\, \mathcal{U}\,, \tag{4.21}$$

and the displacement, velocity of particles, and the pressure caused by the incident and reflected waves, are

$$\mathcal{S}_1 = s_0 \left[\exp(i\, k_1\, z) - \frac{Z_2 - Z_1}{Z_2 + Z_1} \exp(-i\, k_1\, z)\right]$$

$$\mathcal{V}_1 = -i\,\omega\, s_0 \left[\exp(i\, k_1\, z) - \frac{Z_2 - Z_1}{Z_2 + Z_1} \exp(-i\, k_1\, z)\right] \tag{4.22}$$

and

$$\mathcal{P}_1 = -i\,\omega\, Z_1\, s_0 \left[\exp(i\, k_1\, z) + \frac{Z_2 - Z_1}{Z_2 + Z_1} \exp(-i\, k_1\, z)\right]$$

if $z < 0$

In the same manner, we have for the transmitted wave

$$\mathcal{S}_2 = s_0\, \frac{2\, Z_1}{Z_1 + Z_2} \exp(i\, k_2\, z)$$

$$\mathcal{V}_2 = -i\,\omega\, s_0\, \frac{2\, Z_1}{Z_1 + Z_2} \exp(i\, k_2\, z) \tag{4.23}$$

and

$$\mathcal{P}_2 = -i\,\omega\, s_0\, Z_2\, \frac{2\,Z_1}{Z_1 + Z_2}\, \exp(i\,k_2\,z)$$

if $z > 0$

The simplicity of eqs. 4.22 and 4.23 is emphasized by the fact that the coefficients B and C are independent of frequency. This means that the amplitude spectra of reflected and transmitted waves differ from that of the incident wave only by the constant factor. The coefficient

$$R_{12} = \frac{Z_2 - Z_1}{Z_2 + Z_1} \tag{4.24}$$

characterizes the amplitude of the reflected wave and varies in the following limits:

$$-1 \le R_{12} \le 1 \tag{4.25}$$

It is convenient to distinguish two cases:

$$Z_2 < Z_1 \qquad \text{and} \qquad Z_2 > Z_1$$

Case one: $Z_2 < Z_1$ Since the coefficient R_{12} is negative, displacements of incident and reflected waves have the same direction at the boundary, however, their pressures have opposite signs. For instance, a compressional incident wave, which has the same directions of propagation and particle velocity, produces an extensional reflected wave. In fact, it follows from eqs. 4.22 that the pressure in this wave is negative at the boundary, and the directions of propagation and particle velocity differ in sign. Likewise, an extensional incident wave gives rise to a compressional reflected wave. Thus, those waves are always of different types at the boundary $z = 0$.

Suppose that the ratio of impedances is small

$$\frac{Z_2}{Z_1} \ll 1 \tag{4.26}$$

Correspondingly, the coefficient R_{12} is equal to

$$R_{12} = \frac{Z_2/Z_1 \,-\, 1}{Z_2/Z_1 \,+\, 1} \approx -1 + 2\,\frac{Z_2}{Z_1} \tag{4.27}$$

Then superposition of the incident and reflected waves gives at the boundary

$$S_1 \approx 2\,s_0\left(1 - \frac{Z_2}{Z_1}\right) \qquad \mathcal{V}_1 \approx -2\,i\,\omega\,s_0\left(1 - \frac{Z_2}{Z_1}\right)$$

$$\mathcal{P}_1 \approx -2\,i\,\omega\,s_0\,Z_2 = \mathcal{P}_i\,\frac{Z_2}{Z_1} \ll P_i \tag{4.28}$$

The latter shows that the particle displacement and velocity caused by both waves are almost equal to each other, but the pressure is small. In the limit when $Z_2 = 0$, we have at points of the boundary

$$S_1 = 2\, s_0 = 2\, S_i \qquad \text{and} \qquad \mathcal{V}_1 = 2\, \mathcal{V}_i\,, \tag{4.29}$$

that is, the total displacement and the velocity are twice as large as those of the incident wave. At the same time, the pressure vanishes

$$P_1 = 0 \qquad \text{at} \qquad z = 0 \tag{4.30}$$

As is well known, in such a case the boundary is called a free surface and any of eqs. 4.29 or 4.30 represent its boundary condition. It may be proper to notice that they were derived, studying wave propagation along a spring in Part I. Next, consider the transmitted wave. Comparison of eqs. 4.11 and 4.23 shows that the incident and transmitted waves are always of the same type at the boundary.

It is clear that eqs. 4.28 also describe the transmitted wave if

$$\frac{Z_2}{Z_1} \ll 1 \qquad \text{at} \qquad z = 0 \tag{4.31}$$

In the limiting case of free space (when $c_2 \to 0$) the exponent $i\, k_2\, z$ tends to infinity in eqs. 4.23 and therefore the first two of these formulas become meaningless, whereas the last one, $\mathcal{P}_2 \equiv 0$, is satisfied identically.

Case two: $Z_2 > Z_1$ Since the coefficient R_{12} is positive, eq. 4.24, the incident and reflected waves are of the same type at the boundary. Assuming that the ratio Z_1/Z_2 is small, we have

$$\frac{Z_2 - Z_1}{Z_2 + Z_1} \approx 1 - 2\,\frac{Z_1}{Z_2} \tag{4.32}$$

Then

$$\mathcal{S}_1 \approx 2\,\frac{Z_1}{Z_2}\, s_0\,, \qquad \mathcal{V}_1 \approx -2\, i\, \omega\, \frac{Z_1}{Z_2}\, s_0 \tag{4.33}$$

and

$$\mathcal{P}_1 \approx -2\, i\, \omega\, Z_1\, s_0 \left(1 - \frac{Z_1}{Z_2}\right) \tag{4.34}$$

If the lower medium is rigid ($Z_2 \to \infty$), we obtain

$$\mathcal{S}_1 = 0\,, \qquad \mathcal{V}_1 = 0\,, \qquad \text{and} \qquad \mathcal{P}_1 = 2\, \mathcal{P}_i \tag{4.35}$$

Each equation of the set 4.35 can be treated as the boundary condition in this limiting case. Note that due to an oscillatory character of waves, there are always points above the interface where the incident and reflected waves belong to either the same or opposite type regardless of the ratio between Z_1 and Z_2.

If the ratio Z_1/Z_2 is small, eqs. 4.23 yield for transmitted wave

$$\mathcal{S}_2 \approx 2\, s_0 \frac{Z_1}{Z_2}\left(1 - \frac{Z_1}{Z_2}\right) \tag{4.36}$$

$$\mathcal{V}_2 \approx -2\, i\, \omega\, s_0 \frac{Z_1}{Z_2}\left(1 - \frac{Z_1}{Z_2}\right) \tag{4.37}$$

and

$$\mathcal{P}_2 \approx 2\, \mathcal{P}_i \left(1 - \frac{Z_1}{Z_2}\right) \tag{4.38}$$

at $z = 0$. Thus, if Z_2 has a relatively large value, the pressure $\mathcal{P}_2$ is close to $2\,\mathcal{P}_i$.

When velocity c_2 increases, the exponent $i\, k_2\, z$ becomes smaller, and correspondingly eqs. 4.38 correctly describe the wave fields at greater distances from the boundary. In the limit $Z_2 \to \infty$, the transmitted wave vanishes, since

$$\mathcal{S}_2 = \mathcal{V}_2 = 0 \tag{4.39}$$

In other words, the rigid medium is at rest, and the fact that everywhere $\mathcal{P}_2 = 2\,\mathcal{P}_i$ has no any physical meaning.

Transient waves

Now we suppose that the incident wave is an arbitrary function of time and distance:

$$U_i = s_0\, f_i\left[a\left(t - \frac{z}{c_1}\right)\right], \tag{4.40}$$

where a and s_0 are known constants. As was shown in the Part I, functions with an argument

$$a(t \pm z/c_1)$$

satisfy the wave equation

$$\frac{\partial^2 U}{\partial z^2} = \frac{1}{c_1^2}\frac{\partial^2 U}{\partial t^2}, \tag{4.41}$$

provided that the second derivatives exist.

Correspondingly, the potentials of reflected and transmitted waves can be written as

$$U_r = B\, f_r\left[a_1\left(t + \frac{z}{c_1}\right)\right] \tag{4.42}$$

and

$$U_2 = C\, f_2 \left[a_2 \left(t - \frac{z}{c_2} \right) \right] \tag{4.43}$$

Substitution of eqs. 4.40, 4.42, and 4.43 into boundary conditions 4.6 yields

$$\begin{aligned} \rho_1 \left[s_0\, f_i(a\, t) + B\, f_r(a_1\, t) \right] &= \rho_2\, C\, f_2(a_2\, t) \\ \frac{1}{c_1} \left[-a\, s_0\, f_i'(a\, t) + a_1\, B\, f_r'(a_1\, t) \right] &= -\frac{a_2}{c_2}\, C\, f_2'(a_2\, t) \end{aligned} \tag{4.44}$$

Since these equalities have to take place at any time t, we conclude that

$$a_2 = a_1 = a \tag{4.45}$$

and

$$f_i(a\, t) = f_r(a\, t) = f_2(a\, t) \tag{4.46}$$

Then, the set 4.44 is simplified

$$\begin{aligned} \rho_1\, s_0 + \rho_1\, B &= \rho_2\, C \\ -\frac{1}{c_1}\, s_0 + \frac{B}{c_1} &= -\frac{C}{c_2} \end{aligned} \tag{4.47}$$

Whence

$$B = \frac{Z_2 - Z_1}{Z_2 + Z_1}\, s_0\,, \qquad C = \frac{2\, Z_1}{Z_2 + Z_1}\, s_0\,, \tag{4.48}$$

which agrees with eqs. 4.20. Thus, we have for the potential:

$$U_1(z,t) = s_0\, f \left[a \left(t - \frac{z}{c_1} \right) \right] + \frac{Z_2 - Z_1}{Z_2 + Z_1}\, s_0\, f \left[a \left(t + \frac{z}{c_1} \right) \right] \tag{4.49}$$

and

$$U_2(z,t) = \frac{2\, Z_1}{Z_2 + Z_1}\, s_0\, f \left[a \left(t - \frac{z}{c_2} \right) \right] \tag{4.50}$$

We see that the incident, reflected and transmitted waves are described by the same function of time. In other words, the shape of secondary waves coincides with that of the incident wave. Besides, the reflection and transmission coefficients B/A and C/A are independent of the incident wavelet. Later we will observe a similar behavior for oblique incidence, provided that the incident angle does not exceed the so-called critical angle.

It is also useful to derive an expression for $\mathcal{U}_r$, eq. 4.49, in a different way. As we know, the sinusoidal incident wave

$$\mathcal{U}_i = \frac{s_0}{i\,k_1}\,e^{i\,k_1\,z} \tag{4.51}$$

gives rise to the reflected wave

$$\mathcal{U}_r = \frac{B}{i\,k_1}\,e^{-i\,k_1\,z} \tag{4.52}$$

Suppose that the incident wave is an arbitrary function of time:

$$U_i(z,t) = s_0\,f\left[a\left(t - \frac{z}{c_1}\right)\right] \tag{4.53}$$

Then its spectrum is

$$\mathcal{U}_i\,(z,\omega) = s_0\,\mathcal{F}(\omega)\,\frac{e^{i\,k_1\,z}}{i\,k_1} \tag{4.54}$$

This means that

$$U_i(z,t) = \frac{s_0}{2\,\pi}\int\limits_{-\infty}^{\infty}\mathcal{F}(\omega)\,\frac{e^{i\,k_1\,z}}{i\,k_1}\,e^{-i\,\omega\,t}\,d\,\omega \tag{4.55}$$

Each sinusoidal harmonic of this wave generates the reflected wave

$$\frac{1}{2\,\pi}\,B\,\mathcal{F}\,(\omega)\,\frac{e^{-i\,k_1\,z}}{i\,k_1}\,d\,\omega \tag{4.56}$$

The principle of superposition applied to the reflected wave yields

$$U_r(z,t) = \frac{B}{2\,\pi}\int\limits_{-\infty}^{\infty}\mathcal{F}(\omega)\,\frac{e^{-i\,k_1\,z}}{i\,k_1}\,e^{-i\,\omega\,t}\,d\,\omega \tag{4.57}$$

Finally, using eq. 4.55, we arrive at the known formula

$$\mathcal{U}_r(z,t) = B\,f\left[a\left(t + \frac{z}{c_1}\right)\right] \tag{4.58}$$

In a similar manner, we can obtain an expression for the transmitted wave.

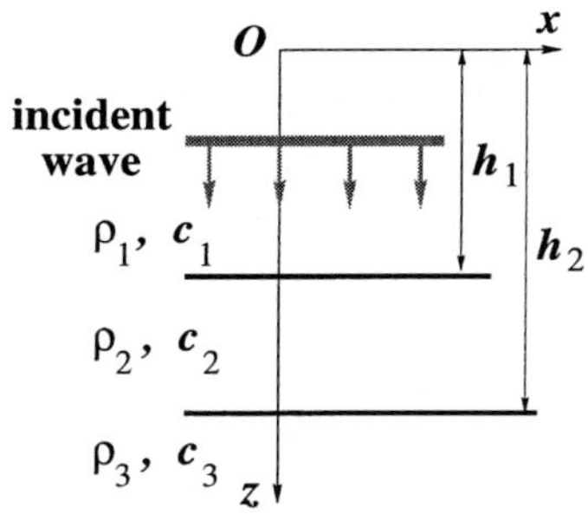

Figure 4.3: Plane incident wave in a medium containing a single homogeneous layer between two homogeneous half-spaces.

4.2 Normal incidence of plane wave in horizontally layered media

It is natural to begin our study of waves in layered media with the simplest case when there is only one layer between two horizontal interfaces, Fig. 4.3. The thickness of the layer is

$$H = h_2 - h_1 \tag{4.59}$$

where h_1 and h_2 are the distances from the coordinate origin to the top and bottom of the layer.

When the incident wave reaches the first boundary $z = h_1$, the reflected and transmitted waves appear. The latter propagates into the layer and, in turn, gives rise to the reflected and transmitted waves at the interface $z = h_2$. The former propagates upward and, again, produces both waves at the top of the layer. It is obvious that this process continues infinitely long, if attenuation is absent. Thus, the reflected wave in the upper medium, as well as the transmitted wave beneath the layer, consists of an infinite number of waves which arise at different times. Similarly, there is infinite number of reflected waves within the layer. Knowing the reflection and transmission coefficients derived in section 2.1, it is not difficult to obtain the transient waves in each medium. For our purpose, however, it is more convenient to study sinusoidal waves first and then apply the Fourier integral to obtain the transient waves.

Sinusoidal waves

Using this approach we formulate the boundary-value problem in terms of the complex amplitude of the potential $\mathcal{U}$ that should satisfy the Helmholtz equation

$$\nabla^2 \mathcal{U} + k^2 \mathcal{U} = 0 \qquad \text{or} \qquad \frac{\partial^2 \mathcal{U}}{\partial z^2} + k^2 \mathcal{U} = 0, \tag{4.60}$$

where $k = \omega/c$ is the wavenumber

$$k = \begin{cases} k_1\,, & \text{if} \quad z < h_1 \\ k_2\,, & \text{if} \quad h_1 \le z \le h_2 \\ k_3\,, & \text{if} \quad z > h_2 \end{cases}$$

Also, the continuity of the pressure and the normal component of displacement yields

$$\rho_1 \mathcal{U}_1 = \rho_2 \mathcal{U}_2\,, \qquad \frac{\partial \mathcal{U}_1}{\partial z} = \frac{\partial \mathcal{U}_2}{\partial z} \qquad \text{if} \quad z = h_1 \tag{4.61}$$

and

$$\rho_2 \mathcal{U}_2 = \rho_3 \mathcal{U}_3\,, \qquad \frac{\partial \mathcal{U}_2}{\partial z} = \frac{\partial \mathcal{U}_3}{\partial z} \qquad \text{if} \quad z = h_2 \tag{4.62}$$

Here $\mathcal{U}_1$, $\mathcal{U}_2$, and $\mathcal{U}_3$ are the complex amplitudes of the potential in the upper, intermediate, and lower parts of the medium, respectively.

Since the incident wave moves vertically along the z-axis, its potential is

$$\mathcal{U}_i = \frac{A}{i\,k_1}\, e^{i\,k_1\,z}\,,$$

where A is a known coefficient.

Correspondingly, for the total potential in each part of the medium, we have

$$\begin{aligned} \mathcal{U}_1 &= \frac{1}{i\,k_1}\left(A\,e^{i\,k_1\,z} + B\,e^{-i\,k_1\,z}\right), \\ \mathcal{U}_2 &= \frac{1}{i\,k_2}\left(C\,e^{i\,k_2\,z} + D\,e^{-i\,k_2\,z}\right), \\ \mathcal{U}_3 &= \frac{1}{i\,k_3}\,E\,e^{i\,k_3\,z}\,, \end{aligned} \tag{4.63}$$

where B, C, D, and E are unknown coefficients.

It is clear that all the functions $\mathcal{U}_1$, $\mathcal{U}_2$, and $\mathcal{U}_3$ satisfy the Helmholtz equation. Substitution of eqs. 4.63 into the boundary conditions 4.61 and 4.62 gives the system of linear equations with respect to our unknowns

$$Z_1\left(A\,e^{i\,k_1\,h_1} + B\,e^{-i\,k_1\,h_1}\right) = Z_2\left(C\,e^{i\,k_2\,h_1} + D\,e^{-i\,k_2\,h_1}\right),$$

$$\begin{aligned} A\,e^{i\,k_1\,h_1} - B\,e^{-i\,k_1\,h_1} &= C\,e^{i\,k_2\,h_1} - D\,e^{-i\,k_2\,h_1}\,, \\ Z_2\left(C\,e^{i\,k_2\,h_2} + D\,e^{-i\,k_2\,h_2}\right) &= Z_3\,E\,e^{i\,k_3\,h_2}\,, \\ C\,e^{i\,k_2\,h_2} - D\,e^{-i\,k_2\,h_2} &= E\,e^{i\,k_3\,h_2} \end{aligned} \tag{4.64}$$

Eliminating E from the last two equations, we obtain

$$Z_2\left(C\,e^{i\,k_2\,h_2} + D\,e^{-i\,k_2\,h_2}\right) = Z_3\left(C\,e^{i\,k_2\,h_2} - D\,e^{-i\,k_2\,h_2}\right)$$

or

$$D = R_{23}\,e^{2\,i\,k_2\,h_2}\,C\,, \tag{4.65}$$

where

$$R_{23} = \frac{Z_3 - Z_2}{Z_3 + Z_2} \tag{4.66}$$

Replacement of D in the first two equations of the set 4.64 yields

$$Z_1\left(A\,e^{i\,k_1\,h_1} + B\,e^{-i\,k_1\,h_1}\right) = Z_2\,e^{i\,k_2\,h_1}\left(1 + R_{23}\,e^{2\,i\,k_2\,H}\right)C$$

and

$$A\,e^{i\,k_1\,h_1} - B\,e^{-i\,k_1\,h_1} = e^{i\,k_2\,h_1}\left(1 - R_{23}\,e^{2\,i\,k_2\,H}\right)C$$

Thus

$$Z_1\,\frac{A\,e^{i\,k_1\,h_1} + B\,e^{-i\,k_1\,h_1}}{A\,e^{i\,k_1\,h_1} - B\,e^{-i\,k_1\,h_1}} = Z_2\,M \tag{4.67}$$

Here

$$M = \frac{1 + R_{23}\,e^{2\,i\,k_2\,H}}{1 - R_{23}\,e^{2\,i\,k_2\,H}} \tag{4.68}$$

Whence

$$B = \frac{Z_2\,M - Z_1}{Z_2\,M + Z_1}\,e^{2\,i\,k_1\,h_1}\,A$$

or

$$B = \frac{R_{12} + R_{23}\, e^{2\,i\,k_2\,H}}{1 + R_{12}\, R_{23}\, e^{2\,i\,k_2\,H}}\, e^{2\,i\,k_1\,h_1}\, A \tag{4.69}$$

and

$$R_{12} = \frac{Z_2 - Z_1}{Z_2 + Z_1} \tag{4.70}$$

Introducing the notation

$$R_{13} = \frac{R_{12} + R_{23}\, e^{2\,i\,k_2\,H}}{1 + R_{12}\, R_{23}\, e^{2\,i\,k_2\,H}}\,, \tag{4.71}$$

we can write the expressions for the coefficients B, C, D, and E in the following form

$$B = R_{13}\, e^{2\,i\,k_1\,h_1}\, A\,, \qquad C = \frac{Z_1}{Z_2}\,(1 + R_{13})\,\frac{e^{i\,(k_1 - k_2)\,h_1}}{1 + R_{23}\, e^{2\,i\,k_2\,H}}\, A\,,$$

$$D = \frac{Z_1}{Z_2}\,(1 + R_{13})\,\frac{R_{23}\, e^{i\,(k_1 - k_2)\,h_1} e^{2i\,k_2\,h_2}}{1 + R_{23}\, e^{2\,i\,k_2\,H}}\, A\,, \tag{4.72}$$

$$E = \frac{Z_1}{Z_2}\,(1 + R_{13})\,\frac{e^{i\,k_1\,h_1}\, e^{i\,k_2\,H}\,(1 - R_{23})\, e^{-i\,k_3\,h_2}}{1 + R_{23}\, e^{2\,i\,k_2\,H}}\, A$$

It is clear that the function R_{13} characterizes the amplitude and phase of the wave reflected from the layer, and it is usually called the reflection coefficient of a layer embedded between two half-spaces. Unlike the case with one boundary, R_{13} is a complex function.

To confirm the derivation, consider two cases.

(a) Suppose that the lower interface is absent ($Z_2 = Z_3$), that is $R_{23} = 0$. Then eq. 4.71 becomes

$$R_{13} = R_{12}$$

and

$$B = R_{12}\, e^{2\,i\,k_1\,h_1}\, A\,, \qquad C = \frac{2\,Z_1}{Z_1 + Z_2}\, e^{i\,(k_1 - k_2)\,h_1}\, A\,,$$

while

$$D = 0\,, \qquad \text{and} \qquad E = \frac{2\,Z_1}{Z_1 + Z_2}\, e^{i\,(k_1 - k_2)\,h_1}\, A = C$$

$$\text{if} \quad k_2 = k_3$$

Correspondingly, the potentials of the reflected and transmitted waves are

$$\mathcal{U}_r = \frac{A}{i\,k_1}\,\frac{Z_2 - Z_1}{Z_2 + Z_1}\, e^{i\,k_1\,h_1}\, e^{-i\,k_1\,(z - h_1)} \tag{4.73}$$

and

$$\mathcal{U}_2 = \frac{A}{i\,k_2}\,\frac{2\,Z_1}{Z_1 + Z_2}\, e^{i\,k_1\,h_1}\, e^{i\,k_2\,(z - h_1)} \tag{4.74}$$

In essence, eqs. 4.73 and 4.74 coincide with those derived in section 2.1. The term $exp\,(i\,k_1\,h_1)$ characterizes the phase of the incident wave at the boundary $z = h_1$, while the exponents with the arguments $k_1\,(z - h_1)$ and $k_2\,(z - h_1)$ describe the phase change of the reflected and transmitted waves, respectively.

(b) Let us assume that the upper boundary is absent

$$Z_1 = Z_2 \qquad \text{or} \qquad R_{12} = 0$$

This yields

$$R_{13} = R_{23}\, e^{2\,i\,k_2\,H}$$

and, therefore,

$$B = R_{23}\, e^{2\,i\,k_2\,h_2}\, A \qquad \text{if} \qquad k_1 = k_2$$

Thus, we have

$$C = A\,, \qquad D = R_{23}\, e^{2\,i\,k_2\,h_2}\, A$$

and

$$E = \frac{2\,Z_2}{Z_2 + Z_3}\, e^{i\,k_2\,h_2}\, e^{-i\,k_3\,h_2}\, A$$

Again we have arrived at equations that describe waves in the presence of the single interface $z = h_2$. Note that coefficients B and D are equal to each other, since they characterize the same reflected wave.

Now we return to the general case and, first, examine the reflected wave in the upper medium, for example, at the plane $z = 0$. Taking into account eqs. 4.63 and 4.69 the potentials of the incident and reflected waves are

$$\mathcal{U}_i = \frac{A}{i\,k_1} \qquad \text{and} \qquad \mathcal{U}_r = \frac{B}{i\,k_1}$$

The ratio between them is the complex number

$$\frac{B}{A} = R_{13}\, e^{2\,i\,k_1\,h_1} \tag{4.75}$$

The reflection coefficient R_{13} depends on three dimensionless parameters: Z_2/Z_1, Z_3/Z_1 and $k_2 H$. The last parameter can be rewritten as

$$k_2\,H = \frac{\omega}{c_2}\,H = 2\,\pi\,\frac{H}{\lambda_2} \tag{4.76}$$

As follows from eqs. 4.75 and 4.76, the function R_{13} is periodic with respect to the ratio H/λ_2. In fact, we have

$$e^{2\,i\,k_2\,H} = \exp\left(4\,i\,\pi\,\frac{H}{\lambda_2}\right) = \exp\left[4\,i\,\pi\,\left(\frac{H}{\lambda_2} + \frac{n}{2}\right)\right],$$

that is, the period is equal to 1/2. Here n is an integer.

It is obvious that the amplitude of the reflected wave decreases when the difference between the impedances becomes smaller. In contrast, if $R_{12} = 1$, the magnitudes of the incident and reflected waves are equal to each other at the upper boundary, because the incident wave does not penetrate into the layer.

Generally speaking, the layer between the two half-spaces acts on the incident wave as a linear frequency filter, changing the spectral content of the reflected and transmitted waves. Fig. 4.4 demonstrates the amplitude and phase responses of such a filter.

Dependence on wavelength λ_2

Inasmuch as the reflected wave is a superposition of waves arising at both boundaries, the phase shift between them plays an important role, which is emphasized by the dependence of the ratio R_{13} on the parameter $k_2\,H$. First, it is instructive to study the asymptotic behavior of the function R_{13} when wavelength λ_2 greatly exceeds the layer thickness H:

$$\lambda_2 \gg H \qquad \text{or} \qquad k_2\,H \ll 1 \tag{4.77}$$

Then, representing the exponent $exp\,(i\,k_2\,H)$ as a power series

$$e^{i\,k_2\,H} = 1 + i\,k_2\,H - \frac{k_2^2\,H^2}{2} + \dots$$

and discarding all terms except the first one, we obtain

$$R_{13} \approx \frac{R_{12} + R_{23}}{1 + R_{12}\, R_{23}} = \frac{(Z_2 - Z_1)\,(Z_2 + Z_3) + (Z_2 + Z_1)\,(Z_3 - Z_2)}{(Z_2 + Z_1)\,(Z_2 + Z_3) + (Z_2 - Z_1)\,(Z_3 - Z_2)}$$

$$= \frac{Z_3 - Z_1}{Z_3 + Z_1} \qquad \text{if} \qquad \lambda_2 \gg H \tag{4.78}$$

The latter coincides with the reflection coefficient from a single interface that separates the media characterized by the impedances Z_1 and Z_3 and is located at the distance h_1 from the coordinate origin. In fact, the complex amplitude of the reflected wave in such a case is

$$\mathcal{U}_r = \frac{A}{i\,k_1}\,\frac{Z_3 - Z_1}{Z_3 + Z_1}\, e^{i\,k_1\,h_1}\, e^{-i\,k_1\,(z\,-\,h_1)} \qquad \text{if} \qquad \lambda_2 \gg H$$

In other words, the influence of the layer thickness H and its impedance Z_2 on the reflected wave is negligible when the condition 4.77 is met. In particular, if the media above and below the layer have equal impedances, $Z_1 = Z_3$, the destructive interference is so strong that the reflected wave disappears

$$\mathcal{U}_r = 0 \qquad \text{if} \qquad \lambda_2 \gg H\,,$$

and the layer cannot be detected. This result is also applied to other frequencies provided that

$$Z_2^2 = Z_1 Z_3 \qquad \text{and} \qquad \frac{H}{\lambda_2} = \frac{1+2n}{4}$$

The behavior of the periodic function R_{13} is shown in Fig. 4.4.

Next, consider the low-frequency spectrum of waves inside and beneath the layer. As follows from eqs. 4.72, if $\lambda_2 \to \infty$ then

$$C = \frac{Z_1}{Z_2}\,(1 + R_{13})\,\frac{e^{i\,k_1\,h_1}}{1 + R_{23}}\, e^{-i\,k_2\,h_1}$$

and

$$D = \frac{Z_1}{Z_2}\,(1 + R_{13})\,\frac{R_{23}\, e^{i\,k_1\,h_1}}{1 + R_{23}}\, e^{i\,k_2\,h_1}$$

Therefore, the potential inside the layer is

$$\mathcal{U}_2 \approx \frac{A}{i\,k_2}\,\frac{Z_1}{Z_2}\,\frac{1 + R_{13}}{1 + R_{23}}\left[e^{i\,k_2\,(z\,-\,h_1)} + R_{23}\, e^{-i\,k_2\,(z\,-\,h_1)}\right]$$

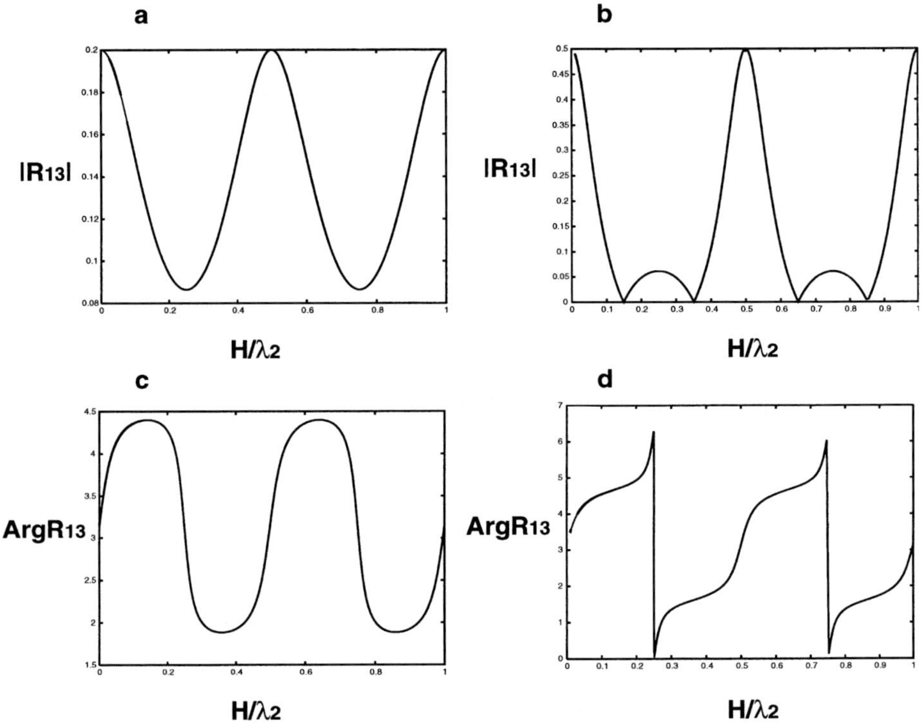

Figure 4.4: Amplitude (a, b) and phase (c, d) responses of the function R_{13} for two different three-layered media. The impedance ratios are $Z_1/Z_2/Z_3 = 3/5/2$ (a, c) and 3/8/1 (b, d).

$$\text{if} \qquad \lambda_2 \gg H \qquad \text{and} \qquad k_1 h_1 < 1$$

Bearing in mind that the exponent $k_2\,(z - h_1) \ll 1$, the complex amplitude $\mathcal{S}_{22}$ of the displacement can be represented as

$$\mathcal{S}_{2z} \approx A\,\frac{Z_1}{Z_2}\,\frac{2\,Z_3}{Z_1 + Z_3}\,\frac{1 - R_{23}}{1 + R_{23}}$$

or

$$\mathcal{S}_{2z} \approx A\,\frac{Z_1}{Z_2}\,\frac{2\,Z_3}{Z_1 + Z_3}\,\frac{Z_2}{Z_3} = A\,\frac{2\,Z_1}{Z_1 + Z_3} \tag{4.79}$$

Thus, the particle displacement inside the layer is independent of its impedance Z_2. For example, if $Z_1 = Z_3$,

$$\mathcal{S}_{2z} = A \tag{4.80}$$

but when $Z_3 = 0$,

$$\mathcal{S}_{2z} = 2\,A\,, \tag{4.81}$$

which corresponds to the free surface.

The former, eq. 4.80, shows that particles inside the layer move at low frequencies with the same velocity as particles of a homogeneous medium.

Let us also examine the low-frequency spectrum of the transmitted wave. The last equation of the set 4.72, gives

$$E \approx A\,\frac{Z_1}{Z_2}\,(1 + R_{13})\,\frac{1 - R_{23}}{1 + R_{23}}\,e^{i\,k_1\,h_1}\,e^{-i\,k_3\,h_2}$$

or

$$\mathcal{U}_3 \approx \frac{A}{i\,k_3}\,e^{i\,k_1\,h_1}\,\frac{2\,Z_1}{Z_1 + Z_3}\,e^{i\,k_3\,(z - h_2)} \qquad \text{if} \qquad \lambda_2 \gg H\,, \tag{4.82}$$

which corresponds to the transmitted wave in a medium with a single interface.

Layer inside a homogeneous space

Next consider one special case of the medium when $Z_1 = Z_3$. Since

$$R_{23} = -R_{12}\,,$$

eq. 4.71 yields

$$R_{13} = R_{12}\,\frac{1 - e^{2\,i\,k_2\,H}}{1 - R_{12}^2\,e^{2\,i\,k_2\,H}}$$

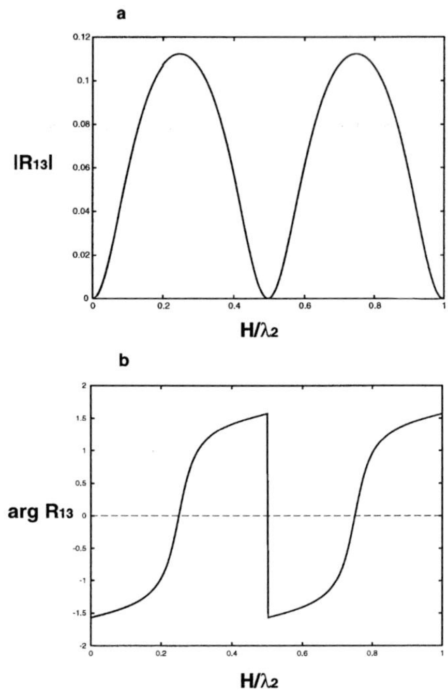

Figure 4.5: Amplitude (a) and phase (b) of the function R_{13} in the case of $Z_1 = Z_3$. The impedance ratios are $Z_2/Z_1 = Z_2/Z_3 = 2.5$.

Then, we can observe an interesting phenomenon, namely, the reflected wave vanishes at certain frequencies. This happens when

$$e^{2\,i\,k_2\,H} = 1 \qquad \text{or} \qquad k_2\,H = \pi\,m\,,$$

where $m = 0,\, 1,\, 2,\, \ldots$

Therefore, the reflected wave is absent due to destructive interference if the wavelength in the layer satisfies the condition

$$\lambda_2 = \frac{2\,H}{m}$$

Note that the same phenomenon can be observed in other cases (Fig. 4.4b). Frequency dependence of the function R_{13} is shown in Fig. 4.5.

Transient waves

For illustration, consider the reflected wave in the upper medium. From eqs. 4.63, 4.71, and 4.72, we have

$$B = R_{13}\,e^{2\,i\,k_1\,h_1}\,A\,, \qquad\qquad R_{13} = \frac{R_{12} + R_{23}\,e^{2\,i\,k_2\,H}}{1 + R_{12}\,R_{23}\,e^{2\,i\,k_2\,H}}$$

and

$$\mathcal{U}_{1r} = \frac{B}{i\,k_1}\, e^{-i\,k_1\,z} \tag{4.83}$$

In order to describe the transient wave, we apply the Fourier transform

$$U_{1r}(t) = \frac{1}{2\,\pi} \int\limits_{-\infty}^{+\infty} \mathcal{F}_0(\omega)\, \mathcal{U}_{1r}(\omega)\, e^{-i\,\omega\,t}\, d\omega\,, \tag{4.84}$$

where $\mathcal{F}_0(\omega)$ is the complex amplitude of the incident wave with the unit amplitude.

As an example, suppose that the incident wave is described by a δ-function. Then its spectrum is

$$\mathcal{F}_0(\omega) = 1$$

Therefore, the reflected wave is

$$U_{1r}(t) = \frac{1}{2\,\pi} \int\limits_{-\infty}^{+\infty} \mathcal{U}_{1r}(\omega)\, e^{-i\,\omega\,t}\, d\omega \tag{4.85}$$

To express the potential $U_{1r}(t)$ in terms of elementary functions, we represent the coefficient B as a series. Since

$$\frac{1}{1+x} = \sum_{n=0}^{\infty} (-1)^n\, x^n\,,$$

the quantity R_{13} becomes

$$\begin{aligned} R_{13} &= R_{12} \sum_{n=0}^{\infty} (-1)^n\, R_{12}^n\, R_{23}^n\, e^{2\,i\,n\,k_2\,H} \\ &\quad + R_{23} \sum_{n=0}^{\infty} (-1)^n\, R_{12}^n\, R_{23}^n\, e^{2\,i\,(n+1)\,k_2\,H} \end{aligned} \tag{4.86}$$

Substituting the latter into eq. 4.85, we obtain two series of Fourier integrals describing the potential of reflected wave. For simplification, we place the coordinate origin on the upper boundary of the layer ($h_1 = 0$) and examine the displacement of the reflected wave:

$$\begin{aligned} s_z = \frac{\partial U_{1r}}{\partial z} &= -\frac{A}{2\,\pi} \sum_{n=0}^{\infty} (-1)^n\, R_{12}^{n+1}\, R_{23}^n \int\limits_{-\infty}^{+\infty} e^{-i\,\omega\,(t - t_{1n})}\, d\omega \\ &\quad - \frac{A}{2\,\pi} \sum_{n=0}^{\infty} (-1)^n\, R_{12}^n\, R_{23}^{n+1} \int\limits_{-\infty}^{+\infty} e^{-i\,\omega\,(t - t_{2n})}\, d\omega\,, \end{aligned} \tag{4.87}$$

where

$$t_{1n} = \frac{|z|}{c_1} + \frac{2\,n\,H}{c_2} \qquad \text{and} \qquad t_{2n} = \frac{|z|}{c_1} + \frac{2\,(n+1)\,H}{c_2}$$

Since

$$\frac{1}{2\,\pi} \int\limits_{-\infty}^{+\infty} e^{-i\,\omega\,(t\,-\,t_n)}\,d\omega \;=\; \delta(t-t_n)\,, \tag{4.88}$$

where $\delta(t-t_n)$ is the δ-function, eq. 4.87 reduces to

$$s_z(t) = -\,A \sum_{n=0}^{\infty} (-1)^n\, R_{12}^n\, R_{23}^n\, \left[R_{12}\,\delta(t-t_{1n}) + R_{23}\,\delta(t-t_{2n})\right] \tag{4.89}$$

This equation shows that the reflected wave in the upper medium is composed of an infinite number of waves arising because of multiple reflections inside the layer. Each of those reflections has the shape of the δ-function. It is convenient to represent eq. 4.89 in a different form:

$$\begin{aligned} s_z(t) \;&=\; -A\,R_{12}\,\delta(t-t_{10}) \\ &\quad + A \sum_{n=1}^{\infty} (-1)^{n+1}\, R_{12}^{n+1}\, R_{23}^{n}\, \delta(t-t_{1n}) \\ &\quad + A \sum_{n=0}^{\infty} (-1)^{n+1}\, R_{12}^{n}\, R_{23}^{n+1}\, \delta(t-t_{2n}) \end{aligned}$$

Replacing the summation index n by m in the first sum and letting $m = n+1$ in the second, we obtain

$$\begin{aligned} s_z(t) \;&=\; -A\,R_{12}\,\delta(t-t_{10}) \\ &\quad + A \sum_{m=1}^{\infty} (-1)^{m+1}\, R_{12}^{m+1}\, R_{23}^{m}\, \delta(t-t_m) \\ &\quad + A \sum_{m=1}^{\infty} (-1)^{m}\, R_{12}^{m-1}\, R_{23}^{m}\, \delta(t-t_m) \end{aligned}$$

or, returning to the index n

$$s_z(t) = -A\,R_{12}\,\delta\left(t-\frac{z}{c_1}\right) + A\,(1-R_{12}^2) \sum_{n=1}^{\infty} (-1)^n\, R_{12}^{n-1}\, R_{23}^n\, \delta(t-t_{1n}) \tag{4.90}$$

For example, the first term of this sum

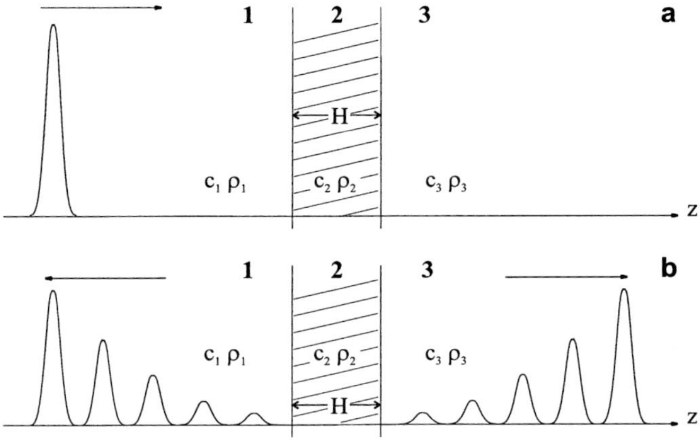

Figure 4.6: Incident, reflected and transmitted waves at given time

$$-A\,(1-R_{12}^2)\,R_{23}\,\delta(t-t_{11}) \;=\; -\,\frac{4\,A\,Z_1\,Z_2}{(Z_1+Z_2)^2}\,R_{23}\,\delta(t-t_{11}) \tag{4.91}$$

describes the reflected wave from the layer bottom which then propagates in the upper medium. In fact, the incident wave gives rise to the transmitted wave in the layer that has the amplitude

$$\frac{2\,Z_1}{Z_1+Z_2}\,A$$

This wave then, generates the reflected wave at the layer bottom with the amplitude

$$\frac{2\,Z_1}{Z_1+Z_2}\,A\,R_{23}$$

When the latter reaches the upper boundary of the layer, it causes the transmitted wave, whose amplitude

$$\frac{4\,Z_1\,Z_2}{(Z_1+Z_2)^2}\,R_{23}\,A$$

coincides with that described by eq. 4.91. The terms with indexes $n = 2, 3$, etc. in eq. 4.90 correspond to the waves that reach an observation point at later times, and in general their amplitudes become smaller as n increases because

$$R_{12}\,R_{23} < 1$$

Let us notice that if the incident wave is an arbitrary function of time, we obtain an equation which is similar to eq. 4.90.

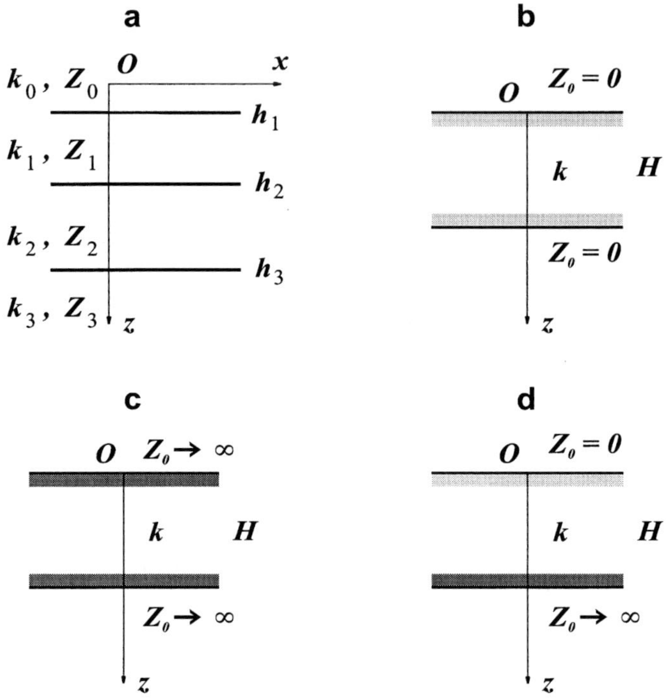

Figure 4.7: Models of layered media.

Suppose that an incident wave is represented by a relatively narrow impulse, that is, the high frequencies in its spectrum play an important role. Then the reflected and transmitted waves are described by a system of separate impulses with decreasing magnitudes, Fig. 4.6. In principle, all impulses except the first contain information about the layer thickness. However, with an increase of observation time, the magnitude of reflected and transmitted waves becomes so small that they cannot be practically detected. Certainly, a decrease of the layer thickness produces overlapping of the reflected impulses.

Recursive formula for function R_n

We have considered in detail the behavior of a plane wave in a medium with two horizontal interfaces. It is obvious that similar features are observed in media containing N layers. In particular, there is an infinite number of waves moving in the opposite directions in each layer. As an example, we derive expression for the potential of the reflected wave in the upper part of the medium with three interfaces, Fig. 4.7a and then extend it to an arbitrary number of layers. By analogy with eqs. 4.64, the system of equations with respect to unknowns A_1, A_2, A_3 and B_1, B_2, B_3 is

$$
\begin{aligned}
Z_0\left(A_0\,e^{i\,k_0\,h_1}+B_0\,e^{-i\,k_0\,h_1}\right) &= Z_1\left(A_1\,e^{i\,k_1\,h_1}+B_1\,e^{-i\,k_1\,h_1}\right),\\
A_0\,e^{i\,k_0\,h_1}-B_0\,e^{-i\,k_0\,h_1} &= A_1\,e^{i\,k_1\,h_1}-B_1\,e^{-i\,k_1\,h_1},\\
Z_1\left(A_1\,e^{i\,k_1\,h_2}+B_1\,e^{-i\,k_1\,h_2}\right) &= Z_2\left(A_2\,e^{i\,k_2\,h_2}+B_2\,e^{-i\,k_2\,h_2}\right),\\
A_1\,e^{i\,k_1\,h_2}-B_1\,e^{-i\,k_1\,h_2} &= A_2\,e^{i\,k_2\,h_2}-B_2\,e^{-i\,k_2\,h_2},\\
Z_2\left(A_2\,e^{i\,k_2\,h_3}+B_2\,e^{-i\,k_2\,h_3}\right) &= Z_3\,E\,e^{i\,k_3\,h_3},\\
A_2\,e^{i\,k_2\,h_3}-B_2\,e^{-i\,k_2\,h_3} &= E\,e^{i\,k_3\,h_3}
\end{aligned}
\tag{4.92}
$$

The last two equations give

$$\frac{A_2\,e^{i\,k_2\,h_3}+B_2\,e^{-i\,k_2\,h_3}}{A_2\,e^{i\,k_2\,h_3}-B_2\,e^{-i\,k_2\,h_3}} = \frac{Z_3}{Z_2}$$

or

$$\frac{1+\exp\left[\,\ln(B_2/A_2)-2ik_2h_3\,\right]}{1-\exp\left[\,\ln(B_2/A_2)-2ik_2h_3\,\right]} = \frac{Z_3}{Z_2}$$

or

$$\coth\left(-\frac{1}{2}\,\ln\frac{B_2}{A_2}+i\,k_2\,h_3\right) = \frac{Z_3}{Z_2}$$

Thus,

$$-\frac{1}{2}\,\ln\frac{B_2}{A_2} = \coth^{-1}\frac{Z_3}{Z_2}-i\,k_2\,h_3 \tag{4.93}$$

From the next two equations of the set 4.92, we have

$$\coth\left(i\,k_1\,h_2-\frac{1}{2}\,\ln\frac{B_1}{A_1}\right) = \frac{Z_2}{Z_1}\,\coth\left(i\,k_2\,h_2-\frac{1}{2}\,\ln\frac{B_2}{A_2}\right) \tag{4.94}$$

Substituting eq. 4.93 into eq. 4.94 yields

$$\coth\left(i\,k_1\,h_2-\frac{1}{2}\,\ln\frac{B_1}{A_1}\right) = \frac{Z_2}{Z_1}\,R_2, \tag{4.95}$$

where

$$R_2 = \coth\left(i\,k_2\,h_2 - i\,k_2\,h_3 + \coth^{-1}\frac{Z_3}{Z_2}\right)$$
$$= \coth\left(-i\,k_2\,H_2 + \coth^{-1}\frac{Z_3}{Z_2}\right) \tag{4.96}$$

As follows from eq. 4.95

$$-\frac{1}{2}\ln\frac{B_1}{A_1} = \coth^{-1}\frac{Z_2}{Z_1}R_2 - i\,k_1\,h_2 \tag{4.97}$$

The first equations of the set 4.92 give

$$\frac{A_0\,e^{i\,k_0\,h_1} + B_0\,e^{-i\,k_0\,h_1}}{A_0\,e^{i\,k_0\,h_1} - B_0\,e^{i\,k_0\,h_1}} = \frac{Z_1}{Z_0}\coth\left(i\,k_1\,h_1 - \frac{1}{2}\ln\frac{B_1}{A_1}\right) \tag{4.98}$$

or, making use of eq. 4.97,

$$\frac{A_0\,e^{i\,k_0\,h_1} + B_0\,e^{-i\,k_0\,h_1}}{A_0\,e^{i\,k_0\,h_1} - B_0\,e^{-i\,k_0\,h_1}} = \frac{Z_1}{Z_0}\coth\left(-i\,k_1\,H_1 + \coth^{-1}\frac{Z_2}{Z_1}R_2\right) = \frac{Z_1}{Z_0}R_3 \tag{4.99}$$

Here

$$R_3 = \coth\left(-i\,k_1\,H_1 + \coth^{-1}\frac{Z_2}{Z_1}R_2\right) \tag{4.100}$$

Correspondingly, for the coefficient B_0 of the reflected wave, we obtain

$$B_0 = \frac{R_3\,Z_1/Z_0 - 1}{R_3\,Z_1/Z_0 + 1}\,e^{2\,i\,k_0\,h_1} \tag{4.101}$$

Knowing B_0 it is a simple matter to determine other coefficients.

Consider two cases that illustrate the recursive character of formulae for R_3.

(a) Suppose that there is only one interface and $Z_1 = Z_2 = Z_3$. Then we have

$$R_2 = \frac{Z_3}{Z_1} = 1 \qquad \text{and} \qquad R_3 = 1$$

Therefore,

$$B_0 = \frac{Z_1 - Z_0}{Z_1 + Z_0}\,A\,,$$

where

$$A = A_0\, e^{2\,i\,k_0\,h_1}$$

and B_0 is the known expression for the reflection coefficient in the medium with one interface.

(b) If there are two boundaries, when $H_2 \to \infty$ and $Z_2 = Z_3$, we obtain

$$R_2 = 1\,, \qquad \text{while} \qquad R_3 = \coth\left(-i\,k_1\,H_1 + \coth^{-1}\frac{Z_2}{Z_1}\right) \tag{4.102}$$

Performing simple algebra, we can see that the coefficient B_0 coincides with that given by eq. 4.69.

This analysis allows us to generalize eq. 4.101 for an N-layered medium:

$$B_0 = \frac{(Z_1/Z_0)\,R_n - 1}{(Z_1/Z_0)\,R_n + 1}\,A_0\,e^{2\,i\,k_0\,h_1}\,, \tag{4.103}$$

where

$$R_n = \coth\left(-i\,k_1\,H_1 + \coth^{-1}\frac{Z_2}{Z_1}\,R_{n-1}\right)\,, \qquad \ldots, \qquad R_1 = 1 \tag{4.104}$$

Normal modes

Next we study wave behavior inside a layer surrounded by homogeneous media and the main attention is paid to the conditions in which normal modes are formed. With this purpose in mind consider three different media. It may be proper to notice that the same phenomenon was already discussed in the case of a spring that has a finite length, (Part I).

Example one Suppose that a layer is bounded by a free space and there is a plane wave propagating along the z-axis, Fig. 4.7b. It is clear that infinite number of reflections arise when the incident wave reaches the top and bottom of the layer. Our goal is to determine the total wave field represented by an infinite sum of waves. The complex amplitude of the potential for each reflected wave can be written as

$$\mathcal{U}_n = \frac{1}{i\,k_n}\left(A_n\,e^{i\,k_n\,z} + B_n\,e^{-i\,k_n\,z}\right)\,, \tag{4.105}$$

where $k_n = \omega_n/c$ is the wavenumber and c is the phase velocity. It is obvious that function $\mathcal{U}_n$ satisfies the Helmholtz equation

$$\nabla^2\,\mathcal{U}_n + k_n^2\,\mathcal{U}_n = 0$$

Let us place the coordinate origin at the upper boundary $z = 0$. Since the pressure vanishes at the free surface, the potential $\mathcal{U}_n$ has to satisfy the following conditions

$$\begin{aligned} A_n + B_n &= 0 \\ A_n\, e^{i\,k_n\,H} + B_n\, e^{-i\,k_n\,H} &= 0 \end{aligned} \tag{4.106}$$

Here H is the layer thickness, Fig. 4.7b. Eliminating the coefficients A_n and B_n, we obtain

$$e^{2\,i\,k_n\,H} = 1 = e^{i\,2\,\pi\,n} \tag{4.107}$$

Thus, in order to satisfy boundary conditions 4.106, the wavenumbers k_n cannot be arbitrary. Instead, they have to be equal to

$$k_n\,H = \pi\,n \qquad \text{or} \qquad k_n = \frac{\pi\,n}{H} \tag{4.108}$$

where n is an integer.

As follows from eqs. 4.106, $B_n = -A_n$ and, correspondingly, eq. 4.105 becomes

$$\mathcal{U}_n = \frac{A_n}{i\,k_n}\left(e^{i\,k_n\,z} - e^{-i\,k_n\,z}\right)$$

or

$$\mathcal{U}_n = \frac{2\,A_n}{k_n}\,\sin\,k_n\,z \tag{4.109}$$

In general, the coefficients A_n are functions of n. Regardless of their values, however, the complex amplitude $\mathcal{U}_n$ satisfies both the Helmholtz equation and the boundary conditions.

Taking into account that

$$U(z,\,t) = \text{Re}\;\mathcal{U}\,e^{-i\,\omega\,t},$$

we have

$$U(z,\,t) = 2\sum_{n=1}^{\infty} A_n\,\frac{\sin\,k_n\,z}{k_n}\,\cos\,\omega_n\,t\,, \tag{4.110}$$

where each term of the sum represents a normal mode.

By definition of the wavenumber,

$$\omega_n = k_n\,c \tag{4.111}$$

are the frequencies of those modes that can be represented as

$$\omega_n = \frac{\pi\,n}{H}\,c \tag{4.112}$$

Therefore the lowest frequency, called the basic or fundamental, is equal to

$$\omega_1 = \frac{c\,\pi}{H} \tag{4.113}$$

Clearly, it becomes smaller as the layer thickness H increases.

The period of this oscillation

$$T = 2\,\frac{H}{c}$$

defines the time interval during which the wave propagates from one boundary to another and back, and shows that the wavelength corresponding to this mode is

$$\lambda_1 = 2\,H$$

As follows from eq. 4.110, the term with index n of this Fourier series describes the standing wave along the z-axis. It has nodes of the pressure at points

$$\frac{\pi\,n}{H}\,z_m = \pi\,m \qquad \text{or} \qquad z_m = \frac{m\,H}{n}, \qquad m = 1,\,2,\,\ldots \tag{4.114}$$

A primary source usually generates a wave with a continuous spectrum that contains all frequencies. However, after the source action ceases the oscillations with frequencies that do not obey eq. 4.112 rapidly decay due to destructive interference. In contrast, the normal modes with frequencies ω_n exist infinitely long in the absence of attenuation. We note that the unknown coefficients A_n may be determined only if the primary source is known.

Example two In this case the layer is surrounded by an absolutely rigid medium, Fig. 4.7c, so the particles of boundaries cannot move. Taking into account a relationship between the potential, eq. 4.105, and the displacement

$$\mathcal{S}_z = \frac{\partial\,\mathcal{U}}{\partial\,z},$$

the boundary conditions give the following system of equations:

$$\begin{aligned} A_n - B_n &= 0 \\ A_n\,e^{i\,k_n\,H} - B_n\,e^{-i\,k_n\,H} &= 0 \end{aligned}$$

Therefore,

$$A_n = B_n \qquad \text{and} \qquad e^{2\,i\,k_n\,H} = 1 \tag{4.115}$$

Since the last equality coincides with eq. 4.107, the frequencies of normal modes are the same as those in the previous example.

The function

$$\mathcal{U}_n = 2\,\frac{A_n}{i\,k_n}\,\cos k_n\,z\,,$$

however, is different. Correspondingly,

$$U_n = -2\,\frac{A_n}{k_n}\,\cos k_n\,z\,\sin\omega_n\,t\,, \tag{4.116}$$

and the total wave is

$$U(z,\,t) = -2\sum_{n=1}^{\infty} A_n\,\frac{\cos k_n\,z}{k_n}\,\sin\omega_n\,t \tag{4.117}$$

Positions of the pressure nodes are now defined as

$$k_n\,z_m = \frac{\pi}{2}\,(2\,m-1) \qquad \text{or} \qquad z_m = \frac{\pi}{2\,k_n}\,(2\,m-1) \tag{4.118}$$

Example three Now we assume that in the upper medium $Z_0 = 0$ while the lower one is rigid, Fig. 4.7d. Since

$$\mathcal{P}(0) = 0 \qquad \text{and} \qquad \mathcal{S}_z(H) = 0\,,$$

the system of equations that defines the frequencies of normal modes is

$$\begin{cases} A_n + B_n = 0 \\ A_n\,e^{i\,k_n\,H} - B_n\,e^{-i\,k_n\,H} = 0 \end{cases}$$

Whence

$$A_n = -B_n \qquad \text{and} \qquad e^{2\,i\,k_n\,H} = -1 \tag{4.119}$$

or

$$e^{2\,i\,k_n\,H} = e^{i\,(2\,n-1)\,\pi}$$

The wavenumbers of those oscillations are equal to

$$k_n = \frac{(2n-1)\,\pi}{2\,H} \qquad \text{or} \qquad f_n = \frac{(2n-1)\,c}{4\,H} \tag{4.120}$$

Respectively, the complex amplitude of the potential for each normal mode is

$$\mathcal{U}_n = \frac{2\,A_n}{k_n}\,\sin\,k_n\,z\,,$$

which leads to the total wave field

$$U(z,\,t) = 2\sum_{n=1}^{\infty} A_n\,\frac{\sin\,k_n\,z}{k_n}\,\cos\,\omega_n\,t \tag{4.121}$$

As we see, eqs. 4.110 and 4.121 coincide, however, they have different values of k_n and A_n. Nevertheless, similarly to the previous cases, each term in the Fourier series 4.121 describes the standing wave along the z-axis.

In all three examples we solved the boundary value problem and found the frequencies of normal modes. It is possible to obtain the same result in a different way. Consider some plane $z =$ const inside the layer. In particular, it can serve as the phase surface of a down-going wave with the initial phase equal to $k\,z$. After one reflection from the bottom and top boundaries, we again deal with the down-going wave whose phase at the plane $z =$ const is

$$k\,z + 2\,k\,H + \varphi_0 \tag{4.122}$$

Here φ_0 is the sum of phases caused by reflections at both interfaces. Thus, the phase difference between two those waves is

$$\Delta\varphi = 2\,k\,H + \varphi_0 \tag{4.123}$$

It is essential that the latter is independent of the z-coordinate and becomes greater when the frequency increases. Inasmuch as the normal modes arise due to constructive interference, the function $\Delta\varphi$ has to be equal to $2\,\pi\,n$:

$$2\,k_n\,H + \varphi_0 = 2\,\pi\,n\,, \tag{4.124}$$

where n is an integer.

For instance, the phase shift at each boundary in the first example is π, and, correspondingly, eq. 4.124 becomes

$$2\,k_n\,H = 2\,\pi\,(n-1) \qquad \text{or} \qquad k_n\,H = \pi\,(n-1)\,, \tag{4.125}$$

which coincides with eq. 4.108. Let us also notice that in all three our examples reflections did not produce waves outside the layer. In other words, the energy of wave does not change which is vital for the existence of normal modes.

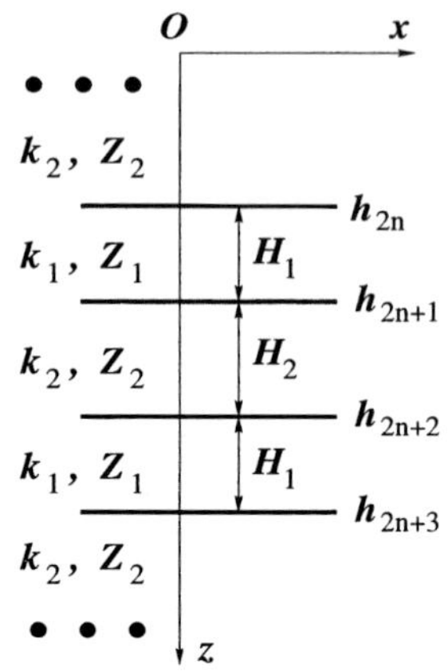

Figure 4.8: Periodic sequence of two layers.

Periodic media

It is also interesting to study propagation of waves for the normal incidence in media containing a periodic sequence of layers. The simplest model consists of only two kinds of layers, and it is shown in Fig. 4.8. Our goal is to determine the reflection coefficient B_0, and with purpose in mind, we proceed from the first two equations of the set 4.104

$$R_n = \coth\left(-i\,k_1\,H_1 + \coth^{-1}\frac{Z_2}{Z_1}\,R_{n-1}\right)\,,$$
$$R_{n-1} = \coth\left(-i\,k_2\,H_2 + \coth^{-1}\frac{Z_3}{Z_2}\,R_{n-2}\right) \qquad (4.126)$$

As follows from the derivation in the previous section, functions R_n depend only on parameters of the media located beneath the plane $z = \text{const}$.

Combining eqs. 4.126, we obtain

$$R_n = \coth\left\{-i\,k_1\,H_1 + \coth^{-1}\left[\frac{Z_2}{Z_1}\coth\left(-i\,k_2\,H_2 + \coth^{-1}\frac{Z_3}{Z_2}\,R_{n-2}\right)\right]\right\}\,, \qquad (4.127)$$

which represents the relation between R_n and R_{n-2} at the planes $z = h_{2n}$ and $z = h_{2n+2} = h_{2n} + H_1 + H_2$.

Inasmuch as the lower medium is periodic

$$R \equiv R_n = R_{n-2} \qquad \text{and} \qquad \frac{Z_3}{Z_2} = \frac{Z_1}{Z_2}$$

Then eq. 4.127 becomes

$$R = \coth\left\{-i\,k_1\,H_1 + \coth^{-1}\left[\frac{Z_2}{Z_1}\coth\left(-i\,k_2\,H_2 + \coth^{-1}\frac{Z_1}{Z_2}\,R\right)\right]\right\} \tag{4.128}$$

Using the equality

$$\coth(x+y) = \frac{1+\coth x\,\coth y}{\coth x + \coth y}\,,$$

and denote

$$\zeta = \frac{Z_2}{Z_1}\coth\left(-i\,k_2\,H_2 + \coth^{-1}\frac{Z_1}{Z_2}\,R\right) = \frac{\dfrac{Z_2}{Z_1} + R\coth(-i\,k_2\,H_2)}{\coth(-i\,k_2\,H_2) + \dfrac{Z_1}{Z_2}\,R}, \tag{4.129}$$

we obtain

$$R = \coth\left(-i\,k_1\,H_1 + \coth^{-1}\zeta\right) = \frac{1+\zeta\,\coth(-i\,k_1\,H_1)}{\zeta + \coth(-i\,k_1\,H_1)} \tag{4.130}$$

Substitution of ζ into eq. 4.130 gives the quadratic equation with respect to R

$$\left[\coth\left(-i\,k_2\,H_2\right) + \frac{Z_1}{Z_2}\coth\left(-i\,k_1\,H_1\right)\right]R^2 + \left(\frac{Z_2}{Z_1} - \frac{Z_1}{Z_2}\right)R -$$

$$-\left[\frac{Z_2}{Z_1}\coth\left(-i\,k_1\,H_1\right) + \coth\left(-i\,k_2\,H_2\right)\right] = 0$$

which can be easily solved.

It is instructive to examine one special case in which

$$k_2\,H_2 = \pi\,m \qquad \text{or} \qquad f_n = \frac{m\,c}{2\,H} \tag{4.131}$$

and m is an integer. Then since

$$\coth(-i\,k_2\,H_2) \to \infty\,,$$

eq. 4.129 yields

$$\zeta = R$$

Therefore, eq. 4.130 reduces to

$$R = \frac{1 + R\coth(-i\,k_1\,H_1)}{R + \coth(-i\,k_1\,H_1)}\,,$$

and its solution is

$$R = \pm 1$$

As follows from eq. 4.103, the reflection coefficient B_0, corresponding to the root $R = 1$ is

$$B_0 = \frac{Z_1 - Z_0}{Z_1 + Z_0} A_0 \, e^{2\, i\, k_0\, h_1}$$

The latter coincides with the expression for media with only one plane boundary at $z = h_1$, that is due to destructive interference the total reflection from other interfaces is absent if the condition 4.131 is met.

Chapter 5

Plane waves in layered media (oblique incidence)

In this chapter we continue our study of plane waves in horizontally layered media when phase surfaces of an incident wave are arbitrarily oriented with respect to the interfaces. Some attention is paid to media in which velocity is a continuous function of the vertical coordinate.

5.1 Reflection and transmission of plane wave $(\alpha < \alpha_{\mathbf{c}})$

At the beginning we assume that the incident wave is sinusoidal and homogeneous. The latter means that the wave amplitude does not change on the phase surface. The direction of wave propagation is characterized by the incidence angle α shown in Fig. 5.1. As before, in order to find reflected and transmitted waves, we have to solve the boundary value problem. With this purpose in mind, let us choose the Cartesian coordinates with the origin O at the interface. The y-axis is tangential to phase surfaces N_i of the incident wave. The z-axis is perpendicular to the boundary and is directed downward, Fig. 5.1.

In Chapter 1 we introduced the critical angle α_c as

$$\sin\alpha_c = \frac{c_1}{c_2} \qquad \text{if} \qquad c_2 > c_1 \tag{5.1}$$

Here we restrict ourselves to the cases in which the incidence angle does not exceed the critical:

$$\alpha \leq \alpha_c \tag{5.2}$$

By definition, the complex amplitude of the potential of the incident wave is

$$\mathcal{U}_i = \frac{A}{i\,k_1}\, e^{i\,k_1\,l}\,, \tag{5.3}$$

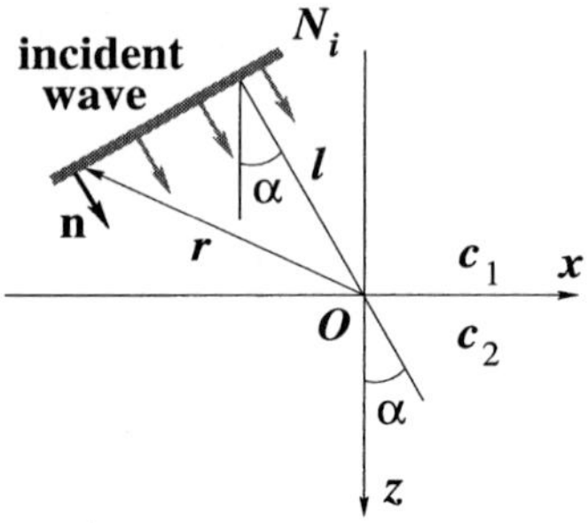

Figure 5.1: Oblique incidence of plane wave.

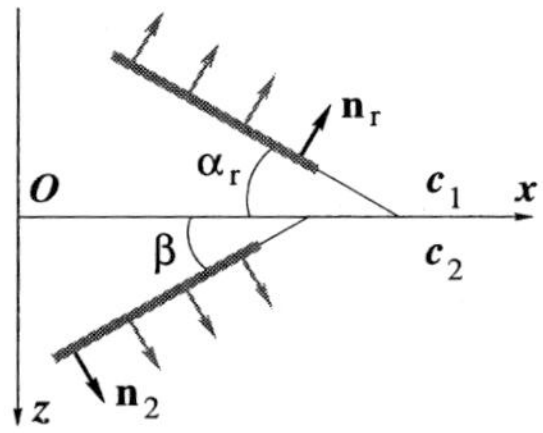

Figure 5.2: Reflected and transmitted plane waves.

where $|l|$ is the distance between the coordinate origin O and the phase surface N_i. It is important to represent l in terms of coordinates x and z. As is seen from Fig. 5.1, the unit vector $\mathbf{n}$, directed along the normal to the planar wavefront, is

$$\mathbf{n} = \sin\alpha\,\mathbf{i}_1 + \cos\alpha\,\mathbf{i}_2\,, \tag{5.4}$$

where $\mathbf{i}_1$ and $\mathbf{i}_2$ are the unit vectors along the x- and z-axes, respectively. Since l can be treated as the scalar component of the radius vector

$$\mathbf{r} = x\,\mathbf{i}_1 + z\,\mathbf{i}_2\,,$$

we have

$$l = \mathbf{r}\cdot\mathbf{n} = x\,\sin\alpha + z\,\cos\alpha$$

Therefore, eq. 5.3 can be written in the form

$$\mathcal{U}_i = \frac{A}{i\,k_1}\,\exp i\,k_1\,(x\,\sin\alpha + z\,\cos\alpha) \tag{5.5}$$

Now we assume that both the reflected and transmitted waves are also plane and homogeneous. Taking into account that unit vectors $\mathbf{n}_r$ and $\mathbf{n}_2$, which characterize wave propagation, are

$$\mathbf{n}_r = x \sin\alpha_r - z \cos\alpha_r \qquad \text{and} \qquad \mathbf{n}_2 = x \sin\beta + z \cos\beta\,, \tag{5.6}$$

we have the following expressions for the potentials of those waves:

$$\mathcal{U}_r = \frac{B}{i\,k_1} \exp i\,k_1\,(x \sin\alpha_r - z \cos\alpha_r) \tag{5.7}$$

and

$$\mathcal{U}_2 = \frac{C}{i\,k_2} \exp i\,k_2\,(x \sin\beta + z \cos\beta) \tag{5.8}$$

Here α_r and β are the angles between the boundary and phase surfaces of the reflected and transmitted waves, Fig. 5.2. It is obvious that the functions $\mathcal{U}_i$, $\mathcal{U}_r$ and $\mathcal{U}_2$ are solutions of the Helmholtz equation. In order to determine the unknown coefficients B and C, we again use boundary conditions that require continuity of the pressure and a normal component of the displacement across the interface:

$$\rho_1\,\mathcal{U}_1 = \rho_2\,\mathcal{U}_2\,, \qquad \frac{\partial\,\mathcal{U}_1}{\partial\,z} = \frac{\partial\,\mathcal{U}_2}{\partial\,z} \qquad \text{if} \qquad z = 0\,, \tag{5.9}$$

where

$$\mathcal{U}_1 = \mathcal{U}_i + \mathcal{U}_r$$

Substitution of eqs. 5.3, 5.7, and 5.8 into the set 5.9 yields

$$Z_1\,[A \exp{(i\,k_1\,x \sin\alpha)} + B \exp{(i\,k_1\,x \sin\alpha_r)}] = Z_2\,C \exp{(i\,k_2\,x \sin\beta)} \tag{5.10}$$

and

$$\begin{aligned} A \cos\alpha \exp{(i\,k_1\,x \sin\alpha)} - B \cos\alpha_r \exp{(i\,k_1\,x \sin\alpha_r)} \\ = C \cos\beta \exp{(i\,k_2\,x \sin\beta)} \end{aligned} \tag{5.11}$$

Since these equalities take place for all values of x, the arguments of the exponential terms have to be equal to each other. Thus we have

$$i\,k_1\,x \sin\alpha = i\,k_1\,x \sin\alpha_r = i\,k_2\,x \sin\beta \tag{5.12}$$

or

$$\alpha = \alpha_r \qquad \text{and} \qquad \frac{\sin\alpha}{c_1} = \frac{\sin\beta}{c_2} \tag{5.13}$$

Correspondingly, in place of eqs. 5.10 and 5.11, we obtain a much simpler system:

$$\begin{aligned} (A+B)\,Z_1 &= C\,Z_2 \\ (A-B)\,\cos\alpha &= C\,\cos\beta \end{aligned} \tag{5.14}$$

Its solution is

$$B = A\,\frac{Z_2\,\cos\alpha - Z_1\,\cos\beta}{Z_2\,\cos\alpha + Z_1\,\cos\beta} \quad \text{and} \quad C = A\,\frac{2\,Z_1\,\cos\alpha}{Z_2\,\cos\alpha + Z_1\,\cos\beta} \tag{5.15}$$

One can represent these coefficients as

$$B = A\,\frac{m\,\cos\alpha - n\,\cos\beta}{m\,\cos\alpha + n\,\cos\beta} \quad \text{and} \quad C = A\,\frac{2\,n\,\cos\alpha}{m\,\cos\alpha + n\,\cos\beta}, \tag{5.16}$$

where

$$m = \frac{\rho_2}{\rho_1} \quad \text{and} \quad n = \frac{c_1}{c_2} \tag{5.17}$$

are the parameters that characterize the ratio of densities of media and propagation velocities.

As follows from eqs. 5.13

$$\cos\beta = \frac{1}{n}\,\sqrt{n^2 - \sin^2\alpha} \tag{5.18}$$

Therefore,

$$B = A\,\frac{m\,\cos\alpha - \sqrt{n^2 - \sin^2\alpha}}{m\,\cos\alpha + \sqrt{n^2 - \sin^2\alpha}} \quad \text{and} \quad C = A\,\frac{2\,n\,\cos\alpha}{m\,\cos\alpha + \sqrt{n^2 - \sin^2\alpha}} \tag{5.19}$$

Thus we have solved the boundary value problem, that is, the potential $\mathcal{U}$ satisfies the Helmholtz equation in the upper and lower parts of the medium as well as the boundary conditions. This means that we were correct in assuming that both the reflected and transmitted waves are homogeneous and plane. The last statement, however, is not always true, and to show that, we rewrite the second equation 5.13 in the form

$$\sin\beta = \frac{\sin\alpha}{n} \tag{5.20}$$

Since the sinusoid of a real argument cannot exceed the unity, we should expect that the transmitted wave is homogeneous only if the velocity ratio n and the incidence angle α satisfy the conditions

$$\sin\alpha \leq 1 \qquad \text{if} \qquad c_2 \leq c_1 \tag{5.21}$$

and

$$\sin\alpha \leq n = \frac{c_1}{c_2} \qquad \text{if} \qquad c_2 > c_1\,, \tag{5.22}$$

which coincide with our assumption, eq. 5.8.

Now we describe the wave behavior, provided that conditions 5.21 and 5.22 are met. Bearing in mind this fact, consider eqs. 5.13. As was shown in Chapter 1 these equalities express Snell's law for arbitrary waves when their wavelength is sufficiently small. Therefore, we conclude that the same law is valid for homogeneous plane waves, regardless of the wavelength, if the interface is planar. The first equation of the set 5.13,

$$\alpha = \alpha_r,$$

states that the incidence and reflection angles are equal. Since this equality is independent of c_1 and c_2, it remains valid even if the condition 5.22 does not take place. As follows from the second equation 5.13, the direction of propagation of a transmitted wave generally differs from that of the incident wave. It is essential that the relationship between angles α and β depends only on the ratio of velocities. Let us illustrate the orientation of phase surfaces of waves in two cases, namely, when

$$n > 1 \qquad \text{and} \qquad n < 1$$

In the first case, when the velocity of the lower medium is smaller, the position of the phase surface is shown in Fig. 5.3a. As is clearly seen the phase surface of the transmitted wave forms with the boundary the angle β, which is smaller than the angle α:

$$\beta < \alpha \tag{5.23}$$

The maximum value of the angle β is defined from eq. 5.20 as

$$\sin\beta = \frac{1}{n} = \frac{c_2}{c_1} < 1 \tag{5.24}$$

Since $n > 1$, this condition is valid for any incidence angle α, that is, the transmitted wave is homogeneous. In the second case, when $c_2 > c_1$, the relationship between angles α and β is different, Fig. 5.3b. It follows from eq. 5.20 that the transmission angle β increases with α and always exceeds it

$$\beta > \alpha \tag{5.25}$$

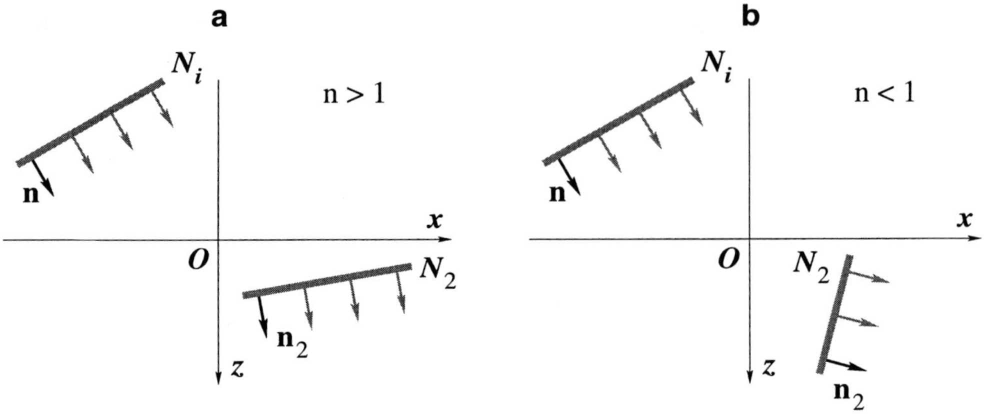

Figure 5.3: Orientation of transmitted plane waves.

Such behavior is observed until the transmitted wave starts to move parallel to the boundary. In this case, $\beta = \pi/2$ and the incidence angle α reaches its critical value α_c. Thus, Snell's law of transmission can be used for homogeneous plane waves, provided that the conditions 5.21 and 5.22 are met, that is

$$0 \leq \alpha \leq \frac{\pi}{2} \quad \text{if} \quad n > 1 \qquad \text{and} \qquad 0 \leq \alpha \leq \alpha_c \quad \text{if} \quad n < 1 \tag{5.26}$$

It may be appropriate to make the following comments

(a) Snell's law was derived from eq. 5.12, which can be rewritten as

$$\frac{x \sin \alpha}{c_1} = \frac{x \sin \alpha_r}{c_1} = \frac{x \sin \beta}{c_2} \tag{5.27}$$

Since the terms $c_1/\sin\alpha$, $c_1/\sin\alpha_r$ and $c_2/\sin\beta$ describe the apparent velocity of each wave along the boundary, we can say that Snell's law means the equivalence of those velocities for incident, reflected, and transmitted waves.

(b) This law was obtained from solution of the boundary value problem, that is, it follows from Newton's laws and the relationship between the pressure and deformation of a medium (Hooke's law). In other words, Snell's law is not independent but follows from other physical principles.

(c) The solution of the boundary value problem also allows us to study wave behavior when the incidence angle exceeds the critical one, α_c. This subject will be discussed in the next section.

Now consider the amplitudes and phases of reflected and transmitted waves. By definition of a homogeneous wave, the potential does not change along a phase surface, and, correspondingly, the particle displacement is directed along the rays, that is perpendicular to the phase surfaces. As follows from eq. 5.7 and 5.8, the horizontal and vertical components of displacements caused by waves are

$$\begin{aligned} \mathcal{S}_{rx} &= B\,\sin\alpha\,\exp\left[i\,k_1\,(x\,\sin\alpha - z\,\cos\alpha)\right] \\ \mathcal{S}_{rz} &= -B\,\cos\alpha\,\exp\left[i\,k_1\,(x\,\sin\alpha - z\,\cos\alpha)\right] \end{aligned} \tag{5.28}$$

and

$$\begin{aligned} \mathcal{S}_{2x} &= C\,\sin\beta\,\exp\left[i\,k_2\,(x\,\sin\beta + z\,\cos\beta)\right] \\ \mathcal{S}_{2z} &= C\,\cos\beta\,\exp\left[i\,k_2\,(x\,\sin\beta + z\,\cos\beta)\right]\,, \end{aligned} \tag{5.29}$$

where

$$B = \frac{\dfrac{Z_2}{Z_1}\,\cos\alpha - \sqrt{1 - \dfrac{\sin^2\alpha}{n^2}}}{\dfrac{Z_2}{Z_1}\,\cos\alpha + \sqrt{1 - \dfrac{\sin^2\alpha}{n^2}}}\,A \tag{5.30}$$

and

$$C = \frac{2\,\cos\alpha}{\dfrac{Z_2}{Z_1}\,\cos\alpha + \sqrt{1 - \dfrac{\sin^2\alpha}{n^2}}}\,A \qquad \text{if} \qquad \frac{\sin\alpha}{n} < 1$$

For comparison we write the expressions for displacement in the incident wave:

$$\begin{aligned} \mathcal{S}_{ix} &= A\,\sin\alpha\,\exp\left[i\,k_1\,(x\,\sin\alpha + z\,\cos\alpha)\right] \\ \mathcal{S}_{iz} &= A\,\cos\alpha\,\exp\left[i\,k_1\,(x\,\sin\alpha + z\,\cos\alpha)\right] \end{aligned} \tag{5.31}$$

As is seen from those equations, the amplitudes of reflected and transmitted waves, normalized by that of the incident wave, depend on three parameters, namely:

(1) the incidence angle α,

(2) the ratio n of propagation velocities and

(3) the ratio Z_2/Z_1 of acoustic impedances.

Before studying the influence of these factors on wave propagation, we describe several general features of reflected waves. First, it is obvious that when the inequality

$$\frac{Z_2}{Z_1}\cos\alpha > \sqrt{1-\frac{\sin^2\alpha}{n^2}} \tag{5.32}$$

takes place, both the incident and reflected waves have the same phase at points of the interface. In the opposite case, the phase shift between them is equal to π. In contrast, the incident and transmitted waves are always in the same phase at the interface.

As follows from the first equation of set 5.30, there may be such incidence angle when the reflected wave is absent everywhere in the upper medium. This happens if

$$\frac{Z_2}{Z_1}\cos\alpha_b = \sqrt{1-\frac{\sin^2\alpha_b}{n^2}}$$

or

$$\sin\alpha_b = \sqrt{\frac{(Z_2/Z_1)^2-1}{(Z_2/Z_1)^2-1/n^2}} \tag{5.33}$$

where α_b is called the Brewster angle . The same effect is observed for plane electromagnetic waves. It is clear that this phenomenon occurs if the following relationships between Z_2/Z_1 and n are satisfied:

$$\frac{Z_2}{Z_1} > 1 \qquad \text{and} \qquad n > 1 \tag{5.34}$$

or

$$\frac{Z_2}{Z_1} < 1 \qquad \text{and} \qquad n < 1 \tag{5.35}$$

Thus, we can say that the intensity of elementary sources of different types on the boundary is such that destructive interference of elementary waves takes place in the medium and the reflected wave is absent when the incidence angle coincides with α_b. In this case in accordance with the second equation of set 5.30,

$$C = \frac{Z_1}{Z_2}A$$

Here we observe an interesting behavior, namely, in spite of the fact that the reflected wave disappears, the amplitude of the transmitted wave is not equal to that of the incident one. This is natural because at the boundary the cross-section of the vector tubes of the Poynting vector changes.

Now suppose that the incidence angle α tends to $\pi/2$. Then, as follows from eqs. 5.28 and 5.29,

$$\mathcal{S}_{rx} = -A\exp\left(i\,k_1\,x\right), \qquad \mathcal{S}_{rz} = 0 \qquad \text{and} \qquad \mathcal{S}_{2x} = \mathcal{S}_{2z} = 0 \tag{5.36}$$

while

$$\mathcal{S}_{ix} = A \exp(i\, k_1\, x)\,, \qquad \mathcal{S}_{iz} = 0$$

Therefore, the transmitted wave is absent, and the reflected wave cancels the incident one. In other words, wave fields are absent everywhere. It is also useful to discuss another interesting feature of wave behavior. Suppose that the propagation velocity is the same everywhere

$$c_1 = c_2 = c \qquad \text{or} \qquad n = 1\,,$$

but the densities above and below the boundary are different. Then, eqs. 5.30 yield

$$B = \frac{Z_2 - Z_1}{Z_2 + Z_1} A = \frac{\rho_2 - \rho_1}{\rho_2 + \rho_1} A \qquad \text{and} \qquad C = \frac{2\, Z_1}{Z_2 + Z_1} A = \frac{2\, \rho_1}{\rho_2 + \rho_1} A \tag{5.37}$$

Thus, regardless of the incidence angle, the amplitudes of reflected and transmitted waves remain the same.

Next, examine the dependence of coefficients B and C on angle α in the case of arbitrary n. At the normal incidence, $\alpha = 0$, we arrive at the known expressions

$$B = \frac{Z_2 - Z_1}{Z_2 + Z_1} A \qquad \text{and} \qquad C = \frac{2\, Z_1}{Z_2 + Z_1} A \tag{5.38}$$

Since eqs. 5.30 do not contain $\sin\alpha$ in the first power and for small α:

$$\sin^2\alpha \approx \alpha^2 \qquad \text{and} \qquad \cos\alpha \approx 1 - \frac{\alpha^2}{2}\,,$$

eqs. 5.38 are reasonable accurate if $\alpha \ll 1$.

The behavior of function $B(\alpha)/A$ is governed by the ratios of impedances and velocities. As is seen from Fig. 5.4a, if $Z_2/Z_1 > 1$ and $n > 1$, the ratio B/A is positive when $\alpha = 0$ and decreases with an increase of α. When $\alpha = \alpha_b$, B/A reaches zero and becomes smaller approaching -1 at large α. At the same time, if $Z_2/Z_1 < 1$, the function $B(\alpha)/A$ is always negative. It gradually decreases and also tends to -1 as α increases. A completely different behavior of $B(\alpha)/A$ is observed if $n < 1$: this ratio monotonically increases and reaches $+1$ at $\alpha = \alpha_c$, Fig. 5.4b. In the case when $Z_2/Z_1 < 1$, the function $B(\alpha)/A$ changes sign at the Brewster angle α_b.

The function $C(\alpha)/A$ is shown in Fig. 5.5a, b. It also depends essentially on the ratios of velocities and impedances. As was pointed our earlier, the range of the incidence angles α that produce homogeneous transmitted wave is restricted to by $0 \le \alpha \le \sin^{-1} n$ when $n < 1$. Note that the amplitude of the transmitted wave can greatly exceed that of incident wave in the vicinity of the critical angle. This fact, however, does not contradict the principle of conservation of energy.

It is also useful to consider wave behavior in two limiting cases. As follows from eqs. 5.30 at the free surface $Z_2 = 0$,

$$B = -A$$

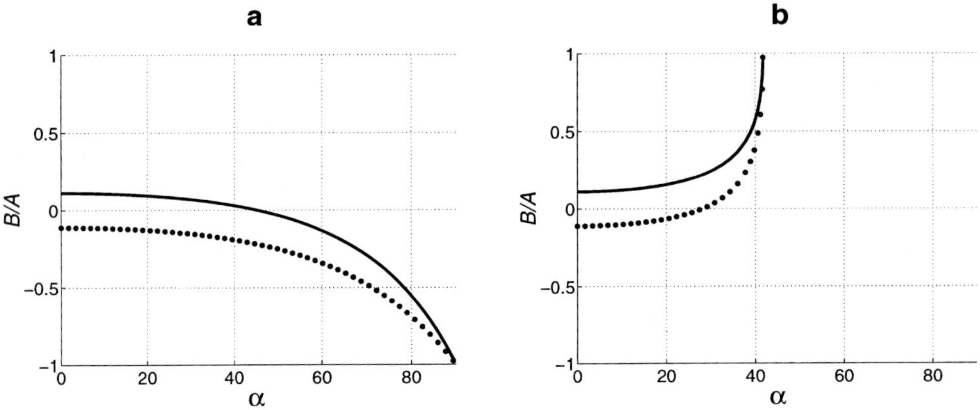

Figure 5.4: Reflection coefficient $B(\alpha)/A$ for velocity ratio $n = 3/2$ (a) and $n = 2/3$ (b). Solid line corresponds to the ratio of impedances $Z_2/Z_1 = 5/4$, dotted – to $Z_2/Z_1 = 4/5$.

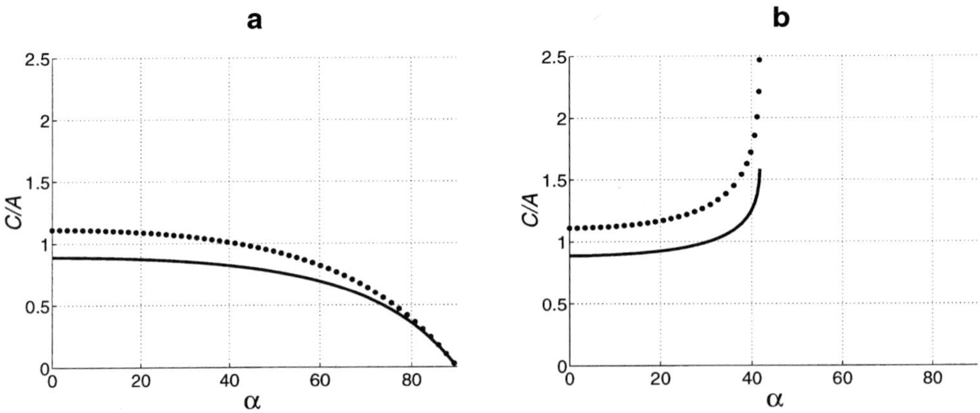

Figure 5.5: Transmission coefficient $C(\alpha)/A$ for the velocity ratio $n = 3/2$ (a) and $n = 2/3$ (b). Solid line corresponds to the ratio $Z_2/Z_1 = 5/4$, dotted – to $Z_2/Z_1 = 4/5$.

Therefore, the reflected and incident waves have equal amplitudes but opposite phases at points of the boundary regardless the angle α. In the opposite case, $Z_2 = \infty$, we obtain

$$B = A$$

We see again that the amplitude of a reflected wave is independent of α.

The derivation of coefficients B and C suggests that intensity of the virtual surface sources is defined only by the boundary conditions. Of course, this intensity depends on the pressure of the incident wave, as well as on its normal component of the displacement. For this reason, the coefficients B and C are, in general, functions of angle α. At the same time, the orientation of the phase surfaces of those waves is determined by the phase shift between the elementary sources. In fact, as follows from the Helmholtz formula, (Chapter 1), for two-dimensional case the reflected and transmitted waves can be expressed in terms of the integrals

$$I_1 = \int_S F_1 e^{i\,k_1\,(r\,+\,x\sin\alpha)}\,dx \qquad \text{and} \qquad I_2 = \int_S F_2\, e^{i\,k_2\,r}\, e^{i\,k_1 x\sin\alpha}\,dx \tag{5.39}$$

Here F_1 and F_2 are some functions and the term $e^{i\,k_1 x\sin\alpha}$ describes the phase of elementary sources that does not change along the y-axis. Thus, the wave behavior is the same in any plane perpendicular to this axis and

$$r = \sqrt{(x_p - x)^2 + z_p^2}$$

where x_p and z_p are the coordinates of the observation point p.

The phase of reflected wave caused by an elementary source is equal to

$$\varphi_r = k_1\,(r + x\,\sin\alpha)$$

Therefore, a stationary point for the first integral in eq. 5.39 is defined as

$$\frac{\partial\,\varphi_r}{\partial\,x} = 0 \qquad \text{or} \qquad -\,k_1\,\frac{x_p - x}{r} + k_1\,\sin\alpha = 0$$

Since

$$\frac{x_p - x}{r} = \sin\alpha_r$$

we again arrive at Snell's law of reflection, and this means that constructive interference takes place if an observation is located on the straight line (ray) which forms with the z-axis the angle equal to the incident one, Fig. 5.6a.

The phase of the transmitted wave is

$$\varphi_2 = k_2\,r + k_1\,x\,\sin\alpha$$

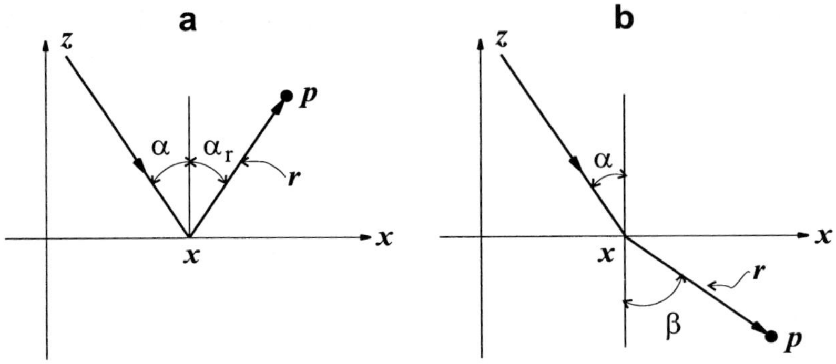

Figure 5.6: Reflected (a) and transmitted (b) rays.

and, hence, the stationary point is defined as

$$-k_2\,\frac{x_p - x}{r} + k_1\,\sin\alpha = 0$$

The latter gives Snell's law of refraction, Fig. 5.6b.

As was mentioned previously, the method of stationary phase is applied when the wavelength is relatively small. In such a case, due to strong oscillations of the integrands in eqs. 5.39, the integrals are mainly determined by some vicinity of the stationary point. However, reflection and transmission of plane waves at plane interface is an exception, since the method of stationary phase gives the correct result (Snell's law) regardless of the wavelength.

Until now we have examined only sinusoidal waves. Inasmuch as coefficients B and C in eqs. 5.30 are real and are independent of frequency, from Fourier's integral follows that they also describe the behavior of transient waves. Thus, we have for their acoustic potentials:

$$\begin{aligned} U_i &= A\,f\left[a\left(t - \frac{x\,\sin\alpha + z\,\cos\alpha}{c_1}\right)\right] \\ U_r &= B\,f\left[a\left(t - \frac{x\,\sin\alpha - z\,\cos\alpha}{c_1}\right)\right] \\ U_2 &= C\,f\left[a\left(t - \frac{x\,\sin\beta + z\,\cos\beta}{c_2}\right)\right], \end{aligned} \tag{5.40}$$

where B and C are given by eqs. 5.30. It is essential that all three waves have the same shape.

5.2 Reflection and refraction of sinusoidal waves $(\alpha > \alpha_c)$

Next we consider an interesting phenomenon called total internal reflection. It may occur when the second medium has higher velocity:

$$c_2 > c_1 \qquad \text{or} \qquad n < 1$$

As was shown earlier, if the incidence angle α is equal to the critical one

$$\sin\alpha_c = n\,,$$

then the transmission angle is $\pi/2$. In other words, this wave moves parallel to the interface and still remains homogeneous, that is, the wave magnitude has the same value on its phase surface at each instant. Let us examine what happens to the reflected and transmitted waves when the incidence angle exceeds the critical.

As demonstrated in the previous section

$$\sin\beta = \frac{\sin\alpha}{n} \qquad \text{or} \qquad \frac{\sin\alpha}{n} > 1 \qquad \text{if} \qquad \alpha > \alpha_c \tag{5.41}$$

Since $\sin\beta$ cannot exceed unity if β is real, we have to conclude that the law of refraction

$$\frac{\sin\alpha}{c_1} = \frac{\sin\beta}{c_2}$$

derived for plane waves from boundary conditions becomes invalid. This may suggest that the transmitted wave is no longer homogeneous if $\alpha > \alpha_c$. To confirm this behavior and find expressions for the reflected and transmitted waves, we solve the boundary value problem again but in a slightly different way. It is essential that this approach allows us to study wave behavior for any value of the incidence angle $0 \le \alpha \le \pi/2$.

As before, when $\alpha \le \alpha_c$, we assume that the reflected plane wave is homogeneous, and the law of reflection $\alpha_1 = \alpha_r = \alpha$ takes place. In accordance with eqs. 5.5 and 5.7, the expressions for complex amplitudes of the potential describing the incident and reflected waves are

$$\mathcal{U}_i = \frac{A}{i\,k_1}\exp\left[i\,k_1\,(x\,\sin\alpha + z\,\cos\alpha)\right] \tag{5.42}$$

and

$$\mathcal{U}_r = \frac{B}{i\,k_1}\exp\left[i\,k_1\,(x\,\sin\alpha - z\,\cos\alpha)\right]$$

To find the potential $\mathcal{U}_2$ we proceed from the Helmholtz equation

$$\frac{\partial^2 \mathcal{U}_2}{\partial x^2} + \frac{\partial^2 \mathcal{U}_2}{\partial z^2} + k_2^2 \mathcal{U}_2 = 0 \tag{5.43}$$

Applying the method of separation of variables the potential can be presented as

$$\mathcal{U}_2 (x,\, z) = T(x)\, P(z) \tag{5.44}$$

Its substitution into eq. 5.43 yields

$$P \frac{d^2 T}{d x^2} + T \frac{d^2 P}{d z^2} + k_2^2\, P T = 0\,,$$

or

$$\frac{1}{T} \frac{d^2 T}{d x^2} + \frac{1}{P} \frac{d^2 P}{d z^2} + k_2^2 = 0 \tag{5.45}$$

This equality takes place if the first two terms are constants, that is,

$$\frac{1}{T} \frac{d^2 T}{d x^2} = -k_2^2\, n_x^2\,, \qquad\qquad \frac{1}{P} \frac{d^2 P}{d z} = -k_2^2\, n_z^2 \tag{5.46}$$

Here n_x and n_z are quantities, that are related to each other. For instance, substitution of eq. 5.46 into eq. 5.45 gives

$$n_x^2 + n_z^2 = 1 \tag{5.47}$$

In particular, n_x and n_z describe directional cosines of the unit vector normal to phase surfaces of the incident and reflected waves. It is obvious that functions

$$\exp\,(\pm i\, k_2\, n_x\, x) \qquad \text{and} \qquad \exp\,(\pm i\, k_2\, n_z\, z)$$

satisfy the first and second equations of the set eq. 5.46, respectively. Bearing in mind eq. 5.44 and the fact that the transient wave is outgoing, its potential can be written in the form

$$\mathcal{U}_2 = \frac{C}{i\, k_2} \exp\left[i\, k_2\, (n_x\, x + n_z\, z)\right] \tag{5.48}$$

To determine unknown coefficients B and C, as well as the quantities n_x and n_z, we use the boundary conditions

$$\rho_1 \mathcal{U}_1 = \rho_2 \mathcal{U}_2 \qquad \text{and} \qquad \frac{\partial \mathcal{U}_1}{\partial z} = \frac{\partial \mathcal{U}_2}{\partial z} \qquad \text{at} \quad z = 0\,,$$

where

$$\mathcal{U}_1 = \mathcal{U}_i + \mathcal{U}_r$$

Substitution of eqs. 5.42 and 5.48 into this set yields

$$\begin{aligned} Z_1\,[A\,\exp(i\,k_1\,x\,\sin\alpha) + B\,\exp(i\,k_1\,x\,\sin\alpha)] &= Z_2\,C\,\exp(i\,k_2\,n_x\,x)\,, \\ A\,\cos\alpha\,\exp(i\,k_1\,x\,\sin\alpha) - B\,\cos\alpha\,\exp(i\,k_1\,x\,\sin\alpha) &= n_z\,C\,\exp(i\,k_2\,n_x\,x) \end{aligned} \tag{5.49}$$

Since these equalities have to take place at any point of the boundary, the arguments of exponents must be equal to each other:

$$k_1\,\sin\alpha = k_2\,n_x \qquad \text{or} \qquad \frac{\sin\alpha}{c_1} = \frac{n_x}{c_2} \tag{5.50}$$

Whence

$$n_x = \frac{\sin\alpha}{n} \tag{5.51}$$

The quantity n_z is defined from eq. 5.47

$$n_z = \sqrt{1 - \frac{\sin^2\alpha}{n^2}} \tag{5.52}$$

We focus here on the case in which

$$\alpha > \alpha_c \qquad \text{or} \qquad \frac{\sin\alpha}{n} > 1$$

For this reason it is convenient to represent n_z as

$$n_z = i\,b_z \tag{5.53}$$

Here

$$b_z = \sqrt{\frac{\sin^2\alpha}{n^2} - 1} \tag{5.54}$$

Correspondingly, in place of eq. 5.48 we have

$$\mathcal{U}_2 = \frac{C}{i\,k_2}\,e^{-k_2\,b_z\,z}e^{i\,k_2\,n_x\,x} \qquad \text{or} \qquad \mathcal{U}_2 = \frac{C}{i\,k_2}\,e^{-k_2\,b_z\,z}e^{i\,k_1\,x\,\sin\alpha} \tag{5.55}$$

Thus, the transmitted wave propagates along the x-axis, but its amplitude changes on its phase surface. Therefore, as in the case of gravitational waves in fluid (Part I), we again deal with inhomogeneous or evanescent waves.

Due to eqs. 5.50, the system 5.49 is greatly simplified, and we have

$$Z_1\,(A+B) = Z_2\,C \qquad \text{and} \qquad \cos\alpha\,(A-B) = i\,b_z\,C$$

Its solution is

$$B = \frac{Z_2\,\cos\alpha - i\,Z_1\sqrt{\dfrac{\sin^2\alpha}{n^2}-1}}{Z_2\,\cos\alpha + i\,Z_1\sqrt{\dfrac{\sin^2\alpha}{n^2}-1}}\,A\,,$$

$$C = \frac{2\,Z_1\,\cos\alpha}{Z_2\,\cos\alpha + i\,Z_1\sqrt{\dfrac{\sin^2\alpha}{n^2}-1}}\,A \qquad \text{if} \qquad \frac{\sin\alpha}{n} > 1 \tag{5.56}$$

Thus, we have solved the boundary value problem, that is we found the functions $\mathcal{U}_r$ and $\mathcal{U}_2$ that satisfy the Helmholtz equation and the boundary conditions. This means that our assumption that the reflected wave is homogeneous when $\alpha > \alpha_c$ was correct, and Snell's law of reflection remains valid. In contrast, the transmitted wave becomes inhomogeneous.

Note that eqs. 5.30 and 5.56 coincide with each other but are applied for different ranges of the incidence angle α. Now we begin to study these waves and, first, consider the wave in the lower part of the medium. If the incident angle α is equal to α_c, then

$$b_z = 0\,,$$

that is, the transmitted wave is still homogeneous and propagates with the velocity c_2 parallel to the interface. The latter follows from the fact that $n_x = 1$ in this case and eq. 5.55 becomes

$$\mathcal{U}_2 = 2\,\frac{Z_1\,A}{Z_2\,i\,k_2}\,e^{i\,k_2\,x}$$

This wave, however, becomes evanescent if $\alpha > \alpha_c$ and moves in the same direction along the x-axis. It is interesting that its propagation velocity depends on the incidence angle. In fact the argument $i\,k_2\,n_x\,x$ of the second exponent in eq. 5.55 can be rewritten as

$$i\,k_2\,n_x\,x \;=\; i\,\frac{w}{c_2}\,n_x\,x \;=\; i\,\frac{w}{c_x}\,x\,,$$

where

$$c_x = \frac{c_2}{n_x}$$

Making use of eq. 5.51, we obtain

$$c_x = \frac{c_1}{\sin\alpha} \qquad \text{if} \qquad \alpha > \alpha_c \tag{5.57}$$

By definition,

$$\sin\alpha_c = \frac{c_1}{c_2},$$

and therefore the velocity of propagation of an inhomogeneous wave lies within the range

$$c_1 \le c_x \le c_2\,, \tag{5.58}$$

while its propagation direction remains the same.

In order to investigate the behavior of this wave, it is convenient to represent the potential $\mathcal{U}_2$ as

$$\mathcal{U}_2 = \frac{M}{i\,k_2}\,\exp(i\,k_2\,n_x\,x)\,, \tag{5.59}$$

where

$$M \;=\; C\,\exp(-k_2\,b_z\,z) \;=\; |M|\;e^{-i\,\varphi_2}$$

Taking into account the second equation of the set 5.56, we can write

$$|M| \;=\; \frac{2\,\cos\alpha\,\exp(-k_2\,b_z\,z)}{\sqrt{\left(\dfrac{Z_2}{Z_1}\right)^2\cos^2\alpha - 1 + \dfrac{\sin^2\alpha}{n^2}}}\;A\,, \tag{5.60}$$

$$\varphi_2 \;=\; \tan^{-1}\frac{Z_1\sqrt{\dfrac{\sin^2\alpha}{n^2} - 1}}{Z_2\,\cos\alpha}$$

Here

$$b_z = \pm\sqrt{\frac{\sin^2\alpha}{n^2} - 1} \qquad \text{if} \qquad \alpha \ge \alpha_c \tag{5.61}$$

We choose the positive value of the radical because in such case the condition at infinity

$$\lim_{z\to\infty} \mathcal{U}_2 = 0$$

is satisfied. As is seen from eq. 5.55 the wave magnitude decreases with distance z when the exponent

$$-k_2\, b_z\, z = -\frac{\omega}{c_2}\, b_z\, z$$

is negative. Since $b_z > 0$, the phase φ_2 is positive, $(\omega > 0)$.

At the same time, it may be proper to notice that later we will use the Fourier integral where integration also includes the negative values of ω. Correspondingly, in order to satisfy the condition at infinity, it is necessary to take the minus sign for the radical 5.61 when $\omega < 0$. Then, due to the term $\exp(-k_2\, b_z\, z)$, the wave again exponentially decays with the distance from the boundary. In such a case, the phase φ_2 is also negative, and therefore it is an odd function of ω

$$\varphi_2(\omega) = \frac{|\omega|}{\omega}\, |\varphi_2(\omega)|$$

and characterizes the phase shift between the transmitted and incident waves at points of the interface.

Consider the magnitude and phase of function M. As follows from eq. 5.60 when the incident angle coincides with the critical one and $z = 0$:

$$|M| = 2\, \frac{Z_1}{Z_2}\, A$$

This function gradually tends to zero when α increases. We demonstrated previously that the transmitted and incident waves are in phase if $\alpha \leq \alpha_c$ and $z = 0$. With further increase of α, however, the phase shift becomes nonzero and grows approaching $\pi/2$. The dependence of $|M|$ and φ_2 on the angle α is shown in Fig. 5.7.

In accordance with eq. 5.60, the magnitude of the transmitted wave exponentially decays along the z-axis, while the phase does not change. It is obvious that the parameter:

$$d = \frac{\omega}{c_2}\, b_z \tag{5.62}$$

characterizes a decrease of the magnitude, and the ratio $1/d$ plays the same role as the skin-depth for the plane electromagnetic field in a conducting medium. For instance, the magnitude decreases by a factor of $e \approx 2.72$ at the distance

$$\delta = \frac{1}{d}$$

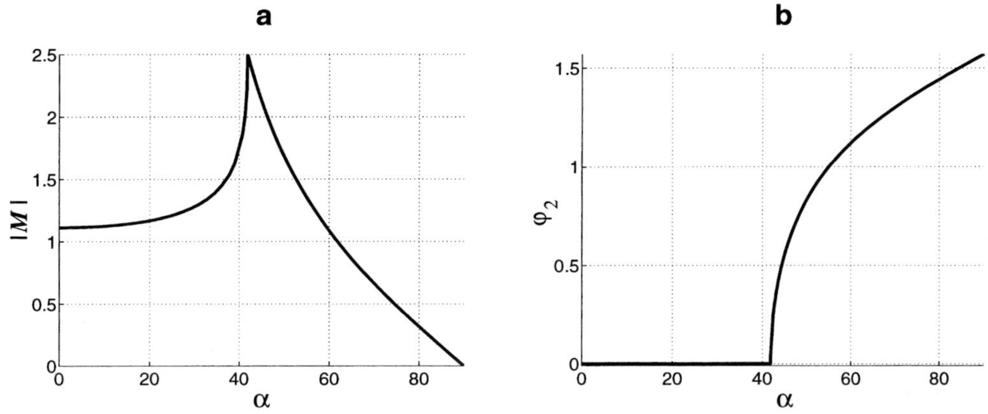

Figure 5.7: Functions $|M|$ and φ_2 in the case of velocity ratio $n = 2/3$ and impedance contrast $Z_2/Z_1 = 4/5$.

from the interface. As follows from eq. 5.54, an increase of the incidence angle α leads to a gradual increase of b_z from zero to its asymptotic value equal to

$$b_z = \sqrt{\frac{c_2^2}{c_1^2} - 1}$$

For illustration, suppose that sound wave moves through air and reaches its boundary with water. In this model

$$\rho_1 = 1.3 \text{ kg/m}^3 \qquad c_1 = 333 \text{ m/s}$$

$$\rho_2 = 1.3 \cdot 10^3 \text{ kg/m}^3 \qquad c_2 = 1500 \text{ m/s}$$

$$f = 2\,\text{Hz}\,, \qquad \alpha = 30^\circ$$

Then

$$Z_1 \approx 443\,\frac{\text{kg}}{\text{m}^2\,\text{sec}}\,, \qquad Z_2 \approx 2 \cdot 10^6\,\frac{\text{kg}}{\text{m}^2\,\text{sec}} \qquad \text{and also} \qquad n \approx 0.22$$

Thus, the critical angle is

$$\sin\alpha_c \approx 0.22 \qquad \text{or} \qquad \alpha_c \approx 13^\circ\,42'$$

Correspondingly, for the incidence angle $\alpha = 30^\circ > \alpha_c$, the parameter characterizing the amplitude decay is

$$d = \frac{2\,\pi \cdot 2}{1.5 \cdot 10^3} \left(20.7 \cdot \frac{1}{4} - 1\right)^{\frac{1}{2}} \approx 0.017\,\mathrm{m}^{-1}\,,$$

that is, the magnitude of the inhomogeneous plane wave decreases by a factor of e when the distance from the interface is around 60 m.

As in the case of water waves (Part I) the depth of penetration of an inhomogeneous wave may be comparable to the wavelength. In fact, the parameter δ can be represented as

$$\delta = \frac{\lambda_2}{2\,\pi\, b_z}$$

Respectively, the low-frequency oscillations can be observed at relatively large distances from the boundary, whereas the high-frequency ones are noticeable only in its vicinity. Because of such behavior, inhomogeneous plane waves are sometimes called the boundary waves. It may be also proper to notice that comparison of the parameter δ with the skin-depth is rather superficial, since a decay of electromagnetic plane waves occurs due to transformation of their energy into heat. In contrast, a decrease of amplitudes of inhomogeneous waves is caused in this case by destructive interference of elementary waves.

Next, consider displacement components of this wave. Since

$$\mathcal{S}_x = \frac{\partial\,\mathcal{U}}{\partial\,x} \qquad \text{and} \qquad \mathcal{S}_z = \frac{\partial\,\mathcal{U}}{\partial\,z}\,,$$

eq. 5.59 gives

$$\begin{aligned} \mathcal{S}_x &= M_1\, n_x\, \exp(-k_2\, b_z\, z)\, \exp\left[i\,(k_2\, n_x\, x - \varphi_2)\right]\,, \\ \mathcal{S}_z &= M_1\, b_z\, \exp(-k_2\, b_z\, z)\, \exp\left[i\, \left(k_2\, n_x\, x - \varphi_2 + \frac{\pi}{2}\right)\right] \end{aligned} \tag{5.63}$$

because

$$-\frac{1}{i} = i = e^{i\,\pi/2} \qquad \text{and} \qquad |M| = M_1\, e^{-k_2\, b_z\, z}$$

Therefore, the particle displacement caused by an inhomogeneous plane wave is

$$\begin{aligned} s_x &= M_1\, n_x\, e^{-k_2\, b_z\, z}\, \cos\left(\omega\, t - k_2\, n_x\, x + \varphi_2\right) \\ s_z &= M_1\, b_z\, e^{-k_2\, b_z\, z}\, \sin\left(\omega\, t - k_2\, n_x\, x + \varphi_2\right)\,, \end{aligned} \tag{5.64}$$

since

$$s = \mathrm{Re}\, \left(\mathcal{S}\, e^{-i\,\omega\, t}\right)$$

As follows from eq. 5.64, the particle motion is elliptically polarized. In fact, we have

$$\frac{s_x}{a^2} + \frac{s_z}{b^2} = 1\,, \tag{5.65}$$

where

$$a = M_1\, n_x\, \exp(-k_2\, b_z\, z) \qquad \text{and} \qquad b = M_1\, b_z\, \exp(-k_2\, b_z\, z)$$

Thus, the end of the vector **s** moves along the ellipse with the semi-axes a and b located in the vertical plane, $y =$ const. This analysis shows that the transmitted wave propagates along the x-axis and its displacement vector returns to the original position after each period. In accordance with eq. 5.65, the ratio of semi-axes is independent of the z-coordinate, that is, the ellipse has the same shape at any distance z from the boundary, but its axes become smaller as z increases. From eq. 5.64, we find

$$\frac{s_z}{s_x} = \frac{b_z}{n_x} \tan\left(\omega\, t - k_2\, n_x\, x + \varphi_2\right) \tag{5.66}$$

The latter shows the particle positions at different times, Fig. 5.8. As was demonstrated in the Part I, propagation of surface waves in water is accompanied by circular motion of particles. Later, studying the surface waves in solids, we will again observe elliptical polarization (Rayleigh and Stoneley waves) .

Assuming that $\alpha > \alpha_c$, consider the reflected wave. From eq. 5.56, it follows that the magnitude of the reflection coefficient B is equal to A

$$|B| = A\,, \tag{5.67}$$

while its phase is

$$\varphi_r = 2\,\varphi_2\,, \tag{5.68}$$

where φ_2 is given by eq. 5.60. Thus, the amplitudes of the reflected and incident waves are equal to each other regardless of the incidence angle α, as long as it exceeds α_c. This indicates that the energy of the incident wave completely transforms into that of the reflected wave during each period. For this reason, such a phenomenon is called the total internal reflection, and it is characterized by following features:

(a) Snell's law of reflection still remains valid.

(b) Amplitudes of the incident and reflected waves are equal to each other.

(c) There is a phase shift between these waves at points of the boundary and it depends on the incident angle $\alpha > \alpha_c$.

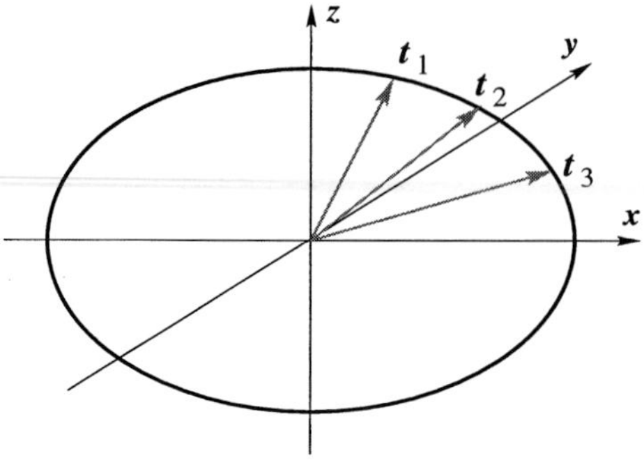

Figure 5.8: Particle motion in inhomogeneous wave.

(d) The transmitted wave is inhomogeneous, and it propagates along the interface with the phase velocity c_x satisfying the inequality

$$c_1 < c_x < c_2$$

Certainly, the transmitted wave, as well as the reflected one, is a result of interference of elementary waves generated by surface sources. It turns out that they are not able to create a homogeneous plane wave in the lower medium. This fact also follows from the Helmholtz formula, since the phase of elementary waves is equal to $k_2\, r + k_1\, x\, \sin\alpha$. As was shown in section 5.1 , the stationary point is defined by the equality

$$\frac{x_p - x_q}{r} = \frac{c_2}{c_1}\,\sin\alpha\,,$$

where the left-hand side is less than unity, while

$$\frac{c_2}{c_1}\,\sin\alpha > 1 \qquad \text{if} \qquad \alpha > \alpha_c$$

Therefore, we have to conclude that the stationary point is absent, that is, the inhomogeneous wave arises due to destructive interference. As a result, its amplitude exponentially decays with the distance from the boundary. We also demonstrated in section 5.1 that the phase of elementary waves in the upper medium is equal to

$$\varphi_r = k_1\, x\, \sin\alpha + k_1\, r$$

In this case the stationary phase exists at points of the ray, where Snell's law of reflection is valid. One can say that the reflected wave is homogeneous due to a constructive interference of elementary waves.

Next, let us summarize the results of our study of the reflection coefficient B/A. With this purpose in mind, it is convenient to use the complex plane, where the real and

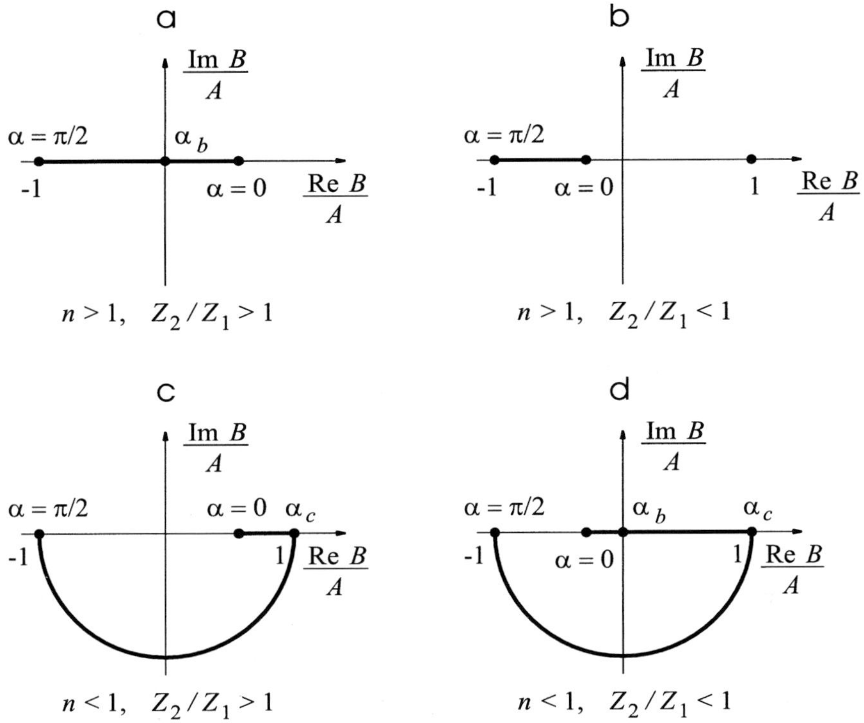

Figure 5.9: Reflection coefficients B/A.

imaginary parts of B/A are plotted along the horizontal and vertical axes, respectively. We distinguish four different cases shown in Fig. 5.9. In the first case

$$n > 1 \qquad \text{and} \qquad \frac{Z_2}{Z_1} > 1\,,$$

which makes the reflection coefficient real and positive within the range

$$0 < \alpha < \alpha_b$$

When the incidence angle coincides with the Brewster angle α_b, the ratio B/A is equal to zero and the reflection is absent. The coefficient B/A is negative at greater values of α and is equal to -1 when the angle α approaches $\pi/2$, Fig. 5.9a.

In the second case, when

$$n > 1 \qquad \text{and} \qquad \frac{Z_2}{Z_1} < 1\,,$$

the function B/A has only negative values, and therefore the reflected wave always exists, Fig. 5.9b.

In the third case

$$n < 1 \qquad \text{and} \qquad \frac{Z_2}{Z_1} > 1,$$

the reflection coefficient is positive when the incident wave propagates along the z-axis ($\alpha = 0$), Fig. 5.9c. Then, with an increase of α, B also increases and equals to A if the angle α coincides with the critical one. Thus, the phase shift between the reflected and incident waves at points of the boundary is absent within the range

$$0 \leq \alpha \leq \alpha_c$$

When $\alpha > \alpha_c$, the total internal reflection is observed. The imaginary part of B/A is negative, and its argument approaches π as α goes to $\pi/2$. In this case, as in the previous one, the reflection coefficient is always nonzero, that is the Brewster angle does not exist.

In the fourth case, when

$$n < 1 \qquad \text{and} \qquad \frac{Z_2}{Z_1} < 1,$$

the reflection coefficient is negative if $\alpha = 0$. The magnitude of B/A decreases with an increase of α, until the reflection vanishes when $\alpha = \alpha_b$. The phase of the reflection wave remains constant and equal to π within the range $0 \leq \alpha \leq \alpha_b$. In the next interval, $\alpha_b \leq \alpha \leq \alpha_c$, the reflection coefficient is positive and approaches 1. Correspondingly, the phase shift is absent. Finally, if the incident angle α exceeds α_c, the total internal reflection takes place, Fig. 5.9d.

For illustration, suppose that

$$n = 0.6 \qquad \text{and} \qquad \frac{Z_2}{Z_1} = 4$$

Such conditions may occur when the incident wave propagating in water reaches the ocean bottom. Behavior of the magnitude and phase of the reflection coefficient B/A is shown in Fig .6.10. In the beginning, when the angle α is small the function $|B/A|$ increases relatively slowly. However, approaching the critical angle $\alpha_c = 36.8°$, it starts to grow rapidly and becomes equal to 1 if $\alpha \geq \alpha_c$. As is seen from Fig .6.10 the phase shift quickly increases and reaches π.

In concluding discussion we note the following:

(a) Considering the principles of geometrical acoustics in Part I, we found out that Fermat's principle predicts the existence of head waves. However, in the case of an incident plane wave, the interference of elementary waves is such that the head waves are not created.

(b) When the wavelength tends to zero, the amplitude of the inhomogeneous wave decreases exponentially and this explains why a Debye series does not describe this type of wave.

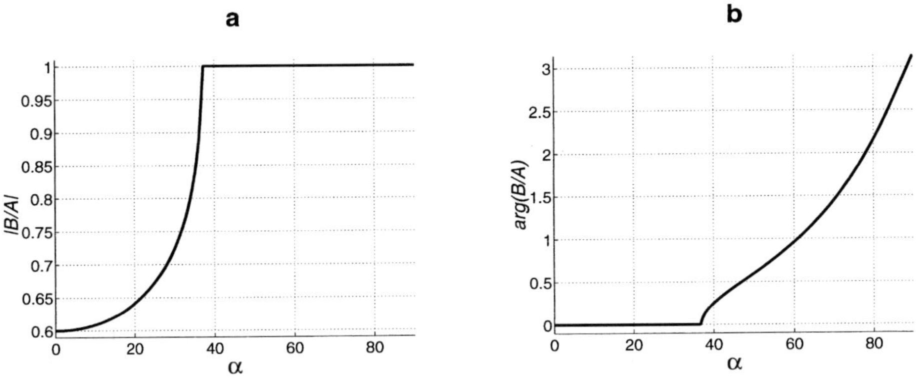

Figure 5.10: $|B/A|$ (a) and phase of B/A (b) when $n = 0.6$ and $Z_2/Z_1 = 4$.

5.3 Distribution of energy between reflected and transmitted waves

It is also useful to consider the fluxes of energy that accompany propagation of the incident, reflected, and transmitted waves. With this purpose in mind we proceed from the Poynting vector

$$\mathbf{Y} = P\,\mathbf{v} \tag{5.69}$$

By definition, it characterizes the density and direction of the energy flux. Here P and $\mathbf{v}$ are the pressure and particle velocity associated with a wave.

In this section we assume that waves are sinusoidal functions of the time. Since the vector $\mathbf{Y}$ is a product of two wave fields, P and $\mathbf{v}$, the use of complex amplitudes requires some precaution. For this reason, instead of these amplitudes, we deal directly with the fields P and $\mathbf{v}$. We will start with the simplest case when the incident wave moves along the z-axis.

Case one: $\alpha = 0$ As was shown earlier the complex amplitudes of the potential are

$$\mathcal{U}_i = \frac{A}{i\,k_1}\exp(i\,k_1\,z) \qquad \mathcal{U}_r = \frac{B}{i\,k_1}\exp(-i\,k_1\,z) \qquad \mathcal{U}_2 = \frac{C}{i\,k_2}\exp(i\,k_2\,z) \tag{5.70}$$

where

$$B = \frac{Z_2 - Z_1}{Z_2 + Z_1}\,A \qquad \text{and} \qquad C = \frac{2\,Z_1}{Z_2 + Z_1}\,A \tag{5.71}$$

Inasmuch as

$$U(z,\,t) = \text{Re}\,\left(\mathcal{U}\,e^{-i\,\omega\,t}\right)\,,$$

we have

$$U_i = -\frac{A}{k_1}\sin(\omega t - k_1 z) \qquad U_r = -\frac{B}{k_1}\sin(\omega t + k_1 z) \qquad , U_2 = -\frac{C}{k_2}\sin(\omega t - k_2 z) \tag{5.72}$$

Bearing in mind that

$$\mathbf{v} = \text{grad}\,\frac{\partial U}{\partial t} \qquad \text{or} \qquad v_z = \frac{\partial^2 U}{\partial z\,\partial t}$$

and

$$P = -\rho\,\frac{\partial^2 U}{\partial t^2}\,,$$

eqs. 5.72 give

$$\begin{aligned} v_{iz} &= \frac{\omega^2}{c_1}\,U_i\,, & P_i &= \rho_1\,\omega^2\,U_i\,, \\ v_{rz} &= -\frac{\omega^2}{c_1}\,U_r\,, & P_r &= \rho_1\,\omega^2\,U_r\,, \\ v_{2z} &= \frac{\omega^2}{c_2}\,U_2\,, & P_2 &= \rho_2\,\omega^2\,U_2\,, \end{aligned} \tag{5.73}$$

since $k_1 = \omega/c_1$ and $k_2 = \omega/c_2$.

As follows from eq. 5.69, the vertical components of the Poynting vector are

$$Y_{iz} = \frac{\rho_1}{c_1}\,\omega^4\,U_i^2\,, \qquad Y_{rz} = -\frac{\rho_1}{c_1}\,\omega^4\,U_r^2$$

and

$$Y_{2z} = \frac{\rho_2}{c_2}\,\omega^4\,U_2^2\,,$$

while the horizontal components are absent.

Using eqs. 5.72 and letting $z = 0$, we have at points of the interface

$$\begin{aligned} Y_{iz} &= \omega^2\,Z_1\,A^2\,\sin^2\omega\,t\,, \\ Y_{rz} &= -\omega^2\,Z_1\,B^2\,\sin^2\omega\,t\,, \\ Y_{2z} &= \omega^2\,Z_2\,C^2\,\sin^2\omega\,t \end{aligned} \tag{5.74}$$

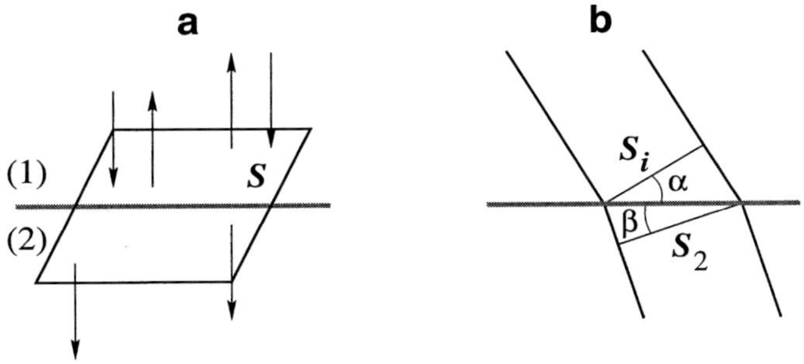

Figure 5.11: Energy fluxes for normal (a) and oblique (b) incidence of plane wave.

The "−" sign in the expression for Y_{rz} indicates that the energy flux of the reflected wave has the direction opposite to the z-axis at each instant.

Let us imagine an elementary vector tube of the fields $\mathbf{Y}_i$ with the cross-section S. When energy inside it reaches the boundary, the reflected and transmitted waves arise. Because of the normal incidence, the cross-sections for those waves are also equal to S, Fig. 5.11a. As is well known, the pressure and normal component of the particle velocity are continuous at the interface. Therefore, as follows from eq. 5.69

$$Y_{1z} = Y_{iz} + Y_{rz} = Y_{2z} \qquad \text{if} \qquad z = 0\,, \tag{5.75}$$

that is, the normal component of the Poynting vectors is a continuous function as well.

This result can be represented in a slightly different way. With this purpose in mind we evaluate the energy flux through a closed surface that surrounds an element of the vector tube, Fig. 5.11a. By definition, the fluxes are

$$N_i = \mathbf{Y}_i \cdot \mathbf{n}\, S\,, \qquad N_r = \mathbf{Y}_r \cdot \mathbf{n}_1\, S\,, \qquad N_2 = \mathbf{Y}_2 \cdot \mathbf{n}\, S$$

or

$$N_i = Y_{iz}\, S\,, \qquad N_r = -Y_{rz}\, S\,, \qquad N_2 = Y_{2z}\, S\,, \tag{5.76}$$

where

$$\mathbf{n} = -\mathbf{n}_1 = \mathbf{k}$$

and $\mathbf{k}$ is the unit vector directed along the z-axis. Then, from eqs. 5.71 and 5.74 we obtain for the flux caused by both the reflected and transmitted waves:

$$N_r + N_2 \;=\; (Y_{2z} - Y_{rz})\, S$$

or

$$N_r + N_2 \;=\; \omega^2 \left[\frac{4\, Z_1^2\, Z_2}{(Z_1 + Z_2)^2} + \frac{Z_1\,(Z_1 - Z_2)^2}{(Z_1 + Z_2)^2}\right] A^2 \sin^2 \omega\, t \;=\; \omega^2\, Z_1\, A^2 \sin^2 \omega\, t$$

On the other hand we have

$$N_i = \omega^2 Z_1 A^2 \sin^2 \omega t$$

and therefore

$$N_i = N_r + N_2 \,, \tag{5.77}$$

that is the energy flux of the incident wave is distributed between the reflected and transmitted waves. We note that the particle velocity and the pressure vary synchronously, changing their signs every half-period. The direction of the Poynting vector, however, remains the same.

Case two: $0 \leq \alpha \leq \alpha_c$ In this more general case, we use eqs. 5.7 and 5.8 to represent the potentials of waves as

$$\begin{aligned} U_i &= -\frac{A}{k_1} \sin(\omega t - k_1 x \sin\alpha - k_1 z \cos\alpha)\,, \\ U_r &= -\frac{B}{k_1} \sin(\omega t - k_1 x \sin\alpha + k_1 z \cos\alpha)\,, \\ U_2 &= -\frac{C}{k_2} \sin(\omega t - k_2 x \sin\beta - k_2 z \cos\beta)\,, \end{aligned} \tag{5.78}$$

where

$$B = \frac{Z_2 \cos\alpha - Z_1 \cos\beta}{Z_2 \cos\alpha + Z_1 \cos\beta} A\,, \qquad C = \frac{2 Z_1 \cos\alpha}{Z_2 \cos\alpha + Z_1 \cos\beta} A$$

and

$$\frac{\sin\alpha}{c_1} = \frac{\sin\beta}{c_2}$$

Correspondingly, the components of the particle velocity and the pressure are

$$\begin{aligned} v_{ix} &= \frac{\omega^2}{c_1} U_i \sin\alpha\,, & v_{iz} &= \frac{\omega^2}{c_1} U_i \cos\alpha\,, & P_i &= \rho_1 \omega^2 U_i\,, \\ v_{rx} &= \frac{\omega^2}{c_1} U_r \sin\alpha\,, & v_{rz} &= -\frac{\omega^2}{c_1} U_r \cos\alpha\,, & P_r &= \rho_1 \omega^2 U_r\,, \\ v_{2x} &= \frac{\omega^2}{c_2} U_2 \sin\beta\,, & v_{2z} &= \frac{\omega^2}{c_2} U_2 \cos\beta\,, & P_2 &= \rho_2 \omega^2 U_2 \end{aligned} \tag{5.79}$$

They obviously coincide with eq. 5.73 if $\alpha = 0$. Thus the vertical components of the Poynting vector are

$$Y_{iz} = \frac{\rho_1}{c_1}\,\omega^4\,U_i^2\,\cos\alpha\,, \qquad Y_{rz} = -\frac{\rho_1}{c_1}\,\omega^4\,U_r^2\,\cos\alpha\,, \qquad Y_{2z} = \frac{\rho_2}{c_2}\,\omega^4\,U_2^2\,\cos\beta \tag{5.80}$$

As in the previous case, the normal component of the Poynting vector is continuous across the boundary. In fact, as follows from eqs. 5.78–5.80 that

$$Y_{1z} = Y_{iz} + Y_{rz} = Z_1\,A^2\omega^2\left[1 - \frac{(Z_2\,\cos\alpha - Z_1\,\cos\beta)^2}{(Z_2\,\cos\alpha + Z_1\,\cos\beta)^2}\right]\cos\alpha\,\sin^2(\omega t - k_1 x\,\sin\alpha)$$

or

$$Y_{1z} = \frac{4\,Z_1^2\,Z_2\,\omega^2\,A^2\cos^2\alpha\,\cos\beta}{(Z_2\,\cos\alpha + Z_1\,\cos\beta)^2}\,\sin^2(\omega\,t - k_1\,x\,\sin\alpha)$$

At the same time the vertical component of the Poynting vector of the transmitted wave at the interface $z = 0$ is

$$Y_{2z} = \frac{4\,Z_1^2\,Z_2\,\omega^2\,A^2\cos^2\alpha\,\cos\beta}{(Z_2\,\cos\alpha + Z_1\,\cos\beta)^2}\,\sin^2(\omega\,t - k_2\,x\,\sin\beta)$$

Taking into account that

$$k_1\,\sin\alpha = k_2\,\sin\beta\,,$$

we prove eq. 5.75 again. In other words, as before, the energy of the incident wave gives rise to the reflected and transmitted waves. In this light, we make three comments.

(a) Tangential components of the Poynting vector are not equal to each other at the boundary:

$$Y_{1x} \neq Y_{2x} \qquad \text{if} \qquad z = 0$$

because the amplitudes of corresponding components of the particle velocity are discontinuous functions. It is obvious that, omitting the sinusoidal term, we have

$$v_{1x} = v_{ix} + v_{rx} = \frac{\omega^2}{c_1}(1 + B)\,\sin\alpha\,A = 2\,\frac{\omega^2}{c_1}\,\frac{Z_2\,\sin\alpha\,\cos\alpha}{Z_2\,\cos\alpha + Z_1\,\cos\beta}\,A$$

At the same time

$$v_{2x} = 2\,\frac{\omega^2}{c_2}\,\frac{Z_1\,\sin\beta\,\cos\alpha}{Z_2\,\cos\alpha + Z_1\,\cos\beta}\,A$$

Whence

$$\frac{v_{1x}}{v_{2x}} = \frac{c_2}{c_1}\,\frac{Z_2}{Z_1}\,\frac{\sin\alpha}{\sin\beta} = \frac{Z_2}{Z_1}$$

Therefore

$$\frac{Y_{1x}}{Y_{2x}} = \frac{Z_2}{Z_1} \tag{5.81}$$

(b) In accordance with Snell's law of reflection $\alpha = \alpha_r$, and therefore the cross-sections of vector tubes of the Poynting vectors for the incident and reflected waves are equal to each other. However, the corresponding cross-section for the transmitted wave differs from two others. As is seen from Fig. 5.11b, the ratio of those areas relates to the angles α and β the following way

$$\frac{S_i}{\cos\alpha} = \frac{S_2}{\cos\beta}$$

As was pointed out earlier there are two reasons why magnitudes of the Poynting vector of the incident and transmitted waves are different at the boundary. First, a part of the energy of the incident wave transforms into the reflected wave. Besides, cross-sections S_i and S_2 are not equal to each other if $\alpha \neq 0$.

(c) If the incidence angle coincides with the Brewster angle then

$$Z_2 \cos\alpha = Z_1 \cos\beta\,,$$

and the normal component of the Poynting vector is

$$Y_{1z} = Y_{iz}\,,$$

while

$$Y_{2z} = \frac{4\,Z_1^2\,Z_2\,\omega^2\cos^2\alpha\,\cos\beta}{4\,Z_1^2\,\cos^2\beta}\,A^2\sin^2(\omega t - k_2 x\sin\beta) \qquad \text{if} \qquad z=0$$

Thus

$$Y_{2z} = Y_{iz} \qquad \text{if} \qquad \alpha = \alpha_b \qquad \text{and} \qquad z = 0 \tag{5.82}$$

By definition, the Poynting vector is parallel to the particle velocity and, correspondingly, it is perpendicular to the phase surfaces. As was pointed out in Chapter 1, this means that the path of propagation of energy coincides with the rays.

Case three: $c_2 > c_1$, $\alpha > \alpha_c$ As we demonstrated in the previous section an inhomogeneous wave arises in the lower medium when $c_2 > c_1$ and α exceeds the critical angle. The complex amplitudes of the potentials in this case are

$$\mathcal{U}_i = \frac{1}{i\,k_1}\exp[i\,k_1\,(x\,\sin\alpha + z\,\cos\alpha)]\,A\,,$$

$$\mathcal{U}_r = \frac{B_1}{i\,k_1}\exp[i\,k_1\,(x\,\sin\alpha - z\,\cos\alpha)]\,A\,,$$

$$\mathcal{U}_2 = \frac{C_1}{i\,k_2}\exp(-k_2\,b_z\,z)\,\exp(i\,k_2\,n_x\,x)\,A\,,$$

where

$$B_1 = \frac{Z_2\,\cos\alpha - i\,Z_1\,\sqrt{n_x^2 - 1}}{Z_2\,\cos\alpha + i\,Z_1\,\sqrt{n_x^2 - 1}} = e^{-i\,\varphi_r}\,,$$

$$C_1 = \frac{2\,Z_1\,\cos\alpha}{Z_2\,\cos\alpha + i\,Z_1\,\sqrt{n_x^2 - 1}} = |C_1|\,e^{-i\,\varphi_2}$$

and

$$n_x = \frac{\sin\alpha}{n}\,, \qquad n = \frac{c_1}{c_2}\,, \qquad b_z = \sqrt{n_x^2 - 1} \qquad \text{and} \qquad \varphi_r = 2\,\varphi_2$$

Respectively, expressions for the potential are

$$\begin{aligned} U_i &= -\frac{A}{k_1}\sin(\omega\,t - k_1\,x\,\sin\alpha - k_1\,z\,\cos\alpha)\,, \\ U_r &= -\frac{A}{k_1}\sin(\omega\,t - k_1\,x\,\sin\alpha + k_1\,z\,\cos\alpha + \varphi_r)\,, \\ U_2 &= -\frac{|C_1|}{k_2}\,A\,e^{-k_2\,b_z\,z}\,\sin(\omega\,t - k_2\,n_x\,x + \varphi_2) \end{aligned} \tag{5.83}$$

Differentiating eqs. 5.83, we obtain for the velocity components and the pressure

$$v_{ix} = -\omega\,A\,\sin\alpha\,\sin(\omega\,t - k_1\,x\,\sin\alpha - k_1\,z\,\cos\alpha)\,,$$

$$v_{iz} = -\omega\,A\,\cos\alpha\,\sin(\omega\,t - k_1\,x\,\sin\alpha - k_1\,z\,\cos\alpha)\,,$$

$$P_i = -\omega\,A\,Z_1\,\sin(\omega\,t - k_1\,x\,\sin\alpha - k_1\,z\,\cos\alpha)\,,$$

$$
\begin{aligned}
v_{rx} &= -\omega\, A\, \sin\alpha\, \sin(\omega\, t - k_1\, x\, \sin\alpha + k_1\, z\, \cos\alpha + \varphi_r)\,, \\
v_{rz} &= \omega\, A\, \cos\alpha\, \sin(\omega\, t - k_1\, x\, \sin\alpha + k_1\, z\, \cos\alpha + \varphi_r)\,, \\
P_r &= -\omega\, A\, Z_1\, \sin(\omega\, t - k_1\, x\, \sin\alpha + k_1\, z\, \cos\alpha + \varphi_r)\,, \\
v_{2x} &= -\omega\, |C_1|\, n_x\, A\, e^{-k_2\, b_z\, z}\, \sin(\omega\, t - k_2\, n_x\, x + \varphi_2)\,, \\
v_{2z} &= \omega\, |C_1|\, b_z\, A\, e^{-k_2\, b_z\, z}\, \cos(\omega\, t - k_2\, n_x\, x + \varphi_2)\,, \\
P_2 &= -\omega\, |C_1|\, Z_2\, A\, e^{-k_2\, b_z\, z}\, \sin(\omega\, t - k_2\, n_x\, x + \varphi_2)
\end{aligned}
$$

Thus, components of the Poynting vector are

$$
\begin{aligned}
Y_{ix} &= \omega^2\, A^2\, Z_1\, \sin\alpha\, \sin^2(\omega\, t - k_1\, x\, \sin\alpha - k_1\, z\, \cos\alpha)\,, \\
Y_{iz} &= \omega^2\, A^2\, Z_1\, \cos\alpha\, \sin^2(\omega\, t - k_1\, x\, \sin\alpha - k_1\, z\, \cos\alpha)\,, \\
Y_{rx} &= \omega^2\, A^2\, Z_1\, \sin\alpha\, \sin^2(\omega\, t - k_1\, x\, \sin\alpha + k_1\, z\, \cos\alpha + \varphi_r)\,, \\
Y_{rz} &= -\omega^2\, A^2\, Z_1\, \cos\alpha\, \sin^2(\omega\, t - k_1\, x\, \sin\alpha + k_1\, z\, \cos\alpha + \varphi_r)\,, \\
Y_{2x} &= \omega^2\, |C_1|^2\, n_x\, A^2\, Z_2\, e^{-2\, k_2\, b_z\, z}\, \sin^2(\omega\, t - k_2\, n_x\, x + \varphi_2)\,, \\
Y_{2z} &= -\omega^2\, |C_1|^2\, b_z\, A^2\, Z_2\, e^{-2\, k_2\, b_z\, z}\, \sin(\omega\, t - k_2\, n_x\, x + \varphi_2)\, \cos(\omega\, t - k_2\, n_x\, x + \varphi_2)
\end{aligned}
\tag{5.84}
$$

First, consider the total normal component Y_{1z} at the boundary $z = 0$

$$Y_{1z} = Y_{iz} + Y_{rz}$$

or

$$Y_{1z} = \omega^2\, A^2\, Z_1\, \cos\alpha\, \left[\sin^2(\omega\, t - k_1\, x\, \sin\alpha) - \sin^2(\omega\, t - k_1\, x\, \sin\alpha + \varphi_r)\right] \tag{5.85}$$

Using the equality

$$\sin^2\theta = \frac{1 - \cos 2\,\theta}{2}\,,$$

eq. 5.85 can be rewritten as

$$Y_{1z} = \frac{\omega^2 A^2 Z_1 \cos\alpha}{2} \left[\cos 2(\omega t - k_1 x \sin\alpha + \varphi_r) - \cos 2(\omega t - k_1 x \sin\alpha)\right]$$

or

$$Y_{1z} = -\omega^2 A^2 Z_1 \cos\alpha \sin\varphi_r \sin 2(\omega t - k_1 x \sin\alpha + \frac{\varphi_r}{2})$$

The period T_* of this function is equal to

$$T_* = \frac{T}{2} = \frac{\pi}{\omega},$$

and we see that

$$\int_0^{T_*} Y_{1z}\, dt = 0 \tag{5.86}$$

Thus, during the period the flux of energy into the lower medium is equal to zero. However, this does not mean that magnitudes of vectors $\mathbf{Y}_i$ and $\mathbf{Y}_r$ are equal to each other at any instant of time. On the contrary, due to the phase shift φ_r, we have

$$|\mathbf{Y}_i(t)| \neq |\mathbf{Y}_r(t)|,$$

that is, the energy moves into the second medium during one-half of the period,

$$0 < t < \frac{T_*}{2}$$

and returns back within the second half.

Also as follows from the continuity of the pressure and the normal component of the particle velocity,

$$Y_{1z}(t) = Y_{2z}(t),$$

that is, energy of the incident wave is distributed between the reflected and inhomogeneous ones. Note that the term "total internal reflection" may cause some confusion since it implies the absence of energy in the lower medium.

Next, consider the energy flux that accompanies propagation of the inhomogeneous plane wave along the boundary. In accordance with eqs. 5.84, the horizontal component of the Poynting vector of this wave,

$$Y_{2x} = \omega^2 |C_1|^2 n_x A^2 Z_2 e^{-2 k_2 b_z z} \sin^2(\omega t - k_2 n_x x + \varphi_2)$$

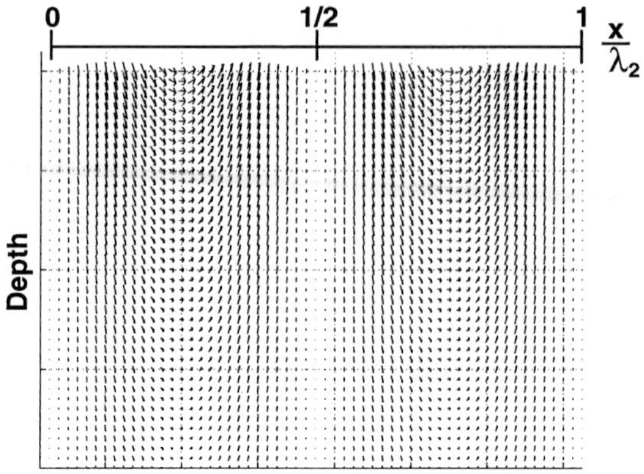

Figure 5.12: Lines of the Poynting vector (the last two equations of set 5.84) of transmitted inhomogeneous wave.

is always positive.

As we know, the phase surfaces of this wave are perpendicular to the boundary and, correspondingly, their normals coincide with the propagation direction. At the same time, vector lines of the Poynting vector have a different orientation, Fig. 5.12. These lines enter the lower medium and, at a distance equal to $\lambda_2/2$, they leave it.

5.4 Reflection and transmission of transient plane wave

Until now we have studied the behavior of the reflected and transmitted waves in the presence of a plane boundary, provided that the incident wave is a sinusoidal function of time. Applying the Fourier transform, we may expect that the main results derived in the previous sections remain valid when the incident wave has arbitrary time dependence. In the beginning, however, we will not use the Fourier integral but, instead, solve the boundary value problem again. As was shown previously, the potential of the incident plane wave can be represented as

$$U_i = A\,f\left[a\left(t - \frac{x\,\sin\alpha + z\,\cos\alpha}{c_1}\right)\right] \tag{5.87}$$

Here A is the known coefficient, α is the incidence angle, and f is an arbitrary function that has second derivatives with respect to its argument.

Solution of the boundary value problem when $\alpha < \alpha_c$

Let us assume that the reflected and transmitted waves have the same time dependence as the incident one. Correspondingly,

$$U_r = B\, f\left[a\left(t - \frac{x\,\sin\alpha_1 - z\,\cos\alpha_1}{c_1}\right)\right],$$

and

$$U_2 = C\, f\left[a\left(t - \frac{x\,\sin\beta + z\,\cos\beta}{c_2}\right)\right] \tag{5.88}$$

By definition, the functions given by eqs. 5.87 and 5.88. are D'Alembert solutions of the wave equation

$$\frac{\partial^2 U}{\partial x^2} + \frac{\partial^2 U}{\partial z^2} = \frac{1}{c^2}\frac{\partial^2 U}{\partial t^2}, \tag{5.89}$$

where

$$c = \begin{cases} c_1 & \text{if} \quad z < 0 \\ c_2 & \text{if} \quad z \geq 0 \end{cases}$$

Since

$$P = -\rho\,\frac{\partial^2 U}{\partial t^2} \qquad \text{and} \qquad \mathbf{s} = \text{grad}\, U\,, \tag{5.90}$$

we have at the boundary

$$\rho_1\, U_1 = \rho_2\, U_2 \qquad \text{and} \qquad \frac{\partial U_1}{\partial z} = \frac{\partial U_2}{\partial z} \qquad \text{if} \quad z = 0 \tag{5.91}$$

Here

$$U_1 = U_i + U_r$$

From the first equality of the set 5.91, it follows that the pressure is continuous

$$\rho_1\,\frac{\partial^2 U_1}{\partial t^2} = \rho_2\,\frac{\partial^2 U_2}{\partial t^2}$$

Substitution of eqs. 5.87–5.88 into the boundary conditions gives

$$\rho_1\, A\, f\left[a\left(t - \frac{x\,\sin\alpha}{c_1}\right)\right] + \rho_1\, B\, f\left[a\left(t - \frac{x\,\sin\alpha_1}{c_1}\right)\right] = \rho_2\, C\, f\left[a\left(t - \frac{x\,\sin\beta}{c_2}\right)\right],$$

and (5.92)

$$-\frac{\cos\alpha}{c_1} A\, f'\left[a\left(t-\frac{x\,\sin\alpha}{c_1}\right)\right] + \frac{\cos\alpha_1}{c_1} B\, f'\left[a\left(t-\frac{x\,\sin\alpha_1}{c_1}\right)\right]$$

$$= -\frac{\cos\beta}{c_2} C\, f'\left[a\left(t-\frac{x\,\sin\beta}{c_2}\right)\right]$$

Here f' is the first derivative of the function f.

As in the case of sinusoidal waves, these equalities have to take place at any time and at each point of the boundary. For this reason, the arguments of the function f must coincide for all three waves. So we again arrive at Snell's law:

$$\alpha = \alpha_1 \qquad \text{and} \qquad \frac{\sin\alpha}{c_1} = \frac{\sin\beta}{c_2}$$

The latter has a physical meaning until the incidence angle α exceeds the critical one, α_c, when the transmitted wave begins to propagate along the boundary. The last equation greatly simplifies the system 5.92, and it reduces to

$$\begin{cases} \rho_1 A + \rho_1 B = \rho_2 C\,, \\ -\cos\alpha\, A + \cos\alpha\, B = -\,n\,\cos\beta\; C\,, \end{cases}$$

The solution of this system,

$$B = \frac{Z_2\,\cos\alpha - Z_1\sqrt{1-\dfrac{\sin^2\alpha}{n^2}}}{Z_2\,\cos\alpha + Z_1\sqrt{1-\dfrac{\sin^2\alpha}{n^2}}}\,A\,,$$

(5.93)

$$C = \frac{2\,Z_1\,\cos\alpha}{Z_2\,\cos\alpha + Z_1\sqrt{1-\dfrac{\sin^2\alpha}{n^2}}}\,A$$

coincides with the corresponding coefficients derived for sinusoidal waves. Thus, we have demonstrated that all three waves are described by the same functions $f(x,\,z,\,t)$. Also Snell's law, as well as the coefficients B and C given by eqs. 5.93 remain valid for any transient waves if $\alpha < \alpha_c$. Of course, this result follows directly from Fourier integrals since B and C are independent of the frequency.

Transient waves beyond the critical angle $\alpha > \alpha_c$

At the same time when the incidence angle exceeds the critical angle, we observe a completely different picture. This can be expected due to the presence of the phase shift in the frequency domain between the reflected, transmitted, and incident waves. Moreover, the magnitude of the evanescent wave decays exponentially with the frequency. Therefore, it is obvious that transient waves behave differently from those in the previous case. For instance, as will be shown, the transient inhomogeneous wave in the lower medium changes its shape depending on the distance from the boundary and this happens because an attenuation of its sinusoidal harmonics is frequency dependent.
Now, proceeding from the Fourier integral, we investigate the transient waves when $\alpha > \alpha_c$. As was derived in section 2.2

$$B = \frac{Z_2\cos\alpha - i\,b_z\,Z_1}{Z_2\cos\alpha + i\,b_z\,Z_1}\,A \qquad \text{and} \qquad C = \frac{2\,Z_1\cos\alpha}{Z_2\cos\alpha + i\,b_z\,Z_1}\,A\,, \tag{5.94}$$

where

$$b_z = \begin{cases} \sqrt{\dfrac{\sin^2\alpha}{n^2} - 1} & \text{if} \quad \omega > 0 \\[2ex] -\sqrt{\dfrac{\sin^2\alpha}{n^2} - 1} & \text{if} \quad \omega < 0 \end{cases}$$

As follows from eqs. 5.94, the arguments of B and C relate to each other as

$$\varphi_r(\omega) = 2\,\varphi_2(\omega)\,,$$

and they are both odd functions of frequency. To emphasize this fact, we write the quantity $\varphi_2(\omega)$ in the form

$$\varphi_2(\omega) = \frac{|\omega|}{\omega}\,|\varphi_2(\omega)| \tag{5.95}$$

Also, the absolute values of the magnitudes of B and C are independent of frequency.

Reflected transient wave

First we demonstrate that even a constant phase shift φ_2 for frequencies with the same sign influences the reflected transient wave. Suppose that the potential of an incident wave is described by the function

$$U_i(x,\,t) = A\,f\left(t - \frac{x\,\sin\alpha}{c_1}\right)$$

at points of the boundary. For simplicity we omit the multiplier a in front of the argument of f.

By definition,

$$f(t_1) = \frac{1}{2\pi} \int_{-\infty}^{\infty} F(\omega)\, e^{-i\,\omega\, t_1}\, d\omega \qquad \text{and} \qquad F(\omega) = \int_{-\infty}^{\infty} f(t_1)\, e^{i\,\omega\, t_1}\, dt_1 \tag{5.96}$$

Here

$$t_1 = t - \frac{x\,\sin\alpha}{c_1}$$

and $F(\omega)$ is the spectrum of the incident wave.

Taking into account eq. 5.95 and the fact that

$$|B| = 1\,,$$

we obtain for the reflected wave

$$U_r(x,\, t_1) = \frac{A}{2\pi} \int_{-\infty}^{\infty} F(\omega)\, \exp\left(-2\,i\,\frac{|\omega|}{\omega}\,|\varphi_2|\right)\, e^{-i\,\omega\, t_1}\, d\omega\,, \tag{5.97}$$

where

$$|\varphi_2| = \tan^{-1} \frac{Z_1 \left|\sqrt{\dfrac{\sin^2\alpha}{n^2} - 1}\right|}{Z_2\,\cos\alpha}$$

The use of Euler's formula,

$$e^{-i\,\theta} = \cos\theta - i\,\sin\theta\,,$$

gives

$$\begin{aligned} U_r(x,\, t_1) \; &= \; \frac{\cos 2\,|\varphi_2|}{2\pi}\, A \int_{-\infty}^{\infty} F(\omega)\, e^{-i\,\omega\, t_1}\, d\omega \\ &\qquad - i\,\frac{\sin 2\,|\varphi_2|}{2\pi}\, A \int_{-\infty}^{\infty} \frac{|\omega|}{\omega}\, F(\omega)\, e^{-i\,\omega\, t_1}\, d\omega\,, \end{aligned} \tag{5.98}$$

because

$$\sin\left(2\,\frac{|\omega|}{\omega}\,|\varphi_2|\right) = \frac{|\omega|}{\omega}\,\sin\left(2\,|\varphi_2|\right)$$

Therefore

$$U_r(t_1) = U_i(t_1)\cos 2\,|\varphi_2| - i\,A\,L(t_1)\,\sin 2\,|\varphi_2| \tag{5.99}$$

Here

$$L(t_1) = \frac{1}{2\,\pi}\int\limits_{-\infty}^{\infty} F(\omega)\,\frac{|\omega|}{\omega}\,e^{-i\,\omega\,t_1}\,d\omega \tag{5.100}$$

These formulae allow us to find the potential $U_r(t_1)$ and they clearly show that the incident and reflected waves depend on time and distance differently. It is also useful to represent the function $L(t_1)$ in terms of the Hilbert transform, (Appendix B). With this purpose in mind, we perform some simple operations. As follows from equation 5.100

$$L(t_1) = -\frac{1}{2\,\pi}\int\limits_{-\infty}^{0} F(\omega)\,e^{-i\,\omega\,t_1}\,d\omega + \frac{1}{2\,\pi}\int\limits_{0}^{\infty} F(\omega)\,e^{-i\,\omega\,t_1}\,d\omega$$

After introducing the variable $u = -\omega$ in the first integral and replacing u with ω, we have

$$L(t_1) = \frac{1}{2\,\pi}\int\limits_{0}^{\infty}\left[F(\omega)\,e^{-i\,\omega\,t_1} - F(-\omega)\,e^{i\,\omega\,t_1}\right]d\omega \tag{5.101}$$

To simplify the integrand, we take into account the fact that the real and imaginary parts of the spectrum

$$F(\omega) = \text{Re}\ F(\omega) + i\,\text{Im}\ F(\omega)$$

are even and odd functions:

$$\text{Re}\ F(\omega) = \text{Re}\ F(-\omega) \qquad \text{and} \qquad \text{Im}\ F(\omega) = -\text{Im}\ F(-\omega)$$

The latter directly follows from the second equation of the set 5.96. Thus, we can write

$$F(-\omega) = F^*(\omega)\,,$$

where

$$F^*(\omega) = \text{Re}\ F(\omega) - i\,\text{Im}\ F(\omega)$$

is the complex conjugate of $F(\omega)$. Correspondingly, eq. 5.101 yields

$$L(t_1) = \frac{1}{2\,\pi}\int\limits_{0}^{\infty}\left[F(\omega)\,e^{-i\,\omega\,t_1} - F^*(\omega)\,e^{i\,\omega\,t_1}\right]d\omega\,, \tag{5.102}$$

and the integrand is the difference between the function $F(\omega)\, e^{-i\omega t_1}$ and its complex conjugate.

Our next step of transformation is based on the second Fourier integral, eq. 5.96. By definition,

$$F(\omega)\, e^{-i\,\omega\, t_1} = \int_{-\infty}^{\infty} f(\tau)\, e^{i\,\omega\,(\tau - t_1)}\, d\tau\,, \tag{5.103}$$

$$F^*(\omega)\, e^{i\,\omega\, t_1} = F(-\omega)\, e^{i\,\omega\, t_1} = \int_{-\infty}^{\infty} f(\tau)\, e^{-i\,\omega\,(\tau - t_1)}\, d\tau$$

Substitution of eqs. 5.103 into eq. 5.101 leads to

$$L = \frac{1}{2\,\pi} \int_0^{\infty} d\omega \int_{-\infty}^{\infty} f(\tau) \left[e^{i\,\omega\,(\tau - t_1)} - e^{-i\,\omega\,(\tau - t_1)} \right] d\tau$$

or

$$L = \frac{i}{\pi} \int_0^{\infty} d\omega \int_{-\infty}^{\infty} f(\tau)\, \sin\omega\,(\tau - t_1)\, d\tau$$

Changing the order of integration gives

$$L = \frac{i}{\pi} \int_{-\infty}^{\infty} f(\tau)\, d\tau \int_0^{\infty} \sin\omega\,(\tau - t_1)\, d\omega \tag{5.104}$$

The integral with respect to ω can be represented as

$$\int_0^{\infty} \sin\omega\,(\tau - t_1)\, d\omega = \lim_{\gamma \to 0} \int_0^{\infty} e^{-\gamma\, t}\, \sin\omega\,(\tau - t_1)\, d\omega$$

This last integral is tabular, and in the limit we have

$$\int_0^{\infty} \sin\omega\,(\tau - t_1)\, d\omega = \frac{1}{\tau - t_1}$$

Thus,

$$L = \frac{i}{\pi} \int_{-\infty}^{\infty} \frac{f(\tau)}{\tau - t_1}\, d\tau = i\, H(t_1)\,,$$

where

$$H(t_1) \equiv \frac{1}{\pi} \int\limits_{-\infty}^{\infty} \frac{f(\tau)}{\tau - t_1}\, d\tau \tag{5.105}$$

is the Hilbert transform of the known function $f(t_1)$.

Respectively, the potential of the reflected wave, eq. 5.99, can be written as

$$U_r(t_1) = A\, f(t_1)\, \cos 2\,|\varphi_2| + A\, H[f(t_1)]\, \sin 2\,|\varphi_2|\,, \tag{5.106}$$

since $|B| = 1$. It is essential that the potential $U_r(t_1)$ is expressed in terms of the known function $f(t_1)$ and its Hilbert transform. The presence of the second term in the right-hand side of eq. 5.106, as well as in eq. 5.99, shows that the incident and reflected waves have different wavelets.

It is clear that, the integrand in eq. 5.105 has a singularity and in order to remove it (Appendix B), we may use the equality

$$\frac{1}{\pi} \int\limits_{-\infty}^{\infty} \frac{d\tau}{\tau - t_1} = 0\,,$$

which is valid because the integrand is an odd function with respect to the point $\tau = t_1$.

Then the Hilbert transform becomes

$$H[f(t_1)] = \frac{1}{\pi} \int\limits_{-\infty}^{\infty} \frac{f(\tau) - f(t_1)}{\tau - t_1}\, d\tau\,, \tag{5.107}$$

and the singularity at $\tau = t_1$ disappears. Certainly a replacement of argument t_1 in eq. 5.99 or 5.106 with $t - z\, \cos\alpha/c_1$ allows us to calculate the reflected wave at any point of the upper medium. To illustrate the influence of the constant phase shift $\varphi_2(\omega)$, consider three different incident waves.

Example one: δ-function Suppose that the incident wavelet has the shape of δ-function

$$U_i(x,\, z,\, t) = A\, \delta\left(t - \frac{x\, \sin\alpha + z\, \cos\alpha}{c_1}\right)$$

By definition of the δ-function, we have

$$\int\limits_a^b f(\tau,\, t_1)\, \delta(\tau)\, d\tau = \begin{cases} f(0,\, t_1) & \text{if} \quad a \le \tau \le b \\ 0 & \text{if} \quad \tau < a \quad \text{or} \quad \tau > b \end{cases}$$

Then, in accordance with eq. 5.105, the Hilbert transform of the δ-function is

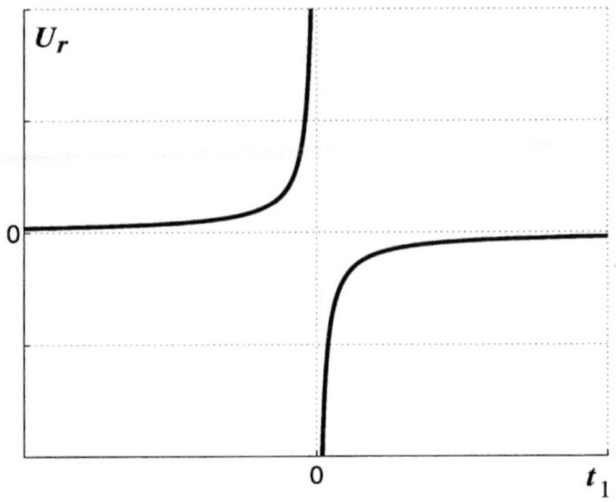

Figure 5.13: The potential $U_r(t_1)$ given by eq. 5.108.

$$H[\delta(t_1)] = -\frac{1}{\pi\, t_1}$$

Correspondingly, the potential of the reflected wave is

$$U_r(t_1) = A\,\delta(t_1)\,\cos 2\,|\varphi_2| - \frac{A}{\pi\, t_1}\,\sin 2\,|\varphi_2| \tag{5.108}$$

Its behavior is shown in Fig. 5.13. Note that knowing this function and applying the Duhamel integral, (Part I), it is not difficult to obtain the reflected wavelet for an incident wave of any arbitrary shape.

Example two: box-car impulse In this case the potential U_i behaves in the following way

$$U_i(t_1) = A\, f(t_1)\,,$$

where

$$f(t_1) = \begin{cases} 1 & \text{if} \quad 0 \le t_1 \le T \\ 0 & \text{if} \quad t_1 < 0 \quad \text{or} \quad t_1 > T \end{cases}$$

Its Hilbert transform is

$$H[f(t_1)] = \frac{1}{\pi}\int_0^T \frac{d\tau}{\tau - t_1} = -\frac{1}{\pi}\,[\ln t_1 - \ln(t_1 - T)\,]$$

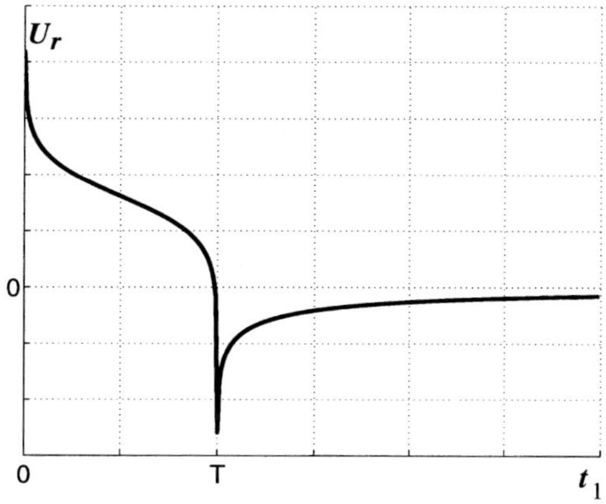

Figure 5.14: The potential $U_r(t_1)$ given by eq. 5.109.

Therefore, the potential of the reflected wave is

$$U_r(t_1) = A\, f(t_1)\, \cos 2\,|\varphi_2| - \frac{A}{\pi}\, \sin 2\,|\varphi_2|\, \ln \frac{t_1}{t_1 - T} \tag{5.109}$$

This function is shown in Fig. 5.14.

Example three: wavelet The Hilbert transform of a more general incident wavelet

$$f(t_1) = \exp\left(-\nu^2\, t_1^2\right)\, \cos\omega\, t_1 \tag{5.110}$$

is shown in Fig. 5.15. Then, using eq. 5.106, we can find the potential of the reflected wave.

Transient inhomogeneous wave

Suppose that the potential of the incident wave is the step-function and $\alpha > \alpha_c$. Then, in order to determine the pressure of an evanescent wave, we proceed from the Fourier integral. Taking into account eqs. 5.60 and 5.61, we find

$$P_2(x,\, z,\, t) \;=\; \frac{2\, A\, Z_2\, \cos\alpha}{\sqrt{\left(\dfrac{Z_2}{Z_1}\right)^2 \cos^2\alpha + \left(\dfrac{\sin\alpha}{n}\right)^2 - 1}} \tag{5.111}$$

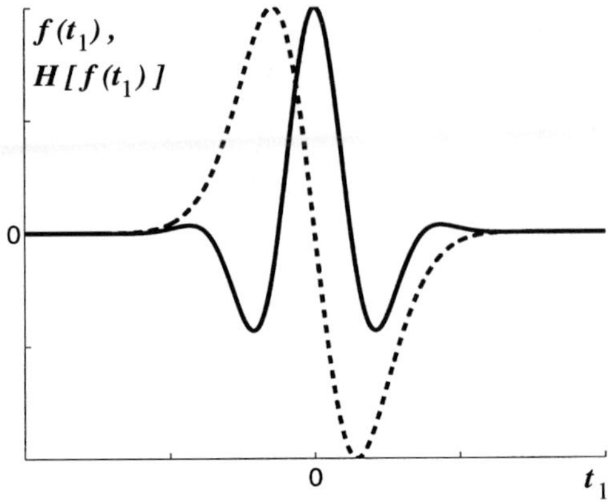

Figure 5.15: Function $f(t_1)$ given by eq. 5.110 (solid) and its Hilbert transform (dashed).

$$\times \frac{1}{2\pi} \int_{-\infty}^{\infty} \exp\left(-\frac{|\omega|}{c_2} |b_z| z\right) \exp\left(-i \frac{|\omega|}{\omega} \varphi_2\right) e^{-i\omega t_1} d\omega ,$$

where

$$t_1 = t - \frac{x \sin\alpha}{c_1}$$

Behavior of the pressure at different distances from the interface is demonstrated in Fig. 5.16 and it clearly shows the change of shape of wave with increase of the depth, where the low-frequency harmonics play a dominant role.

5.5 Reflection and transmission of inhomogeneous plane waves

It is also interesting to study the behavior of reflected and transmitted waves when the incident wave is inhomogeneous and propagates along a plane boundary. Such a wave can arise somewhere in the upper medium. As follows from eq. 5.55, its potential is

$$\mathcal{U}_i = \frac{A}{i\,k_1} \exp(-k_1\, b_{1z}\, z) \exp(i\, k_1\, n_{1x}\, x) , \tag{5.112}$$

where A, n_{1x} and b_{1z} are known, and

$$n_{1x} > 1 , \qquad b_{1z} = \sqrt{n_{1x}^2 - 1} \tag{5.113}$$

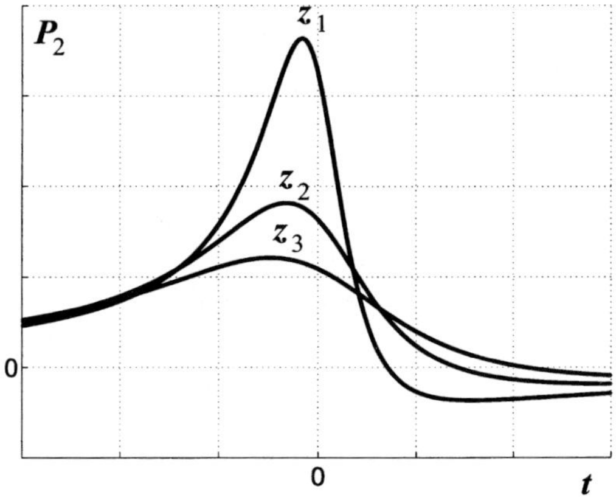

Figure 5.16: The potential of pressure $P_2(x = 0,\, z,\, t)$, eq. 5.111, as a function of time t at different depths $z_3 > z_2 > z_1$ below the interface. The dimensionless model parameters are $n = 1/2$, $n_x = \sqrt{2}$, $b_z = 1$.

First, we assume that both the reflected and transmitted waves are homogeneous and plane so their potentials are

$$\mathcal{U}_r = \frac{B}{i\,k_1} \exp\left[i\,k_1\,(x\,\sin\alpha_r - z\,\cos\alpha_r)\right] ,$$
$$\mathcal{U}_2 = \frac{C}{i\,k_2} \exp\left[i\,k_2\,(x\,\sin\beta + z\,\cos\beta)\right] \tag{5.114}$$

As we know, the boundary conditions are

$$\rho_1 \mathcal{U}_1 = \rho_2 \mathcal{U}_2\,, \qquad \frac{\partial \mathcal{U}_1}{\partial z} = \frac{\partial \mathcal{U}_2}{\partial z} \qquad \text{at} \qquad z = 0 \tag{5.115}$$

Here

$$\mathcal{U}_1 = \mathcal{U}_i + \mathcal{U}_r$$

For instance, substitution of eq. 5.112 and 5.114 into the first equality of the set 5.115 gives

$$A\,Z_1\,\exp(i\,k_1\,n_{1x}\,x) + B\,Z_1\,\exp(i\,k_1\,x\,\sin\alpha_r) = C\,Z_2\,\exp(i\,k_2\,x\,\sin\beta)$$

This equality occurs for all x if the arguments of exponents are equal to each other, that is

$$k_1 n_{1x} = k_1 \sin\alpha_r = k_2 \sin\beta$$

or

$$\sin\alpha_r = n_{1x} \qquad \text{and} \qquad \frac{\sin\alpha_r}{c_1} = \frac{\sin\beta}{c_2}$$

Because $n_{1x} > 1$, the first equality cannot be satisfied, and therefore our assumption about the reflected wave is incorrect. For this reason, we suppose that the reflected wave is also inhomogeneous and propagates along the boundary decaying, with distance z ($z < 0$). Then, the first equation of the set 5.114 is replaced with

$$\mathcal{U}_r = \frac{B}{i\,k_1} \exp(k_1\, b_{1z}\, z) \exp(i\, k_1\, n_{1x}\, x) \tag{5.116}$$

Now, making use of the boundary conditions and eqs. 5.112, 5.114 and 5.116, we arrive at the system

$$\begin{aligned} A\,Z_1 \exp(i\,k_1\,n_{1x}\,x) + B\,Z_1 \exp(i\,k_1\,n_{1x}\,x) &= C\,Z_2 \exp(i\,k_2\,x\,\sin\beta)\,, \\ -A\,b_{1z} \exp(i\,k_1\,n_{1x}\,x) + B\,b_{1z} \exp(i\,k_1\,n_{1x}\,x) &= i\,C\,\cos\beta \exp(i\,k_2\,x\,\sin\beta) \end{aligned}$$

It follows from these equations that

$$k_1\,n_{1x} = k_2 \sin\beta \qquad \text{or} \qquad \frac{n_{1x}}{c_1} = \frac{\sin\beta}{c_2} \tag{5.117}$$

and

$$\begin{aligned} A\,Z_1 + B\,Z_1 &= C\,Z_2\,, \\ -A\,b_{1z} + B\,b_{1z} &= i\,C\,\cos\beta \end{aligned}$$

The solution of this system is

$$B = \frac{Z_2\,b_{1z} + i\,Z_1\,\cos\beta}{Z_2\,b_{1z} - i\,Z_1\,\cos\beta}\,A \qquad \text{and} \qquad C = \frac{2\,Z_1\,b_{1z}}{Z_2\,b_{1z} - i\,Z_1\,\cos\beta}\,A \tag{5.118}$$

Thus, we have demonstrated that our last assumptions are correct, provided that the equalities 5.117 are met. In other words, if the second medium has a smaller velocity and

$$\sin\beta = n_{1x}\,\frac{c_2}{c_1} \leq 1\,, \tag{5.119}$$

the incident and reflected waves are inhomogeneous, while the transmitted one is homogeneous.

The condition 5.119 can be written as

$$\sin\beta = \frac{c_2}{c_1^\star} \le 1\,,$$

where

$$c_1^\star = \frac{c_1}{n_{1x}} < 1$$

is the phase velocity of the incident and reflected waves.

As follows from eqs. 5.118, these waves have the same amplitude and propagate along the boundary with equal phase velocities $c_1^\star$, but there is a phase shift between them. At the same time, the amplitude and phase of the transmitted wave differ from those of the incident one.

Next, let us assume that inequality 5.119 is invalid:

$$\sin\beta = n_{1x}\,\frac{c_2}{c_1} > 1 \tag{5.120}$$

It is natural to expect in such a case that the transmitted wave, as well as the other two types, is inhomogeneous, and therefore

$$\mathcal{U}_2 = \frac{C}{i\,k_2}\,\exp(-k_2\,b_{2z}\,z)\exp(i\,k_2\,n_{2x}\,x) \qquad \text{if} \qquad z > 0 \tag{5.121}$$

Applying the boundary conditions again we obtain a new system for the unknown coefficients B and C

$$\begin{aligned} A\,Z_1\,\exp(i\,k_1\,n_{1x}\,x) + B\,Z_1\,\exp(i\,k_1\,n_{1x}\,x) &= C\,Z_2\,\exp(i\,k_2\,n_{2x}\,x)\,, \\ -A\,b_{1z}\,\exp(i\,k_1\,n_{1x}\,x) + B\,b_{1z}\,\exp(i\,k_1\,n_{1x}\,x) &= -C\,b_{2z}\,\exp(i\,k_2\,n_{2x}\,x) \end{aligned} \tag{5.122}$$

The equality of exponential arguments yields

$$k_1\,n_{1x} = k_2\,n_{2x} \qquad \text{or} \qquad n_{2x} = \frac{c_2}{c_1}\,n_{1x} = \frac{n_{1x}}{n} \tag{5.123}$$

and, correspondingly,

$$b_{2z} = \sqrt{n_{2x}^2 - 1} \tag{5.124}$$

Because of eqs. 5.123, the system 5.122 becomes

$$\begin{aligned} A\,Z_1 + B\,Z_1 &= C\,Z_2\,, \\ -A\,b_{1z} + B\,b_{1z} &= -b_{2z}\,C \end{aligned}$$

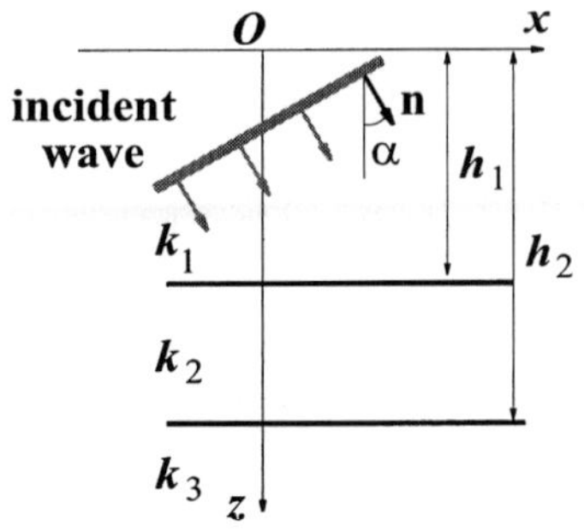

Figure 5.17: Oblique plane incident wave in a medium containing a single homogeneous layer between two homogeneous half-spaces.

Then

$$B = \frac{Z_2\, b_{1z} - Z_1\, b_{2z}}{Z_2\, b_{1z} + Z_1\, b_{2z}}\, A \qquad \text{and} \qquad C = \frac{2\, Z_1\, b_{1z}}{Z_2\, b_{1z} + Z_1\, b_{2z}}\, A \tag{5.125}$$

Thus, as in the case of homogeneous waves, the incident and transmitted waves are always in phase at $z = 0$, while the phase of the reflected wave either coincides with the other two if $Z_2\, b_{1z} \geq Z_1\, b_{2z}$, or differs by π if $Z_2\, b_{1z} < Z_1\, b_{2z}$.

5.6 Oblique incidence of plane waves in media with two interfaces

As before, suppose that the propagation direction of the incident plane wave makes an arbitrary angle α with the z-axis, Fig. 5.17. The y-axis is parallel to the phase surfaces of the wave, that is, neither the displacement nor the pressure changes in the direction of y. The complex amplitude of the potential of the incident wave can be represented as

$$\mathcal{U}_i = \frac{A}{i\, k_1}\, e^{i\, k_1\, x\, \sin\alpha}\, e^{i\, k_1\, z\, \cos\alpha}, \tag{5.126}$$

where A is the known constant.

Homogeneous waves

We assume in the beginning that the reflected and transmitted waves arising at both interfaces are homogeneous. Then, the potentials of waves in this media are

$$\mathcal{U}_1 = \frac{1}{i\, k_1} \left(A\, e^{i\, k_1\, z\, \cos\alpha} + B\, e^{-i\, k_1\, z\, \cos\alpha} \right) e^{i\, k_1\, x\, \sin\alpha},$$

$$\mathcal{U}_2 = \frac{1}{i\,k_2}\left(C\,e^{i\,k_2\,z\,\cos\beta} + D\,e^{-i\,k_2\,z\,\cos\beta}\right)e^{i\,k_2\,x\,\sin\beta}, \tag{5.127}$$

$$\mathcal{U}_3 = \frac{1}{i\,k_3}E\,e^{i\,k_3\,z\,\cos\gamma}\,e^{i\,k_3\,x\,\sin\gamma}$$

Here the coefficients B, C, D and E are unknowns, and the angles α, β, and γ relate to each other through Snell's law:

$$\frac{\sin\alpha}{c_1} = \frac{\sin\beta}{c_2} = \frac{\sin\gamma}{c_3} \tag{5.128}$$

It is clear that the functions $\mathcal{U}_1$, $\mathcal{U}_2$ and $\mathcal{U}_3$ are solutions of Helmholtz equation

$$\nabla^2\mathcal{U} + k^2\mathcal{U} = 0$$

Also, from the continuity of the pressure and the normal component of the displacement at each boundary it follows that

$$\begin{aligned} \rho_1\,\mathcal{U}_1 = \rho_2\,\mathcal{U}_2\,, \qquad & \frac{\partial\,\mathcal{U}_1}{\partial\,z} = \frac{\partial\,\mathcal{U}_2}{\partial\,z} \qquad \text{at} \quad z = h_1 \\ \rho_2\,\mathcal{U}_2 = \rho_3\,\mathcal{U}_3\,, \qquad & \frac{\partial\,\mathcal{U}_2}{\partial\,z} = \frac{\partial\,\mathcal{U}_3}{\partial\,z} \qquad \text{at} \quad z = h_2 \end{aligned} \tag{5.129}$$

Substitution of eqs. 5.127 into the boundary conditions 5.129 leads to the following system of equations for four unknowns:

$$\begin{aligned} Z_1\left(A\,e^{i\,k_1 h_1\cos\alpha} + B\,e^{-i\,k_1 h_1\cos\alpha}\right) &= Z_2\left(C\,e^{i\,k_2 h_1\cos\beta} + D\,e^{-i\,k_2 h_1\cos\beta}\right), \\ \cos\alpha\left(A\,e^{i\,k_1 h_1\cos\alpha} - B\,e^{-i\,k_1 h_1\cos\alpha}\right) &= \cos\beta\left(C\,e^{i\,k_2 h_1\cos\beta} - D\,e^{-i\,k_2 h_1\cos\beta}\right), \\ Z_2\left(C\,e^{i\,k_2 h_2\cos\beta} + D\,e^{-i\,k_2 h_2\cos\beta}\right) &= Z_3\,E\,e^{i\,k_3 h_2\cos\gamma}, \\ \cos\beta\left(C\,e^{i\,k_2 h_2\cos\beta} - D\,e^{-i\,k_2 h_2\cos\beta}\right) &= \cos\gamma\;E\,e^{i\,k_3 h_2\cos\gamma} \end{aligned} \tag{5.130}$$

Eliminating E from the last two equations, we obtain

$$\frac{C\, e^{i\,k_2 h_2 \cos\beta} + D\, e^{-i\,k_2 h_2 \cos\beta}}{C\, e^{i\,k_2 h_2 \cos\beta} - D\, e^{-i\,k_2 h_2 \cos\beta}} = \frac{Z_3 \cos\beta}{Z_2 \cos\gamma} = \frac{Z_3^{(1)}}{Z_2^{(1)}}, \tag{5.131}$$

where

$$Z_2^{(1)} = \frac{Z_2}{\cos\beta} \qquad \text{and} \qquad Z_3^{(1)} = \frac{Z_3}{\cos\gamma} \tag{5.132}$$

As follows from eqs. 5.131

$$D = R_{23}^{(1)}\, e^{2\,i\,k_2\,h_2\,\cos\beta}\, C \tag{5.133}$$

Here

$$R_{23}^{(1)} = \frac{Z_3^{(1)} - Z_2^{(1)}}{Z_3^{(1)} + Z_2^{(1)}} \tag{5.134}$$

Making use of eq. 5.133, we can rewrite the first two equations of the set 5.130 in the form

$$\begin{aligned} &Z_1 \left(A\, e^{i\,k_1\,h_1\,\cos\alpha} + B\, e^{-i\,k_1\,h_1\,\cos\alpha}\right) \\ &\quad = Z_2\, C\left(e^{i\,k_2\,h_1\,\cos\beta} + R_{23}^{(1)}\, e^{2\,i\,k_2\,h_2\,\cos\beta}\, e^{-i\,k_2\,h_1\,\cos\beta}\right), \\ &\cos\alpha \left(A\, e^{i\,k_1\,h_1\,\cos\alpha} - B\, e^{-i\,k_1\,h_1\,\cos\alpha}\right) \\ &\quad = C\,\cos\beta\left(e^{i\,k_2\,h_1\,\cos\beta} - R_{23}^{(1)}\, e^{2\,i\,k_2\,h_2\,\cos\beta}\, e^{-i\,k_2\,h_1\,\cos\beta}\right) \end{aligned} \tag{5.135}$$

Whence

$$\frac{A\, e^{i\,k_1\,h_1\,\cos\alpha} + B\, e^{-i\,k_1\,h_1\,\cos\alpha}}{A\, e^{i\,k_1\,h_1\,\cos\alpha} - B\, e^{-i\,k_1\,h_1\,\cos\alpha}} = \frac{Z_2^{(1)}}{Z_1^{(1)}}\, R_{13}^{(1)} \tag{5.136}$$

and

$$R_{13}^{(1)} = \frac{1 + R_{23}^{(1)}\, e^{2\,i\,k_2\,H\,\cos\beta}}{1 - R_{23}^{(1)}\, e^{2\,i\,k_2\,H\,\cos\beta}} \tag{5.137}$$

Here

$$Z_1^{(1)} = \frac{Z_1}{\cos\alpha} \qquad \text{and} \qquad H = h_2 - h_1$$

is the layer thickness.

From eq. 5.136 we arrive at the expression of coefficient B that characterizes the reflected wave in the upper half-space:

$$B = \frac{Z_2^{(1)}\, R_{13}^{(1)} - Z_1^{(1)}}{Z_2^{(1)}\, R_{13}^{(1)} + Z_1^{(1)}}\, A\, e^{2\,i\,k_1\,h_1\,\cos\alpha} \tag{5.138}$$

It is a simple matter to verify that if $\alpha = 0$, we obtain the known coefficient B for the normal incidence.

Because of multiple reflections inside the layer, there is an infinite number of waves approaching the upper boundary $z = h_1$. It is essential that all these waves make the same angle β with the interface. For this reason, the reflected wave field in the upper medium consists of waves characterized by equal reflection angles α. Unlike the case of a single boundary the coefficient $|B|$ depends on frequency. When wavelength λ_2 greatly exceeds layer thickness H, the asymptotic value of the ratio $R_{13}^{(1)}$ is

$$R_{13}^{(1)} \to \frac{1 + R_{23}^{(1)}}{1 - R_{23}^{(1)}} = \frac{Z_3^{(1)}}{Z_2^{(1)}}$$

Therefore,

$$B \to \frac{Z_3^{(1)} - Z_1^{(1)}}{Z_3^{(1)} + Z_1^{(1)}}\, A\, e^{2\,i\,k_1\,h_1\,\cos\alpha}\,, \tag{5.139}$$

that is, the influence of the layer parameters: H and Z_2 vanishes due to interference of waves arising at both interfaces. Since the exponent $2\,i\,k_2\,H\cos\beta$ in eq. 5.137 becomes smaller as α increases, the same effect is observed at relatively higher frequencies.

Note that the function $|B/A|$ is periodical, and its behavior is shown in Fig. 5.18. In particular, there is a frequency, at which the reflection coefficient amplitude has a minimum. As is seen from Fig. 5.19 the influence of a change of the incident angle is relatively small except in the vicinity of the critical one. Also, since the incidence angle α defines two others: β and γ, all quantities $Z_1^{(1)}$, $Z_2^{(1)}$ and $Z_3^{(1)}$ are functions of α.

Now consider two special cases.

Case one Suppose that layer thickness H and the wavelength λ_2 relate as

$$H\,\cos\beta = \frac{m}{2}\,\lambda_2\,, \tag{5.140}$$

where m is an integer. Then we have

$$2\,i\,k_2\,H\,\cos\beta = 2\,i\,\frac{\omega}{c_2}\,m\,\frac{\lambda_2}{2} = 2\,\pi\,i\,m$$

Thus

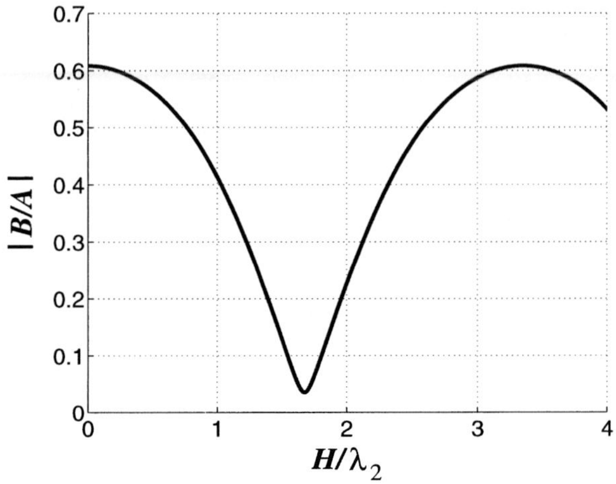

Figure 5.18: $|B/A|$ as a function of ratio H/λ_2 in medium with parameters: $c_1/c_3 = 1/4$, $c_2/c_3 = 1/2$, $Z_1/Z_3 = 1/3$, $Z_2/Z_3 = 2/3$. The incidence angle $\alpha = 10°$.

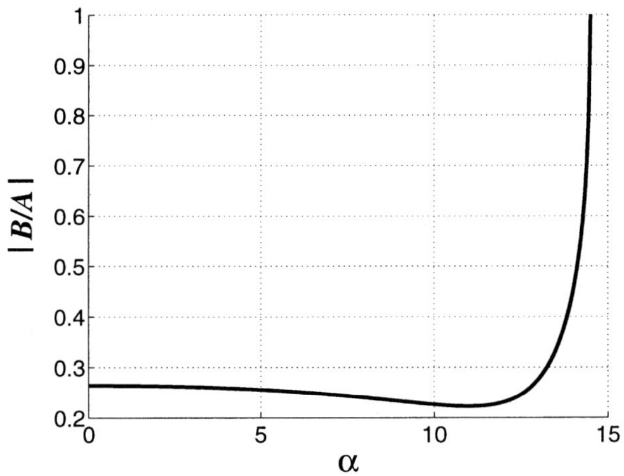

Figure 5.19: Dependence $|B/A|$ on the incidence angle α for the medium specified in Fig. 5.18 ($H/\lambda_2 = 2$).

$$R_{13}^{(1)} = \frac{1+R_{23}^{(1)}}{1-R_{23}^{(1)}} = \frac{Z_3^{(1)}}{Z_2^{(1)}} \qquad \text{and} \qquad B = \frac{Z_3^{(1)} - Z_1^{(1)}}{Z_3^{(1)} + Z_1^{(1)}} A\, e^{2\,i\,k_1\,h_1\,\cos\alpha}$$

The latter shows that the reflected wave does not contain information about the layer if the condition 5.140 is met, and such a situation is caused by the interference of waves.

Case two Next we assume that

$$H\,\cos\beta = \frac{\lambda_2}{4}\,(2\,m-1) \qquad \text{or} \qquad 2\,k_2\,H\,\cos\beta = (2\,m-1)\,\pi \tag{5.141}$$

Then eq. 5.137 becomes

$$R_{13}^{(1)} = \frac{Z_2^{(1)}}{Z_3^{(1)}}$$

Respectively, the reflection coefficient, eq. 5.138, is

$$B = \frac{\left(Z_2^{(1)}\right)^2 - Z_1^{(1)}\,Z_3^{(1)}}{\left(Z_2^{(1)}\right)^2 + Z_1^{(1)}\,Z_3^{(1)}}\,A\,e^{2\,i\,k_1\,h_1\,\cos\alpha}\,, \tag{5.142}$$

and it is equal to zero when

$$Z_2^{(1)} = \sqrt{Z_1^{(1)}\,Z_3^{(1)}} \tag{5.143}$$

Thus, we have demonstrated that the reflected wave is absent in the upper medium if the condition 5.141 is satisfied and the impedance $Z_2^{(1)}$ is the geometric mean of the impedances $Z_1^{(1)}$ and $Z_3^{(1)}$. As in the case of Brewster angle, section 5.1, we observe the maximum of the destructive interference.

Inasmuch as the product $H\cos\beta$ plays an important role, let us discuss its geometrical meaning. Suppose that a phase surface N intersects the upper boundary of the layer at the point with coordinates x and z, Fig. 5.20a. Then, the product $H\cos\beta$ is equal to the distance d traversed by this phase surface until it intersects the lower boundary at the point with the same coordinate x. As will be shown later the value of $H\cos\beta$ determines the frequencies of normal modes.

Inhomogeneous waves

We may observe the total internal reflection at the top or bottom interface as the incidence angle α increases, which leads to the appearance of inhomogeneous waves. Bearing this in mind, let us represent the potentials in the form

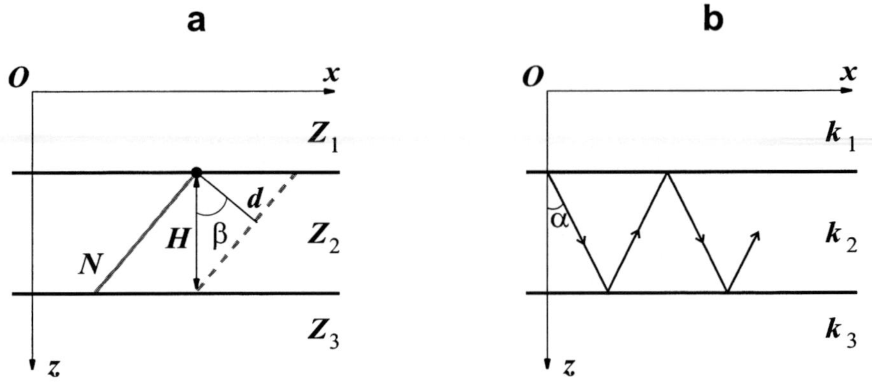

Figure 5.20: Geometrical meaning of $H\cos\beta$ (a). Waves inside the layer (b).

$$
\begin{aligned}
\mathcal{U}_1 &= \frac{1}{i\,k_1}\left(A\,e^{i\,k_1\,z\,\cos\alpha} + B\,e^{-i\,k_1\,z\,\cos\alpha}\right)e^{i\,k_1\,x\,\sin\alpha} && \text{for} \quad z < h_1\,, \\
\mathcal{U}_2 &= \frac{1}{i\,k_2}\left(C\,e^{i\,k_2\,b_{2z}\,z} + D\,e^{-i\,k_2\,b_{2z}\,z}\right)e^{i\,k_2\,n_{2x}\,x} && \text{for} \quad h_1 < z < h_2\,, \\
\mathcal{U}_3 &= \frac{1}{i\,k_3}\,E\,e^{i\,k_3\,b_{3z}\,z}\,e^{i\,k_3\,n_{3x}\,x} && \text{for} \quad z > h_2
\end{aligned}
\tag{5.144}
$$

Unlike the examples in section 5.2, the quantities b_{2z} and b_{3z} can be either real or imaginary.

As was shown earlier from the Helmholtz equation

$$\frac{\partial^2\mathcal{U}}{\partial x^2} + \frac{\partial^2\mathcal{U}}{\partial z^2} + k^2\,\mathcal{U} = 0$$

follows the relationship between n_x and b_z:

$$-k^2\,b_z^2 - k^2\,n_x^2 + k^2 = 0$$

and therefore

$$n_x^2 + b_z^2 = 1 \qquad \text{or} \qquad b_z = \sqrt{1 - n_x^2}$$

In accordance with eqs. 5.144, the equality

$$i\,k_1\,\sin\alpha = i\,k_2\,n_{2x}$$

has to take place in order to satisfy the boundary conditions at interface $z = h_1$. Thus

$$n_{2x} = \frac{k_1}{k_2}\,\sin\alpha = \frac{\sin\alpha}{n_1}\,,$$

where

$$n_1 = \frac{c_1}{c_2}$$

We further assume that the incidence angle α exceeds the critical one:

$$\sin\alpha > \sin\alpha_c = n_1$$

Respectively, the quantity b_{2z} is purely imaginary

$$b_{2z} = i\,\sqrt{n_{2x}^2 - 1} = i\,\sqrt{\frac{\sin^2\alpha}{n_1^2} - 1}$$

Thus, we see from the second equation of the set 5.144 that the wave field inside the layer is represented as a superposition of two inhomogeneous plane waves traveling along the x-axis. One of them exponentially decays when the z-coordinate increases, the other grows in the same direction.

In the same manner, the condition at boundary $z = h_2$ gives

$$k_3\,n_{3x} = k_2\,n_{2x}$$

or

$$n_{3x} = \frac{k_2}{k_3}\,n_{2x} = \frac{c_3}{c_2}\,\frac{\sin\alpha}{n_1} = \frac{\sin\alpha}{n_1\,n_2}\,, \tag{5.145}$$

where

$$n_2 = \frac{c_2}{c_3}$$

Therefore,

$$b_{3z} = \sqrt{1 - n_{3x}^2} = \sqrt{1 - \left(\frac{\sin\alpha}{n_1\,n_2}\right)^2} \tag{5.146}$$

and it can be either real or imaginary. In the first case the plane wave $\mathcal{U}_3$ is homogeneous, and it propagates away from interface $z = h_2$, whereas in the second case the inhomogeneous plane wave $\mathcal{U}_3$ travels along this boundary.

Wave amplitudes

Having determined n_x and b_z, it is a simple matter to find the wave amplitudes. The boundary conditions give the system of equations is similar to the set 5.130

$$\begin{aligned}
Z_1 \left(A\, e^{i\,k_1\,h_1\,\cos\alpha} + B\, e^{-i\,k_1\,h_1\,\cos\alpha}\right) &= Z_2 \left(C\, e^{i\,k_2\,h_1\,b_{2z}} + D\, e^{-i\,k_2\,h_1\,b_{2z}}\right), \\
\cos\alpha \left(A\, e^{i\,k_1\,h_1\,\cos\alpha} - B\, e^{-i\,k_1\,h_1\,\cos\alpha}\right) &= b_{2z} \left(C\, e^{i\,k_2\,h_1\,b_{2z}} - D\, e^{-i\,k_2\,h_1\,b_{2z}}\right), \\
Z_2 \left(C\, e^{i\,k_2\,h_2\,b_{2z}} + D\, e^{-i\,k_2\,h_2\,b_{2z}}\right) &= Z_3\, E\, e^{i\,k_3\,h_2\,b_{3z}}, \\
b_{2z} \left(C\, e^{i\,k_2\,h_2\,b_{2z}} - D\, e^{-i\,k_2\,h_2\,b_{2z}}\right) &= b_{3z}\, E\, e^{i\,k_3\,h_2\,b_{3z}}
\end{aligned} \tag{5.147}$$

The solution of this system gives

$$\begin{aligned}
B &= \frac{Z_2^{(2)}\,R_{13}^{(2)} - Z_1^{(2)}}{Z_2^{(2)}\,R_{13}^{(2)} + Z_1^{(2)}}\, A\, e^{2\,i\,k_1\,h_1\,\cos\alpha}, \\
C &= \frac{Z_1}{Z_2}\,\frac{2\,Z_2^{(2)}\,R_{13}^{(2)}}{L}\, A\, e^{-i\,k_2\,h_1\,b_{2z}}\, e^{i\,k_1\,h_1\,\cos\alpha}
\end{aligned} \tag{5.148}$$

and

$$\begin{aligned}
D &= \frac{Z_1}{Z_2}\,\frac{2\,Z_2^{(2)}\,R_{13}^{(2)}\,R_{23}^{(2)}}{L}\, A\, e^{i\,k_2\,h_2\,b_{2z}}\, e^{i\,k_2\,b_{2z}\,H}\, e^{i\,k_1\,h_1\,\cos\alpha}, \\
E &= \frac{Z_1}{Z_3}\,\frac{2\,Z_2^{(2)}\,R_{13}^{(2)}\,(1 + R_{23}^{(2)})}{L}\, A\, e^{i\,k_2\,b_{2z}\,H}\, e^{-i\,k_3\,b_{3z}\,h_2}\, e^{i\,k_1\,h_1\,\cos\alpha}
\end{aligned} \tag{5.149}$$

Here

$$Z_1^{(2)} = Z_1^{(1)} = \frac{Z_1}{\cos\alpha}, \qquad Z_2^{(2)} = \frac{Z_2}{b_{2z}}, \qquad Z_3^{(2)} = \frac{Z_3}{b_{3z}},$$

$$R_{23}^{(2)} = \frac{Z_3^{(2)} - Z_2^{(2)}}{Z_3^{(2)} + Z_2^{(2)}}, \qquad R_{13}^{(2)} = \frac{1 + R_{23}^{(2)}\, e^{2\,i\,k_2\,H\,b_{2z}}}{1 - R_{23}^{(2)}\, e^{2\,i\,k_2\,H\,b_{2z}}}$$

and

$$L = \left(Z_2^{(2)}\, R_{13}^{(2)} + Z_1^{(2)}\right)\left(1 + R_{23}^{(2)}\, e^{2\,i\,k_2\,H\,b_{2z}}\right)$$

Note that in deriving reflection coefficient B in a layered medium, we can use the same recursive relationship as in the case of the normal incidence. To illustrate the wave behavior and eqs. 5.148–5.149, we consider several cases.

Case one Suppose that

$$Z_2 = Z_3 \qquad \text{and} \qquad c_2 = c_3$$

Then

$$n_{2x} = n_{3x}\,, \qquad b_{2z} = b_{3z} = i\sqrt{\frac{\sin^2\alpha}{n_1^2} - 1}$$

and

$$R_{13}^{(2)} = 1$$

Therefore,

$$B = \frac{Z_2^{(2)} - Z_1^{(2)}}{Z_2^{(2)} + Z_1^{(2)}}\, A\, e^{2\,i\,k_1\,h_1\,\cos\alpha}$$

or

$$B = \frac{Z_2\,\cos\alpha - i\,|b_{2z}|\,Z_1}{Z_2\,\cos\alpha + i\,|b_{2z}|\,Z_1}\, A\, e^{2\,i\,k_1\,h_1\,\cos\alpha}\,,$$

that is, the total internal reflection is observed.

Correspondingly

$$C = \frac{2\,Z_1}{Z_2\,\cos\alpha + i\,|b_{2z}|\,Z_1}\, A\, e^{-i\,k_1\,h_1\,\cos\alpha}\, e^{-i\,k_2\,h_1\,b_{2z}}$$

Thus, we have obtained the expressions for the reflected and transmitted waves in a medium with a single interface when $\alpha > \alpha_c$. Assuming that the layer thickness H tends to infinity, we arrive at the same result. In fact, since $b_{2z} = i\,|b_{2z}|$, the quantity $R_{13}^{(2)}$ is equal to unity again.

Case two If $c_1 < c_2 < c_3$, then $b_{2z} = i\,|b_{2z}|$ and $b_{3z} = i\,|b_{3z}|$. Correspondingly, factors $R_{23}^{(2)}$ and $R_{13}^{(2)}$ are real, and this means that the amplitudes of the incident and reflected waves are equal to each other at the boundary $z = h_1$. At the same time inhomogeneous waves propagate along the x-axis inside the layer and in the lower medium. In accordance with eqs. 5.144, the amplitudes of these waves inside the layer decrease with the distance from the boundaries. Thus, the layer plays the role of a screen, whose influence depends on the media parameters, as well as on the frequency and the incidence angle. Since the exponent in the term $exp\,(-k_2\,|b_{2z}|\,z)$ is directly proportional to the frequency, the attenuation becomes stronger as ω increases. For instance, if the condition

$$|b_{2z}|\,H \geq \lambda_2$$

is met, the waves only slightly penetrate into the lower medium. In contrast, when

$$|b_{2z}|\,H \ll \lambda_2$$

the attenuation or, more precisely, the destructive interference is small. In other words, the layer behaves as a low-pass filter. This analysis shows that the time dependence of the waves in the layer and lower half-space is generally different from that of the incident wave. The Rayleigh and Love waves exhibit a similar behavior as functions of frequency.

Case three Suppose that

$$n_1\, n_2 > 1 \qquad \text{or} \qquad \frac{c_1}{c_3} > 1$$

Then, as follows from eqs. 5.146, the parameter b_{3z} is real and, correspondingly, the wave is homogeneous in the lower medium. Letting $n_{3x} = \sin\gamma$ and $b_{3z} = \cos\gamma$, we see that this waves forms angle γ with the z-axis.

Wave tunneling

An understanding of the total internal reflection allows us to begin studying a phenomenon called wave tunneling. Consider wave propagation inside a layer surrounded by half-spaces with greater velocity, Fig. 5.20b. Let us assume that the propagation angle exceeds the critical angle, that is, $\alpha > \alpha_c$. Therefore, the total internal reflection is observed at all points of the layer boundaries. Correspondingly, the wave energy remains constant within the layer, which plays the role of a waveguide. At the same time, inhomogeneous waves travel outside the layer along the x-axis and decay exponentially with the distance from the interfaces. The propagation of waves inside the layer is called the tunneling effect. This, may happen, for instance, in the ocean where there is a low-velocity water layer that serves as the waveguide. As is seen from Fig. 5.20b, the velocity component of the tunneled wave along the x-axis is

$$c_x = \frac{c_1}{\sin\alpha} \tag{5.150}$$

In particular, when

$$\alpha = \alpha_c \qquad \text{and} \qquad \sin\alpha_c = \frac{c_1}{c_2},$$

we have

$$c_x = c_2$$

If angle α, exceeds the critical one, then

$$c_x < c_2$$

and, therefore, the velocity of the tunneled wave may vary as

$$c_1 \leq c_x \leq c_2$$

At the same time our study of waves performed in this section shows the following. If the medium surrounding the waveguide is layered and its velocity decreases with distance from the waveguide, the total internal reflection may no longer be observed. Because in this case the wave energy escapes from the waveguide into another low-velocity medium, the amplitude of the tunneled wave would decrease along the layer.

5.7 Wave propagation in waveguides

In this section we continue the discussion of waves propagating along a layer, assuming that their direction makes some angle θ with the z-axis, Fig. 5.21. Waves in the layer may exist for several reasons. For instance, a plane wave can travel through the upper medium and partially penetrate into the layer. When the wave reaches the bottom of the layer the reflected wave appears. It travels upward, reaches the layer boundary, and causes another reflected wave that goes downward. It is obvious that reflections from both interfaces give rise to two families of waves, namely, the down- and up-going ones. As we see from Fig. 5.21, each wave of these families makes the same angle θ with the z-axis. We can also imagine a source of primary waves located inside the layer. At a sufficiently great distance from the source the resulting wave field can be approximately described by a system of plane waves propagating along the layer and making different angles with its interfaces. Selecting plane waves with any angle θ, we obtain the wave family that is similar to that in the first case. This discussion suggests that each group of waves, up-going and down-going, undergoes multiple reflections at the boundaries and advances within the layer. The interference of these waves may be either destructive or constructive, and our goal is to examine the superposition of these waves when they interfere in a constructive way. With this purpose in mind, we consider two examples illustrating propagation of plane waves inside the layer. The latter has density ρ, thickness H, and velocity of wave propagation c_1.

As is well known, the partial solution of the Helmholtz equation inside the layer is

$$\mathcal{U} = \frac{1}{i\,k_1}\left(A_0 e^{i\,k_1\,z\,\cos\theta} + B_0\,e^{-i\,k_1\,z\,\cos\theta}\right)e^{i\,k_1\,x\,\sin\theta}\,, \tag{5.151}$$

where the terms $A_0 e^{i\,k_1\,z\cos\theta}$ and $B_0 e^{-i\,k_1\,z\cos\theta}$ describe the down-going and up-going wave fields, respectively.

First, we consider sinusoidal waves and start with a relatively simple case.

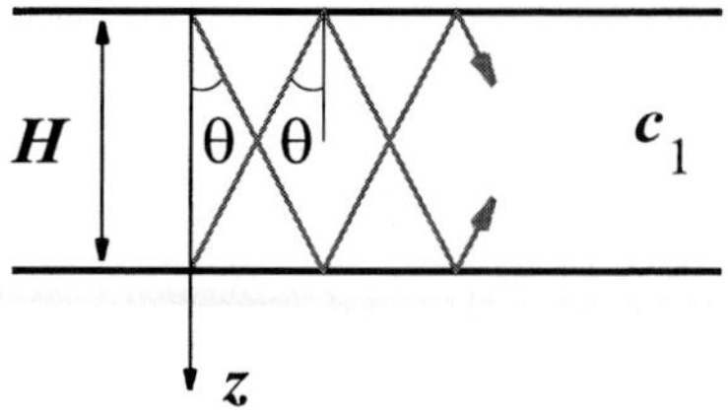

Figure 5.21: Waveguide with plane boundaries.

Example one: The layer is surrounded by ideally rigid media In this case, the vertical component of displacement disappears at each point of both boundaries. Therefore, the boundary conditions are

$$\frac{\partial \mathcal{U}}{\partial z} = 0 \quad \text{at} \quad z = 0 \qquad \text{and} \qquad \frac{\partial \mathcal{U}}{\partial z} = 0 \quad \text{at} \quad z = H \tag{5.152}$$

Substitution of eq. 5.151 into eqs. 5.152 gives

$$\begin{cases} A_0 - B_0 = 0\,, \\ A_0\, e^{i\,k_1\,H\,\cos\theta} - B_0\, e^{-i\,k_1\,H\,\cos\theta} = 0 \end{cases} \tag{5.153}$$

Eliminating the unknown coefficient, we arrive at the equation with respect to the wavenumber k_1

$$e^{2i\,k_1\,H\,\cos\theta} = 1$$

Since

$$k_1 = \frac{\omega}{c_1} \qquad \text{and} \qquad e^{\,2\pi\,i\,n} = 1\,,$$

we obtain

$$2\,i\,\frac{\omega_n}{c_1}\,H\,\cos\theta = 2\,\pi\,i\,n \qquad \text{or} \qquad f_n = \frac{n\,c_1}{2\,H\,\cos\theta}\,, \tag{5.154}$$

where n is an integer.

Respectively, the wavenumbers are

$$k_n = \frac{\omega_n}{c_1} = \frac{\pi\,n}{H\,\cos\theta} \tag{5.155}$$

Substituting eq. 5.155 into eq. 5.151 and taking into account that $A_0 = B_0$, eqs. 5.153, we find for the potential corresponding to k_n:

$$\mathcal{U}_n = \frac{A_0}{i\,k_n}\left(e^{\pi\,i\,n\,z/H} + e^{-\pi\,i\,n\,z/H}\right)\,\exp\left(i\,\frac{\omega_n}{c_1}\,x\,\sin\theta\right) \tag{5.156}$$

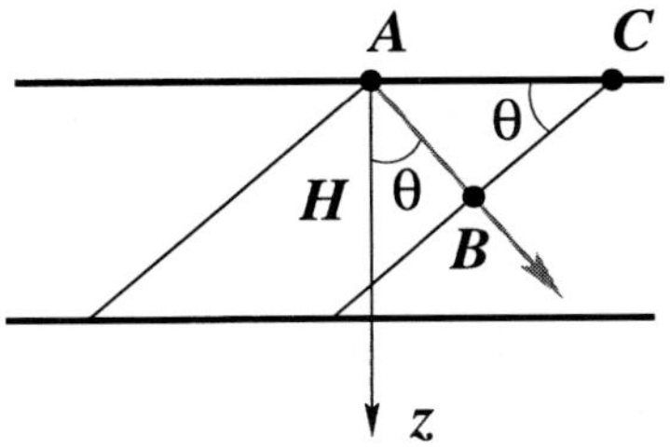

Figure 5.22: Illustration of eq. 5.158.

Since

$$U = \text{Re}\, \mathcal{U} e^{-i\omega t},$$

we have

$$U_n = \frac{2A_0}{k_n} \cos\left(\pi n \frac{z}{H}\right) \cos\left(\omega_n t + \frac{\pi}{2} - x \frac{\omega_n}{c_1} \sin\theta\right) \tag{5.157}$$

or

$$U_n = -\frac{2A_0}{k_n} \cos\left(\pi n \frac{z}{H}\right) \sin\omega_n \left(t - \frac{x \sin\theta}{c_1}\right),$$

where

$$\omega_n = \frac{\pi n c_1}{H \cos\theta}$$

Equation 5.157 describes the wave that is caused by constructive interference of reflected waves and propagates parallel to the layer boundaries with the velocity

$$c = \frac{c_1}{\sin\theta} \tag{5.158}$$

Such a dependence can be easily explained. In fact, when point A of the phase surface of reflected plane wave advances at the distance AB along the ray, the wave reaches point C in the horizontal direction. As follows from Fig. 5.22,

$$AB = AC \sin\theta \qquad \text{or} \qquad \frac{AB}{\Delta t} = \frac{AC}{\Delta t} \sin\theta,$$

that is

$$c_1 = c \sin\theta,$$

which coincides with eq. 5.158. When θ increases, the phase velocity c along the x-axis becomes smaller and approaches acoustic velocity c_1.

In accordance with eq. 5.157 the amplitude of this wave is

$$A(z) = \frac{2A_0}{k_n} \cos\left(\pi n \frac{z}{H}\right), \tag{5.159}$$

that is, it is a sinusoidal function of the vertical coordinate z and is independent of time.

We see that the wave is traveling along the layer in the horizontal direction but has the appearance of a standing wave in the vertical direction. In fact, the amplitude remains constant at all points of any horizontal plane $z =$ const inside the layer. In particular, the integer values of m satisfying the conditions

$$\pi\, n\, \frac{z}{H} = \frac{\pi}{2}\,(2\,m-1) \qquad \text{or} \qquad \frac{z}{H} = \frac{2\,m-1}{2\,n} \leq 1$$

determine the planes of nodes where the pressure vanishes for each given n.

Bearing in mind that the horizontal and vertical components of the particle displacement relate to the potential U as

$$s_x = \frac{\partial U}{\partial x} \qquad \text{and} \qquad s_z = \frac{\partial U}{\partial z},$$

we find

$$s_x = 2\,A_0\,\sin\theta\,\cos\left(\pi\, n\,\frac{z}{H}\right)\,\cos\omega_n\left(t - \frac{x\,\sin\theta}{c_1}\right)$$

and

$$s_z = 2\,A_0\,\cos\theta\,\sin\left(\pi\, n\,\frac{z}{H}\right)\,\sin\omega_n\left(t - \frac{x\,\sin\theta}{c_1}\right)$$

Therefore, the wave propagation inside the layer is accompanied by elliptical particle motion, and the aspect ratio of the ellipse depends on the angle θ and the z-coordinate.

We have demonstrated that interference of homogeneous plane waves with linear particle motion may produce a system of elliptically polarized inhomogeneous waves traveling in the horizontal direction with velocity c. As in the case of normal incidence Chapter 4, wave propagation inside the layer is possible only for certain frequencies corresponding to a given incidence angle θ. Those frequencies are

$$f_1 = \frac{c_1}{2\,H\,\cos\theta}, \qquad f_2 = \frac{2\,c_1}{2\,H\,\cos\theta}, \qquad f_3 = \frac{3\,c_1}{2\,H\,\cos\theta}, \qquad \ldots \tag{5.160}$$

Otherwise, the boundary conditions cannot be satisfied.

We mentioned previously that the product

$$l = H\,\cos\theta$$

characterizes the distance between the same phase surface when it reaches the top and bottom of the layer, Fig. 5.20a. Thus, frequencies f_n, eq. 5.160, are directly proportional to the velocity c_1 and inversely proportional to separation $2\,l$. We see that the largest possible period corresponds to the normal incidence, and it is equal to the time during which the wave travels along the z-axis from the top to the bottom of the layer and back. As is well known a wave with some index n is usually called the normal mode of the nth order. It is clear from eq. 5.158 that all normal modes that have different amplitudes and frequencies move with the same velocity c as long as they are formed by plane waves with the same incidence angle θ. By definition, with an increase of n the frequency of these waves becomes higher:

$$\omega_n = \frac{c_1 \pi}{H \cos\theta} n \qquad \text{and} \qquad \lambda_n = \frac{c}{f_n}$$

Also, as follows from eq. 5.159, the larger the order of the normal mode, the more oscillations of its amplitude are observed along the z-axis. It is obvious that for the up-going and down-going plane waves and normal modes, we have

$$\omega = c_1 k_1 \qquad \text{and} \qquad \omega_n = c_1 k_n \tag{5.161}$$

Due to the constructive interference, the latter travels without attenuation inside the layer, that is, the tunneling effect is observed, and for this reason, such a layer is called the waveguide. Now let us imagine a source of primary waves and consider several different situations.

Case one Suppose that a source "generates" a plane wave with an incident angle θ and the frequency f that does not correspond to those of normal modes. Then the superposition of reflected waves has a destructive character and the tunneling effect along the layer is absent. In contrast, if the source produces the plane wave with a frequency of any normal mode the interference of reflected waves becomes constructive and the resulting wave propagates along the x-axis without attenuation.

Case two Consider an example when the source gives rise to a system of normal modes that are formed by plane waves with the same incident angle θ. Then the total wave propagates along the layer without attenuation, and it represents a sum of modes of different amplitudes and frequencies. All of them, however, have the same velocity c, and therefore the shape of the resulting wave does not change with distance.

Case three Next we assume that the source "generates" a transient plane wave with a continuous spectrum and the incident angle θ. Let us treat this wave as an infinite sum of sinusoidal waves with different frequencies but the same angle θ. By definition, the waves with frequencies of normal modes propagate along a layer of a given thickness H, whereas a superposition of plane waves with other frequencies has a destructive character. In particular, it is true for waves with frequencies which are lower than that of the first normal mode:

$$f < f_1 = \frac{c_1}{2 H \cos\theta}$$

This is why f_1 is often called the cut-off frequency. Thus, as in the previous case, the total wave propagating along the layer is the superposition of normal modes. In accordance with eq. 5.157, its potential is

$$U = -2 \sum_{n=1}^{\infty} \frac{A_{0n}}{k_n} \cos\left(\pi n \frac{z}{H}\right) \sin \omega_n \left(t - \frac{x \sin\theta}{c_1}\right) \tag{5.162}$$

Of course, the shape of this wave does not change with distance x.

Case four Suppose that the source "generates" two homogeneous plane waves with the same frequency f but different incident angles: θ_1 and θ_2. As we already know, they define two sets of frequencies of normal modes:

$$f_{1n} = \frac{n\, c_1}{2\,H\,\cos\theta_1} \qquad \text{and} \qquad f_{2n} = \frac{n\, c_1}{2\,H\,\cos\theta_2}$$

It is obvious that if the frequency f does not coincide with any frequency of either set, propagation is almost absent. In contrast, if f coincides with that of one set, for instance, $f = f_{1n}$, then there is only a single inhomogeneous plane wave (normal mode) traveling along the layer with the velocity

$$c^{(1)} = \frac{c_1}{\sin\theta_1}$$

When the frequency f coincides with one frequency of each set

$$f = f_{1n_1} = f_{2n_2}\,,$$

the total wave consists of two inhomogeneous waves of the same frequency. They have, however, different amplitudes, initial phases and velocities. Correspondingly, the shape of this composite wave varies with the distance from the source. A similar behavior is observed when there are N homogeneous plane waves with different incident angles, and their frequency f corresponds to that of normal modes of different sets. Note that the superposition of waves with the same frequency and different velocities produces a sinusoidal function of time with the given frequency. Its amplitude changes with distance x also as a sinusoidal function.

Case five If the source generates two transient plane waves with different incident angles θ_1 and θ_2, then, due to interference of reflected waves, there are two infinite sets of inhomogeneous sinusoidal waves. They are characterized by two sets of frequencies f_{1n} and f_{2n} and propagate along the layer. Each group of these waves has a certain velocity c. Since frequencies f_{1n} or f_{2n} differ by an integer, for example,

$$f_{1n} = n\, f_{11} \qquad \text{and} \qquad f_{2n} = n\, f_{21}\,,$$

we can expect that the sum of waves corresponding to each group is described by a periodic function with respect to both arguments x and t, that is,

$$U_1 = F_1(x\,,z)\,g_1\left[a\left(t - \frac{x}{c^{(1)}}\right)\right] \qquad \text{and} \qquad U_2 = F_1(x\,,z)\,g_2\left[a\left(t - \frac{x}{c^{(2)}}\right)\right]$$

Inasmuch as those two waves move with different velocities, the total wave

$$U = U_1 + U_2$$

changes its shape as the distance from the source increases. Of course, a similar behavior occurs in the real case when a source gives rise to an arbitrary number of transient waves with various incidence angles.

Velocity of normal modes as a function of frequency

We have derived simple expressions for phase velocity c and frequencies f_n of normal modes:

$$c = \frac{c_1}{\sin\theta} \qquad \text{and} \qquad f_n = \frac{n\,c_1}{2\,H\,\cos\theta} \tag{5.163}$$

Thus

$$\sin\theta = \frac{c_1}{c} < 1 \qquad \text{or} \qquad \cos\theta = \sqrt{1 - \frac{c_1^2}{c^2}} < 1\,,$$

and the second equation of the set 5.163 yields

$$f_n = \frac{n\,c_1}{2\,H\sqrt{1 - \frac{c_1^2}{c^2}}} \tag{5.164}$$

Certainly, the phase velocity c of the normal mode and its frequency f_n are functions of the incident angle θ of the plane waves that form this mode. At the same time, eq. 5.164 can be treated as the relationship between the phase velocity and the frequency for each normal mode. Let us now denote the phase velocity of the n-th mode by c_{pn} and consider its dependence on the frequencies f. In fact, we have from eq. 5.164:

$$\frac{c_{pn}}{c_1} = \frac{1}{\sqrt{1 - \frac{n^2}{q^2}}} \tag{5.165}$$

where the dimensionless parameter q is

$$q = \frac{2\,H\,f}{c_1} = \frac{2\,H}{\lambda_1}$$

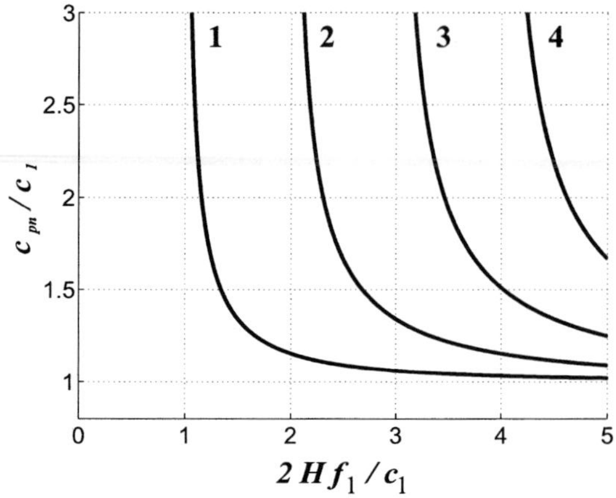

Figure 5.23: The function $c_{pn}/c_1\,(q)$, n is the curve index. Curves are labeled with the values of n.

and

$$\lambda_1 = \frac{c_1}{f}$$

The behavior of the ratio c_{pn}/c_1 as a function of parameter q for any normal mode is shown in Fig. 5.23. Such dependences, usually called the dispersion curves, illustrate the following. With an increase of the frequency f, when the angle θ tends to $\pi/2$, the phase surfaces of reflected waves become almost normal to the boundaries. Therefore, velocity c of normal modes approaches acoustic velocity c_1. In contrast, if frequency f is relatively small, the phase surfaces of plane waves are nearly parallel to the interfaces, that is, $\theta \to 0$. Correspondingly, the velocity of a normal mode along the waveguide tends to infinity.

Example two: The layer is bounded by the free surface at $z = 0$ and the rigid basement at $z = h$, (Fig. 5.24).

Case one By analogy with the previous example, we first assume that a source "generates" a single plane wave the frequency ω and incident angle θ. Taking into account the fact that the pressure and the normal component of displacement disappear at the top an bottom of the layer, respectively, we have

$$\mathcal{U} = 0 \quad \text{at} \quad z = 0 \qquad \text{and} \qquad \frac{\partial \mathcal{U}}{\partial z} = 0 \quad \text{at} \quad z = H \tag{5.166}$$

Substitution of eq. 5.151 into eq. 5.166 gives

$$\begin{cases} A_0 + B_0 = 0\,, \\ A_0\, e^{i\,k_1\,H\,\cos\theta} - B_0\, e^{-i\,k_1\,H\,\cos\theta} = 0 \end{cases}$$

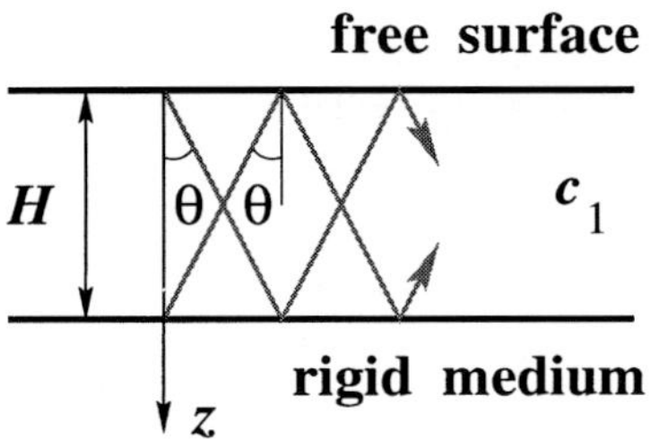

Figure 5.24: Waveguide with free and rigid boundaries.

Therefore

$$A_0 = -B_0$$

and

$$e^{2\,i\,k_1\,H\,\cos\theta} = -1 = e^{i\,\pi(2n-1)} \qquad \text{or} \qquad 2\,i\,k_1\,H\,\cos\theta = i\,\pi(2n-1)\,,$$

that is,

$$f_n = \frac{(2\,n-1)\,c_1}{4\,H\,\cos\theta} \tag{5.167}$$

Thus we have found again that only waves with certain frequencies can propagate along the layer without attenuation. Those frequencies depend on angle θ, layer thickness H, and acoustic velocity c_1. From eqs. 5.151 and 5.167, we obtain

$$\mathcal{U}_n = \frac{A_0}{i\,k_n}\left\{\exp\left[i\,\frac{\pi}{2}\,(2\,n-1)\,\frac{z}{H}\right] - \exp\left[-i\,\frac{\pi}{2}\,(2\,n-1)\,\frac{z}{H}\right]\right\}\,e^{i\,k_n\,x\,\sin\theta}$$

or

$$U_n = \frac{2\,A_0}{k_n}\,\sin\left[\frac{\pi}{2}\,(2\,n-1)\,\frac{z}{H}\right]\,\cos\omega_n\left(t-\frac{x\,\sin\theta}{c_1}\right) \tag{5.168}$$

As before, we see that constructive interference of reflected waves produces an inhomogeneous plane wave propagating along the waveguide with the phase velocity given by eq. 5.158. The term

$$\frac{2\,A_0}{k_n}\,\sin\left[\frac{\pi}{2}\,\frac{z}{H}\,(2\,n-1)\right]$$

describes the standing wave along the z-axis.

Case two Now suppose that the source generates a finite set of plane transient waves with different incident angles, so that

$$0 \le \theta \le \frac{\pi}{2} \qquad \text{and} \qquad 0 < \omega < \infty$$

Each transient wave with the given angle θ can be represented as an infinite sum of sinusoidal plane waves, including those with frequencies f_n. Since only the latter can propagate along the layer, the transient wave is described by the sum of normal modes. It is essential that such a wave is periodical in time and does not change its shape with the distance. This happens because all sinusoidal waves have the same phase velocity c_p when $\theta = \text{const}$. Correspondingly, the resultant wave caused by the source consists of the sum of such periodic waves. They are characterized by different sets of frequencies f_n and propagate with different velocities. For this reason, the resultant wave is a transient one, and its shape changes with the distance from the source.

Case three Next we assume that angle θ of a plane transient waves, caused by a source, continuously varies within the interval

$$0 \leq \theta \leq \frac{\pi}{2}$$

Therefore, the resulting wave also has a continuous frequency spectrum. In such a case, it is natural to characterize the motion of this wave using the concept of the group velocity c_{gn} , Part I. With this purpose in mind, we first establish the dispersion relation $c_{pn}(\omega)$.

As follows from eq. 5.167

$$\sqrt{\left(1 - \sin^2 \theta\right)} = \frac{\pi\,(2\,n - 1)\,c_1}{2\,\omega_n\,H}$$

or

$$\sin \theta = \sqrt{1 - \frac{\pi^2\,(2\,n - 1)^2\,c_1^2}{4\,H^2\,\omega_n^2}}$$

Taking into account eq. 5.158, we obtain for the phase velocity of the normal mode:

$$c_{pn}(\omega) = \frac{c_1}{\sqrt{1 - \dfrac{\pi^2\,(2\,n - 1)^2\,c_1^2}{4\,H^2\,\omega^2}}} \tag{5.169}$$

By definition, the group velocity c_{gn} is

$$c_{gn}(\omega) = \left(\frac{d\,k_n}{d\,\omega}\right)^{-1}$$

Because

$$k_n = \frac{\omega}{c_{pn}},$$

we have

$$\frac{d\,k_n}{d\,\omega} = \frac{d}{d\,\omega}\left(\frac{\omega}{c_{pn}}\right) = \frac{1}{c_{pn}} - \frac{\omega}{c_{pn}^2}\,\frac{d\,c_{pn}}{d\,\omega}$$

Thus,

$$c_{gn}(\omega) = c_{pn}\left(1 - \frac{\omega}{c_{pn}}\,\frac{d\,c_{pn}}{d\,\omega}\right)^{-1} \tag{5.170}$$

Differentiation of eq. 5.169 gives

$$\frac{d\,c_{pn}(\omega)}{d\,\omega} = -\,c_1\left[1 - \frac{\pi^2\,(2\,n-1)^2\,c_1^2}{4\,H^2\,\omega^2}\right]^{-3/2}\,\frac{\pi^2\,(2\,n-1)^2\,c_1^2}{4\,H^2\,\omega^3}$$

or

$$\frac{d\,c_{pn}(\omega)}{d\,\omega} = -\,\frac{\pi^2\,(2\,n-1)^2\,c_{pn}^3}{4\,H^2\,\omega^3}$$

Then

$$\left(1 - \frac{\omega}{c_{pn}}\,\frac{d\,c_{pn}}{d\,\omega}\right)^{-1} = \left[1 + \frac{\pi^2\,(2\,n-1)^2\,c_{pn}^2}{4\,H^2\,\omega^2}\right]^{-1}$$

Finally, we obtain

$$c_{gn}(\omega) = c_{pn}\left[1 + \frac{\pi^2\,(2\,n-1)^2\,c_{pn}^2}{4\,H^2\,\omega^2}\right]^{-1}, \tag{5.171}$$

where phase velocity c_{pn} is given by eq. 5.169. Behavior of the group and phase velocities as functions of a non-dimensional parameter

$$\frac{H\,f}{c_1} = \frac{H}{\lambda_1}$$

for two first normal modes is shown in Fig. 5.25. As we see from these curves, velocities $c_{pn}(\omega)$ and $c_{gn}(\omega)$ approach each other and become equal to c_1 when frequencies are sufficiently large

$$\frac{H\,f}{c_1} > 1\,.$$

This corresponds to almost horizontal propagation of the waves, that is, $\theta \to \pi/2$. In other words, if frequencies satisfy the condition

$$\frac{H}{\lambda_1} > 1\,,$$

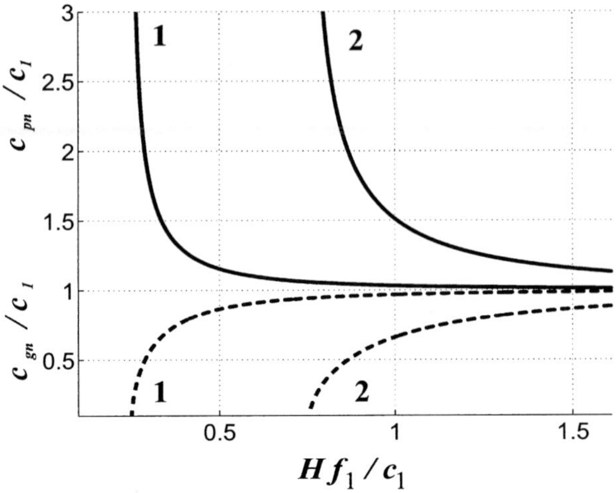

Figure 5.25: The ratios $c_{pn}(\omega)/c_1$ (solid) and $c_{gn}(\omega)/c_1$ (dashed). The numbers on curves label the orders of normal modes.

the normal modes are formed by reflected waves with large values of angle θ. In contrast, with a decrease of frequency the difference between those velocities becomes greater and, as we already know, angle θ tends to zero. Thus, proceeding from solutions of the boundary value problem, we have demonstrated that plane sinusoidal waves characterized by certain frequencies can propagate along the layer bounded by either two rigid media or by rigid and free half-spaces. At the same time the system of plane waves with other frequencies cannot exist in the layer. From the mathematical point of view this means that such waves do not satisfy boundary conditions. This behavior of both groups of plane waves suggests that either constructive or destructive interference takes place.

Phase shift between plane waves

By analogy with the case of normal incidence, let us study the process of interference in some detail, and with this purpose in mind consider the phase surface N of the downgoing wave, Fig. 5.26. As we know, the reflected up-going and down-going waves appear as a result of the action of this wave, and this phenomenon may be continuously observed. It is also clear that both the up-going and down-going waves make the same angle θ with the boundaries, and, correspondingly, their phase surfaces are parallel to each other. If the frequencies or wavelengths are such that the phase difference between these waves is equal to $2\pi n$, their superposition is constructive. The waves with other frequencies have a phase shift different from $2\pi n$, and they also interfere, but they cancel each other after a relatively small number of reflections.

Note that the phase difference of two down-going or up-going waves is defined by

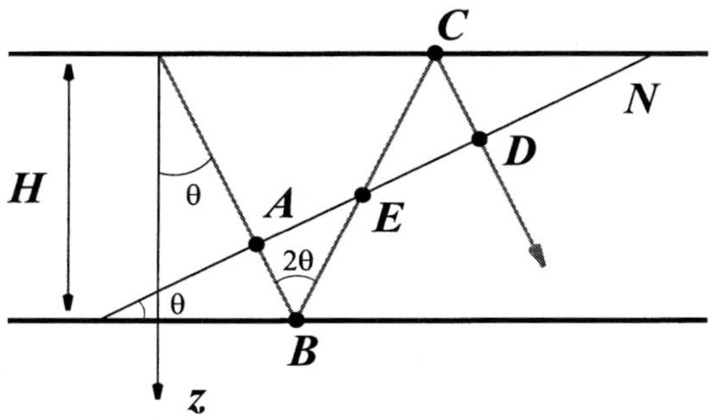

Figure 5.26: Illustration of interference.

the following factors. One is the phase shift between the incident and reflected waves arising at the layer top. It is equal to π, since this boundary is free. The second factor is defined by the distance $ABECD$, Fig. 5.26. Taking into account that the argument of the sinusoidal function is

$$\omega t - k_1 l = \omega t - \frac{2\pi}{\lambda_1} l ,$$

the condition for the constructive interference can be written as

$$k_1 \, |ABCD| + \pi = 2\,\pi n$$

or

$$\frac{2\,\pi}{\lambda_1} \left(|AB| + |BC| + |CD| \right) = (2\,n - 1)\,\pi , \qquad (5.172)$$

where n is an integer and λ_1 is the wavelength of the reflected wave.

Since

$$\lambda_1 = \frac{c_1}{f_n} ,$$

the result of interference essentially depends on frequency. As is seen from Fig. 5.26 $AB = BE \cos 2\,\theta$, $CD = CE \cos 2\,\theta$ and $BC = H / \cos\theta$. Substitution of these terms into eq. 5.172 gives

$$\frac{2\,\pi\,H}{\lambda_1} \left(\frac{\cos 2\,\theta}{\cos\theta} + \frac{1}{\cos\theta} \right) = (2\,n - 1)\,\pi$$

Making use of the equality

$$\cos 2\,\theta + 1 = 2\,\cos^2\theta ,$$

we finally obtain

$$\frac{f_n\,H\,\cos\theta}{c_1} = \frac{2\,n - 1}{4} ,$$

which coincides with eq. 5.167 which was derived from solution of the boundary value problem. Of course, the same process of interference is observed in the previous case, when the layer was surrounded by rigid half-spaces. We have considered two models of media where the embedded layer behaves as a waveguide. This may also happen in other cases, for instance, if surrounding medium has a higher velocity than that in the layer and the incidence angle of the plane waves exceeds the critical. Then, for certain frequencies constructive interference takes place and the normal modes propagate along the layer.

5.8 Waves in media with continuously changing velocity

So far we studied plane waves mainly in piecewise homogeneous media. Next consider a more complicated model in which both wave propagation velocity c and density ρ are continuous functions of spatial coordinates. First, we again discuss a relationship between the pressure and the medium parameters.

Equation for the pressure

As was already shown in Chapter 1, there are two equations describing the behavior of the displacement $\mathbf{s}$ and the pressure P:

$$\rho c^2 \,\text{div}\, \mathbf{s} = -P \qquad \text{and} \qquad \frac{\partial^2 \mathbf{s}}{\partial t^2} = -\frac{1}{\rho}\,\text{grad}\, P \tag{5.173}$$

Here P is the additional pressure caused by a wave.

Differentiation of the first equation 5.173 twice with respect to time gives

$$\rho c^2 \,\text{div}\, \frac{\partial^2 \mathbf{s}}{\partial t^2} = -\frac{\partial^2 P}{\partial t^2} \tag{5.174}$$

Then, making use of the second relation, we arrive at the following differential equation of the second order for the pressure:

$$\rho c^2 \,\text{div}\left(\frac{1}{\rho}\,\text{grad}\, P\right) = \frac{\partial^2 P}{\partial t^2}$$

Since

$$\text{div}\,(\varphi\, \text{grad}\, \psi) = \varphi \nabla^2 \psi + \nabla\varphi \nabla\psi$$

and

$$\text{grad}\, \frac{1}{\rho} = -\frac{1}{\rho^2}\,\text{grad}\, \rho\,,$$

we have

$$c^2\,\nabla^2 P - \frac{c^2}{\rho}\,\nabla\rho\cdot\nabla P = \frac{\partial^2 P}{\partial t^2}$$

or

$$\nabla^2 P - \frac{\operatorname{grad}\rho\cdot\operatorname{grad}P}{\rho} = \frac{1}{c^2}\frac{\partial^2 P}{\partial t^2} \tag{5.175}$$

Assuming that waves are sinusoidal functions of time, we can write

$$P = \operatorname{Re}\,\mathcal{P}\,e^{-i\omega t}, \tag{5.176}$$

where $\mathcal{P}$ is the complex amplitude of the pressure. Substitution of eq. 5.176 into eq. 5.175 yields

$$\nabla^2\mathcal{P} + k^2\,\mathcal{P} - \frac{\nabla\rho\cdot\nabla\mathcal{P}}{\rho} = 0 \tag{5.177}$$

Here $k = \omega/c$ is the wavenumber.

Suppose that the pressure and displacement caused by a wave depend on two spatial coordinates x and z, while the medium parameters ρ and c are the functions of the z-coordinate only:

$$\rho = \rho(z) \qquad \text{and} \qquad c = c(z)$$

Then in place of eq. 5.177, we obtain

$$\frac{\partial^2\mathcal{P}}{\partial x^2} + \frac{\partial^2\mathcal{P}}{\partial z^2} + k^2\,\mathcal{P} - \frac{1}{\rho}\frac{d\rho}{dz}\frac{\partial\mathcal{P}}{\partial z} = 0 \tag{5.178}$$

This equation essentially differs from the Helmholtz equation because it contains the product of the derivatives of the density and pressure. To simplify this equation we introduce the function

$$\Psi \equiv \frac{\mathcal{P}}{\sqrt{\rho}} \tag{5.179}$$

Thus,

$$\mathcal{P} = \sqrt{\rho}\,\Psi \qquad \text{and} \qquad \frac{\partial^2\mathcal{P}}{\partial x^2} = \sqrt{\rho}\,\frac{\partial^2\Psi}{\partial x^2},$$

$$\frac{\partial\mathcal{P}}{\partial z} = \frac{\Psi}{2}\left(\frac{1}{\sqrt{\rho}}\frac{d\rho}{dz}\right) + \sqrt{\rho}\,\frac{\partial\Psi}{\partial z}$$

and

$$\frac{\partial^2 \mathcal{P}}{\partial z^2} = \frac{1}{\sqrt{\rho}} \frac{d\rho}{dz} \frac{\partial \Psi}{\partial z} + \frac{\Psi}{2} \frac{d}{dz} \left(\frac{1}{\sqrt{\rho}} \frac{d\rho}{dz} \right) + \sqrt{\rho} \frac{\partial^2 \Psi}{\partial z^2}$$

$$= \frac{1}{\sqrt{\rho}} \frac{d\rho}{dz} \frac{\partial \Psi}{\partial z} + \sqrt{\rho} \frac{\partial^2 \Psi}{\partial z^2} + \frac{\Psi}{2\sqrt{\rho}} \frac{d^2\rho}{dz^2} - \frac{\Psi}{4\sqrt{\rho^3}} \left(\frac{d\rho}{dz} \right)^2$$

Eq. 5.178 then becomes

$$\sqrt{\rho} \frac{\partial^2 \Psi}{\partial x^2} + \sqrt{\rho} \frac{\partial^2 \Psi}{\partial z^2} + \frac{1}{\sqrt{\rho}} \frac{d\rho}{dz} \frac{\partial \Psi}{\partial z} + \frac{\Psi}{2\sqrt{\rho}} \frac{d^2\rho}{dz^2} - \frac{\Psi}{4\sqrt{\rho^3}} \left(\frac{d\rho}{dz} \right)^2 +$$

$$+ k^2 \sqrt{\rho}\, \Psi - \frac{1}{\rho} \frac{d\rho}{dz} \left(\frac{\Psi}{2\sqrt{\rho}} \frac{d\rho}{dz} + \sqrt{\rho} \frac{\partial \Psi}{\partial z} \right) = 0$$

or

$$\sqrt{\rho} \frac{\partial^2 \Psi}{\partial x^2} + \sqrt{\rho} \frac{\partial^2 \Psi}{\partial z^2} + \sqrt{\rho}\, \Psi \left[k^2 + \frac{1}{2\rho} \frac{d^2\rho}{dz^2} - \frac{3}{4} \left(\frac{1}{\rho} \frac{d\rho}{dz} \right)^2 \right] = 0\,,$$

and eq. 5.177 can be written as:

$$\nabla^2 \Psi + \mathcal{K}^2\, \Psi = 0\,, \tag{5.180}$$

where

$$\mathcal{K}^2 = k^2 + \frac{1}{2\rho} \frac{d^2\rho}{dz^2} - \frac{3}{4} \left(\frac{1}{\rho} \frac{d\rho}{dz} \right)^2 \tag{5.181}$$

The similarity with the Helmholtz equation is obvious. The difference, however, is that instead of the single coefficient k^2 we now have a sum of three terms. The first one depends on the velocity which changes with the z-coordinate. The last two are the functions of the density and they also vary with z. Let us notice that in general, when

$$c = c(x,\, y,\, z) \qquad \text{and} \qquad \rho = \rho(x,\, y,\, z)\,,$$

the same transformation again leads to eq. 5.180 with the coefficient $\mathcal{K}^2$ given by

$$\mathcal{K}^2 = k^2 + \frac{\nabla^2 \rho}{2\rho} - \frac{3}{4} \left(\frac{\nabla \rho}{\rho} \right)^2$$

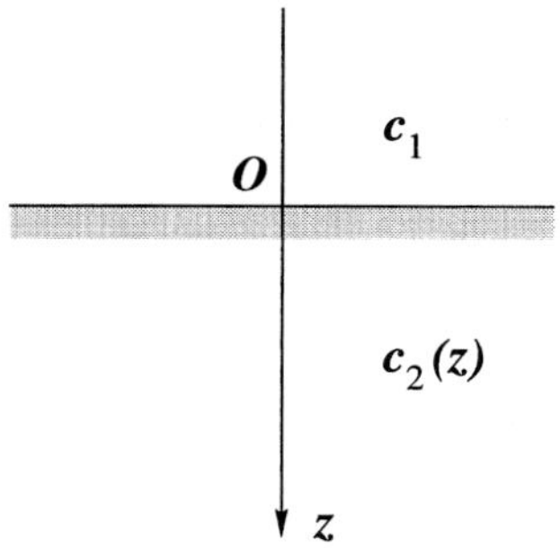

Figure 5.27: Medium with continuous change of velocity.

Our main goal in this section is to study a wave behavior when only velocity c varies. For this reason we assume that

$$\frac{\partial \rho}{\partial z} = 0 \,,$$

and eq. 5.180 becomes

$$\nabla^2 \Psi + k^2(z)\, \Psi = 0 \tag{5.182}$$

Thus, the function $\Psi(x,\, z)$ satisfies the Helmholtz equation where the wavenumber k depends on z.

Boundary conditions

Inasmuch as the wave equation is valid at regular points only, consider the behavior of the function $\Psi(x,\, z)$ at the interface between two media, Fig. 5.27. First, taking into account eq. 5.179 and the continuity of pressure, we obtain

$$\Psi_1 = \Psi_2 \,,$$

provided that the density is equal in both media. As follows from the second equation of the set 5.173

$$\omega^2 \mathcal{S} = \frac{1}{\rho} \operatorname{grad} \mathcal{P}$$

and, in particular,

$$\omega^2 \mathcal{S}_z = \frac{1}{\rho} \frac{\partial \mathcal{P}}{\partial z} \qquad \text{or} \qquad \omega^2 \mathcal{S}_z = \frac{1}{\sqrt{\rho}} \frac{\partial \Psi}{\partial z}$$

Therefore, from the continuity of the normal component of displacement, we conclude that the derivative $\partial \Psi / \partial z$ is also continuous. Correspondingly, the boundary conditions for the function Ψ at any point of the plane $z = 0$ are

$$\Psi_1 = \Psi_2 \qquad \text{and} \qquad \frac{\partial \Psi_1}{\partial z} = \frac{\partial \Psi_2}{\partial z} \tag{5.183}$$

Note that in the accuracy of a constant the function, Ψ plays the role of the complex amplitude of the pressure, because $\rho =$ const.

Solution of the wave equation

Similarly to the case of constant wavenumber, we apply the method of separation of variables and represent the function $\Psi(x,\, z)$ as

$$\Psi(x,\, z) = \mathcal{X}(x)\, \mathcal{Z}(z) \tag{5.184}$$

Its substitution into eq. 5.182 yields

$$\frac{1}{\mathcal{X}(x)} \frac{d^2\mathcal{X}}{d\,x^2} + \frac{1}{\mathcal{Z}(z)} \frac{d^2\mathcal{Z}}{d\,z^2} + k^2(z) = 0$$

or

$$-\frac{1}{\mathcal{X}(x)} \frac{d^2\mathcal{X}}{d\,x^2} = \frac{1}{\mathcal{Z}(z)} \frac{d^2\mathcal{Z}}{d\,z^2} + k^2(z)$$

This equality occurs only when both the left-hand and right-hand sides are equal to the same constant. Denoting it as m^2, we obtain two ordinary differential equations:

$$\frac{d^2\mathcal{X}}{dx^2} + m^2\mathcal{X} = 0 \tag{5.185}$$

and

$$\frac{d^2\mathcal{Z}}{dz^2} + \left[k^2(z) - m^2\right] \mathcal{Z} = 0 \tag{5.186}$$

The first one, eq. 5.185, has the well known solution

$$\mathcal{X}_m = A_m\, e^{i\,m\,x} + B_m\, e^{-i\,m\,x}\,, \tag{5.187}$$

which is independent of the spatially varying propagation velocity. In contrast, the solution of eq. 5.186 depends on $c(z)$ and can be expressed explicitly in only a few cases, and one of them is given below. In general, numerical methods have to be used to determine behavior of the function $\mathcal{Z}(z)$.

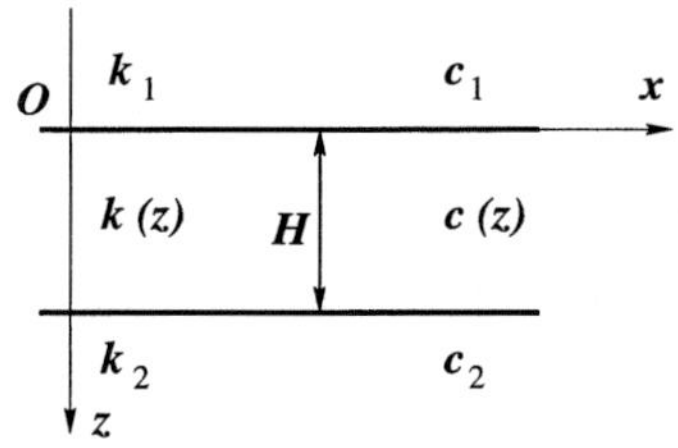

Figure 5.28: Layer with continuous change of velocity.

Function $k^2(z)$ is linear

Suppose that the medium consists of three elements. Its upper and lower half-spaces have constant velocities c_1 and c_2, while the layer between them is inhomogeneous in such a way that $k^2(z)$ is a linear function of z, Fig. 5.28. Then, for the refraction index n.

$$k(z) = k_1\, n(z)\,, \tag{5.188}$$

we have

$$n^2(z) = \begin{cases} 1 & z \le 0 \\ 1 + \alpha\, z & 0 \le z \le H \\ k_2^2/k_1^2 = c_1^2/c_2^2 & z \ge H \end{cases} \tag{5.189}$$

Here

$$\alpha = \frac{k_2^2 - k_1^2}{k_1^2\, H} = \frac{c_1^2 - c_2^2}{c_2^2\, H} \tag{5.190}$$

Correspondingly, the velocity $c(z)$ is a continuous function of z:

$$c(z) = \begin{cases} c_1 & z \le 0 \\ \dfrac{c_1}{\sqrt{1 + \alpha\, z}} & 0 \le z \le H \\ c_2 & z \ge H \end{cases} \tag{5.191}$$

At the same time, the derivatives $d^n c(z)/dz^n$ of any order n are discontinuous at the top and bottom interfaces.

Since the upper and lower parts of the medium are homogeneous, the solutions of eq. 5.186 in those regions are obtained by analogy with that of eq. 5.185:

$$\mathcal{Z}_m = C_m^{(1)}\, e^{i\, m_1\, z} + D_m^{(1)}\, e^{-i\, m_1\, z} \qquad \text{if} \qquad z \le 0\,,$$

$$\mathcal{Z}_m = C_m^{(2)}\, e^{i\,m_2\,z} + D_m^{(2)}\, e^{-i\,m_2\,z} \qquad \text{if} \qquad z \geq H\,, \tag{5.192}$$

where

$$m_1 = \sqrt{k_1^2 - m^2} \qquad \text{and} \qquad m_2 = \sqrt{k_2^2 - m^2} \tag{5.193}$$

Solution of equation 5.186

Now we find a solution of eq. 5.186 inside the layer. Substitution of eqs. 5.188–5.189 into eq. 5.186 gives

$$\frac{d^2\mathcal{Z}}{dz^2} + \left[k_1^2\,(1+\alpha\,z) - m^2\right]\mathcal{Z} = 0$$

or

$$\frac{d^2\mathcal{Z}}{dz^2} + \left[k_1^2 - m^2 + k_1^2\,\alpha\,z\right]\mathcal{Z} = 0 \tag{5.194}$$

First introducing the variable

$$u_1 = k_1^2 - m^2 + k_1^2\,\alpha\,z\,, \tag{5.195}$$

we have

$$\frac{d\,\mathcal{Z}}{dz} = \frac{d\,\mathcal{Z}}{du_1}\,\frac{du_1}{dz} = k_1^2\,\alpha\,\frac{d\,\mathcal{Z}}{du_1}$$

and

$$\frac{d^2\mathcal{Z}}{dz^2} = k_1^4\,\alpha^2\,\frac{d^2\mathcal{Z}}{du_1^2}$$

Then, eq. 5.194 becomes

$$\frac{d^2\mathcal{Z}}{du_1^2} + \frac{u_1}{k_1^4\,\alpha^2}\,\mathcal{Z} = 0 \tag{5.196}$$

It is useful to make one more change of variables

$$u_1 = u\left(k_1^4\,\alpha^2\right)^{1/3} \tag{5.197}$$

This gives

$$\frac{d\,\mathcal{Z}}{du_1} = \frac{d\,\mathcal{Z}}{du}\,\frac{du}{du_1} = \left(k_1^4\,\alpha^2\right)^{-1/3}\,\frac{d\,\mathcal{Z}}{du}$$

and

$$\frac{d^2\mathcal{Z}}{du_1^2} = \left(k_1^4\,\alpha^2\right)^{-2/3}\frac{d^2\mathcal{Z}}{du^2}$$

Therefore, in place of eq. 5.196 we have

$$\left(k_1^4\,\alpha^2\right)^{-2/3}\frac{d^2\mathcal{Z}}{du^2} + \frac{(k_1^4\,\alpha^2)^{1/3}}{k_1^4\,\alpha^2}u\,\mathcal{Z} = 0$$

or

$$\frac{d^2\mathcal{Z}}{du^2} + u\,\mathcal{Z} = 0\,, \tag{5.198}$$

where

$$u = \frac{k_1^2 - m^2 + k_1^2\,\alpha\,z}{(k_1^4\,\alpha^2)^{1/3}} \tag{5.199}$$

The solutions of eq. 5.198 are well known, and they have the the form

$$\mathcal{Z} = u^{1/2}\,N_{1/3}\left(\frac{2}{3}\,u^{3/2}\right) \tag{5.200}$$

Here $N_{1/3}$ is either a Bessel function of the first or second kind, $J_{1/3}(v)$ and $Y_{1/3}(v)$, respectively, or combination of them; the Hankel function $H_{1/3}^{(1)}(v)$ or $H_{1/3}^{(2)}(v)$. By definition,

$$v = \frac{2}{3}\,u^{3/2} \tag{5.201}$$

Thus, we can use for partial solutions of eq. 5.198 one of the following two expressions:

$$\begin{aligned} \mathcal{Z}(v) &= v^{1/3}\left[C_m\,J_{1/3}(v) + D_m\,Y_{1/3}(v)\right] \\ &\text{or} \\ \mathcal{Z}(v) &= v^{1/3}\left[C_m\,H_{1/3}^{(1)}(v) + D_m\,H_{1/3}^{(2)}(v)\right] \end{aligned} \tag{5.202}$$

Now we are ready to study some interesting features of wave behavior in media with continuous changes of velocity. With this purpose in mind consider three cases.

Example one Assume that the upper half-space is homogeneous, whereas the parameter $k^2(z)$ linearly increases in the lower half-space when $z > 0$, Fig. 5.29. In other words, coefficient α, eq. 5.189, is positive, and the velocity decreases with z. We also suppose

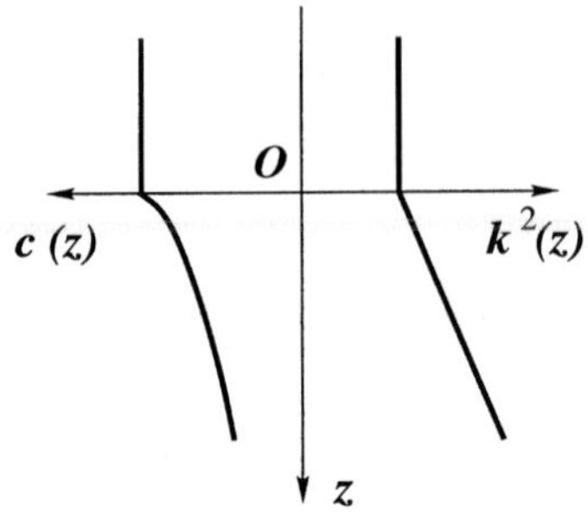

Figure 5.29: Velocity and wavenumber as functions of z.

that the incident plane wave propagates in the upper half-space toward the boundary. As we mentioned earlier, the functions Ψ and $\mathcal{P}$ differ by a constant factor. Correspondingly, we can use the expressions for the complex amplitude of the pressure. In particular, for the plane incident wave we have

$$\Psi_i = A_1 \, e^{i\,k_1\,z\,\cos\theta} \, e^{i\,k_1\,x\,\sin\theta} \tag{5.203}$$

Here θ is the incidence angle and A_1 is the known magnitude.

The reflected wave is absent in such a medium if we consider the high frequency limit, (Chapter 1), but with a decrease of frequency we may expect its appearance. This wave in the upper half-space is caused by the superposition of elementary waves due to secondary sources inside the lower medium, as well as at the interface (the last example of this section). Then, the function Ψ_1 should be written in the form

$$\Psi_1 = \left(A_1 \, e^{i\,k_1\,z\,\cos\theta} + B_1 \, e^{-i\,k_1\,z\,\cos\theta}\right) e^{i\,k_1\,x\,\sin\theta} \qquad \text{if} \qquad z \le 0 \tag{5.204}$$

The latter implies that Snell's law of reflection is valid, and solution of the boundary value problem shows that this assumption is correct. In accordance with eqs. 5.184, 5.187, and 5.202, the expression for the function Ψ_2 in the lower medium may have the form

$$\Psi_2 = v^{1/3} \left[C_m \, H^{(1)}_{1/3}(v) + D_m \, H^{(2)}_{1/3}(v)\right] \left[A_m \, e^{i\,m\,x} + B_m \, e^{-i\,m\,x}\right] \quad \text{if} \quad z \ge 0 \tag{5.205}$$

Fortunately, the right-hand side of this equation can be greatly simplified. First, as follows from the boundary conditions 5.183, the functions Ψ_1 and Ψ_2 are equal to each other at each point of the plane $z = 0$. This may happen only if they have the same dependence on x. Therefore, comparison of eqs. 5.204 and 5.205 gives

$$B_m = 0 \qquad \text{and} \qquad m = k_1 \sin\theta \tag{5.206}$$

Respectively, eq. 5.205 reduces to

$$\Psi_2 = v^{1/3} \left[C_m H^{(1)}_{1/3}(v) + D_m H^{(2)}_{1/3}(v) \right] e^{i k_1 x \sin\theta} \tag{5.207}$$

The argument u is also simplified. In fact, eq.5.199 yields

$$u = \frac{\cos^2\theta + \alpha z}{\alpha^{2/3}} k_1^{2/3} \tag{5.208}$$

The next simplification relates to the asymptotic behavior of Hankel functions. As is known, for large arguments these function can be represented as

$$H^{(1)}_{1/3}(v) \approx \sqrt{\frac{2}{\pi v}}\, e^{i\,(v - 5\pi/12)}, \qquad H^{(2)}_{1/3}(v) \approx \sqrt{\frac{2}{\pi v}}\, e^{-i\,(v - 5\pi/12)} \tag{5.209}$$

if $v \to \infty$.

Bearing in mind that the time dependence is described by the function $exp\,(-i\,\omega\,t)$, we conclude that the term

$$v^{1/3} D_m H^{(2)}_{1/3}(v)\, e^{i k_1 x \sin\theta}$$

in the right-hand side of eq. 5.207 describes the up-going wave that travels from $z = \infty$ toward the plane $z = 0$. Since the sources are absent at $z = \infty$, we have to set

$$D_m = 0\,,$$

and eq. 5.207 becomes

$$\Psi_2 = v^{\frac{1}{3}} C_m H^{(1)}_{\frac{1}{3}}(v)\, e^{i k_1 x \sin\theta} \qquad \text{if} \qquad z \geq 0$$

Thus, the expressions for function Ψ are

$$\Psi_1 = \left(A\, e^{i k_1 z \cos\theta} + B\, e^{-i k_1 z \cos\theta} \right) e^{i k_1 x \sin\theta},$$
$$\Psi_2 = v^{1/3}\, C\, H^{(1)}_{1/3}(v)\, e^{i k_1 x \sin\theta} \tag{5.210}$$

Since coefficients B and C depend on angle θ, the index m is omitted. Also, we have from eqs. 5.201 and 5.208

$$v = \frac{2\,k_1}{3\,\alpha}\left(\cos^2\theta + \alpha\,z\right)^{3/2} \tag{5.211}$$

To determine unknown coefficients, we make use of the boundary conditions 5.183. They give

$$\begin{aligned} A + B &= C\, v_0^{1/3}\, H^{(1)}_{1/3}(v_0)\,, \\ i\,k_1\cos\theta\,(A - B) &= C\,\frac{\partial}{\partial v}\left[v^{1/3}\,H^{(1)}_{1/3}(v)\right]\frac{\partial v}{\partial z} \qquad \text{if} \qquad z = 0 \end{aligned} \tag{5.212}$$

Here

$$v_0 = \frac{2\,k_1}{3\,\alpha}\cos^3\theta \tag{5.213}$$

and

$$\frac{\partial v}{\partial z} = k_1\cos\theta \qquad \text{if} \qquad z = 0 \tag{5.214}$$

As follows from the theory of Bessel functions,

$$\frac{\partial}{\partial v}\left[v^{1/3}\,H^{(1)}_{1/3}(v)\right] = v^{1/3}\,H^{(1)}_{-2/3}(v) \tag{5.215}$$

Thus, in place of eqs. 5.212 we have

$$\begin{aligned} A + B &= C\, v_0^{1/3}\, H^{(1)}_{1/3}(v_0)\,, \\ i\,(A - B) &= C\, v_0^{1/3}\, H^{(1)}_{-2/3}(v_0) \end{aligned} \tag{5.216}$$

This yields

$$\frac{A - B}{A + B} = -i\,\frac{H^{(1)}_{-2/3}(v_0)}{H^{(1)}_{1/3}(v_0)} \tag{5.217}$$

Then

$$B = \frac{i\,H^{(1)}_{1/3}(v_0) - H^{(1)}_{-2/3}(v_0)}{i\,H^{(1)}_{1/3}(v_0) + H^{(1)}_{-2/3}(v_0)}\,A\,,$$

and

$$C = \frac{2\,i\,v_0^{-1/3}}{i\,H^{(1)}_{1/3}(v_0) + H^{(1)}_{-2/3}(v_0)}\,A \tag{5.218}$$

Thus, we have solved the boundary value problem and have found the wave fields in both parts of the medium. In particular, it was shown that Snell's law of reflection is still valid at any frequency. Also, it is interesting to notice that the dependence of our solutions on the x-coordinate is the same in both lower and upper media. As follows from eqs. 5.210 and 5.218, the reflected and transmitted waves are expressed in terms of Hankel functions which, in general, have complex values.

The asymptotic wave behavior

It is useful to study the reflection and transmission coefficients in two limiting cases, namely, $v_0 \gg 1$ and $v_0 \ll 1$. In accordance with eq. 5.213 they describe the high- and low-frequency spectra, respectively.

The parameter α, which enters into the refraction index n in the form

$$n^2 = 1 + \alpha\, z\,,$$

characterizes the rate of change of n^2 with distance z. From the last equality we have

$$n = \sqrt{1+\alpha\, z} \qquad \text{and} \qquad \frac{d\,n}{d\,z} = \frac{\alpha}{2\sqrt{1+\alpha\, z}} \tag{5.219}$$

This means that the maximum rate of change in index n occurs at the plane $z = 0$, and it is equal to $\alpha/2$, that is

$$\alpha = 2\,\frac{d\,n}{d\,z} \qquad \text{if} \qquad z = 0 \tag{5.220}$$

High-frequency spectrum

In this case it is natural to make use of asymptotic formulae for the Hankel functions:

$$H^{(1)}_{1/3}(v_0) \approx \sqrt{\frac{2}{\pi\, v_0}}\, e^{i\,(v_0 - 5\pi/12)} \left(1 + \frac{5}{72\,i\,v_0} + \ldots\right),$$

$$H^{(1)}_{-2/3}(v_0) \approx \sqrt{\frac{2}{\pi\, v_0}}\, e^{i\,(v_0 + \pi/12)} \left(1 - \frac{7}{72\,i\,v_0} + \ldots\right) \tag{5.221}$$

Substitution of the latter into eq. 5.217 yields

$$\frac{A-B}{A+B} \approx -\,i\,e^{i\pi/2}\,\frac{1 - \dfrac{7}{72\,i\,v_0}}{1 + \dfrac{5}{72\,i\,v_0}}$$

or

$$\frac{A-B}{A+B} \approx 1 - \frac{1}{6\,i\,v_0} = 1 + \frac{i}{6\,v_0}$$

As follows from this approximation

$$B \approx -\frac{i}{12\,v_0}\,A = -\frac{i\,\alpha}{8\,k_1\,cos^3\theta}\,A \tag{5.222}$$

or

$$|B| \approx \frac{\lambda_1}{8\,\pi\,cos^3\theta}\,\frac{d\,n}{d\,z}\bigg|_{z=0} A \tag{5.223}$$

In accordance with eq. 5.223, the reflection coefficient decreases as the wavelength becomes smaller and it tends to zero when $v \to \infty$ or $\lambda_1 \to 0$. This means that the asymptotic series describing waves at high frequencies does not contain the zero-order term. In reality, the reflected wave always exists, however, its high-frequency spectrum is proportional to a wavelength λ_1. The reflections are absent only in the case of homogeneous media, where $d\,n/d\,z = 0$.

Next consider the transmission coefficient C/A when $v_0 \gg 1$. The denominator in eq. 5.218 can be represented as

$$\begin{aligned} i\,H^{(1)}_{1/3}(v_0) + H^{(1)}_{-2/3}(v_0) &\approx \sqrt{\frac{2}{\pi\,v_0}}\,e^{i\,v_0}\left[e^{i\left(\frac{\pi}{2}-\frac{5\pi}{12}\right)} + e^{i\,\frac{\pi}{12}}\right] \\ &= 2\sqrt{\frac{2}{\pi\,v_0}}\,e^{i\,v_0}\,e^{i\pi/12} \end{aligned}$$

Thus

$$C \approx \frac{2\,i\,v_0^{-1/3}\,e^{-i\,v_0}\,\sqrt{\pi\,v_0}}{2\sqrt{2}}\,e^{-i\,\frac{\pi}{12}}\,A \approx i\,v_0^{1/6}\sqrt{\frac{\pi}{2}}\,e^{-i\,\frac{\pi}{12}}\,e^{-i\,v_0}\,A$$

Correspondingly, we obtain for the function Ψ_2,

$$\Psi_2 \approx v^{1/3}\,i\,v_0^{1/6}\sqrt{\frac{\pi}{2}}\,e^{-i\,\frac{\pi}{12}}\,e^{-i\,v_0}\sqrt{\frac{2}{\pi}}\,v^{-1/2}\,e^{-i\,\frac{5\pi}{12}}\,e^{i\,v}\,e^{i\,k_1\,x\,\sin\theta}\,A,$$

or

$$\Psi_2 \approx \left(\frac{v_0}{v}\right)^{1/6} e^{i\,(v-v_0)}\,e^{i\,k_1\,x\,\sin\theta}\,A \tag{5.224}$$

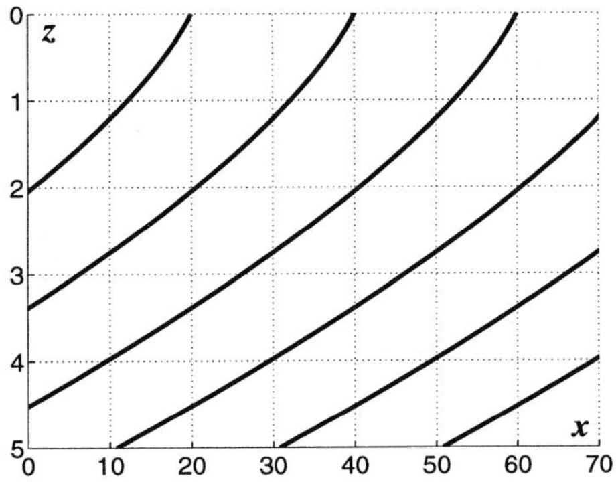

Figure 5.30: Phase surfaces in the medium with $\alpha = 2\ \text{km}^{-1}$. The angle of the incident wave is $\theta = 30°$.

In particular, we have $\Psi_1 = \Psi_2$ at the plane $z = 0$. As follows from eq. 5.224, the phase surface of the transmitted wave in the high-frequency limit is defined as

$$(\cos^2\theta + \alpha\, z)^{3/2} - \cos^3\theta + \frac{3}{2}\,\alpha\, x\, \sin\theta = \text{const} \ ,$$

and, in general, this wave is not plane, Fig. 5.30. The exception is the normal incidence, when the phase surfaces are horizontal. In this case $\theta = 0$ and

$$\Psi_2 = (1 + \alpha\, z)^{-1/4}\, e^{i\,(v\, -\, v_0)}\, A\,,$$

where

$$v = \frac{2\,k_1}{3\,\alpha}\,(1 + \alpha\, z)^{3/2}$$

Therefore, the amplitude $|\Psi_2|$ decreases with distance z relatively slowly. It is clear that, the amplitude of the particle velocity and the pressure vary with the depth even when $\theta = 0$. Since acoustic energy is not transformed into heat, such behavior of the wave is a result of interference. For this reason the energy flux along the z-axis during each period is independent of the coordinate z. Of course, this follows from the fact that the Poynting vector does not have a horizontal component.

Low-frequency spectrum

The wave behavior is completely different when the parameter v_0 is small. First, consider reflection coefficient B/A. With this purpose in mind, we make use of the asymptotic formula for the Bessel function

$$J_\nu(v_0) \approx \left(\frac{v_0}{2}\right)^\nu \frac{1}{\Gamma(1+\nu)} \qquad \text{if} \qquad v_0 \ll 1 \tag{5.225}$$

and the relationship

$$H_\nu^{(1)}(v_0) = \frac{i}{\sin \pi \nu} \left[e^{-i\pi\nu} J_\nu(v_0) - J_{-\nu}(v_0) \right] \tag{5.226}$$

Here $\Gamma(1+\nu)$ is the gamma function of the argument $(1+\nu)$ and ν is the order of the Bessel and Hankel functions.

Taking into account that $v_0 \to 0$, we have

$$\begin{aligned} H_{\frac{1}{3}}^{(1)}(v_0) &\approx -\frac{i}{\sin \pi/3} J_{-1/3}(v_0), \\ H_{-2/3}^{(1)}(v_0) &\approx -\frac{i}{\sin 2\pi/3} e^{2i\pi/3} J_{-2/3}(v_0) \end{aligned} \tag{5.227}$$

Since

$$J_{-1/3}(v_0) \approx \left(\frac{v_0}{2}\right)^{-1/3} \frac{1}{\Gamma(2/3)} \qquad \text{and} \qquad J_{-2/3}(v_0) \approx \left(\frac{v_0}{2}\right)^{-2/3} \frac{1}{\Gamma(1/3)},$$

the Hankel functions are

$$H_{1/3}^{(1)}(v_0) \approx -\frac{i}{\sin \pi/3} \left(\frac{v_0}{2}\right)^{-1/3} \frac{1}{\Gamma(2/3)},$$

and

$$H_{-2/3}^{(1)}(v_0) \approx -\frac{i}{\sin 2\pi/3} e^{2i\pi/3} \left(\frac{v_0}{2}\right)^{-2/3} \frac{1}{\Gamma(1/3)}$$

Then, as follows from eq. 5.217,

$$\frac{A-B}{A+B} = -i\, \frac{2^{1/3} e^{2i\pi/3}\, \Gamma(2/3)}{v_0^{1/3}\, \Gamma(1/3)}$$

Bearing in mind that

$$-i = e^{-i\pi/2} \qquad \text{and} \qquad -i\, e^{i2\pi/3} = e^{i\pi/6},$$

we obtain

$$\frac{A-B}{A+B} = \frac{d}{v_0^{1/3}}, \tag{5.228}$$

where

$$d = e^{i\pi/6} \frac{2^{1/3}\,\Gamma(2/3)}{\Gamma(1/3)} \tag{5.229}$$

Thus,

$$B = -\frac{d - v_0^{1/3}}{d + v_0^{1/3}} A \qquad \text{or} \qquad B \approx -\left(1 - 2\frac{v_0^{1/3}}{d}\right) A \quad \text{if} \quad v_0 \ll 1 \tag{5.230}$$

From eqs. 5.213 and 5.229, we have

$$v_0 = \frac{2\,k_1}{3\,\alpha} \cos^3\theta \qquad \text{and} \qquad d = (i + \sqrt{3})\,\frac{\Gamma(2/3)}{2^{2/3}\,\Gamma(1/3)}$$

Thus, the reflection coefficient B/A tends to -1 as the frequency decreases. As a consequence, the superposition of both waves,

$$\Psi_1 = A\left(e^{i\,k_1\,z\,\cos\theta} - e^{-i\,k_1\,z\,\cos\theta}\right) e^{i\,k_1\,x\,\sin\theta} \qquad \text{if} \qquad z \le 0$$

forms the standing wave along the z-axis,

$$\Psi_1 = 2\,i\,A \sin\left(k_1\,z\,\cos\theta\right) e^{i\,k_1\,x\,\sin\theta} \tag{5.231}$$

In particular, the incident and reflected wave cancel each other at the plane $z = 0$. Correspondingly, the pressure vanishes at this plane, like at a free surface. This may suggest that the wave field also disappears inside the lower medium in the limit $v \to 0$.

To demonstrate such behavior, consider the coefficient C given by eq. 5.218. Substitution of the Hankel functions into the latter yields

$$C \approx -2^{1/3}\,\Gamma\left(\frac{1}{3}\right) \sin\frac{2\,\pi}{3}\, e^{-2\,i\,\pi/3}\, v_0^{1/3}\, A$$

Here the term $H_{1/3}^{(1)}(v_0)$ is disregarded. Correspondingly, the function Ψ_2 can be represented as

$$\Psi_2 \approx v^{1/3}\,C\,H_{1/3}^{(1)}(v)\,e^{\,i\,k_1 x \sin\theta} \approx 2^{2/3}\,\frac{\Gamma(1/3)}{\Gamma(2/3)}\,e^{-i\,\frac{\pi}{6}}\,v_0^{1/3}\,A\,e^{\,i\,k_1 x \sin\theta}$$

Thus with the decrease of frequency the amplitude of a transmitted wave becomes smaller, and in the limit $v_0 \to 0$ it vanishes. This result is intuitively obvious. Since the wave with almost zero frequency or almost infinite wavelength senses the presence of a medium with zero velocity at $z \to \infty$, its amplitude has to behave according to the conditions at infinity that formally coincide with the conditions at the free surface.

The analysis of waves at high and low frequencies allows us to make several comments.

(a) The total wave field in the upper homogeneous half-space consists of the incident and reflected waves. The amplitude of the latter essentially depends on frequency. This suggests that interference of elementary reflected waves arising at different points in the lower medium plays a dominant role. In particular, the superposition at high frequencies is such that the reflected wave in the upper medium almost disappears, and all energy of the incident wave propagates through the interface into the lower half-space. The Poynting vector, which is tangential to transmitted rays in the zero-approximation of geometrical acoustics, is directed away from the boundary at each point.

(b) In contrast, at low frequencies the spectrum of a reflected wave is almost equal to that of the incident one. Also, as a result of interference, the total wave field in the medium tends to zero.

(c) The fact that the pressure is equal to zero at boundary $z = 0$ when $\lambda_1 \to \infty$ can be interpreted in the following way. Since the wave velocity decreases along the z-axis, we can imagine such a large distance H that the velocity is almost zero if $z > H$. In other words, the plane $z = H$ can be treated as free. Therefore, in this approximation, we deal with the layer of a finite thickness H overlying the free space. Our study of the low-frequency spectrum indicates that such an inhomogeneous layer becomes transparent if the wavelength greatly exceeds its thickness. Correspondingly, the incident and reflected waves cancel each other at the plane $z = 0$.

Example two Next consider a medium with two horizontal interfaces, Fig. 5.28, and demonstrate that the inhomogeneous layer has no influence on the reflected wave at low frequencies. The velocity $c(z)$ is defined by eq. 5.191 and, as before, it is a continuous function of coordinate z. Taking into account the results obtained in the previous case, we can write the function Ψ in the following form

$$
\begin{aligned}
\Psi_1 &= \left[A\, e^{i\,k_1\,z\,\cos\theta_1} + B\, e^{-i\,k_1\,z\,\cos\theta_1}\right] e^{i\,k_1\,x\,\sin\theta_1} && \text{if} \quad z < 0\,, \\
\Psi_2 &= v^{1/3}\left[C\, H^{(1)}_{1/3}(v) + D\, H^{(2)}_{1/3}(v)\right] e^{i\,k_1\,x\,\sin\theta_1} && \text{if} \quad 0 \geq z \geq H\,, \qquad (5.232) \\
\Psi_3 &= E\, e^{i\,k_2\,z\,\cos\theta_2}\, e^{i\,k_2\,x\,\sin\theta_2} && \text{if} \quad z \geq H
\end{aligned}
$$

Here θ_1 is the incidence angle and B, C, D and E are unknown amplitude coefficients.

As follows from eq. 5.211

$$
v = \frac{2\,k_1}{3\,\alpha}\left(\cos^2\theta_1 + \alpha\,z\right)^{3/2} \qquad \text{and} \qquad \frac{dv}{dz} = k_1\,\sqrt{\cos^2\theta_1 + \alpha\,z}
$$

Further, we make use of the known equalities

$$\frac{d}{dv}\left[v^{1/3}\,H^{(1)}_{1/3}(v)\right] = v^{1/3}\,H^{(1)}_{-2/3}(v) \qquad \text{and} \qquad \frac{d}{dv}\left[v^{1/3}\,H^{(2)}_{1/3}(v)\right] = v^{1/3}\,H^{(2)}_{-2/3}(v)$$

Also, the continuity of the function Ψ at plane $z = H$ requires that

$$k_1\,x\,\sin\theta_1 = k_2\,x\,\sin\theta_2$$

or

$$\frac{\sin\theta_1}{c_1} = \frac{\sin\theta_2}{c_2} \tag{5.233}$$

At upper boundary $z = 0$ we have

$$v_1 = \frac{2\,k_1}{3\,\alpha}\cos^3\theta_1 \qquad \text{and} \qquad \frac{dv}{dz} = k_1\,\cos\theta_1\,, \tag{5.234}$$

while at the layer bottom, $z = H$ eq (5.190):

$$v_2 = \frac{2\,k_1}{3\,\alpha}\left(\cos^2\theta_1 + \frac{c_1^2 - c_2^2}{c_2^2}\right)^{3/2}$$

or

$$v_2 = \frac{2\,k_1}{3\,\alpha}\left(\frac{c_1^2}{c_2^2} - \sin^2\theta_1\right)^{3/2}$$

Taking into account Snell's law and the equality

$$\alpha\,H = \frac{c_1^2}{c_2^2} - 1\,,$$

we find

$$v_2 = \frac{2\,k_2}{3\,\alpha}\,\frac{c_1^2}{c_2^2}\,\cos^3\theta_2 \qquad \text{and} \qquad \frac{dv}{dz} = k_2\,\cos\theta_2 \tag{5.235}$$

Now, applying the boundary conditions at both interfaces 5.183 and making use of the eq. 5.15, we obtain

$$\begin{aligned}
A + B &= v_1^{1/3}\left[C\,H^{(1)}_{1/3}(v_1) + D\,H^{(2)}_{1/3}(v_1)\right], \\
i\,(A - B) &= v_1^{1/3}\left[C\,H^{(1)}_{-2/3}(v_1) + D\,H^{(2)}_{-2/3}(v_1)\right], \\
E\,e^{i\,k_2\,H\,\cos\theta_2} &= v_2^{1/3}\left[C\,H^{(1)}_{1/3}(v_2) + D\,H^{(2)}_{1/3}(v_2)\right], \\
i\,E\,e^{i\,k_2\,H\,\cos\theta_2} &= v_2^{1/3}\left[C\,H^{(1)}_{-2/3}(v_2) + D\,H^{(2)}_{-2/3}(v_2)\right]
\end{aligned} \tag{5.236}$$

Since we are interested in the low-frequency spectrum only, consider the solution of this system when arguments v_1 and v_2 are both small.

In accordance with eqs. 5.227

$$H^{(1)}_{1/3}(v) \approx -\frac{i}{\sin \pi/3}\left(\frac{v}{2}\right)^{-1/3}\frac{1}{\Gamma(2/3)}$$

and

$$H^{(1)}_{-2/3}(v) \approx -\frac{i}{\sin \pi/3}\, e^{2i\pi/3}\left(\frac{v}{2}\right)^{-2/3}\frac{1}{\Gamma(1/3)} \qquad \text{if} \qquad v \to 0$$

Taking into account the relationship

$$H^{(2)}_{\nu}(v) = -\frac{i}{\sin \pi\,\nu}\left[e^{i\,\pi\,\nu}\, J_{\nu}(v) - J_{-\nu}(v)\right]$$

and eq. 5.225,

$$J_{\nu}(v) \approx \left(\frac{v}{2}\right)^{\nu}\frac{1}{\Gamma(1+\nu)},$$

we have for the Hankel functions of the second order

$$H^{(2)}_{1/3}(v) \approx \frac{i}{\sin \pi/3}\left(\frac{v}{2}\right)^{-1/3}\frac{1}{\Gamma(2/3)}$$

and

$$H^{(2)}_{-2/3}(v) \approx \frac{i}{\sin 2\,\pi/3}\, e^{-2\,i\,\pi/3}\left(\frac{v}{2}\right)^{-2/3}\frac{1}{\Gamma(1/3)}$$

Thus,

$$H^{(2)}_{1/3}(v) \approx -\,H^{(1)}_{1/3}(v) \qquad \text{and} \qquad H^{(2)}_{-2/3}(v) \approx e^{-i\,\frac{\pi}{3}}\, H^{(1)}_{-2/3}(v) \tag{5.237}$$

Correspondingly, the system 5.236 becomes

$$\begin{aligned}
A + B &= v_1^{1/3}\,(C - D)\, H^{(1)}_{1/3}(v_1)\,, \\
i\,(A - B) &= v_1^{1/3}\left(C + e^{-i\,\frac{\pi}{3}}\, D\right) H^{(1)}_{-2/3}(v_1)\,, \\
E\, e^{i\,k_2\,H\,\cos\theta_2} &= v_2^{1/3}\,(C - D)\, H^{(1)}_{1/3}(v_2)\,, \\
i\,E\, e^{i\,k_2\,H\,\cos\theta_2} &= v_2^{1/3}\left(C + e^{-i\,\frac{\pi}{3}}\, D\right) H^{(1)}_{-2/3}(v_2)
\end{aligned}$$

Elimination of E gives

$$i = \frac{(C + e^{-i\frac{\pi}{3}} D)}{(C - D)} \frac{H^{(1)}_{-2/3}(v_2)}{H^{(1)}_{1/3}(v_2)} \tag{5.238}$$

From the first two equations we have

$$i\frac{A - B}{A + B} = \frac{(C + e^{-i\frac{\pi}{3}} D)}{(C - D)} \frac{H^{(1)}_{-2/3}(v_1)}{H^{(1)}_{1/3}(v_1)} \tag{5.239}$$

Whence

$$\frac{A - B}{A + B} = \frac{H^{(1)}_{1/3}(v_2)\, H^{(1)}_{-2/3}(v_1)}{H^{(1)}_{1/3}(v_1)\, H^{(1)}_{-2/3}(v_2)} \quad \text{if} \quad v_1 \ll 1 \quad \text{and} \quad v_2 \ll 1 \tag{5.240}$$

Substitution of eqs. 5.227 into eq. 5.240 gives

$$\frac{A - B}{A + B} = \left(\frac{v_2}{v_1}\right)^{-1/3} \left(\frac{v_1}{v_2}\right)^{-2/3} = \left(\frac{v_2}{v_1}\right)^{1/3}$$

and taking into account eqs. 5.234 – 5.235, we have

$$\frac{A - B}{A + B} = \left(\frac{c_1}{c_2}\right)^{2/3} \left(\frac{k_2}{k_1}\right)^{1/3} \frac{\cos\theta_2}{\cos\theta_1} = \frac{c_1}{c_2}\frac{\cos\theta_2}{\cos\theta_1} = \frac{Z_1^\star}{Z_2^\star}, \tag{5.241}$$

where

$$Z_1^\star = \frac{c_1\rho_1}{\cos\theta_1} \quad \text{and} \quad Z_2^\star = \frac{c_2\rho_2}{\cos\theta_2}, \quad \rho_1 = \rho_2$$

Whence

$$B = \frac{Z_2^\star - Z_1^\star}{Z_2^\star + Z_1^\star} A \tag{5.242}$$

The latter is the well-known expression for the reflection coefficient of the plane wave in a medium with a single planar interface. Thus, we have proved that the intermediate layer becomes transparent, provided that its thickness is much smaller than the wavelength.

It is clear from the physical point of view that this result is independent of the velocity $c(z)$ inside the layer. Our study illustrates the inability of seismic waves of a certain frequency to detect relatively thin layers. This type of wave behavior is, however, instructive. Indeed, in reality a boundary between two media is not a zero-thickness mathematical surface but a very thin intermediate layer with some finite thickness Δh

where velocity varies rapidly. The influence of such a transition layer on wave propagation, however, becomes negligible as long as the wavelength λ greatly exceeds its thickness. Correspondingly, the amplitudes of reflected and transmitted waves are indistinguishable from those that would be obtained in the case of absence of the layer. As a consequence, the wave fields almost coincide with those of the case in which the velocity $c(z)$ abruptly changes at the boundary.

Example three Let us consider again a medium with one horizontal interface, Fig. 5.27. Unlike the first example suppose that the refraction index n, defined from the equality

$$k(z) = k_1\, n(z)\,,$$

behaves as

$$n(z) = \begin{cases} 1 & z < 0 \\ \dfrac{k_2}{k_1}\sqrt{1+\alpha\, z} & z \geq 0 \end{cases} \tag{5.243}$$

Here

$$k_1 = \frac{\omega}{c_1} \qquad \text{and} \qquad k_2 = \frac{\omega}{c_2}$$

Correspondingly, the velocity $c(z)$ is also discontinuous at interface $z = 0$:

$$c(z) = \begin{cases} c_1 & z < 0 \\ c_2/\sqrt{1+\alpha\, z} & z \geq 0 \end{cases} \tag{5.244}$$

Here c_2 is the velocity in the lower medium in the vicinity of the boundary.

Then, in place of eqs. 5.210, we obtain the following expressions for the function Ψ:

$$\begin{aligned} \Psi_1 &= \left[A\, e^{i\, k_1 z \cos\theta} + B\, e^{-i\, k_1 z \cos\theta}\right] e^{i\, k_1 x \sin\theta} && \text{if} \quad z < 0\,, \\ \Psi_2 &= v^{1/3}\, C\, H^{(1)}_{1/3}(v)\, e^{i\, k_2 x \sin\theta_1} && \text{if} \quad z \geq 0 \end{aligned} \tag{5.245}$$

Here θ is an incidence angle that does not exceed the critical one. In order to satisfy the boundary conditions the equality

$$k_1 \sin\theta = k_2 \sin\theta_1 \tag{5.246}$$

has to take place. This means that Snell's law of refraction remains valid even though the transmitted wave is not plane due to heterogeneity of the lower medium.

As follows from eq. 5.199, we have

$$v = \frac{2\,k_2}{3\,\alpha}\left(\cos^2\theta_1 + \alpha\,z\right)^{3/2} \tag{5.247}$$

The boundary conditions (eq. 5.183) give

$$\begin{aligned} A + B &= C\,v_0^{1/3}\,H^{(1)}_{1/3}(v_0)\,, \\ i\,k_1\,\cos\theta\,(A - B) &= C\,v_0^{1/3}\,H^{(1)}_{-2/3}(v_0)\,\frac{\partial v}{\partial z}\,, \end{aligned} \tag{5.248}$$

where

$$v_0 = \frac{2\,k_2}{3\,\alpha}\cos^3\theta_1 \qquad \text{and} \qquad \frac{\partial v}{\partial z} = k_2\,\cos\theta_1$$

Thus, the system of equations becomes

$$\begin{aligned} A + B &= C\,v_0^{1/3}\,H^{(1)}_{1/3}(v_0) \\ \frac{i}{Z_1^\star}\,(A - B) &= \frac{C\,v_0^{1/3}\,H^{(1)}_{-2/3}(v_0)}{Z_2^\star}\,, \end{aligned} \tag{5.249}$$

where

$$Z_1^\star = \frac{c_1\,\rho_1}{\cos\theta} \qquad \text{and} \qquad Z_2^\star = \frac{c_2\,\rho_2}{\cos\theta_1}\,, \qquad \rho_1 = \rho_2$$

Then,

$$\frac{A - B}{A + B} = \frac{Z_1^\star}{i\,Z_2^\star}\,\frac{H^{(1)}_{-2/3}(v_0)}{H^{(1)}_{1/3}(v_0)} .$$

The latter yields

$$B = \frac{i\,Z_2^\star\,H^{(1)}_{1/3}(v_0) - Z_1^\star\,H^{(1)}_{-2/3}(v_0)}{i\,Z_2^\star\,H^{(1)}_{1/3}(v_0) + Z_1^\star\,H^{(1)}_{-2/3}(v_0)}\,A \tag{5.250}$$

and

$$C = \frac{2\,i\,Z_2^\star\,v_0^{-1/3}}{i\,Z_2^\star\,H^{(1)}_{1/3}(v_0) + Z_1^\star\,H^{(1)}_{-2/3}(v_0)}\,A \tag{5.251}$$

From a physical point of view, it is clear that the reflected and transmitted waves are caused by the surface and volume sources, and the contributions of each of them depend on the wavelength.

First, as before, consider the asymptotic behavior of these waves. As follows from eq. 5.221 at the high-frequency spectrum

$$H^{(1)}_{1/3}(v_0) \approx \sqrt{\frac{2}{\pi v_0}}\, e^{i\,(v_0 - 5\,\pi/12)} \qquad \text{and} \qquad H^{(2)}_{-2/3}(v_0) \approx \sqrt{\frac{2}{\pi v}}\, e^{i\,(v + \pi/12)}$$

Therefore eq. 5.250 yields

$$B \approx \frac{Z_2^\star - Z_1^\star}{Z_2^\star + Z_1^\star}\, A\,,$$

which coincides with the reflection coefficient if the lower and upper parts of the medium were homogeneous. In other words, the interference of elementary waves traveling upward and caused by the volume sources is destructive. Correspondingly, the reflected wave in the upper medium is defined mainly by the surface sources.

In accordance with eqs. 5.228, we have at the low-frequency limit

$$\frac{A - B}{A + B} = \frac{Z_1^\star}{i\,Z_2^\star}\,\frac{d}{v_0^{1/3}}\,,$$

where d is given by eq. 5.229. This means that $B \to -A$ when the frequency tends to zero. This behavior was already explained in the example one.

Applying the same approach, we can find asymptotic formulae for the function Ψ_2, which describes the wave field in the lower medium. From eqs. 5.224 and 5.251 we obtain

$$\Psi_2 = \frac{2\,Z_2^\star\,A}{Z_2^\star + Z_1^\star}\left(\frac{v_0}{v}\right)^{1/6} e^{i\,(v - v_0)}\, e^{i\,k_2\,\sin\theta_1} \qquad \text{if} \qquad v \gg 1$$

In particular, at the boundary

$$\Psi_2 = \frac{2\,Z_2^\star\,A}{Z_2^\star + Z_1^\star}\, e^{i\,k_2\,\sin\theta_1}\,,$$

which corresponds to a medium containing two homogeneous half-spaces. The wave magnitude decreases relatively slowly with an increase of the distance from the boundary.

In the low-frequency limit the function Ψ_2 becomes

$$\Psi_2 = 2^{2/3}\,\frac{\Gamma(1/3)}{\Gamma(2/3)}\,\frac{Z_2^\star}{Z_1^\star}\, v_0^{1/3}\, e^{i\,\frac{\pi}{6}}\, e^{i\,k_2 \sin\theta_1}\,,$$

which indicates that when the frequency decreases, the wave field tends to zero at any distance from the boundary.

Chapter 6

Spherical waves in the presence of horizontal interface

In this chapter we describe the behavior of waves arising at the horizontal interface when the incident wave is caused by an elementary spherical source. We derive exact formulas for the solution of this problem and describe asymptotic approaches leading to closed expressions for various waves. It is natural to start this study from two limiting cases when the medium is relatively simple and the wave fields can be expressed in terms of elementary functions. Latter in this Chapter, we discuss more general cases.

6.1 Spherical source in a half-space with a free or rigid surface

Free surface

At the beginning suppose that the point source of a spherical wave is located in the half-space at the distance h from the free surface, Fig. 6.1a. As is well known, (Part I), the potential of this wave is

$$U_i(R,\,t) = \frac{A}{R}\, f\left[a\left(t - \frac{R}{c_1}\right)\right], \tag{6.1}$$

where A and a are given constants, R is the distance between the source and the observation point and c_1 is the wave velocity in the lower medium. The function $f[a\,(t - R/c_1)]$ is defined by the source and it is practically arbitrary.

Further we introduce the cylindrical system of coordinates r, φ, and z with the origin at the boundary directly above the source and the z-axis directed downward. It is clear that, due to the axial symmetry, the wave field is independent of φ. As in the case of plane waves, we should observe a reflection and correspondingly the potential of the total wave field can be written as

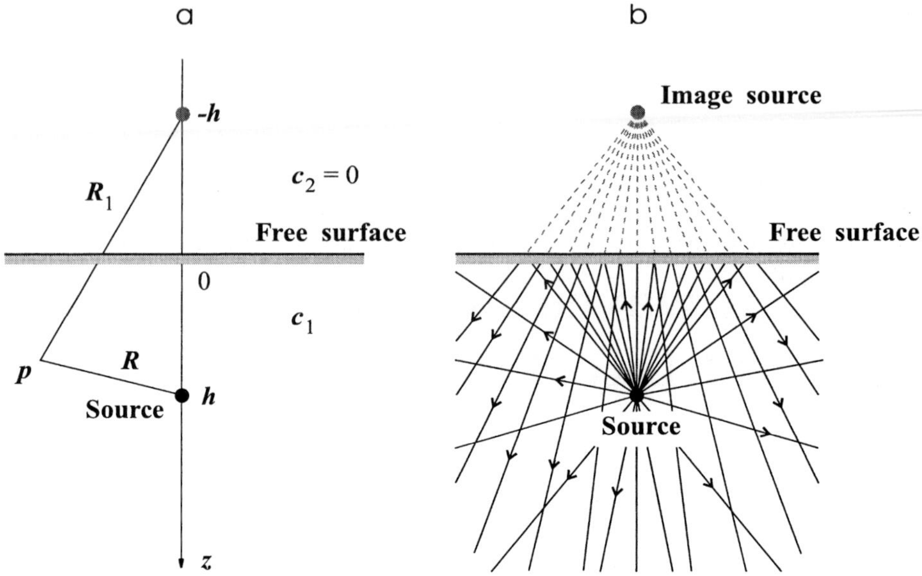

Figure 6.1: Reflection of spherical waves from the free surface.

$$U(r,\ z) = U_i(r,\ z) + U_r(r,\ z)\,, \tag{6.2}$$

where $U_r(r,\ z)$ is the potential of the reflected wave.

By definition, both functions U_i and U_r are solutions of the wave equation and therefore at regular points of the medium

$$\nabla^2 U = \frac{1}{c_1}\frac{\partial^2 U}{\partial t^2} \tag{6.3}$$

Also, at each point of the free surface the pressure, caused by both waves, is zero:

$$P = 0 \tag{6.4}$$

Since

$$P = -\rho\,\frac{\partial^2 U}{\partial t^2}\,, \tag{6.5}$$

where ρ is the density, the boundary condition 6.4 at all points of this surface takes the form

$$U(p,\ t) = 0 \qquad \text{if} \quad z = 0 \tag{6.6}$$

Here p denotes coordinates of the observation point. Thus, in order to determine the reflected wave, it is necessary to satisfy the initial and boundary conditions. The former requires that

$$U(p,\,t) \equiv 0 \qquad \text{and} \qquad \frac{\partial U(p,\,t)}{\partial t} \equiv 0 \qquad \text{if} \qquad t < 0 \tag{6.7}$$

In essence, we have formulated Dirichlet's boundary value problem. The remarkable simplicity of the boundary condition 6.6 allows us to find the solution almost instantly. It is obvious that the reflected wave travels away from the boundary, where its potential satisfies the condition

$$U_r(p,\,t) = -U_i(p,\,t) \tag{6.8}$$

Bearing this in mind, we assume that the secondary wave is also the spherical one, and its potential is described as

$$U_r(p,\,t) = -\frac{A}{R_1}\, f\left[a\left(t - \frac{R_1}{c_1}\right)\right], \tag{6.9}$$

where

$$R_1 = \sqrt{r^2 + (z+h)^2} \qquad \text{and} \qquad R = \sqrt{r^2 + (z-h)^2} \tag{6.10}$$

It is obvious that the function U satisfies the wave equation, as well as the boundary conditions. In particular,

$$U(r,\,0,\,t) = 0$$

at the plane $z = 0$ because $R = R_1$. Thus, we have solved the boundary value problem and, therefore, the potential at each point of the medium is

$$U(p,\,t) = \frac{A}{R}\, f\left[a\left(t - \frac{R}{c_1}\right)\right] - \frac{A}{R_1}\, f\left[a\left(t - \frac{R_1}{c_1}\right)\right] \tag{6.11}$$

The latter clearly shows that the reflected wave is also spherical. Its fictitious source is the mirror reflection of the real one, Fig. 6.1b.

It may be proper to notice that a similar situation sometimes occurs when the electric and electromagnetic fields are examined in media with a single plane interface.

Snell's law

In the previous chapter we studied the reflection of plane waves at a plane interface and demonstrated that the incidence and reflection angles are equal to each other regardless of the acoustic impedances of both media. Next consider those angles at the surface $z = 0$ for the incident spherical wave. As we see from Fig. 6.2, the directions of propagation of both waves form equal angles with the normal to the free surface ($\theta_i = \theta_r$). Therefore, Snell's law is valid not only for plane waves but also for spherical waves reflected at the

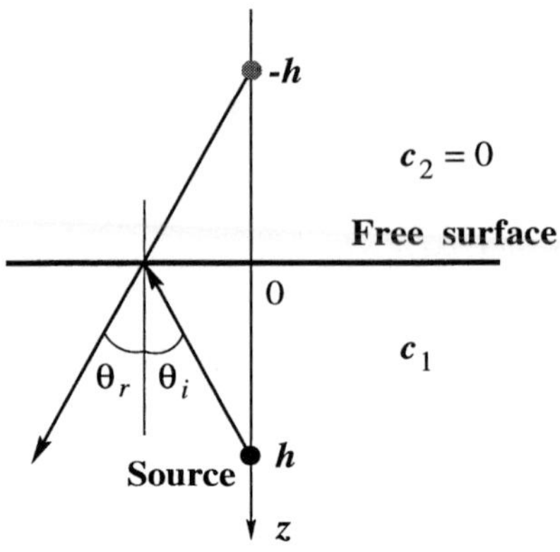

Figure 6.2: Snell's law for spherical waves.

free surface. This is an important generalization of the law, since in such a case it can be applied to any transient wave regardless of its frequency spectrum and the distance between the source and the boundary.

Superposition of incident and reflected waves

Let us study an interference of the incident and reflected waves near the boundary and, first, consider the stationary case when the source generates sinusoidal waves. The complex amplitudes of the potential can be written in the form

$$\mathcal{U}_i = \frac{A}{i\,k_1\,R}\, e^{i\,\omega\,R/c_1}\,, \qquad \mathcal{U}_r = -\frac{A}{i\,k_1\,R_1}\, e^{i\,\omega\,R_1/c_1}\,,$$

where ω is the frequency and $k_1 = \omega/c_1$ is the wavenumber. Therefore, we have

$$U(p,\,t) = \text{Re}\,(\mathcal{U}\,e^{-i\,\omega\,t})$$

or

$$U = -\frac{A}{k_1}\left[\frac{1}{R}\,\sin\omega\left(t - \frac{R}{c_1}\right) - \frac{1}{R_1}\,\sin\omega\left(t - \frac{R_1}{c_1}\right)\right] \tag{6.12}$$

Suppose that observations are performed at such distances from the surface that the inequality

$$\frac{z+h}{r} \ll 1 \tag{6.13}$$

is met. Then

$$R = r\sqrt{1 + \frac{(z-h)^2}{r^2}} \approx r + \frac{1}{2}\,\frac{(z-h)^2}{r}\,,$$

$$R_1 = r\sqrt{1+\frac{(z+h)^2}{r^2}} \approx r + \frac{1}{2}\,\frac{(z+h)^2}{r} \tag{6.14}$$

Substitution of eqs. 6.14 into eq. 6.12 yields

$$U(p,\,t) \approx -\frac{A}{k_1\,r}\left[\sin\omega\left(t-\frac{R}{c_1}\right) - \sin\omega\left(t-\frac{R_1}{c_1}\right)\right]$$

$$= -\frac{2\,A}{k_1\,r}\,\sin\left(\omega\,\frac{R_1-R}{2\,c_1}\right)\cos\omega\left(t-\frac{R_1+R}{2\,c_1}\right)$$

Since

$$\frac{R_1+R}{2} \approx r$$

and

$$R_1 - R \approx \frac{1}{2r}\left[(z+h)^2-(z-h)^2\right] = \frac{2\,z\,h}{r},$$

we have

$$U(p,\,t) \approx -\frac{2\,A}{k_1\,r}\,\sin\left(\omega\,\frac{z\,h}{c_1\,r}\right)\cos\omega\left(t-\frac{r}{c_1}\right) \tag{6.15}$$

Correspondingly, the magnitude of U is zero at the points, where

$$2\,\frac{z\,h}{\lambda_1\,r} = 0,\ 1,\ 2,\ \ldots,$$

and reaches its maximum if

$$4\,\frac{z\,h}{\lambda_1\,r} = 1,\ 3,\ 5,\ \ldots$$

Assuming that

$$\omega\,\frac{z\,h}{c_1\,r} \ll 1, \tag{6.16}$$

eq. 6.15 becomes

$$U(p,\,t) \approx \frac{2\,A\,z\,h}{r^2}\,\cos\omega\left(t-\frac{r}{c_1}-\pi\right) \tag{6.17}$$

Thus, the amplitude of the potential of the total wave field decreases in the horizontal direction as $1/r^2$, that is, more rapidly than the amplitude of the spherical wave. This effect is sometimes called the Lloyd mirror reflection.

As we should expect, the amplitude also decreases when either h or z tends to zero. In accordance with eq. 6.5, the pressure depends on coordinates r and z in the same manner as the potential, and such behavior is called the pressure-release effect. Since the condition 6.16 can be rewritten as

$$2\pi \frac{z\,h}{\lambda_1\, r} \ll 1\,,$$

with an increase of wavelength the same tendency is observed at greater distances from the boundary.

Next, consider the interference of transient waves. It is clear that if the source begins to act at the moment $t=0$, the incident and reflected waves arrive at the observation point below the boundary at instants $t = R/c_1$ and $t = R_1/c_1$, respectively. Therefore, since $R_1 > R$, the interference may occur only when $t \geq R_1/c_1$. Note also that, as in the case of plane waves, the compressional incident wave gives rise to the extensional reflected wave at points of the free surface and vice versa. For illustration, suppose that the pressure of the source varies as

$$P(t) = P_0\, e^{-m\,t}$$

and correspondingly, the potential at each point is

$$U = -\frac{P_0}{m^2\,\rho_1} \begin{cases} 0\,, & t < \dfrac{R}{c_1} \\[2ex] \dfrac{e^{-m\,(t-R/c_1)}}{R}\,, & \dfrac{R}{c_1} \leq t < \dfrac{R_1}{c_1} \\[2ex] \dfrac{e^{-m\,(t-R/c_1)}}{R} - \dfrac{e^{-m\,(t-R_1/c_1)}}{R_1}\,, & t \geq \dfrac{R_1}{c_1} \end{cases} \tag{6.18}$$

Here ρ_1 is the density of the lower medium. The curves in Fig. 6.3 show the result of interference of the incident and reflected waves.

Spectrum of the total potential

As we know, both the incident and reflected waves are described by the same function f with different arguments. Let us study the spectrum of the resultant wave field. Applying the Fourier transform, we obtain

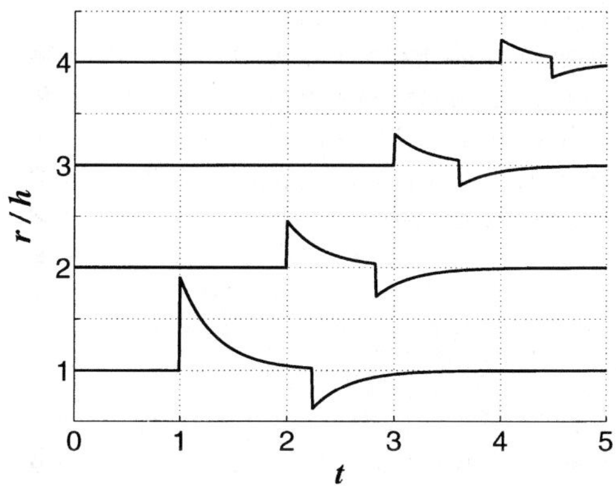

Figure 6.3: The potential U, eq. 6.18, at observation points with coordinates $(h,\ r)$.

$$\mathcal{U}(\omega) = \frac{1}{R}\int_0^\infty f\left[a\left(t-\frac{R}{c_1}\right)\right]e^{i\omega t}\,dt - \frac{1}{R_1}\int_{\Delta t}^\infty f\left[a\left(t-\frac{R}{c_1}-\Delta t\right)\right]e^{i\omega t}\,dt \qquad (6.19)$$

The low limit in the second integral is

$$\Delta t = \frac{R_1 - R}{c_1},$$

because its integrand vanishes when $t < \Delta t$. A change of variables in the second integral of eq. 6.19 gives

$$\mathcal{U}(\omega) = \frac{1}{R}\int_0^\infty f\left[a\left(t-\frac{R}{c_1}\right)\right]e^{i\omega t}\,dt - \frac{1}{R_1}e^{i\omega\Delta t}\int_0^\infty f\left[a\left(t-\frac{R}{c_1}\right)\right]e^{i\omega t}\,dt$$

or

$$\mathcal{U}(\omega) = \left(1 - \frac{R}{R_1}e^{i\omega\Delta t}\right)\mathcal{U}_i(\omega)\,, \qquad (6.20)$$

where

$$\mathcal{U}_i(\omega) = \frac{1}{R}\int_0^\infty f\left[a\left(t-\frac{R}{c_1}\right)\right]e^{i\omega t}\,dt$$

is the spectrum of the incident wave.

In particular, the amplitude spectrum of the total wave

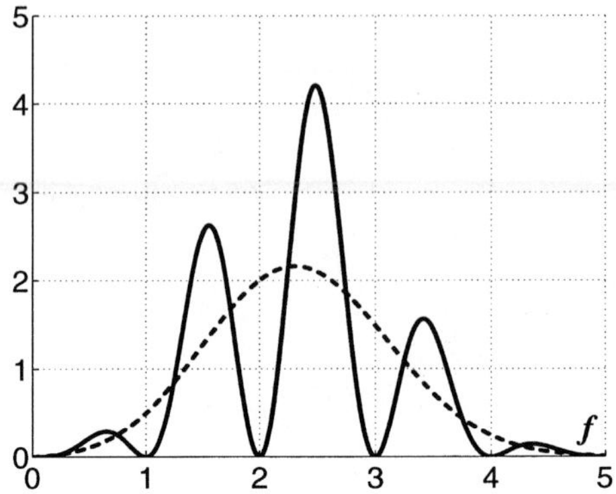

Figure 6.4: Amplitude spectra of the incident (dashed line) and total (solid line) wave fields.

$$|\mathcal{U}(\omega)| = \left|1 - \frac{R}{R_1}\, e^{i\,\omega\,\Delta t}\right|\ |\mathcal{U}_i(\omega)|$$

is equal to the amplitude spectrum of the incident wave multiplied by the factor

$$N = \left[\left(1 - \frac{R}{R_1}\ \cos\omega\,\Delta t\right)^2 + \frac{R^2}{R_1^2}\ \sin^2\omega\,\Delta t\right]^{1/2} \tag{6.21}$$

If we assume that $R \approx R_1$ but $\Delta t \neq 0$, the latter becomes

$$N = \sqrt{2\,(1 - \cos\omega\,\Delta t)} \tag{6.22}$$

Thus, the amplitude spectrum of the total wave field is equal to zero at frequencies

$$f = 0\,, \quad \frac{1}{\Delta t}\,, \quad \frac{2}{\Delta t}\,, \quad \ldots$$

and reaches its maxima when

$$f = \frac{1}{2\,\Delta t}\,, \quad \frac{3}{2\,\Delta t}\,, \quad \frac{5}{2\,\Delta t}\,, \quad \ldots$$

The behavior of the amplitude spectra of the incident and total wave fields are shown in Fig. 6.4.

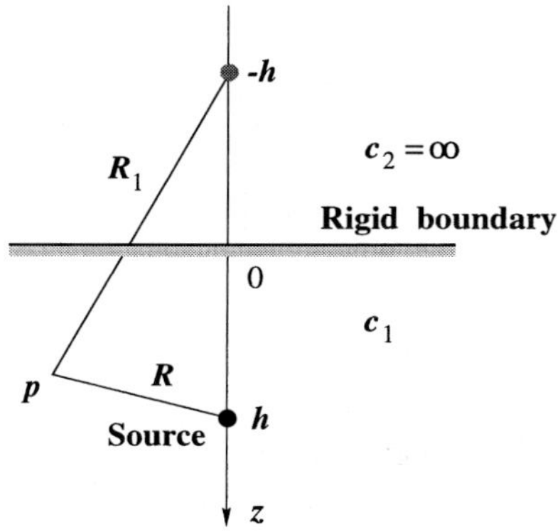

Figure 6.5: Reflection of spherical waves from a rigid boundary.

Rigid surface

Now suppose that the source of the spherical wave is located beneath a rigid surface, Fig. 6.5. As before, we attempt to solve the boundary value problem, assuming that the reflected wave is also spherical and that the position of a fictitious source is again the mirror reflection of the real one. Since the surface is rigid, the vertical component of the displacement at all points of the boundary is equal to zero

$$s_z = 0 \qquad \text{if} \qquad z = 0 \tag{6.23}$$

Therefore, in terms of the potential the latter can be written in the form

$$\frac{\partial U}{\partial z} = 0 \qquad \text{or} \qquad \frac{\partial U_r}{\partial z} = -\frac{\partial U_i}{\partial z}, \tag{6.24}$$

where

$$U = U_i + U_r$$

Let us represent the potential as

$$U(r,\, z,\, t) = \frac{A}{R}\, f\left[a\left(t - \frac{R}{c_1}\right)\right] + \frac{A}{R_1}\, f\left[a\left(t - \frac{R_1}{c_1}\right)\right] \tag{6.25}$$

Here R and R_1 are given by eq. 6.10. Comparison of eqs. 6.11 and 6.25 shows that the potentials of reflected waves in both cases differ by sign only. This means that at a rigid surface the incident wave generates the reflected wave of the same type, that is both waves are either compressional or extensional if $z = 0$.

Taking the derivative of the potential with respect to z, we obtain

$$\frac{\partial U_i}{\partial z} = -\frac{A}{R^2}\frac{z-h}{R}\, f\left[a\left(t-\frac{R}{c_1}\right)\right] - \frac{A\,a}{c_1\,R}\frac{z-h}{R}\, f'\left[a\left(t-\frac{R}{c_1}\right)\right]$$

and

$$\frac{\partial U_r}{\partial z} = -\frac{A}{R_1^2}\frac{z+h}{R_1}\, f\left[a\left(t-\frac{R_1}{c_1}\right)\right] - \frac{A\,a}{c_1\,R_1}\frac{z+h}{R_1}\, f'\left[a\left(t-\frac{R_1}{c_1}\right)\right]$$

Since at the boundary $R = R_1$, the condition 6.24, is satisfied as well as other conditions of the boundary value problem. Thus, our assumption was correct, and the reflected wave is spherical. The center of its phase surfaces is situated at the point with coordinates $0, -h$. Similarly to the case of a free surface, Fig. 6.2, Snell's law of reflection correctly describes a ray direction.

In accordance with eq. 6.25, at all points of the boundary $z = 0$, the potential of the total wave field is

$$U(r,\,0,\,t) = \frac{2\,A}{R}\, f\left[a\left(t-\frac{R}{c_1}\right)\right] \tag{6.26}$$

Correspondingly, the pressure, P, and the horizontal component of the displacement also double. A similar behavior is observed beneath the boundary when the distance from the source R greatly exceeds h. At the same time, the vertical component s_z of the displacement is close to zero. Note that, unlike the case of the free surface, total potential U does not vanish if the source approaches the boundary. Rather, it is

$$U(p,\,t) \to 2\,U_i(p,\,t) \qquad \text{if} \qquad h \to 0$$

We have demonstrated that the reflected wave is spherical in the two extreme cases of the free and rigid boundaries, and the incident and reflection angles are equal to each other at all points of the boundaries. Thus, Snell's law, which is valid in the limit of geometrical acoustics and for plane waves propagating in media with plane interfaces, is also applicable in the two cases discussed above. It may be proper to point out that the same approach was used to solve the boundary value problems in both examples.

General case and the mirror-reflection method

Also it is useful to demonstrate that in the general case when the medium consists of two half-spaces with finite nonzero values of the impedance Z, this method for solving the boundary value problem cannot be applied. In other words, the waves arising at the interface may differ from the spherical ones. In order to prove this fact, let us assume, first, that the reflected and transmitted waves are both spherical and that their fictitious sources are located on the z-axis, Fig. 6.6. Then, for the potentials in each medium, we have

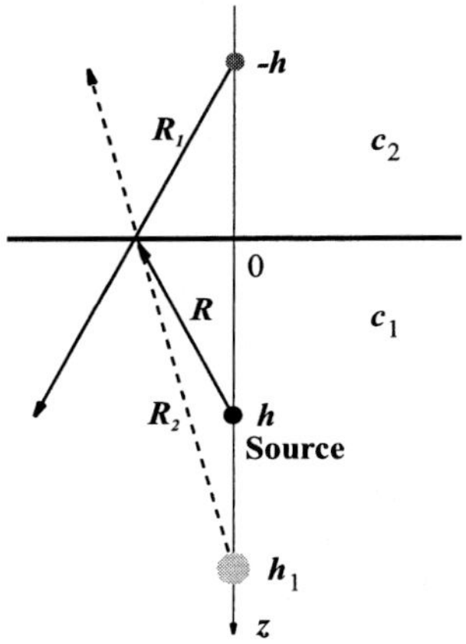

Figure 6.6: The mirror-reflection method in the case of finite nonzero velocities c_1 and c_2.

$$U_1=\frac{A}{R}\,f\left[a\left(t-\frac{R}{c_1}\right)\right]+\frac{B}{R_1}\,f_1\left[a_1\left(t-\frac{R_1}{c_1}\right)\right]\qquad \text{if}\qquad z\geq 0 \tag{6.27}$$

and

$$U_2=\frac{C}{R_2}\,f_2\left[a_2\left(t-\frac{R_2}{c_2}\right)\right]\qquad \text{if}\qquad z<0\,, \tag{6.28}$$

where

$$R=\sqrt{r^2+(z-h)^2}\,,\qquad R_1=\sqrt{r^2+(z+h)^2}\qquad \text{and}\qquad R_2=\sqrt{r^2+(z-h_1)^2}$$

Here h and h_1 denote the positions of the real and fictitious sources, Fig. 6.6.

As is well known, the potential has to satisfy the conditions

$$\rho_1\,U_1=\rho_2\,U_2\qquad \text{and}\qquad \frac{\partial U_1}{\partial z}=\frac{\partial U_2}{\partial z}\qquad \text{if}\quad z=0 \tag{6.29}$$

Substitution of eqs. 6.27 and 6.28 into the first equation of the last set gives

$$\rho_1\,\frac{A}{R}\,f\left[a\left(t-\frac{R}{c_1}\right)\right]+\rho_1\,\frac{B}{R}\,f_1\left[a_1\left(t-\frac{R}{c_1}\right)\right]=\rho_2\,\frac{C}{R_2}\,f_2\left[a_2\left(t-\frac{R_2}{c_2}\right)\right],$$

since $R=R_1$ if $z=0$. This equality has to take place at any instant which means that

$$f = f_1 = f_2\,, \qquad a = a_1 = a_2 \qquad \text{and} \qquad t - \frac{R}{c_1} = t - \frac{R_2}{c_2}$$

or

$$R_2 = \frac{R}{n} \qquad \text{and} \qquad n = \frac{c_1}{c_2}$$

Whence

$$h_1 = \sqrt{\left(\frac{1}{n^2} - 1\right) r^2 + \frac{h^2}{n^2}} \tag{6.30}$$

The dependence of h_1 on r indicates that the position of the fictitious source for the transmitted waves has to change for different observation points. In other words, the transmitted wave is not spherical. Also if

$$\left|\frac{1}{n^2} - 1\right| r^2 > \frac{h^2}{n^2} \qquad \text{and} \qquad \frac{1}{n^2} < 1\,,$$

h_1 becomes complex. Moreover, the function

$$\frac{C}{R_2}\, f_2 \left[a_2 \left(t - \frac{R_2}{c_2}\right)\right]$$

no longer satisfies the wave equation because the coordinate h_1 depends on r. Therefore, our assumptions were incorrect and, correspondingly, the method of mirror reflection cannot be used. For this reason, in the next section we solve this boundary problem applying the method of separation of variables.

6.2 Spherical source in medium with a plane boundary (general case)

Boundary value problem

Suppose that the source of the acoustic wave is located at the distance d above the boundary between media with different densities and wave velocities, Fig. 6.7. As before, we formulate the boundary value problem in order to determine the wave field. Taking into account the axial symmetry of the problem, it is convenient to choose the cylindrical coordinate system and place its origin at the boundary. Here, c_1, ρ_1 and c_2, ρ_2 are the velocities and densities in the upper and lower media, respectively, Fig. 6.7. Let us assume at the beginning that the source generates the sinusoidal wave with the frequency ω and then, using the Fourier integral, we can study the transient waves. It is natural to formulate the boundary value problem in terms of the complex amplitude of the potential $\mathcal{U}$, which relates to pressure $\mathcal{P}$ and displacement $\mathcal{S}$ as

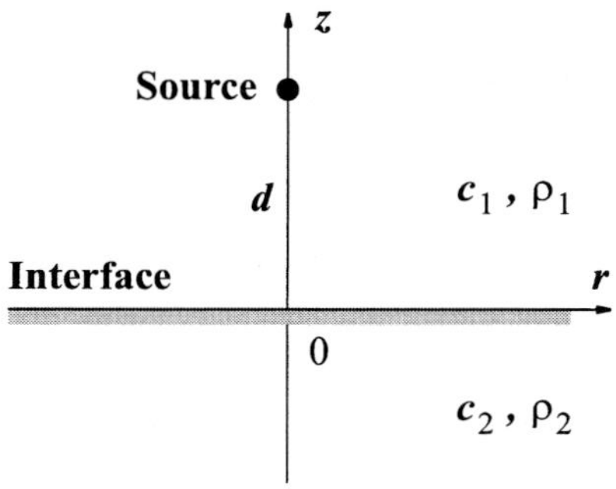

Figure 6.7: Elementary source in the presence of plane interface.

$$\mathcal{P} = \omega^2 \rho \mathcal{U} \qquad \text{and} \qquad \mathcal{S} = \text{grad}\,\mathcal{U}\,, \tag{6.31}$$

where $\mathcal{P}$ and each component of $\mathcal{S}$ are the complex amplitudes.

As we already know, the potential $\mathcal{U}$ should satisfy the following conditions. At regular points in both media, it is a solution of the Helmholtz equation

$$\nabla^2\mathcal{U} + k^2\mathcal{U} = 0\,, \tag{6.32}$$

where

$$k = \begin{cases} k_1 = \dfrac{\omega}{c_1} & \text{if} \quad z > 0 \\ \\ k_2 = \dfrac{\omega}{c_2} & \text{if} \quad z < 0 \end{cases} \tag{6.33}$$

The pressure and the vertical component of the displacement are continuous functions at the media interface. In accordance with eqs. 6.31 we have

$$\rho_1 \mathcal{U}_1 = \rho_2 \mathcal{U}_2 \qquad \text{and} \qquad \frac{\partial \mathcal{U}_1}{\partial z} = \frac{\partial \mathcal{U}_2}{\partial z} \tag{6.34}$$

Here $\mathcal{U}_1$ and $\mathcal{U}_2$ are the complex amplitudes of the potential in the upper and lower media, respectively.

The potential of the incident wave caused by the source can be represented as

$$\mathcal{U}_i = C\,\frac{e^{i\,k_1 R}}{R}\,, \tag{6.35}$$

where C is a given constant and

$$R = \sqrt{r^2 + (z-d)^2} \tag{6.36}$$

is the distance from the source to an observation point with coordinates r and z.

The potential in the upper medium is a sum of two terms:

$$\mathcal{U}_1 = \mathcal{U}_i + \mathcal{U}_r \tag{6.37}$$

and $\mathcal{U}_r$ is the potential of the reflected wave field which has a finite value everywhere. It is clear that the total potential $\mathcal{U}_1$ tends to $\mathcal{U}_i$ in approaching the source:

$$\mathcal{U}_1 \to \mathcal{U}_i \qquad \text{if} \qquad R \to 0 \tag{6.38}$$

The latter characterizes the condition near the source.

The spherical wave generated by the source, as well as the reflected and transmitted waves, decrease with the distance from the source. In other words, the waves have to obey the condition

$$\mathcal{U}_1 \to 0 \qquad \text{and} \qquad \mathcal{U}_2 \to 0 \qquad \text{if} \qquad R \to \infty \tag{6.39}$$

Thus, we have formulated the boundary value problem which requires the potential $\mathcal{U}$ to satisfy the following conditions:

(a) At regular points

$$\nabla^2 \mathcal{U}_1 + k_1^2 \mathcal{U}_1 = 0 \qquad \text{if} \qquad z > 0$$

and

$$\nabla^2 \mathcal{U}_2 + k_2^2 \mathcal{U}_2 = 0 \qquad \text{if} \qquad z < 0$$

(b) Near the source

$$\mathcal{U}_1 \to \mathcal{U}_i \qquad \text{if} \qquad R \to 0$$

(c) At infinity

$$\mathcal{U}_1 \to 0 \qquad \text{and} \qquad \mathcal{U}_2 \to 0 \qquad \text{if} \qquad R \to \infty$$

(d) At the boundary

$$\rho_1 \mathcal{U}_1 = \rho_2 \mathcal{U}_2 \qquad \text{and} \qquad \frac{\partial \mathcal{U}_1}{\partial z} = \frac{\partial \mathcal{U}_2}{\partial z} \qquad \text{if} \qquad z = 0$$

As follows from the physical point of view, as well as from the theorem of uniqueness, there is only one field of the pressure and the displacement which satisfies all of those conditions. Since we consider Dirichlet's boundary value problem, (Part I), the potential $\mathcal{U}$ is also defined uniquely. The same results follow from the first equation of the set 6.31.

Solution of the boundary problem

First, we find the solution of the Helmholtz equation. Bearing in mind the axial symmetry, we represent this equation in the cylindrical coordinates, (Part I):

$$\frac{\partial^2 \mathcal{U}}{\partial r^2} + \frac{1}{r}\frac{\partial \mathcal{U}}{\partial r} + \frac{\partial^2 \mathcal{U}}{\partial z^2} + k^2 \mathcal{U} = 0\,, \tag{6.40}$$

since the derivative with respect to the azimuthal angle φ vanishes:

$$\frac{\partial^2 \mathcal{U}}{\partial \varphi^2} = 0$$

Applying the method of separation of variables, the potential $\mathcal{U}$ is written in the form

$$\mathcal{U}(r,\, z,\, \omega) = \mathcal{T}(r)\, \mathcal{Z}(z,\, \omega) \tag{6.41}$$

Substitution of eq. 6.41 into eq. 6.40 and dividing both its sides by the product $\mathcal{T}\mathcal{Z}$ gives

$$\frac{1}{\mathcal{T}}\frac{d^2 \mathcal{T}}{d r^2} + \frac{1}{r\mathcal{T}}\frac{d\mathcal{T}}{d r} + \frac{1}{\mathcal{Z}}\frac{d^2 \mathcal{Z}}{d z^2} + k^2 = 0 \tag{6.42}$$

This equality indicates that the sum of the first two terms, as well as that of the last two, are constants which differ by sign only. Therefore, in place of eq. 6.42, we can write

$$\begin{aligned} \frac{1}{\mathcal{T}}\frac{d^2 \mathcal{T}}{d r^2} + \frac{1}{r\mathcal{T}}\frac{d\mathcal{T}}{d r} &= \pm m^2\,, \\ \frac{1}{\mathcal{Z}}\frac{d^2 \mathcal{Z}}{d z^2} + k^2 &= \mp m^2\,, \end{aligned} \tag{6.43}$$

where m is an arbitrary number.

Thus, instead of the partial differential equation, we have arrived at two ordinary differential equations, whose solutions are well known. Selecting the sign in the right-hand side of eqs. 6.43, we have to take into account the fact that the wave field has a finite amplitude everywhere except the source location. Suppose we choose the sign "+" in the first equation of this set. Then it gives

$$\frac{d^2 \mathcal{T}}{d r^2} + \frac{d\mathcal{T}}{r\, d r} - m^2 \mathcal{T} = 0$$

The solution of this equation is expressed in terms of the modified Bessel functions of the zero order:

$$I_0\,(mr) \qquad \text{and} \qquad K_0\,(mr)$$

Inasmuch as $I_0(mr)$ increases unlimitedly when the distance $r \to \infty$, while $K_0(mr)$ becomes infinitely large at all points of the z-axis, neither of those functions can describe the reflected and transmitted wave fields. For this reason, we choose the signs "−" and "+" in the first and second equations, respectively. Therefore, eqs. 6.43 become

$$\begin{aligned} &\frac{d^2\mathcal{T}}{dr^2}+\frac{d\mathcal{T}}{r\,dr}+m^2\,\mathcal{T}=0\,,\\ &\frac{d^2\mathcal{Z}}{dz^2}-(m^2-k^2)\,\mathcal{Z}=0 \end{aligned} \tag{6.44}$$

The solution of the first equation is expressed in terms of the Bessel functions

$$J_0\,(mr) \qquad \text{and} \qquad Y_0\,(mr)\,,$$

whose behavior has been studied in detail. Since the function $Y_0(mr)$ is infinitely large at the z-axis, it cannot describe the waves.

The solution of the second equation is the exponent

$$\exp\left[\pm\sqrt{(m^2-k^2)}z\right]$$

Thus, in accordance with eq. 6.41, we have

$$\mathcal{U}_m=\left(A_m\,e^{-m_n z}+B_m\,e^{m_n z}\right)J_0(mr) \tag{6.45}$$

Here

$$m_n=\sqrt{m^2-k_n^2} \tag{6.46}$$

and $n=1,\,2$.

We would expect that the potential $\mathcal{U}$, describing the real physical field of waves, is independent of the variable m which appears due to the algebraic transformations. To remove an influence of m, we make use of the general solution of the Helmholtz equation

$$\mathcal{U}(r,\,z,\,\omega)=\int\limits_0^\infty\left(A_m\,e^{-m_n z}+B_m\,e^{m_n z}\right)J_0(mr)\,dm\,, \tag{6.47}$$

where A_m and B_m are the unknown coefficients and they do not depend on the coordinates r and z of the observation point. Note that the integrand in eq. 6.47 is the function of m, but the definite integral is independent of m.

It is clear that the function $\mathcal{U}_m$ (eq. 6.45) satisfies the Helmholtz equation regardless the values of A_m and B_m. In other words, the Helmholtz equation has an infinite number of solutions, and our goal is to choose such A_m and B_m that $\mathcal{U}(r,z,\omega)$ would also satisfy other conditions of the boundary value problem. In order to accomplish this task we represent the primary potential $\mathcal{U}_i$ in the same manner as the potentials of the secondary waves. This can be done with the help of the Sommerfeld integral

$$\frac{e^{i\,k_1 R}}{R}=\int\limits_0^\infty\frac{m}{m_1}\,e^{-m_1\,|z-d\,|}\,J_0(mr)\,dm\,, \tag{6.48}$$

where

$$m_1 = \sqrt{m^2 - k_1^2}$$

Then the expressions for the complex amplitudes of the potential in both media are

$$\mathcal{U}_1 = C \int_0^\infty \left(\frac{m}{m_1} e^{-m_1 |z-d|} + A_m e^{-m_1 z} \right) J_0(mr)\, dm \qquad \text{if} \qquad z > 0\,,$$

$$\mathcal{U}_2 = C \int_0^\infty B_m e^{m_2 z} J_0(mr)\, dm \qquad \text{if} \qquad z < 0 \tag{6.49}$$

Here

$$m_2 = \sqrt{m^2 - k_2^2}$$

We note the following.

(a) The real part of the radicals m_1 and m_2 is chosen to be positive,

$$\text{Re } m_1 > 0 \qquad \text{and} \qquad \text{Re } m_2 > 0 \tag{6.50}$$

to ensure a convergence of integrals 6.49.

(b) The term containing the exponent $e^{m_1 z}$ was discarded in the expression for the potential in the upper medium. Otherwise, the wave magnitude would increase with the distance from the source, $R \to \infty$, which contradicts the condition at infinity. For the same reason, the term with the exponent $e^{-m_2 z}$ does not describe the potential in the lower medium.

(c) It is obvious that the potential $\mathcal{U}_1$ tends to $\mathcal{U}_i$ as the distance from the source goes to zero. Thus the functions $\mathcal{U}_1$ and $\mathcal{U}_2$ given by eqs. 6.49 obey the Helmholtz equation, as well as the conditions near the source and at infinity.

Now we find such coefficients A_m and B_m that satisfy the conditions at the boundary $z = 0$. Substitution of eqs. 6.49 into eqs. 6.34 gives the system of linear equations

$$\begin{aligned} \rho_1 \left(\frac{m}{m_1} e^{-m_1 d} + A_m \right) &= \rho_2 B_m \\ m\, e^{-m_1 d} - m_1 A_m &= m_2 B_m \end{aligned} \tag{6.51}$$

This remarkable transition from the equality of integrals to that of their integrands is based on fundamental features of Bessel functions, which can be written as if

$$\int_0^\infty F_1(m)\, J_0(mr)\, dm = \int_0^\infty F_2(m)\, J_0(mr)\, dm\,,$$

then

$$F_1(m) = F_2(m)$$

The solution of the system 6.51 is

$$A_m = \frac{m}{m_1}\,\frac{m_1 - b\,m_2}{m_1 + b\,m_2}\, e^{-m_1 d} \tag{6.52}$$

and

$$B_m = \frac{2\,m\,b}{m_1 + b\,m_2}\, e^{-m_1 d}\,, \tag{6.53}$$

where

$$b = \frac{\rho_1}{\rho_2}$$

Thus, if $z > 0$,

$$\mathcal{U}_1 = C\int_0^\infty \frac{m}{m_1}\, e^{-m_1|z-d|}\, J_0(mr)\, dm + C\int_0^\infty \frac{m}{m_1}\,\frac{m_1 - b\,m_2}{m_1 + b\,m_2}\, e^{-m_1(z+d)}\, J_0(mr)\, dm \tag{6.54}$$

and, if $z < 0$,

$$\mathcal{U}_2 = 2\,b\,C\int_0^\infty \frac{m}{m_1 + b\,m_2}\, e^{-m_1 d}\, e^{m_2 z}\, J_0(mr)\, dm \tag{6.55}$$

As follows from equality 6.48, the potential $\mathcal{U}_1$ can be rewritten as

$$\mathcal{U}_1 = C\,\frac{e^{i\,k_1 R}}{R} + C\int_0^\infty \frac{m}{m_1}\,\frac{m_1 - b\,m_2}{m_1 + b\,m_2}\, e^{-m_1(z+d)}\, J_0(mr)\, dm \tag{6.56}$$

We see that the second term in the right-hand side of eq. 6.54 and the function $\mathcal{U}_2$ characterize the reflected and transmitted waves, respectively. Those equations also describe the behavior of the head wave, which exists if $c_2 > c_1$, as well as evanescent ones.

It is simple to obtain again the expressions for the potential $\mathcal{U}_1$ in the two extreme cases examined in the previous section. In fact, suppose that the boundary is free, that is

$$\rho_2 = 0 \qquad \text{or} \qquad b \to \infty$$

Then, in accordance with eq. 6.52, we have

$$A_m = -\frac{m}{m_1}\,e^{-m_1 d}$$

and therefore, eq. 6.56 becomes

$$\mathcal{U}_1 = C\,\frac{e^{i\,k_1 R}}{R} - C\int\limits_0^\infty \frac{m}{m_1}\,e^{-m_1(z+d)}\,J_0(mr)\,dm$$

The use of the Sommerfeld integral gives

$$\mathcal{U}_1 = C\,\frac{e^{i\,k_1 R}}{R} - C\,\frac{e^{i\,k_1 R_1}}{R_1}\,, \tag{6.57}$$

where

$$R_1 = \sqrt{r^2 + (z+d)^2}$$

In contrast, if the lower medium is rigid

$$\rho_2 \to \infty \qquad \text{and} \qquad b = 0$$

Whence

$$A_m = \frac{m}{m_1}\,e^{-m_1 d}\,,$$

and the potential $\mathcal{U}_1$ is

$$\mathcal{U}_1 = C\,\frac{e^{i\,k_1 R}}{R} + C\,\frac{e^{i\,k_1 R_1}}{R_1} \tag{6.58}$$

These waves, eqs. 6.57–6.58, were studied in detail previously.

Also it is useful to consider the case where the density changes at the boundary, $\rho_1 \neq \rho_2$, but the velocity of propagation is constant everywhere: $c_1 = c_2 = c$. Then for the reflected and transmitted waves, eqs. 6.54–6.55, we obtain

$$\mathcal{U}_r = C\,\frac{1-b}{1+b}\,\frac{e^{ik_1 R_1}}{R_1} \qquad \text{and} \qquad \mathcal{U}_2 = C\,\frac{2b}{1+b}\,\frac{e^{ik_1 R_2}}{R_2}$$

since $m_1 = m_2$. Here

$$R_2 = \sqrt{r^2 + (d + |z|)^2}$$

is the distance between the source and an observation point in the lower medium.

Thus, we see that both waves described by the functions $\mathcal{U}_r$ and $\mathcal{U}_2$ are spherical and the transmitted wave differs from the incident one by the term $2b/(1+b)$ only.

6.3 Analysis of wave fields by the stationary phase method

In the previous chapter, we have studied the reflected, transmitted, and evanescent waves in the presence of a planar boundary, provided that the incident wave is also a plane one. In reality, however, such a wave does not exist; it only serves as a very useful approximation of the actual wave. Besides, having examined the limiting case of the geometrical acoustics it was found that the head wave may arise too. For this reason it is natural to make use of the solution of the boundary value problem for an elementary spherical source and describe the behavior of these waves. As will be demonstrated in this section, the method of stationary phase allows us to study most of them in the wave zone, $(r > \lambda)$, which is of great practical interest. First, we review main features of this approach (Part I).

The method of stationary phase

Consider the integral

$$I = \int_{-\infty}^{\infty} f(m)\, e^{i\alpha h(m)}\, dm \tag{6.59}$$

The integrand is the product of two terms: $f(m)$ and $exp[i\alpha h(\omega)]$. The first one usually changes relatively slowly with m, but the exponential term can vary rapidly when α is sufficiently large. Correspondingly, in such a case the oscillating character of the integrand is due to the second term. Regardless of the meaning of the function $I(\alpha)$, it is convenient to treat the integrand as a sinusoidal wave with spatial frequency m. Its amplitude and initial phase, that is the argument of $f(m)$, are both functions of m in general. Therefore, the integral $I(\alpha)$ describes a superposition of these waves. As is well known (Part I) within the range of frequencies m, where the phase

$$\varphi(m) = \alpha\, h(m) \tag{6.60}$$

varies only slightly, constructive interference occurs and the sum of such sinusoids may become rather large. In contrast, outside this range the phase $\varphi(m)$ can change very quickly and for this reason sinusoids cancel each other (destructive interference), and their sum is relatively small. Therefore, the integral $I(\alpha)$ is mainly defined by the first range. Of course, it is impossible to draw the exact boundary between these ranges, since in reality there is always an intermediate range. This analysis shows that our goal is to determine a position of the first range and then perform an evaluation of the integral over this range of m. In essence, the method of stationary phase consists of these two steps. It is natural to characterize a position of this range with the help of the point m_0, where the change of phase $\varphi(m)$ is small. Therefore, it can be determined from the equation

$$\frac{\partial\, h(m)}{\partial\, m} = 0 \tag{6.61}$$

As we know, (Part I), such point m_0 is called the stationary one, and it corresponds to either a maximum or a minimum of the function $h(m)$. It is obvious that

$$\Delta\varphi = \varphi(m) - \varphi(m_0) = \alpha[h(m) - h(m_0)]$$

in the vicinity of m_0 essentially depends on the value of the parameter α. For example, if it is relatively small, then even sufficiently strong deviations from the stationary point m_0 may cause a small change in the phase. In contrast, when the parameter α is rather large, an insignificant difference, Δm, may produce a great change of the phase. In other words, in this case the width of the first range of frequencies m becomes small. Thus, only for a relatively large value of α can we assume that the phase is nearly constant within this interval, while outside it varies very rapidly. These two important features of the function $\varphi(m)$ allow us to perform great simplifications in equation 6.59. First, the function $f(m)$ is replaced by its value at the stationary point

$$f(m) = f(m_0)\,, \tag{6.62}$$

since its change within the narrow interval is usually small. Also, we expand $h(m)$ in Taylor's series and restrict ourselves by the first three terms:

$$h(m) = h(m_0) + h'(m_0)\,(m - m_0) + \frac{h''(m_0)}{2}\,(m - m_0)^2 + ...$$

Since $h'(m_0) = 0$, we have

$$h(m) = h(m_0) + \frac{h''(m_0)}{2}\,(m - m_0)^2 + ... \tag{6.63}$$

Substitution of eqs. 6.62–6.63 into eq. 6.59 yields

$$I = f(m_0)\,e^{i\,\alpha\,h(m_0)} \int\limits_{-\infty}^{\infty} \exp\left[i\,\alpha\,\frac{h''(m_0)}{2}\,(m - m_0)^2\right] dm \tag{6.64}$$

Let us introduce the new variable

$$v^2 = \frac{\alpha\,|h''(m_0)|}{2}\,(m - m_0)^2$$

Hence

$$v = \sqrt{\frac{\alpha\,|h''(m_0)|}{2}}\,(m - m_0) \qquad \text{and} \qquad dv = \sqrt{\frac{\alpha\,|h''(m_0)|}{2}}\,dm\,,$$

that is

$$dm = \sqrt{\frac{2}{\alpha\,|h''(m_0)|}}\,dv$$

Whence

$$I(\alpha) = \frac{\sqrt{2}\,e^{i\,\alpha\,h(m_0)}}{\sqrt{\alpha\,|h''(m_0)|}}\,f(m_0)\int\limits_{-\infty}^{\infty} e^{\pm i\,v^2}\,dv \tag{6.65}$$

The appearance of different signs in the exponent is related to the fact that

$$h''(m_0) = |h''(m_0)| \qquad \text{if} \qquad h''(m_0) > 0$$

and

$$h''(m_0) = -|h''(m_0)| \qquad \text{if} \qquad h''(m_0) < 0$$

Inasmuch as

$$\int\limits_{-\infty}^{\infty} e^{\pm iv^2}\,dv = \sqrt{\frac{\pi}{2}}\,(1 \pm i)\,,$$

we have

$$I(\alpha) = f(m_0)\,(1 \pm i)\,\sqrt{\frac{\pi}{\alpha\,|h''(m_0)|}}\,e^{i\alpha\,h(m_0)}$$

or

$$I(\alpha) = f(m_0)\,\sqrt{\frac{2\pi}{\alpha\,|h''(m_0)|}}\,e^{i\,[\alpha\,h(m_0)\,\pm\,\pi/4]} \tag{6.66}$$

Bearing in mind that $\varphi''(m) = \alpha\,h''(m)$, the last equation gives

$$I(\alpha) = f(m_0)\left[\frac{2\pi}{|\varphi''(m_0)|}\right]^{1/2} \exp i\,[\varphi(m_0) \pm \frac{\pi}{4}] \tag{6.67}$$

and the sign in front of $\pi/4$ corresponds to that of the second derivative, $h''(m_0)$.

Thus, in place of the exact expression, eq. 6.59, we arrive at the approximate one, and its accuracy becomes higher with an increase of the parameter α. It may be appropriate to emphasize again that in deriving eq. 6.67, it was assumed that the amplitude and phase of the function $f(m)$ change rather slowly around the stationary point. As will be shown later, this condition is not always met. Now we begin to apply the method of stationary phase to eqs. 6.54 – 6.55, describing waves of the elementary spherical source and they are

$$\mathcal{U}_1 = C\int\limits_0^\infty \frac{m}{m_1}\, e^{-m_1|z-d|}\, J_0(mr)\, dm + C\int\limits_0^\infty \frac{m}{m_1}\,\frac{m_1 - b\,m_2}{m_1 + b\,m_2}\, e^{-m_1(z+d)}\, J_0(mr)\, dm \tag{6.68}$$

if $z > 0$, and

$$\mathcal{U}_2 = 2\,b\,C\int\limits_0^\infty \frac{m}{m_1 + b\,m_2}\, e^{-m_1 d}\, e^{m_2 z}\, J_0(mr)\, dm \tag{6.69}$$

if $z < 0$, where

$$m_1 = \sqrt{m^2 - k_1^2}\,, \qquad m_2 = \sqrt{m^2 - k_2^2}\,, \qquad b = \frac{\rho_1}{\rho_2}$$

Before we derive asymptotic expressions for the waves, consider again the Sommerfeld integral representing the potential of the incident wave, eq. 6.68:

$$\mathcal{U}_i = C\,\frac{e^{-ik_1 R}}{R} = C\int\limits_0^\infty \frac{m}{m_1}\, e^{-m_1|z-d|}\, J_0(mr)\, dm \tag{6.70}$$

and

$$R = \sqrt{r^2 + (z-d)^2}$$

Here it is assumed that $e^{i\omega t}$ describes the time dependence of the wave fields.

Cylindrical harmonics

It is instructive to interpret eq. 6.70 in the following way. The real spherical wave described by the potential

$$C\,\frac{e^{-ik_1 R}}{R}$$

and propagating away from the source is represented as an infinite sum of harmonics (elementary waves)

$$\mathcal{U}_{im} = C\,\frac{m}{m_1}\, e^{-m_1|z-d|}\, J_0(mr)\, dm \tag{6.71}$$

which are symmetrical with respect to the z-axis. Their amplitudes depend on the variable of integration m. Also, with an increase of m the harmonics change more rapidly in the horizontal and vertical directions. In contrast, they vary relatively slowly if m is small. This consideration permits us to predict that at sufficiently large distances from the source, where the incident wave changes only slightly, the harmonics with small m are dominant. At the same time, in the vicinity of the source and at some moderate distances from it harmonics with rather large m can give a significant contribution. The parameter m, which has the dimension [m^{-1}], can be treated as the spatial frequency of the harmonics and it characterizes their change in both directions. Since the harmonics of the incident wave decay more rapidly as m increases, we may say that harmonics corresponding to small m have a greater depth of penetration.

It is convenient to distinguish two types of harmonics, namely, the "diffusive" and the "wave" harmonics. First, suppose that parameter m exceeds k_1, that is

$$m_1 = \sqrt{m^2 - k_1^2} > 0 \tag{6.72}$$

is a real number. Then, harmonics of the potential U_{im}, which satisfy this condition, decay exponentially along the z-axis in both directions from the plane $z = d$. They can be written as

$$U_{im} = \text{Re}\, \mathcal{U}_{im}\, e^{i\omega t}$$

or

$$U_{im} = C\,\frac{m}{m_1}\, e^{-m_1|z-d|}\, J_0(mr)\,\cos(\omega t)\, dm \tag{6.73}$$

This means that the latter vary synchronously with the source at all points of the medium. Such behavior is also observed in the case of electromagnetic fields in a nonconducting medium when the effect of propagation is neglected. In other words, it can be assumed that the velocity of propagation is infinitely large, which is an essential feature of diffusion. Taking this into account, the harmonics of the incident wave with $m > k_1$ are sometimes called diffusive harmonics. Let us notice that the spatial spectrum of these waves depends on the wavenumber k_1. In particular, the influence of such spatial harmonics becomes stronger as k_1 decreases.

Next, assume that

$$m < k_1 \tag{6.74}$$

Since $m_1 = i\sqrt{k_1^2 - m^2} = in_1$, the complex amplitude of the potential of these harmonics is

$$\mathcal{U}_{im} = C\,\frac{m}{in_1}\, e^{-in_1|z-d|}\, J_0(mr)\, dm$$

or

$$U_{im} = C\,\frac{m}{n_1}\, J_0(mr)\, dm\, \sin\left[\omega t - n_1|z-d|\right] \tag{6.75}$$

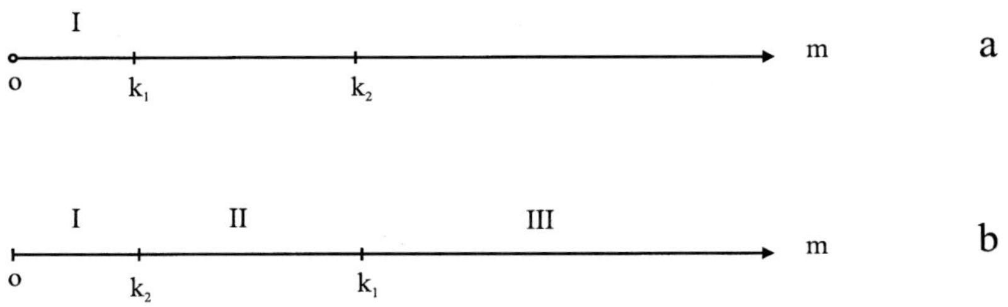

Figure 6.8: Three intervals of integration.

The latter describes the wave traveling along the z-axis away from the plane $z = d$ with the velocity

$$c_2 = \frac{\omega}{n_1} = \frac{\omega}{\sqrt{k_1^2 - m^2}} \tag{6.76}$$

To some extent, such harmonics are similar to inhomogeneous plane waves, because their phase surfaces are planes given by $\omega t - n_1|z - d| = \text{const}$. At the same time, the wave magnitude does not decay exponentially , but it changes at this surface along the radial direction as the Bessel function $J_0(mr)$. For example, the magnitude remains constant along any horizontal circle with the radius r. As follows from eq. 6.76, phase velocity c_2 increases with an increase of m and tends to infinity when $m = k_1$. In the opposite case, $m = 0$, the velocity c_2 is equal to that of a plane homogeneous wave

$$c_2 = \frac{\omega}{k_1} = c_1 \,,$$

and the amplitude becomes constant along the phase surface because $J_0(0) = 1$.

Thus, we have represented the spherical incident wave by a system of diffusive and wave harmonics. The latter propagate along the z-axis with different phase velocities and, unlike the diffusive harmonics, do not decay in this direction. For this reason, it is proper to expect that the contribution of wave harmonics, $(m < k_1)$, into the incident wave U_i becomes greater as the distance z increases.

It is essential that each harmonic of the incident wave gives rise to that of the secondary waves with the same spatial frequency m_1, eq. 6.68-6.69. In order to study the relationship between them, it is useful to distinguish two cases: $c_1 > c_2$ and $c_2 > c_1$. Correspondingly, there are three intervals of integration, shown in Fig. 6.8.

As an example, consider the harmonics of the transmitted wave $\mathcal{U}_2$, provided that $c_2 < c_1$. Then we observe the following:

1. If $m < k_1$, the wave harmonics of the incident wave produce the wave harmonics of the secondary wave $\mathcal{U}_2$.

2. If $k_1 < m < k_2$, the diffusive harmonics $\mathcal{U}_{im}$ cause the wave harmonics of $\mathcal{U}_{2m}$ which propagates without decay along the z-axis.

3. If $m > k_2$, both the incident and transmitted harmonics have the diffusion type.

The relationship between the incident and secondary harmonics is simpler in the upper medium. In fact, they both have either the wave ($m < k_1$) or diffusive ($m > k_1$) character. This analysis may allow us to reach some useful conclusions. In particular, the interval of integration

$$m > k_1 \tag{6.77}$$

does not make a significant contribution to the magnitude of the secondary waves when the source is located far away from the boundary. This happens because of the rapid decay of diffusion harmonics.

Reflected wave

Now we derive an approximate formula for the potential of the reflected wave provided that the product of distance r and spatial frequency m satisfies an inequality

$$mr \gg 1 \tag{6.78}$$

In such a case, the Bessel function $J_0(mr)$ can be represented as

$$J_0(mr) \approx \sqrt{\frac{2}{\pi mr}}\, \cos\left(mr - \frac{\pi}{4}\right)$$

or

$$J_0(mr) \approx \sqrt{\frac{1}{2\pi mr}}\left[e^{i\,(mr - \pi/4)} + e^{-i\,(mr - \pi/4)}\right] \qquad \text{if} \qquad mr \gg 1 \tag{6.79}$$

Our first goal is to obtain an approximate expression for the secondary wave in the upper medium. Substitution of eq. 6.79 into eq. 6.68 gives

$$\mathcal{U}_s \approx \frac{C}{\sqrt{2\pi r}} \int\limits_0^{\infty} \frac{\sqrt{m}}{m_1}\,\frac{m_1 - bm_2}{m_1 + bm_2}\left[\exp\left(i\left(mr - n_1(z+d) - \frac{\pi}{4}\right)\right)\right.$$
$$\left. + \exp\left(-i\left(mr + n_1(z+d) - \frac{\pi}{4}\right)\right)\right] dm \tag{6.80}$$

Here $m_1 = in_1$ and $n_1 = \sqrt{k_1^2 - m^2}$. Thus, the integrand contains rapidly oscillating functions which allow us to apply the method of stationary phase to evaluate this integral.

Consider the first exponent,

$$\varphi_1 = mr - n_1(z+d) - \frac{\pi}{4}$$

Its derivative with respect to m is

$$\frac{d\varphi_1}{dm} = r + \frac{m}{n_1}(z+d)\,,$$

since

$$\frac{dn_1}{dm} = -\frac{m}{n_1}$$

Correspondingly, the stationary point m_0 should obey the equality

$$\frac{d\varphi_1}{dm} = r + \frac{m_0}{n_1}(z+d) = 0$$

Whence

$$m_0 = -\frac{n_1 r}{z+d} \qquad \text{and} \qquad n_1 = \sqrt{k_1^2 - m_0^2} \tag{6.81}$$

Inasmuch as n_1 is either a positive, if $k_1 > m_0$, or an imaginary number when $k_1 < m_0$, the stationary point for the first exponent is located outside the interval of integration $0 \leq m < \infty$. In other words, the term of the integrand which contains $\exp i[mr - n_1(z+d) - \pi/4]$ does not make a noticeable contribution, and it can be discarded in our approximation. Therefore, in place of eq. 6.80 we have

$$\mathcal{U}_s \approx \frac{C}{\sqrt{2\pi r}} \int\limits_0^\infty \frac{\sqrt{m}}{m_1} \frac{m_1 - bm_2}{m_1 + bm_2}\, e^{-i(mr + n_1(z+d) - \pi/4)}\, dm \tag{6.82}$$

or

$$\mathcal{U}_s = \frac{C\, e^{i\pi/4}}{\sqrt{2\pi r}}\, I\,, \tag{6.83}$$

where

$$I = \int\limits_0^\infty f(m)\, e^{i\varphi(m)}\, dm\,, \tag{6.84}$$

and the latter coincides with eq. 6.59, except for the lower limit. However, this difference is not important because the integral is mainly defined by the small interval around the stationary point m_0. Of course, this conclusion is valid as soon as m_0 is not equal to zero.

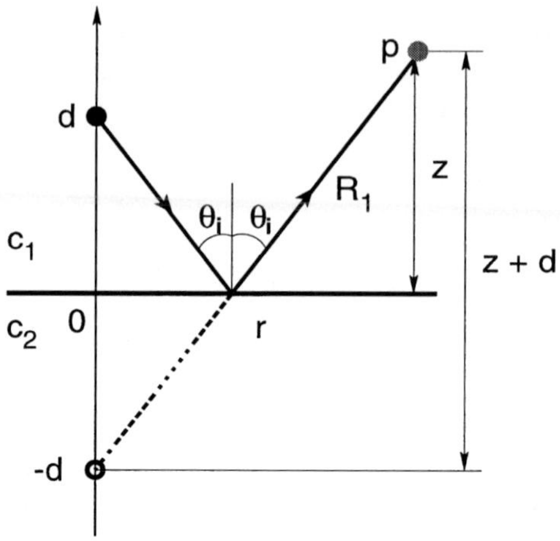

Figure 6.9: Illustration of eq. 6.86.

Now we make use of eq. 6.67 and with this purpose in mind let us find $\varphi(m_0)$, $\varphi''(m_0)$, and $f(m_0)$. As follows from eq. 6.82

$$\varphi(m) = -mr - n_1\,(z+d)\,, \qquad \varphi'(m) = -r + \frac{m}{n_1}\,(z+d)\,,$$
$$\varphi''(m) = \frac{z+d}{n_1} + \frac{m^2}{n_1^3}\,(z+d) \tag{6.85}$$

The value of the stationary point, m_0, is found from the equation:

$$\varphi'(m_0) = 0 \qquad \text{or} \qquad r = \frac{m_0\,(z+d)}{\sqrt{k_1^2 - m_0^2}}\,.$$

Hence

$$m_0 = \frac{k_1 r}{\sqrt{r^2 + (z+d)^2}}\,, \tag{6.86}$$

which clearly shows that the stationary point is defined by the wave number k_1 and the position of the observation point. For example, if $r = 0$, then $m_0 = 0$. With an increase of r the stationary point approaches k_1 and, therefore,

$$0 \leq m_0 < k_1 \tag{6.87}$$

Inasmuch as $\text{Re}\, m_1 = 0$ within this range, the incident and secondary waves are described by wave harmonics.

Now we take an important step and introduce the incident angle θ_i. Then, as is seen from Fig. 6.9

$$m_0 = k_1 \sin\theta_i \tag{6.88}$$

Here θ_i is the angle between the normal to the boundary and the line that connects the former with the source.

Thus, observation points which are located along the line R_1 are characterized by the same stationary point m_0.

By definition, we have

$$\varphi(m_0) = -m_0 r - \sqrt{k_1^2 - m_0^2}\,(z+d)$$

or

$$\varphi(m_0) = -k_1\left[r\sin\theta_i + (z+d)\cos\theta_i\right]$$

Since

$$r = R_1\sin\theta_i \qquad\qquad (z+d) = R_1\cos\theta_i\,,$$

where R_1 is the distance from the observation point p to the fictitious source, Fig. 6.9, we obtain

$$\varphi(m_0) = -k_1 R_1 \tag{6.89}$$

In accordance with eq. 6.80

$$f(m) = \frac{\sqrt{m}}{m_1}\,\frac{m_1 - bm_2}{m_1 + bm_2}$$

We also have

$$m_1 = \sqrt{m_0^2 - k_1^2} = k_1\sqrt{\sin^2\theta_i - 1} = ik_1\cos\theta_i$$

and

$$m_2 = \sqrt{m_0^2 - k_2^2} = \sqrt{k_1^2\sin^2\theta_i - k_2^2} = ik_1\sqrt{n^2 - \sin^2\theta_i}$$

Thus

$$f(m_0) = \frac{\sqrt{k_1\sin\theta_i}}{ik_1\cos\theta_i}\,\frac{\cos\theta_i - b\sqrt{n^2-\sin^2\theta_i}}{\cos\theta_i + b\sqrt{n^2-\sin^2\theta_i}}$$

or

$$f(m_0) = \frac{\beta\, e^{-i\pi/2}\sqrt{\sin\theta_i}}{\sqrt{k_1}\cos\theta_i}\,, \tag{6.90}$$

where

$$\beta = \frac{\cos\theta_i - b\sqrt{n^2 - \sin^2\theta_i}}{\cos\theta_i + b\sqrt{n^2 - \sin^2\theta_i}} \qquad \text{and} \qquad n = c_1/c_2\,, \qquad b = \rho_1/\rho_2\,.$$

Taking into account that $Z = \rho\, c$, we can write

$$\beta = \frac{Z_2\cos\theta_i - Z_1\sqrt{1 - \sin^2\theta_i/n^2}}{Z_2\cos\theta_i + Z_1\sqrt{1 - \sin^2\theta_i/n^2}} \tag{6.91}$$

In accordance with eq. 6.85, the second derivative of $\varphi(m_0)$ is

$$\varphi''(m_0) = \frac{z+d}{k_1\cos\theta_i} + \frac{k_1^2\sin^2\theta_i}{k_1^3\cos^3\theta_i}(z+d) = \frac{z+d}{k_1\cos^3\theta_i}$$

or

$$\varphi''(m_0) = \frac{R_1^3}{k_1(z+d)^2}\,, \tag{6.92}$$

and it is always positive. Then a substitution of eqs. 6.89-6.92 into eq. 6.67 yields

$$I = \frac{\sqrt{2\pi}\, e^{-i\pi/4}\, \beta\, \sqrt{\sin\theta_i}\,(z+d)}{\cos\theta_i\, R_1^{3/2}}\, e^{-i\,k_1 R_1}$$

Correspondingly, the expression for the potential, eq. 6.83, becomes

$$\mathcal{U}_s = C\,\beta\,\frac{e^{-ik_1R_1}}{R_1} \tag{6.93}$$

In general, the wave field in the upper medium may consist of the reflected and head waves. At the same time, eq. 6.93 describes the former only.

Before we continue, let us represent the phase, eq. 6.85, as

$$\varphi(m) = -k_1\, R_1\, \bar{\varphi}(m)\,,$$

where

$$\bar{\varphi}(m) = \frac{mr + n_1\,(z+d)}{k_1 R_1}$$

and

$$\text{Re}\ \bar{\varphi}(m) \leq 1\,.$$

For instance, at the stationary point $\bar{\varphi}(m_0) = 1$. This consideration shows that the product $k_1 R_1$ plays the role of a large parameter and, correspondingly, eq. 6.93 describes the wave with reasonable accuracy if

$$k_1 R_1 > 1\,, \tag{6.94}$$

that is, at the wave zone where distance R_1 exceeds the wave length.

Bearing this fact in mind, let us discuss the main features of the reflected wave. First, it is a spherical one and its center is located at the point, that is the mirror reflection of the source with respect to the interface. The term β, eq. 6.91, coincides with the reflection coefficient of the plane wave, reflected at the boundary (Chapter 5), and it depends on the ratios c_1/c_2 and ρ_1/ρ_2 and the angle θ_i. The latter is formed by the z-axis and the line connecting the fictitious source with an observation point. Therefore, unlike the amplitude of incident wave, that of the reflected wave varies on its phase surface. It is convenient to introduce the spherical system of coordinates, R_1 and θ. Then at each point components of the particle displacement are written as

$$S_{R_1} = C\,\beta\,\frac{\partial}{\partial R_1}\left(\frac{e^{-ik_1R_1}}{R_1}\right) \qquad \text{and} \qquad S_\theta = C\,\frac{e^{-ik_1R_1}}{R_1^2}\,\frac{\partial\beta}{\partial\theta}$$

This means that in general they have different magnitudes and phases. Respectively, during each period the particles of the medium move along an ellipse. Its shape, as well as an orientation, depend on the observation point and the frequency. Inasmuch as the pressure varies on the phase surface, the Poynting vector $\mathbf{Y} = P\mathbf{v}$ has both the normal and tangential components, Y_{R_1} and Y_θ. At the same time it is a simple matter to demonstrate that within each period the energy flux along the phase surface is equal to zero.

There is another remarkable feature of this wave if $kR > 1$ (Chapter 2). As is seen from Fig. 6.9, the incident and reflected angles obey Snell's law:

$$\theta_r = \theta_i \tag{6.95}$$

Now it may be appropriate to make several comments related to the behavior of the reflected wave. as well as the stationary phase method.

1. If the velocity of the lower medium is smaller, $c_2 < c_1$, the stationary point m_0 is located only in the first interval of integration, Fig. 6.8a, where the incident and reflected waves are represented by wave harmonics only.

2. In this case the amplitude of the reflected wave, βC, varies rather slowly with the angle θ and, correspondingly, the derivative $\partial\beta/\partial\theta$ has a finite value everywhere. As a result, eq. 6.93 describes correctly the reflected wave at all points of the upper medium, regardless of the value of θ_i.

3. In the case when $c_2 > c_1$ the stationary point may be located either in the first or second intervals of integration, Fig. 6.8b, and its value is defined by the position of the observation point. In particular, if it is situated along line R_1, which forms the critical angle θ_c with the z-axis, eq. 6.88 gives $m_0 = k_2$. Correspondingly, the reflection coefficient β is real, as in the previous case, provided that $\theta_i \le \theta_c$. However it becomes complex when $\theta_i > \theta_c$. In fact we have from eq. 6.91

$$\beta = \frac{Z_2 \cos\theta_i - i\sqrt{\sin^2\theta_i/n^2 - 1}}{Z_2 \cos\theta_i + i\sqrt{\sin^2\theta_i/n^2 - 1}}, \tag{6.96}$$

that is $|\beta| = 1$, and the internal reflection is observed.

4. At any position of the stationary point $m_0 \le k_2$ or $m_0 > k_2$ waves in the upper medium are represented, as before, by wave harmonics only.

5. In deriving eq. 6.93 it was assumed that the function $f(m)$ varies rather slowly. However, if the stationary point is equal to k_2, the rate of change of $f(m)$ with respect to m or $\partial\beta/\partial\theta$ tends to infinity. In other words, in such a case the method of stationary phase cannot be applied. This means that if the position of the observation point is characterized by the critical angle or close to it eq. 6.93 becomes invalid.

6. By definition, the potential $\mathcal{U}_s$ represents the total wave field. However, eq. 6.93 describes only one part, namely, the reflected wave.

7. In Chapter 2 we derived exactly the same expression of the wave field $\mathcal{U}_s$, using the Kirchhoff approach and the transport equation for the wave amplitude in the zero approximation. It may be useful to remind that this equation is based on the principle of locality. In other words, the flux of the Poynting vector inside its vector tube remains constant. Besides, it moves to an observation point from a certain zone of the boundary, which is defined by Snell's law.

Transmitted wave

Next, consider the asymptotic behavior of the transmitted wave in the lower medium. First, by analogy with the previous case of the reflected wave, we assume that $mr \gg 1$, and, correspondingly, the Bessel function $J_0(mr)$ can be replaced by its asymptotical expression, eq. 6.79. Substitution of the latter into eq. 6.69 yields

$$\mathcal{U}_2 = \frac{C}{\sqrt{2\pi r}} \left[\int_0^\infty f(m)\, e^{i\,\varphi_1(m)}\, dm + \int_0^\infty f(m)\, e^{i\,\varphi_2(m)}\, dm \right] \tag{6.97}$$

Here

$$f(m) = \frac{F_2(m)}{\sqrt{m}}\,, \tag{6.98}$$

$$F_2(m) = \frac{2\,b\,m}{m_1 + b\,m_2}\,, \tag{6.99}$$

$$\varphi_1(m) = mr - n_2|z| - n_1 d - \frac{\pi}{4}\,, \tag{6.100}$$

and

$$\varphi_2(m) = -mr - n_2|z| - n_1 d + \frac{\pi}{4}\,, \tag{6.101}$$

where $n_1 = \sqrt{k_1^2 - m^2}$ and $n_2 = \sqrt{k_2^2 - m^2}$. Performing differentiation of the functions $\varphi_1(m)$ and $\varphi_2(m)$ and equating them to zero, we arrive at equations that define the stationary point m_0:

$$\frac{d\varphi_1}{dm} = r + \frac{m_0}{n_2}|z| + \frac{m_0}{n_1}d \;=\; 0 \tag{6.102}$$

and

$$\frac{d\varphi_2}{dm} = -r + \frac{m_0}{n_2}|z| + \frac{m_0}{n_1}d \;=\; 0 \tag{6.103}$$

First, assume that the point m_0 satisfies the condition: $m_0 < k_1$ and $m_0 < k_2$. Then, terms $m_0|z|/n_2$ and $m_0 d/n_1$ are real positive numbers, and therefore function $\varphi_1(m)$ does not have a stationary point. At the same time, for the phase $\varphi_2(m)$ we have

$$\frac{m_0\,d}{\sqrt{k_1^2 - m_0^2}} + \frac{m_0\,|z|}{\sqrt{k_2^2 - m_0^2}} = r \tag{6.104}$$

As is seen from Fig. 6.10, the stationary point m_0 can be expressed in terms of the incident and transmitted angles in the following way

$$\frac{m_0}{\sqrt{k_1^2 - m_0^2}} = \tan\theta_i\,, \qquad \frac{m_0}{\sqrt{k_2^2 - m_0^2}} = \tan\theta_2 \tag{6.105}$$

The first equation from this set shows that, as in the case of the reflected wave,

$$m_0 = k_1 \sin\theta_i\,, \tag{6.106}$$

while the last equality indicates that angles θ_i and θ_2 obey Snell's law of refraction. In fact, the substitution of the ratio

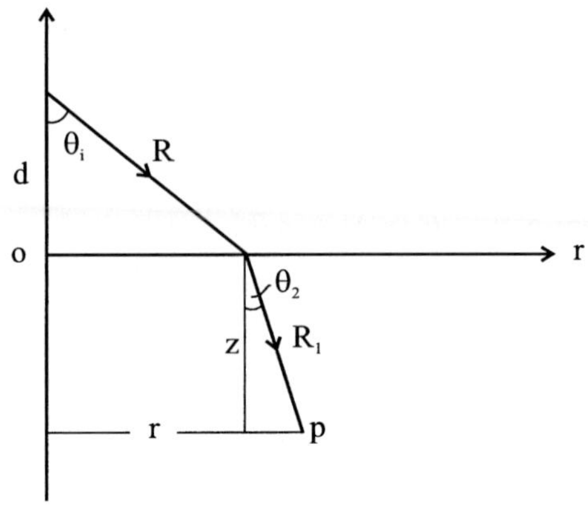

Figure 6.10: Illustration of eq. 6.104

$$\frac{\sin\theta_i}{c_1} = \frac{\sin\theta_2}{c_2}$$

gives

$$\frac{m_0}{\sqrt{k_2^2 - m_0^2}} = \frac{\sin\theta_i}{\sqrt{n^2 - \sin^2\theta_i}} = \tan\theta_2$$

It is useful to rewrite eq. 6.104 as

$$d\tan\theta_i + |z|\tan\theta_2 = r \tag{6.107}$$

Thus, discarding the first integral at the right-hand side of eq. 6.97, we obtain

$$\mathcal{U}_2 = \frac{C\,e^{i\pi/4}}{\sqrt{2\pi r}} \int\limits_0^\infty f(m)\,e^{i\,\varphi(m)}\,dm \tag{6.108}$$

where

$$\varphi(m) = -mr - n_2|z| - n_1 d \tag{6.109}$$

Before we apply eq. 6.67, it is proper to notice that the value of stationary point m_0 essentially depends on the position of the observation point. It is a simple matter to see that the function $f(m)$ varies rather slowly in the vicinity of m_0 except in the case when θ_2 tends to $\pi/2$. This means that the incident angle coincides with the critical one.

In accordance with eq. 6.101, the second derivative of the phase is

$$\frac{d^2\varphi}{dm^2} = n_2^{-1}|z| + m_0^2 n_2^{-3}|z| + n_1^{-1}d + m_0^2 n_1^{-3}d = \tag{6.110}$$

$$= n_2^{-3}|z|(n_2^2 + m_0^2) + n_1^{-3}d(n_1^2 + m_0^2) = k_2^2 n_2^{-3}|z| + k_1^2 n_1^{-3}d$$

Since $n_1 = k_1 \cos\theta_i$ and $n_2 = k_2 \cos\theta_2$, we have

$$\frac{d^2\varphi(m)}{dm^2} = k_2^{-1}\frac{|z|}{\cos^3\theta_2} + k_1^{-1}\frac{d}{\cos^3\theta_i} \tag{6.111}$$

The substitution of eq. 6.106 into eq. 6.99 yields

$$F_2(m_0) = \frac{2\,b\,\sin\theta_i\,e^{-i\pi/2}}{(\cos\theta_i + b\,n\cos\theta_2)}, \tag{6.112}$$

The expression of the function $\varphi(m_0)$ is

$$\varphi(m_0) = -k_1\,r\sin\theta_i - k_2\,|z|\cos\theta_2 - k_1\,d\cos\theta_i =$$

$$= -k_1\,(r\sin\theta_i + d\cos\theta_i) - k_2\,|z|\cos\theta_2$$

Then, making use of eq. 6.107, we have

$$\varphi(m_0) = -k_1\,[\sin\theta_i\,(d\tan\theta_i + |z|\tan\theta_2) + d\,\cos\theta_i\,] - k_2\,|z|\,\cos\theta_2 =$$

$$= -k_1\,d\,(\sin\theta_i\tan\theta_i + \cos\theta_i) - k_2\,|z|\,(\cos\theta_2 + \frac{k_1}{k_2}\sin\theta_i\tan\theta_2) =$$

$$= -k_1\,\frac{d}{\cos\theta_i} - k_2\,\frac{|z|}{\cos\theta_2}$$

As is seen from Fig. 6.10, the latter becomes

$$\varphi(m_0) = -k_1 R - k_2 R_1 \tag{6.113}$$

Substitution of $\varphi(m_0)$, $\varphi''(m_0)$, and $f(m_0)$ into eq. 6.67 and use of eq. 6.108 gives an approximate expression for the transmitted wave

$$\mathcal{U}_2 = \frac{2\,C\,e^{-ik_1R}\,e^{-ik_2R_1}}{\left(\dfrac{\cos\theta_i}{b} + n\cos\theta_2\right)} \left[\frac{r}{\sin\theta_i}\left(\frac{d}{\cos^3\theta_i} + \frac{|z|}{n\,\cos^3\theta_2}\right)\right]^{-1/2} \tag{6.114}$$

In deriving this potential, we found that the stationary point m_0 coincides with the such point for the reflected wave (eq. 6.106). For this reason, it is not surprising that the same expression of $\mathcal{U}_2$ was obtained from the principles of the geometrical acoustics.

It is clear that this wave is not spherical and its amplitude varies along the phase surface $k_1R + k_2R_1 = \text{const}$. Also, as follows from eq. 6.105, if the second medium has lower velocity, $c_2 < c_1$, the transmitted wave given by eq. 6.114 arises at all points of the boundary. However, in the opposite case, $c_2 > c_1$, this wave appears only at the points of the interface where the incident angle θ_i does not exceed the critical one. In particular, in accordance with eq. 6.114, the potential $\mathcal{U}_2$ is equal to zero when $\theta_i = \theta_c$. Of course, as was mentioned earlier, this equation becomes invalid in the vicinity of this point.

Now we investigate the behavior of transmitted waves of different types.

Evanescent transmitted waves

Until now we have considered stationary points, that are situated at the initial part of integration, shown in Fig. 6.8. Next let us pay attention to cases when they are located in a following interval of m. With this purpose in mind it is convenient to distinguish two cases: $c_2 > c_1$ and $c_2 < c_1$.

Case one First, suppose that the second medium has a higher velocity and that the stationary point m_0 is present within the interval $k_2 < m < k_1$. Since the radical m_2 becomes real, $(m > k_2)$, in place of eq. 6.109 we have

$$\varphi(m) = -mr - n_1 d \tag{6.115}$$

Therefore, the point m_0 is defined from the equation

$$\frac{d\varphi}{dm} = -r + \frac{m_0}{n_1}\,d = 0 \qquad \text{or} \qquad \frac{m_0}{\sqrt{k_1^2 - m_0^2}} = \frac{r}{d} \tag{6.116}$$

Whence

$$k_2 < m_0 = \frac{k_1}{\sqrt{1 + (d/r)^2}} < k_1$$

Bearing in mind that $\cot\theta_i = d/r$, we have again

$$m_0 = k_1\,\sin\theta_i \tag{6.117}$$

However, in this case $m_0 > k_2$, and therefore the incident angle exceeds the critical one.

Inasmuch as the function $\varphi_1(m)$, eq. 6.100, still does not have the stationary point, instead of eq. 6.108 we obtain

$$\mathcal{U}_2 = \frac{C\, e^{i\pi/4}}{\sqrt{2\pi r}} \int\limits_0^\infty \frac{F_2(m)}{\sqrt{m}}\, e^{-m_2|z|}\, e^{i\,\varphi(m)}\, dm \tag{6.118}$$

At the point m_0 the phase $\varphi(m)$ can be represented as

$$\varphi(m_0) = -k_1\, r\, \sin\theta_i - k_1\, d\, \cos\theta_i = -k_1 R \tag{6.119}$$

As follows from eq. 6.111 and 6.98

$$\frac{d^2\varphi(m_0)}{dm^2} = \frac{k_1^{-1}\, d}{\cos^3\theta_i} = \frac{k_1^{-1}\, R}{\cos^2\theta_i} \tag{6.120}$$

and

$$\frac{F_2(m_0)}{\sqrt{m_0}}\, e^{-m_2|z|} = \frac{2b\sqrt{\sin\theta_i}}{\sqrt{k_1}}\, \frac{e^{-k_1|z|\sqrt{\sin^2\theta_i - n^2}}}{\left(i\cos\theta_i + b\sqrt{\sin^2\theta_i - n^2}\right)} \tag{6.121}$$

where $\sin\theta_i > n$. Then eq. 6.67 yields

$$\mathcal{U}_2 = \frac{2\,b\,Ci}{R}\, \frac{\cos\theta_i\, e^{-k_1|z|\sqrt{\sin^2\theta_i - n^2}}}{\left(i\cos\theta_i + b\sqrt{\sin^2\theta_i - n^2}\right)}\, e^{-ik_1 R} \tag{6.122}$$

This describes the wave field beneath the boundary, sometimes called the evanescent wave. Inasmuch as the term $e^{-ik_1 R}$ can be written in the form

$$\exp\left(-i\,\frac{\omega}{c_1}\,\frac{r}{\sin\theta_i}\right),$$

the phase velocity of such "wave" varies from point to point of the boundary and it is equal to

$$c = c_1 \sin\theta_i \tag{6.123}$$

With an increase of the incident angle it tends to c_1, but near the critical angle $c = c_1^2/c_2$. In this light we note that the behavior of the displacement and the pressure, described by eq. 6.122, has a rather diffusive character. This field simply accompanies the reflected wave beyond the critical angle. A decay of the field $\mathcal{U}_2$ along the z-axis is defined by the term $\exp(-k_1|z|\sqrt{\sin^2\theta_i - n^2})$ and it becomes greater with an increase of the incident angle. Unlike in the previous case, where we derived an expression for the transmitted wave, the function F_2 is complex and, correspondingly, it influences the phase of the integrand. In our approximation, this fact was not taken into account. Nevertheless, it was possible to demonstrate that at the boundary the incident wave generates simultaneously the reflected wave as well as the diffusive field in the lower medium. Thus, the total internal reflection takes place.

It is obvious, eq. 6.101, that the stationary point is absent within the last interval of integration, $m > k_1$.

Case two Now suppose that $c_2 < c_1$. As was demonstrated previously the stationary point is within the first range, $(m < k_1)$, at $m_0 = k_1 \sin\theta_i$. An integration in its vicinity gave an expression for the transmitted wave traveling through the lower medium. In the second interval $k_1 < m < k_2$ the phase $\varphi_2(m)$, eq. 6.101, becomes equal to

$$\varphi_2(m) = -mr - n_2|z| + \frac{\pi}{4} \tag{6.124}$$

Respectively, we have

$$\frac{d\varphi_2}{dm} = -r + \frac{m_0}{n_2}|z| = 0$$

Then,

$$\frac{m_0}{\sqrt{k_2^2 - m_0^2}} = \frac{r}{|z|} \qquad \text{and} \qquad m_0 = \frac{k_2}{\sqrt{1 + z^2/r^2}} = \frac{k_2\, r}{R_2} \tag{6.125}$$

Here R_2 is the distance between the origin of coordinates, O, and the observation point. As before,

$$\varphi_2(m) = \varphi(m) + \frac{\pi}{4}$$

and

$$\varphi(m_0) = -m_0 r - n_2|z| = -\frac{k_2\, r^2}{R_2} - \frac{k_2\, z^2}{R_2} = -k_2 R_2 \tag{6.126}$$

For the second derivative of $\varphi(m_0)$, eq. 6.110, we have

$$\frac{d^2\varphi(m_0)}{dm^2} = k_2^2\, n_2^{-3}\, |z|$$

Since

$$n_2 = \sqrt{k_2^2 - m_0^2} = k_2\, \frac{|z|}{r}\, \frac{1}{\sqrt{1 + \left(\frac{z}{r}\right)^2}},$$

the latter gives

$$\varphi''(m_0) = \frac{r^3}{k_2\, z^2}\left(1 + \frac{z^2}{r^2}\right)^{3/2}$$

Introducing angle θ,

$$z = R_2 \sin\theta \qquad \text{and} \qquad r = R_2 \cos\theta\,,$$

we obtain

$$\varphi''(m_0) = \frac{R_2}{k_2 \sin^2\theta} \tag{6.127}$$

Correspondingly, the potential $\mathcal{U}_2$ is equal to

$$\mathcal{U}_2 = \frac{2\,i\,b\,C\,e^{-m_1 d}\,k_2\,\sin\theta}{(m_1 + i\,b\,n_2)}\,\frac{e^{-ik_2R_2}}{R_2} \tag{6.128}$$

Here $m_1 = \sqrt{m_0^2 - k_1^2}$ and $n_2 = \sqrt{k_2^2 - m_0^2}$.

Bearing in mind that $m_0 > k_1$, the appearance of this spherical wave can be described in the following way. The diffusive harmonics of the incident wave reach the boundary and give rise to the transmitted wave in the lower medium. The center of this spherical wave is located at the origin of the coordinates. It is obvious that with an increase of the distance between the primary source and the boundary the influence of this wave becomes smaller.

In conclusion of this section let us notice that the method of stationary phase is an approximate one and for this reason it is natural that we obtain an incomplete picture of wave fields. For instance, this approach did not allow us to study the head wave, as well as some of the evanescent waves.

6.4 Head wave and integration along branch cuts

Now we derive an asymptotic expression for a different part of the potential. For this purpose, let us use the following equality

$$J_0(mr) = \frac{1}{2}\left[H_0^{(1)}(mr) + H_0^{(2)}(mr)\right] \tag{6.129}$$

Here $H_0^{(1)}(mr)$ and $H_0^{(2)}(mr)$ are the Hankel functions of the first and second kind. As is well known their asymptotic expressions are

$$H_0^{(1)}(mr) \approx \sqrt{\frac{2}{\pi\,m\,r}}\;e^{i\,(m\,r\,-\,\pi/4)}\,, \qquad H_0^{(2)}(mr) \approx \sqrt{\frac{2}{\pi\,m\,r}}\;e^{-i\,(m\,r\,-\,\pi/4)} \tag{6.130}$$

if $mr \gg 1$.

Substitution of eqs. 6.129 into eq. 6.54 gives for the secondary wave

$$\mathcal{U}_s = \frac{C}{2} \int_0^\infty F(m)\, m\, H_0^{(1)}(mr)\, dm + \frac{C}{2} \int_0^\infty F(m)\, m\, H_0^{(2)}(mr)\, dm\,, \tag{6.131}$$

where

$$F(m) = \frac{1}{m_1} \frac{m_1 - b\,m_2}{m_1 + b\,m_2}\, e^{-m_1\,(z+d)} \tag{6.132}$$

Consider the behavior of this function on complex plane m, even though the integration in eq. 6.131 is performed along the real axis of m. As is seen from eq. 6.132, the function $F(m)$ has four singularities, namely, the branch points

$$m = \pm\, k_1 \qquad \text{and} \qquad m = \pm\, k_2$$

located on the real axis, where

$$m_1 = 0 \qquad \text{and} \qquad m_2 = 0\,,$$

respectively.

Note that $F(m)$ is finite everywhere, except at the point $m_1 = 0$, since

$$\text{Re}\; m_1 \geq 0 \qquad \text{and} \qquad \text{Re}\; m_2 \geq 0$$

Thus, $F(m)$ is analytical function on the complex plane m, except the branch points. As is well known, the integral of any analytical function along any closed path is zero, (Appendix A), and this remarkable feature is used in deriving asymptotic formulae. From eqs. 6.130 we see that the Hankel functions of the first and second order decay exponentially with an increase of m if

$$\text{Im}\; m > 0 \qquad \text{and} \qquad \text{Im}\; m < 0\,,$$

respectively. Taking into account this fact, consider the integral

$$\oint_{C_1} F(m)\, m\, H_0^{(1)}(mr)\, dm \tag{6.133}$$

along the closed path C_1 located in the first quadrant of the complex plane, Fig. 6.11a. Due to the exponential decay of $H_0^{(1)}(mr)$ for large arguments, the integral along the portion of the path with infinitely large radius, Im $m \to \infty$, tends to zero (Appendix A), and we have

$$\int_0^\infty F(m)\, m\, H_0^{(1)}(mr)\, dm + \int_{i\infty}^0 F(m)\, m\, H_0^{(1)}(mr)\, dm = 0$$

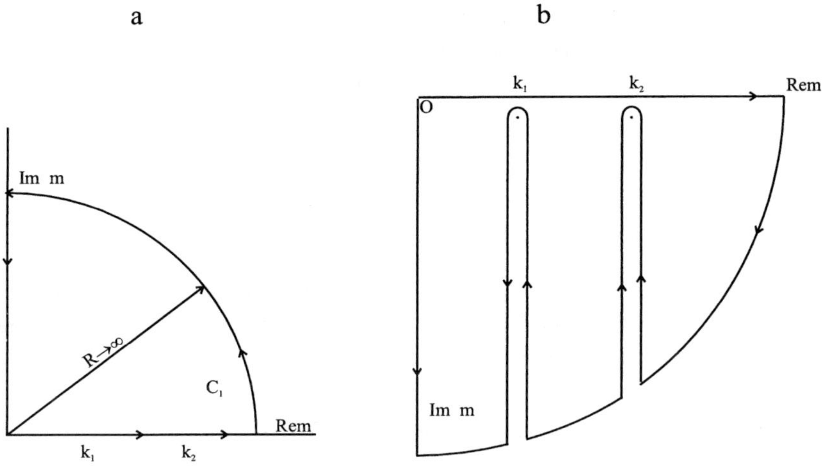

Figure 6.11: (a), (b) Contours of integration.

or

$$\int_0^{\infty} F(m)\, m\, H_0^{(1)}(mr)\, dm = \int_0^{i\infty} F(m)\, m\, H_0^{(1)}(mr)\, dm \tag{6.134}$$

Note that in deriving the asymptotical formulas we assume at the beginning that branch points have a very small imaginary part.

Next, consider the closed path C_2 located in the fourth quadrant in Fig. 6.11b. In essence, it consists of three closed contours, so that the integrand is an analytical function inside each of them. Then we also have

$$\oint_{C_2} F(m)\, m\, H_0^{(2)}(mr)\, dm = 0 \tag{6.135}$$

It is assumed that the branch points are situated slightly beneath the real axis of m. Bearing in mind the asymptotic behavior of $H_0^{(2)}(mr)$, the integral along the quarter of circle with an infinitely large radius, Im $m \to \infty$, can be neglected. This gives

$$\oint_{C_2} F(m)\, m\, H_0^{(2)}(mr)\, dm = \int_0^{\infty} \dots + \int_{k_2 - i\infty}^{k_2} \dots + \int_{k_2}^{k_2 - i\infty} \dots$$

$$+ \int_{k_1 - i\infty}^{k_1} \dots + \int_{k_1}^{k_1 - i\infty} \dots + \int_{-i\infty}^{0} \dots = 0 \tag{6.136}$$

The four integrals in the right-hand side of this equality are taken along branch lines corresponding to the branch points k_2 and k_1. Since

$$F(m) = F(-m) \qquad \text{and} \qquad H_0^{(2)}(-mr) = -\,H_0^{(1)}(mr)\,,$$

the last integral can be written as

$$\int\limits_{-i\infty}^{0} F(m)\, m\, H_0^{(2)}(mr)\, dm = \int\limits_{i\infty}^{0} F(m)\, m\, H_0^{(2)}(-mr)\, dm$$

$$= \int\limits_{0}^{i\infty} F(m)\, m\, H_0^{(1)}(mr)\, dm \tag{6.137}$$

Substituting eqs. 6.134–6.137 into eqs. 6.131, we see that the potential $\mathcal{U}_s$ is expressed in terms of integrals along the branch cuts:

$$\begin{aligned} \mathcal{U}_s &= \frac{C}{2} \int\limits_{k_1 - i\infty}^{k_1} m\, H_0^{(2)}(mr)\,[\,F(m_1,\, m_2) - F(-m_1,\, m_2)\,]\, dm\, + \\ &+ \frac{C}{2} \int\limits_{k_2 - i\infty}^{k_2} m\, H_0^{(2)}(mr)\,[\,F(m_1,\, m_2) - F(m_1,\, -m_2)\,]\, dm \end{aligned} \tag{6.138}$$

Here we used of the fact that radicals m_1 and m_2 change their signs around the branch points k_1 and k_2, respectively.

In deriving asymptotic expressions of integrals 6.138, we take into account the exponential decay of function $H_0^{(2)}(mr)$ along the cut lines, provided that

$$|m|\, r \gg 1 \qquad \text{and} \qquad \text{Im}\, m < 0$$

In other words, those integrals are defined mainly by the behavior of their integrands in the vicinity of branch points and because of this the integration paths were deformed.

Contribution of branch cut near point k_1, (Re $m = k_1$)

Let us introduce a new variable

$$m = k_1\,(1 - i\,x)\,, \tag{6.139}$$

Then

$$dm = -i\, k_1\, dx \tag{6.140}$$

and for small x, corresponding to the vicinity of the branch point, we have

$$m \approx k_1 \tag{6.141}$$

Also

$$m_1 = \sqrt{m^2 - k_1^2} = k_1 \sqrt{(1 - i\,x)^2 - 1} \approx k_1 \sqrt{-2\,i\,x} = k_1 \sqrt{2\,x}\, e^{-i\,\pi/4} \tag{6.142}$$

and

$$m_2 = \sqrt{m^2 - k_2^2} \approx \sqrt{k_1^2 - k_2^2} = k_1 \sqrt{1 - \frac{c_1^2}{c_2^2}} \qquad \text{if} \qquad x \to 0$$

or

$$m_2 \approx k_1\, \mu\,, \tag{6.143}$$

where

$$\mu = \sqrt{1 - \frac{c_1^2}{c_2^2}} \tag{6.144}$$

As follows from eq. 6.130

$$H_0^{(2)}(mr) \approx \sqrt{\frac{2}{\pi\, m\, r}}\, e^{-i\,(k_1\, r\, -\, \pi/4)}\, e^{-k_1\, r\, x} \tag{6.145}$$

The difference of function F, eq. 6.138, can be represented as

$$\Delta F = F(m_1,\, m_2) - F(-m_1,\, m_2)$$

$$= \frac{1}{m_1} \left[\frac{m_1 - b\, m_2}{m_1 + b\, m_2}\, e^{-m_1\,(z\,+\,d)} + \frac{m_1 + b\, m_2}{m_1 - b\, m_2}\, e^{m_1\,(z\,+\,d)} \right],$$

because m_1 has opposite signs at both branch cuts. After simple manipulations we obtain

$$\Delta F = \frac{2\left[\,(m_1^2 + b^2\, m_2^2)\, \cosh m_1\,(z + d) + 2\, b\, m_1\, m_2\, \sinh m_1\,(z + d)\,\right]}{m_1\,(m_1^2 - b^2\, m_2^2)} \tag{6.146}$$

As follows from eqs. 6.142-6.144

$$m_1^2 - b^2\, m_2^2 = 2\, k_1^2\, x\, e^{-i\pi/2} - k_1^2\, b^2\, \mu^2 = k_1^2 \left(2\, x\, e^{-i\pi/2} - b^2\, \mu^2 \right)$$

Also

$$m_1^2 + b^2\, m_2^2 = k_1^2 \left(2\, x\, e^{-i\,\pi/2} + b^2\, \mu^2 \right)$$

and

$$2\, b\, m_1\, m_2 \; = \; 2\, b\, k_1^2\, \mu\, \sqrt{2\, x}\, e^{-i\pi/4}$$

Correspondingly, function ΔF becomes

$$\begin{aligned} \Delta F \;&\approx\; \frac{2\, e^{i\pi/4}}{-k_1^3\, b^2\, \mu^2\, \sqrt{2\, x}} \left[k_1^2\, b^2\, \mu^2\, \cosh \alpha \sqrt{x} + 2\, k_1^2\, b\, \mu\, \sqrt{2\, x}\, e^{-i\pi/4}\, \sinh \alpha \sqrt{x} \right] \\ &= \; -\frac{2\, e^{i\pi/4}}{k_1\, \sqrt{2\, x}} \left(\cosh \alpha \sqrt{x} + \frac{2\, \sqrt{2\, x}\, e^{-i\pi/4}}{b\, \mu}\, \sinh \alpha \sqrt{x} \right) \end{aligned} \qquad (6.147)$$

$$\text{if} \quad x \rightarrow 0\,,$$

where

$$\alpha = \sqrt{2}\, k_1\, e^{-i\pi/4}\, (z + d) \qquad (6.148)$$

In accordance with eq. 6.138

$$\mathcal{U}_s = \mathcal{U}_s^{(1)} + \mathcal{U}_s^{(2)} \qquad (6.149)$$

Here

$$\mathcal{U}_s^{(1)} = \frac{C}{2} \int\limits_{k_1 - i\,\infty}^{k_1} m\, \Delta F\, H_0^{(2)}(mr)\, dm\,, \qquad \mathcal{U}_s^{(2)} = \frac{C}{2} \int\limits_{k_2 - i\,\infty}^{k_2} m\, \Delta F\, H_0^{(2)}(mr)\, dm \qquad (6.150)$$

From eqs. 6.139, and 6.142 - 6.145 we obtain

$$\mathcal{U}_s^{(1)} \approx -\frac{C\, i\, k_1^2}{2} \int\limits_{\infty}^{0} \Delta F\, H_0^{(2)}(xr)\, dx \approx \frac{C\, i\, k_1^{3/2}}{\sqrt{2\, \pi\, r}}\, e^{-i\, k_1\, r}\, e^{i\, \pi/4} \int\limits_0^{\infty} \Delta F\, e^{-k_1\, r\, x}\, dx$$

Substitution of eqs. 6.147 gives

$$\mathcal{U}_s^{(1)} \approx -\frac{\sqrt{k_1}\, e^{-ik_1 r}}{\sqrt{\pi\, r}}\, C \left[\int\limits_0^{\infty} \frac{\cosh \alpha \sqrt{x}}{\sqrt{x}}\, e^{-k_1 r x}\, dx + \frac{2^{3/2}\, e^{-i\,\frac{\pi}{4}}}{b\, \mu} \int\limits_0^{\infty} e^{-k_1 r x}\, \sinh \alpha \sqrt{x}\, dx \right] \qquad (6.151)$$

Introducing another new variable, $y = \sqrt{x}$, we obtain

$$dx = 2\sqrt{x}\,dy = 2\,y\,dy\,,$$

Therefore,

$$\mathcal{U}_s^{(1)} \approx -\frac{2\,C\,\sqrt{k_1}\,e^{-i\,k_1\,r}}{\sqrt{\pi\,r}}\left(I_1 + \frac{2^{3/2}\,e^{-i\,\frac{\pi}{4}}}{b\,\mu}\,I_2\right) \tag{6.152}$$

Here

$$I_1 = \int_0^\infty e^{-k_1\,r\,y^2}\cosh\alpha y\,dy \qquad \text{and} \qquad I_2 = \int_0^\infty y\,e^{-k_1\,r\,y^2}\sinh\alpha y\,dy$$

The value of the first integral is well known

$$I_1 = \frac{1}{2}\sqrt{\frac{\pi}{k_1\,r}}\,\exp\left(\frac{\alpha^2}{4\,k_1\,r}\right) \tag{6.153}$$

At the same time the second integral can be represented as

$$I_2 = \frac{\partial\,I_1}{\partial\,\alpha}\,,$$

that is

$$I_2 = \sqrt{\frac{\pi}{k_1\,r}}\,\frac{\alpha}{4\,k_1\,r}\,\exp\left(\frac{\alpha^2}{4\,k_1\,r}\right),$$

where α is given by eq. 6.148.

Thus, we have

$$\mathcal{U}_s^{(1)} \approx -C\sqrt{\frac{k_1}{\pi\,r}}\,e^{-i\,k_1\,r}\sqrt{\frac{\pi}{k_1\,r}}\,\exp\left(\frac{\alpha^2}{4\,k_1\,r}\right)\left(1 + \frac{\alpha}{k_1\,r}\,\frac{\sqrt{2}}{b\,\mu}\,e^{-i\,\frac{\pi}{4}}\right)$$

$$= -C\,\frac{e^{-i\,k_1\,r}}{r}\left[1 - \frac{2\,i\,(z+d)}{r\,b\,\mu}\right]\exp\left(\frac{\alpha^2}{4\,k_1\,r}\right) \tag{6.154}$$

if $\quad k_1 r \gg 1$

Also that

$$\frac{z+d}{r} \ll 1\,, \tag{6.155}$$

we can disregard the second term in parenthesis in the right-hand side of eq. 6.154 and obtain

$$\mathcal{U}_s^{(1)} \approx -\frac{C}{r}\,\exp\left\{-i\,k_1\left[r+\frac{(z+d)^2}{2\,r}\right]\right\} \tag{6.156}$$

The latter describes the reflected wave when condition 6.155 is met. In fact, the distance from the mirror reflection of the source, R_1, is

$$R_1 = \sqrt{r^2+(z+d)^2} \approx r+\frac{(z+d)^2}{2\,r}$$

As we might expect, this wave obeys Snell's law and its amplitude decreases as R^{-1}, since r is approximately equal to R_1 in this case. We see that the reflected and incident waves are shifted in phase by π. The same result follows from the reflection coefficient for large incidence angles θ. Comparison of eq. 6.93 and 6.156 shows that the method of stationary phase allows us to describe the reflected wave correctly in a more broad range of the incidence angles than does the approach based on evaluation of integrals along the branch cut Re $m = k_1$.

Contribution of branch point k_2

Now we evaluate the integral given by the second equation of set 6.150, that is, the function $\mathcal{U}_s^{(2)}$. As before, we assume that the integral is mainly defined by the behavior of the integrand near the branch point. By analogy with the previous case, let us introduce a new variable x:

$$m = k_2 - i\,k_1\,x \tag{6.157}$$

Whence

$$dm = -i\,k_1\,dx \tag{6.158}$$

and for small values of x

$$m \approx k_2$$

Also,

$$m_1 = \sqrt{m^2-k_1^2} \approx \sqrt{k_2^2-k_1^2} = i\,k_1\,\mu\,, \tag{6.159}$$

where μ is given by eq. 6.144. At the same time,

$$m_2 = \sqrt{m^2-k_2^2} = \sqrt{(k_2-i\,k_1\,x)^2-k_2^2} \approx \sqrt{-2\,i\,k_1\,k_2\,x} = k_1\sqrt{2\,x\,\frac{c_1}{c_2}}\,e^{-i\pi/4} \tag{6.160}$$

and

$$H_0^{(2)}(mr) \approx \sqrt{\frac{2\,c_2}{c_1\,\pi\,k_1\,r}}\; e^{-i\,(k_2\,r\,-\,\pi/4)}\, e^{-k_1\,r\,x} \tag{6.161}$$

By definition the function ΔF is equal to

$$\begin{aligned} \Delta F &= \frac{e^{-m_1\,(z\,+\,d)}}{m_1}\left[\frac{m_1 - b\,m_2}{m_1 + b\,m_2} - \frac{m_1 + b\,m_2}{m_1 - b\,m_2}\right] \\ &= -e^{-m_1\,(z\,+\,d)}\,\frac{4\,b\,m_2}{m_1^2 - b^2\,m_2^2} \end{aligned} \tag{6.162}$$

Making use of eqs. 6.157–6.162, we have

$$\begin{aligned} \Delta F &\approx -\frac{4\,b\sqrt{2\,x\,\dfrac{c_1}{c_2}}\;e^{-i\,\pi/4}\,e^{-i\,k_1\,\mu\,(z\,+\,d)}\,k_1}{-k_1^2\,\mu^2 - b^2\,k_1^2\left(2\,x\,\dfrac{c_1}{c_2}\right)e^{-i\pi/2}} \\ &\approx \frac{4\,b\sqrt{2\,x\,\dfrac{c_1}{c_2}}\;e^{-i\pi/4}\,e^{-i\,k_1\,\mu\,(z\,+\,d)}}{k_1\,\mu^2} \end{aligned} \tag{6.163}$$

Substitution of eqs. 6.161 and 6.163 into the expression for $\mathcal{U}_s^{(2)}$ gives

$$\begin{aligned} \mathcal{U}_s^{(2)} &\approx -\frac{C}{2}\,i\,k_1\,k_2\,\frac{4\,b\sqrt{2\,\dfrac{c_1}{c_2}}\;e^{i\,k_1\,\mu\,(z\,+\,d)}}{k_1\,\mu^2}\sqrt{\frac{2\,c_2}{c_1\,\pi\,k_1\,r}}\;e^{i\,k_2\,r}\int\limits_0^\infty \sqrt{x}\,e^{-k_1\,r\,x}\,dx \\ &= -\frac{4\,C\,i\,b\,k_2}{\sqrt{\pi\,k_1\,r}}\,\frac{\exp\{i\,[k_2\,r + k_1\,\mu\,(z+d)]\}}{\mu^2}\int\limits_0^\infty \sqrt{x}\,e^{-k_1\,r\,x}\,dx \end{aligned} \tag{6.164}$$

Letting

$$\sqrt{x} = y\,,$$

the integral in the last equation becomes

$$I_3 = 2 \int_0^\infty y^2 e^{-k_1 r y^2}\, dy\,, \tag{6.165}$$

because

$$dx = 2\, y\, dy$$

This integral is tabular

$$I_3 = \frac{\sqrt{\pi}}{2\,(k_1\, r)^{3/2}} \tag{6.166}$$

Whence

$$\mathcal{U}_s^{(2)} \approx -\frac{2\, i\, C\, b\, k_2}{k_1^2\, \mu^2\, r^2}\, \exp\{-i\,[k_2\, r + k_1\, \mu\,(z+d)]\} \tag{6.167}$$

The latter characterizes the wave that differs essentially from the reflected one. First, the exponential function depends on both wavenumbers k_1 and k_2. This indicates that the wave path is located in the upper and lower media, where the wave propagates with the velocities c_1 and c_2, respectively. To understand such behavior, it is reasonable to consider two cases, namely,

$$c_1 > c_2 \qquad \text{and} \qquad c_1 < c_2$$

In the first case, parameter μ becomes imaginary, and letting

$$\mu = \sqrt{1 - \frac{c_1^2}{c_2^2}} = -i\,|\mu|\,,$$

we obtain

$$\mathcal{U}_s^{(2)} \approx \frac{2\, i\, C\, b\, k_2\, e^{i\, k_2\, r}}{k_1^2\, |\mu^2|\, r^2}\, e^{-k_1\, |\mu|\, (z+d)} \tag{6.168}$$

This means that the wave, propagating in the lower medium and decreasing with the distance from the source as $1/r^2$, generates an evanescent wave in the upper medium. The latter exponentially decays with an increase of z.

In the second case, $c_1 < c_2$, we represent the exponent

$$k_2\, r + k_1\, \mu\,(z+d)$$

with real μ in the following form:

$$k_2\, r + k_1\, \mu\,(z+d) \;=\; \omega \left[\frac{r}{c_2} + \frac{z+d}{c_1}\sqrt{1 - \frac{c_1^2}{c_2^2}}\,\right] = \omega \left[\frac{r}{c_2} + \frac{z+d}{c_1}\,\cos\theta_c\right]$$

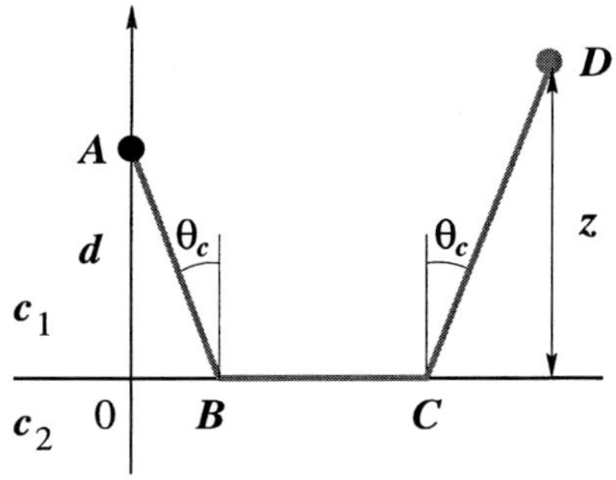

Figure 6.12: Geometry of rays of the head wave.

$$= \omega \left[\frac{r-(z+d)\,\tan\theta_c}{c_2} + \frac{z+d}{c_1}\left(\cos\theta_c + \frac{c_1}{c_2}\tan\theta_c\right)\right]$$

$$= \omega \left[\frac{r-(z+d)\,\tan\theta_c}{c_2} + \frac{(z+d)}{c_1}\,\frac{\cos^2\theta_c+\sin^2\theta_c}{\cos\theta_c}\right] \qquad (6.169)$$

$$= \omega \left[\frac{r-(z+d)\,\tan\theta_c}{c_2} + \frac{z+d}{c_1\,\cos\theta_c}\right]$$

The expression in brackets describes the time the wave takes to travel along the path shown in Fig.6.12. In fact, we have $AB = d/\cos\theta_c$ and $CD = z/\cos\theta_c$, that is, the path in the upper medium is

$$AB + CD = \frac{z+d}{\cos\theta_c}$$

The length of the path in the lower medium is

$$BC = r - (d+z)\,\tan\theta_c$$

Here θ_c is the critical angle

$$\theta_c = \sin^{-1}\frac{c_1}{c_2} \qquad \text{if} \qquad c_2 > c_1$$

As follows from eq. 6.167, the amplitude of the wave decreases as $1/r^2$, that is, more rapidly than those of the incident and reflected waves. In this sense, we see some analogy with Lloyd mirror effect described in section 6.1. As is well known, this wave is called the head wave. In order to visualize its appearance, it may be useful to apply the Huygens principle (Part I). Suppose that the wave propagates along the boundary in the lower medium. In accordance with this principle, each point of the boundary can be treated

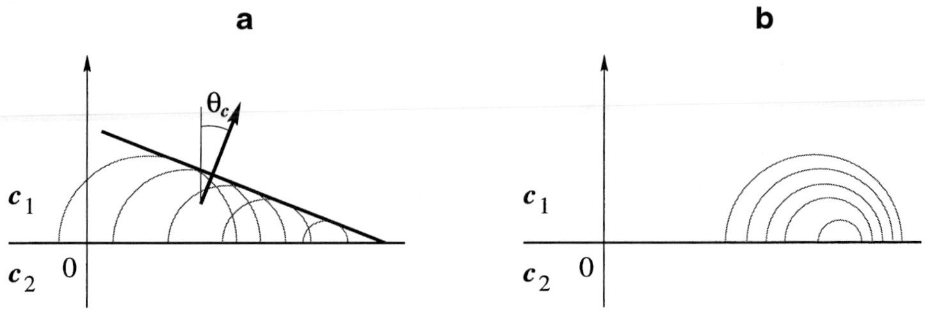

Figure 6.13: Wavefront in the case $c_1 < c_2$ and $c_1 > c_2$.

as the source of the secondary spherical wave in the upper medium. When the velocity c_2 exceeds c_1, the elementary wavefronts overlap and form the envelope that represent the front of the head wave. Its ray direction is defined by critical angle θ_c, Fig. 6.13a. If $c_1 > c_2$, however, the wavefronts do not overlap each other and their envelope is not formed, Fig. 6.13b. In such a case, wave propagation in the upper medium is almost absent; instead, an inhomogeneous wave appears, and it exponentially decays with the distance from the interface.

In accordance with eq. 6.167, rays of the head wave are parallel to each other and they form with the z-axis angle equal to θ_c. Correspondingly, taking into account the axial symmetry, the phase surfaces of this wave are conical and their apexes are situated at the z-axis, Fig. 6.14b. Inasmuch as the potential is inversely proportional to the square of r, the wave fields vary on each phase surface and decrease with an increase of the distance from the z-axis. In particular, this means that a motion of particles is characterized, as in the case of the reflected and transmitted waves, by elliptical polarization. However, with an increase of the distance or frequency, this effect becomes weak and the linear polarization is practically observed. At the same time, the vertical coordinate, z, makes an influence on the phase only.

We note the following.

(a) We were able to predict the existence of the head waves in Part I, proceeding from the principles of the geometrical acoustics. The solution of the boundary value problem, however, allows us to prove their presence and to find the approximate relationship between the amplitude of head waves on one hand and the distance r and the wave frequency ω on the other.

(b) Inasmuch as the amplitude is inversely proportional to the frequency it is natural that the head waves cannot be described by the zero term of the asymptotic series.

(c) All asymptotic formulae derived in this section imply that the observation point is located in the wave zone, $R > \lambda$.

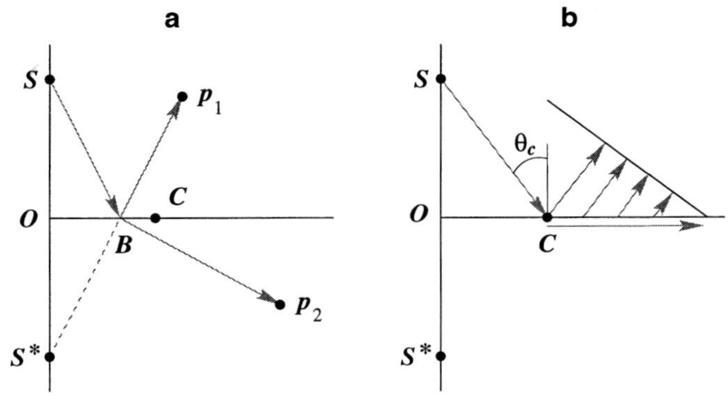

Figure 6.14: (a) Two areas of interface (b) Head wave.

6.5 Wave and near zones

First, proceeding from asymptotic formulae for potential $\mathcal{U}_s$, we describe the wave behavior at sufficiently large distances, $R > \lambda$, from the source. The most important result of this study is the fact that it is possible to make use to a great extent the concepts of the geometrical acoustics and the theory of plane waves. For instance, these equations suggest that, except in some cases, the energy propagates along the directions defined by the rays. As usually consider the two cases separately, namely,

$$c_2 > c_1 \qquad \text{and} \qquad c_1 > c_2$$

Case one: $c_2 > c_1$ Rays of the incident wave arrive at the boundary at different angles and it is convenient to distinguish two areas of the interface, located inside and outside the circle OC respectively. At points of this circle, the incident angle coincides with the critical one, Fig. 6.14. In accordance with the direction of rays of the incident wave the apparent velocity c_a along the boundary varies within the first and second intervals as

$$\infty > c_a > c_2 \qquad \text{and} \qquad c_2 > c_a > c_1$$

At the beginning consider the ray intersecting the boundary at some point B, Fig. 6.14a. In this case, it gives rise to the reflected and transmitted waves. The direction of their rays is defined by Snell's law, and their amplitudes can be evaluated by using reflection and transmission coefficients derived for plane waves. Thus, at point p_1 in the upper medium, the spherical direct (incident) and reflected waves arrive. In contrast, the transmitted wave in the lower medium, which is represented by the ray Bp_2 in Fig. 6.14a, is not spherical. We remember that amplitudes of the reflected and transmitted waves vary on the corresponding phase surfaces (wavefronts). Of course, all three waves have the same apparent velocity c_a at each point of the boundary and it varies with a change in direction of the incident ray.

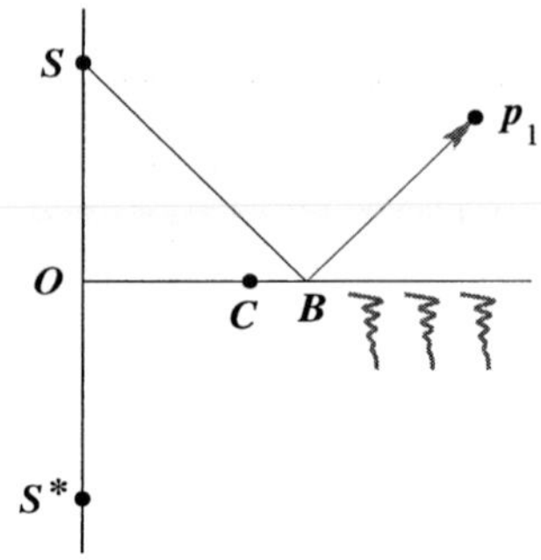

Figure 6.15: The internal reflection.

Now suppose that the point B coincides with C, that is, the incident ray makes the critical angle θ_c with the normal to the boundary. Then, as follows from Snell's law, this wave also generates the reflected and transmitted waves. It is essential that the phase surface of the latter is normal to the boundary, and its apparent velocity coincides with c_2. Wave propagation along the boundary is accompanied by vibrations of particles near the interface, which generates the head wave in the upper medium, Fig. 6.14b. As follows from Snell's law, the angles between all rays of this wave and the z-axis are equal to θ_c, that is they are parallel to each other. As was pointed out earlier, the head wave has conical phase surfaces with the apexes located at the z-axis. It is clear that the apparent velocity of the head wave along the boundary is equal to c_2.

Note that our description of the wave behavior at points of the boundary where the angle of incidence is close to the critical one, is very approximate and it is related to several factors. First of all, in such a place the zero term of the Debye series is not sufficient to characterize the reflected and transmitted waves. Secondary, the expression for the head wave is valid only at relatively large distances from the z-axis. Finally, in the vicinity of points where $\theta_i = \theta_c$ the evanescent wave arises in the lower medium and it cannot be described by the Debye series.

Next, consider the effect caused by rays of the incident wave crossing the second interval, that is when $OB > OC$. Since $\theta > \theta_c$, the internal reflection is observed, whereas the reflected wave is still a spherical one, Fig. 6.15. Amplitudes of the incident and reflected waves are equal to each other at the boundary, that is all incident energy transforms to the reflected one during one period. Also, there is a wave field which decays exponentially with the distance from the boundary, and it is caused by the part of the source energy that penetrates into the lower medium within the interval $OB > OC$. Moreover, the head wave is accompanied by a transmitted wave, which propagates in the second medium with velocity c_2. Thus we have two groups of waves. The first group includes the incident and reflected waves in the upper medium as well as the evanescent wave in the lower medium. The second group consists of the transmitted wave propagating in the lower medium along the boundary and the head wave in the upper medium. As an

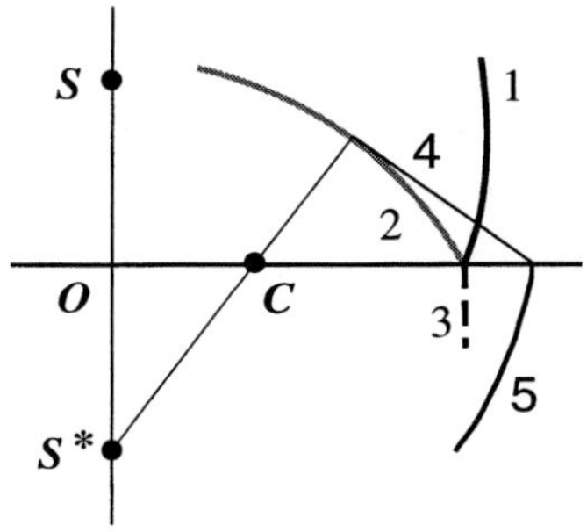

Figure 6.16: Phase surfaces: 1 – direct, 2 – reflected, 3 – evanescent 4 – head, 5 – transmitted waves.

illustration, the positions of the phase surfaces of waves are shown in Fig. 6.16. The lower edge of the front of the head wave coincides with that of the wave propagating in the lower medium. At the same time, its upper edge is formed by the intersection with the line S^*C.

Case two: $c_1 > c_2$ In this case, we observe again the reflected and transmitted waves, however, the head wave is absent. Instead, an evanescent wave arises in the upper medium, and it decays exponentially with the distance from the boundary. The existence of this wave may be explained in the following way. The diffusive harmonics of the incident spherical wave which have phase velocities between c_1 and c_2 generate the wave harmonics in the lower medium. Propagating along the boundary, these harmonics excite evanescent waves in the upper medium. These waves cannot be described, of course, by the leading term of the Debye expansion.

Near zone

Until now we examined waves at distances from the source that exceed the wavelength in both media. Next consider the opposite case and suppose that $R \ll \lambda$. In order to find asymptotic expression for the potential, let us assume that

$$k_1 \to 0 \qquad \text{and} \qquad k_2 \to 0$$

Then, in accordance with eqs. 6.54 and 6.55, we have

$$\mathcal{U}_{1s} \approx \frac{C}{R}\frac{1-b}{1+b} \qquad \text{and} \qquad \mathcal{U}_2 \approx \frac{2b}{1+b}\frac{C}{R_1} \tag{6.170}$$

where

$$R_1 = \sqrt{(z+d)^2 + r^2}$$

These equations clearly show that in the near zone, $R \ll \lambda$, the secondary fields change synchronously with the primary one, and the effect of propagation is negligible.

6.6 Wave asymptotics and the steepest descent method

Now we improve some asymptotic formulae derived earlier for wave fields by using the saddle point method (the steepest descent one). At the beginning we study the wave behavior in the upper part of a medium, where the primary source is located.

Reflected wave

In solving the boundary value problem (section 4.2), the function describing the incident wave was represented as

$$\frac{e^{i\,k_1 R}}{R} = \int\limits_0^\infty \frac{m}{m_1}\, e^{-m_1\,|z-d|}\, J_0(mr)\, dm$$

It is also convenient to apply a different form of Sommerfeld integral, namely

$$\frac{e^{i\,k_1 R}}{R} = \frac{i}{2}\int\limits_{-\infty}^{\infty} H_0^{(1)}(mr)\,\exp(\,i\,n_1\,|z-d|\,)\,\frac{m}{n_1}\,dm\,, \tag{6.171}$$

where $H_0^{(1)}(mr)$ is the Hankel function of the zero order and

$$n_1 = (k_1^2 - m^2)^{1/2}$$

Let us assume that the pressure of the incident wave, P_i, is given by eq. 6.171. Then, in accordance with eq. 6.54 the pressure of the reflected wave is

$$P_r = \frac{i}{2}\int\limits_{-\infty}^{\infty} V(m)\,H_0^{(1)}(mr)\,\exp\left[\,i\,n_1\,(z+d)\,\right]\frac{m}{n_1}\,dm\,, \tag{6.172}$$

where

$$V(m) = \frac{n_1 - b\,n_2}{n_1 + b\,n_2} \tag{6.173}$$

and

$$b = \frac{\rho_1}{\rho_2}\,, \qquad n_2 = (k_2^2 - m^2)^{1/2}$$

In the following discussion we limit ourselves to finding the leading and correction terms which describe the wave field in the wave zone. For this reason, we make use of the asymptotic expression of the Hankel function for large arguments, containing only two terms:

$$H_0^{(1)}(u) = \left(\frac{2}{\pi u}\right)^{1/2} e^{iu}\, e^{-i\pi/4}\left[1 - \frac{i}{8u} + O\left(\frac{1}{u^2}\right)\right], \tag{6.174}$$

where $O\left(\frac{1}{u^2}\right)$ is a small quantity of the order u^{-2}. Introducing a dimensionless variable of integration

$$q = \frac{m}{k_1}, \tag{6.175}$$

eq. 6.172 becomes

$$P_r = \left(\frac{k_1}{2\pi r}\right)^{1/2} e^{i\pi/4} \int_{-\infty}^{\infty} F(q) \exp\left[i\, k_1 R_1\, f(q)\right] dq \tag{6.176}$$

Here

$$R_1 = \left[r^2 + (z+d)^2\right]^{1/2}, \tag{6.177}$$

$$f(q) = q \sin\theta_0 + \sqrt{1-q^2}\,\cos\theta_0\,, \tag{6.178}$$

$$\sin\theta_0 = \frac{r}{R_1}, \qquad \cos\theta_0 = \frac{z+d}{R_1},$$

and

$$F(q) = \sqrt{\frac{q}{1-q^2}}\left[1 - \frac{i}{8k_1 r q} + O\left(\frac{1}{k_1^2 r_1^2}\right)\right] V(q)$$

or

$$F(q) = \sqrt{\frac{q}{1-q^2}}\left[1 - \frac{i}{8k_1 r q} + O\left(\frac{1}{k_1^2 r_1^2}\right)\right] \frac{\sqrt{1-q^2} - b\sqrt{n^2-q^2}}{\sqrt{1-q^2} + b\sqrt{n^2-q^2}} \tag{6.179}$$

and, as before,

$$n = \frac{c_1}{c_2}$$

Thus, the Fresnel coefficient, eq. 6.173, is expressed in terms of density ratio, b, and refraction index n, as well as the variable of integration, q. The angle θ_0 has a meaning of a specular angle, and it is formed by the normal to the boundary and the line, connecting the mirror reflection of the source with the observation point, Fig. 6.2. From the conditions

$$\operatorname{Im}\, n_1 \geq 0 \qquad \text{and} \qquad \operatorname{Im}\, n_2 \geq 0\,,$$

formulated in section 4.2, it follows that the values of square roots of $1-q^2$ and n^2-q^2

in eq. 6.179 are uniquely determined for any q by the requirement that imaginary parts of the square roots are non-negative.

With the large parameter, $k_1 R_1 \gg 1$, entering the integrand as a factor in the exponential, the integral in eq. 6.176 has the standard form, which allows one to apply the saddle point method (Appendix C). By definition, the stationary points are determined by the equation

$$\frac{d\,f}{d\,q} = 0 \tag{6.180}$$

Differentiation of eq. 6.178 gives

$$\sin\theta_0 - \frac{q}{\sqrt{1-q^2}}\cos\theta_0 = 0$$

Whence, the saddle point is equal to

$$q_0 = \sin\theta_0 \tag{6.181}$$

and is located on the path of integration in eq. 6.176. Note that at the stationary point q_0 we have

$$\frac{d^2 f}{d\,q^2} = -\cos^{-2}\theta_0 \neq 0$$

and such points are called simple.

Deformation of the integration path

Our next goal is to find on the complex plane of the variable q the path of the steepest descent, γ, which has three important features, namely (Appendix C),

(a) it passes through the saddle point,

(b) the real part of $f(q)$ remains constant along this contour,

(c) the imaginary part of $f(q)$ increases with the distance from the saddle point,

$$\text{Im } f(q) \geq \text{Im } f(\sin\theta_0)$$

It is obvious that along such a path the exponent

$$\exp\left[\,i\,k_1\,R_1\,f(q)\,\right]$$

rapidly becomes smaller with the increase of the distance from the saddle point and at the same time it does not oscillate. Inasmuch as at the stationary point, $q = \sin\theta_0$, we have

$$f(\sin\theta_0) = 1$$

or

$$\text{Re } f(\sin\theta_0) = 1 \qquad \text{and} \qquad \text{Im } f(\sin\theta_0) = 0\,,$$

the equation of the steepest descent path, γ, can be written as

$$q \sin\theta_0 + \sqrt{1-q^2}\cos\theta_0 \;=\; 1 + i\,s^2\,, \tag{6.182}$$

where q is a complex number and

$$-\infty \,<\, s \,<\, \infty$$

It is useful to visualize the shape of the path γ in the complex plane. First, let us find possible intersections of γ with the imaginary axis, where $q = \pm i\,|q|$. Its substitution into eq. 6.182 yields

$$\pm i\,|q|\,\sin\theta_0 + \sqrt{1+|q|^2}\,\cos\theta_0 \;=\; 1 + i\,s^2\,,$$

that is,

$$\sqrt{1+|q|^2}\,\cos\theta_0 = 1$$

Then

$$|q| = \tan\theta_0$$

Taking into account that $s^2 > 0$, we have

$$\text{Im } q\,\sin\theta_0 \,>\, 0\,,$$

and therefore the path γ intersects the imaginary axis only once at the point

$$q = i\,\tan\theta_0 \tag{6.183}$$

As follows from eq. 6.181 this path intersects the real axis at the saddle point

$$q_0 = \sin\theta_0 < 1$$

Besides, there is another point of intersection when $q > 1$. In fact, from eq. 6.182 we have

$$q\,\sin\theta_0 \pm i\,\sqrt{q^2-1}\cos\theta_0 \;=\; 1 + i\,s^2\,,$$

Whence

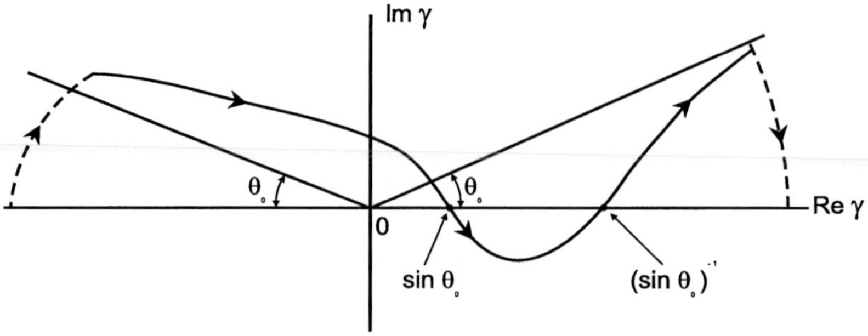

Figure 6.17: Original integration path and the steepest descent path γ.

$$q = (\sin\theta_0)^{-1} > 1 \tag{6.184}$$

Finally, at large values of $|q|$

$$\sqrt{1-q^2} \approx \pm i\,q \qquad \text{and} \qquad s \gg 1$$

Respectively, eq. 6.182 becomes

$$q\,\sin\theta_0 \pm i\,q\,\cos\theta_0 = i\,s^2 \qquad \text{or} \qquad \pm\,q\,\cos\theta_0 - i\,q\,\sin\theta_0 = s^2$$

Thus,

$$q\,e^{-i\theta_0} = s^2 \qquad \text{or} \qquad q\,e^{i\theta_0} = s^2\,e^{i\pi}$$

Since

$$q = |q|\,e^{i\,\arg q}\,,$$

we obtain

$$\arg q = \theta_0 \qquad \text{or} \qquad \arg q = \pi - \theta_0 \qquad \text{if} \quad |q| \gg 1$$

and they define the asymptotes of the path γ. Its sketch is shown in Fig. 6.17.

Next we need to carefully determine when the integral along the initial contour (the line Im $q = 0$ of the complex plane) can be replaced by the integral along γ. According to Cauchy's theorem, an integration contour can be deformed in the complex plane without a change of the integral of an analytical function, if the end points of the original and the deformed paths coincide and there are no singularities between these paths (Appendix A). To meet the first requirement, we connect the original integration contour and the steepest descent path γ by circular arcs in the first and second quadrants of the γ plane, as is shown in Fig. 6.17 (dashed lines). Now we demonstrate that the integrals along the two arcs tend to zero when their radius increases unlimitedly. Bearing in mind that

$$\text{Im}\,(1-q^2)^{1/2} \geq 0\,,$$

we have

$$(1-q^2)^{1/2} \approx i\,q \qquad \text{and} \qquad (1-q^2)^{1/2} \approx -i\,q \qquad \text{if} \qquad q \to \infty$$

in the first and second quadrants, respectively. From eq. 6.178 it follows that in these quadrants

$$f(q) = i\,q\,\exp(-i\,\theta_0) \qquad \text{and} \qquad f(q) = -i\,q\,\exp(i\,\theta_0)$$

Inasmuch as the argument of q varies from θ_0 to 0 on the arc in the first quadrant and from π to $\pi - \theta_0$ on the arc in the second quadrant, the imaginary part of $f(q)$ is positive. Therefore, it tends to infinity as $|q| \to \infty$ and, respectively, adding the two infinitely large arcs to the steepest descent path does not change the value of the integral. In other words, we can consider that the original integration contour and γ have common end points. The issue of singularities encountered in the deformation of the original contour is a more complicated subject and will be discussed later. For now, we will ignore their influence and evaluate the integral along the path γ.

Derivative dq/ds

Before we begin this procedure, it is useful to determine the derivative dq/ds at the saddle point, since its value is used later. From eq. 6.182 we have

$$\frac{df}{dq}\frac{dq}{ds} = 2\,i\,s$$

or

$$\frac{dq}{ds}\left(\sin\theta_0 - \frac{q}{\sqrt{1-q^2}}\,\cos\theta_0\right) = 2\,i\,s\,,$$

that is,

$$\frac{dq}{ds} = \frac{2\,i\,s}{\sin\theta_0 - \cos\theta_0\,q/\sqrt{1-q^2}} \tag{6.185}$$

In approaching the saddle point, the numerator and the denominator of eq. 6.185 tend to zero, since

$$s \to 0 \qquad \text{and} \qquad q_0 = \sin\theta_0$$

To resolve this indeterminacy we apply L'Hopital rule, and it gives

$$\frac{d\,q}{d\,s} = \frac{2\,i}{\dfrac{d}{dq}\left[-q\,(1-q^2)^{-1/2}\right]\,\cos\theta_0\,\dfrac{dq}{ds}}$$

Thus, at $s = 0$,

$$\frac{d\,q}{d\,s} = \sqrt{\frac{-2\,i}{\dfrac{d}{dq}\left[q\,(1-q^2)^{-1/2}\right]\,\cos\theta_0}}$$

Since

$$\frac{d}{dq}\left(\frac{q}{\sqrt{1-q^2}}\right) = (1-q^2)^{-3/2}\,,$$

we obtain at the saddle point ($q_0 = \sin\theta_0$):

$$\frac{dq}{ds} = \sqrt{\frac{-2\,i\,\cos^3\theta_0}{\cos\theta_0}} = 2^{1/2}\cos\theta_0\,e^{-i\pi/4} \tag{6.186}$$

Evaluation of the integral along path γ

In accordance with eq. 6.176 we have

$$P_s = \left(\frac{k_1}{2\pi r}\right)^{1/2} e^{i\pi/4}\,I \tag{6.187}$$

where

$$I = \int_\gamma F(q)\,\exp\left[i\,a\,f(q)\right]dq\,, \tag{6.188}$$

$$a = k_1 R_1 \gg 1$$

and P_s is the pressure of the wave related to an integration along the path γ.

Taking into account eqs. 6.178 and 6.182, the integral I can be represented as

$$I = e^{i\,k_1 R_1}\int_{-\infty}^{\infty} e^{-a\,s^2}\,\phi(s)\,ds \tag{6.189}$$

Inasmuch as

$$dq = \frac{dq}{ds}\,ds$$

we have

$$\phi(s) = F(q)\,\frac{dq}{ds} \tag{6.190}$$

Let us assume that the parameter a is large and, correspondingly, the integral is defined by small values of s only, that is, the vicinity of the saddle point gives the main contribution. Because of this, it is natural to represent the function $\phi(s)$ as the Taylor series

$$\phi(s) = \phi(0) + \phi'(0)s + \phi''(0)\frac{s^2}{2} + ... \tag{6.191}$$

Substitution of the latter into eq. 6.189 and the use of equalities

$$\int_{-\infty}^{\infty} e^{-as^2}\,ds = \sqrt{\frac{\pi}{a}}, \qquad \int_{-\infty}^{\infty} e^{-as^2}\,s^2\,ds = \frac{1}{2}\sqrt{\frac{\pi}{a^3}} \tag{6.192}$$

gives

$$I = e^{ik_1R_1}\left(\frac{\pi}{k_1R_1}\right)^{1/2}\left[\phi(0) + \frac{1}{4k_1R_1}\,\phi''(0) + ...\right] \tag{6.193}$$

Note that the integrand

$$s\,e^{-a\,s^2}$$

is an odd function, and therefore the integral containing the first derivative, $\phi'(0)$, vanishes. Thus, the steepest descent method allows us to represent the value of the integral as a series by inverse powers of the large parameter a if the function $\phi(s)$ varies sufficiently slowly with respect to $exp\left[-a\,s^2\right]$. This means that the $|s|$ values are small compared to the region significant for integration. Therefore, we can restrict ourselves to one or two terms of the series, eq. 6.193.

In order to find $\phi(0)$ and $\phi''(0)$, we express the function $\phi(s)$ in terms of $F(q)$ and $f(q)$. Since, in general,

$$f(q) = f(q_0) + i\,s^2,$$

we have

$$f'(q)\,\frac{dq}{ds} = 2\,i\,s$$

Whence

$$\frac{dq}{ds} = \frac{2\,i\,s}{f'(q)}$$

Then eq. 6.190 gives

$$\phi(s) = 2\,i\,s\,\frac{F(q)}{f'(q)} \tag{6.194}$$

Finally, after using rather cumbersome algebra, we obtain

$$\phi(0) = \sqrt{-\frac{2}{f''(q_0)}}\,F(q_0) \tag{6.195}$$

and

$$\frac{\phi''(0)}{2} = \phi(0)\left[\frac{f'''}{(f'')^2}\frac{F'}{F} + \frac{1}{4}\frac{f^{IV}}{(f'')^2} - \frac{5}{12}\frac{(f''')^2}{(f'')^3} - \frac{F''}{F\,f''}\right]_{q=q_0}$$

Respectively, the integral I, eq. 6.188, becomes

$$I = \sqrt{\frac{2\pi i}{k_1 R_1 f''(q_0)}}\,e^{ik_1 R_1}\,F(q_0)\left[1 + \frac{1}{2ik_1 R_1}\left(\frac{F' f'''}{F(f'')^2} + \frac{f^{IV}}{4(f'')^2} - \frac{5(f''')^2}{12(f'')^3} - \frac{F''}{F f''}\right)_{q=q_0}\right] \tag{6.196}$$

The sign of the square root of the right-hand side is determined by the following condition: the argument of the square root equals the angle that the tangent to the steepest descent path forms with the real axis of q at $q = q_0$. It is related to the fact that after transition to the new variable the integrand contains the term dq/ds, which characterizes the slope of γ.

Specular portion of the reflected wave

Now we apply this general result to the wave field in the upper half space assuming that the integrals along the real axis of q and the steepest descent path are equal. Inasmuch as singularities are not taken into account, it is natural that we are going to obtain information about some part of the reflected field that is often called the specular wave.

After some cumbersome but simple algebra involving repeated differentiation of the functions $f(s)$ and $F(s)$, we obtain

$$P_s = \frac{e^{i\,k_1 R_1}}{R_1}\left[V(\sin\theta_0) - \frac{iN}{k_1 R_1}\right] \tag{6.197}$$

where

$$N = \left[\frac{1-q^2}{2}\frac{\partial^2 V}{\partial q^2} + \frac{1-2q^2}{2q}\frac{\partial V}{\partial q}\right]_{q\,=\,\sin\theta_0} = \frac{1}{2}\left(\frac{\partial^2 V}{\partial\theta^2} + \frac{\partial V}{\partial\theta}\cot\theta\right)_{\theta\,=\,\theta_0} \tag{6.198}$$

Here we neglected terms $O(k_1^{-2}R^{-2})$ compared to unity.

Note that these results hold for an arbitrary analytical dependence of the reflection coefficient V on q or the incidence angle $\theta = \sin^{-1} q$. The use of the rightmost side of eq. 6.198 is convenient when the reflection coefficient is given as a function of the incident angle, rather than q, and also for θ_0 close to $\pi/2$. The derivatives of V with respect to q in eq. 6.198 generally tend to infinity at $q \to 1$, but their combination remains finite as it follows from the right-hand side of this equation.

For the case of an interface between two fluids, substitution of the Fresnel reflection coefficient V in eq. 6.198 gives an explicit expression for N:

$$N = \frac{1-n^2}{b} \frac{2\left(n^2-1\right)+3\cos^2\theta_0-\cos^4\theta_0+b\cos\theta_0\left(2n^2+\sin^2\theta_0\right)\sqrt{n^2-\sin^2\theta_0}}{\left(n^2-\sin^2\theta_0\right)^{3/2}\left(\cos\theta_0+b\sqrt{n^2-\sin^2\theta_0}\right)^3} \tag{6.199}$$

In the limiting case when $k_1 R_1 \to \infty$, we arrive at the expression for a spherical wave

$$P_s = V(\sin\theta_0)\,\frac{e^{i\,k_1 R_1}}{R_1} \tag{6.200}$$

"caused" by a fictitious source, located at the point with coordinates $0, -d$. This is the reason why this portion of the reflected wave is called specular. It is important to emphasize that the amplitude of this wave depends not only on R_1, but on the specular angle θ_0 through the value $V(\theta_0)$.

Now let us make several comments, namely:

(a) Equation 6.200 was derived earlier using the stationary phase method (section 6.3).

(b) The right-hand side of eq. 6.197 represents the sum of the first two terms of the Debye series, describing a specular wave at the high frequencies.

(c) As follows from eq. 6.197, the second term decreases with distance more rapidly than the first one and they depend differently on angle θ_0. In particular, near the critical angle $\sin\theta_0 = n$, the second term becomes infinitely large and, correspondingly, this asymptotics becomes invalid.

(d) This analysis of the far-field asymptotics of the integral representation of the reflected and other waves is based on the asymptotic expansion of the Hankel function, eq. 6.174, which is not justified when $\theta_0 \to 0$. Indeed, the argument of the Hankel function, $k_1 r q$, is not large in the vicinity of the stationary point, $q_0 = \sin\theta_0$. Moreover, the integral representation, eq. 6.172, does not have physical meaning when $r = 0$ (that is, when the observation point is on the same vertical as the primary source), since the Hankel function of zero argument is not defined. A different approach is required to find the reflected wave asymptotics at the line $r = 0$ and in its vicinity. However, it turns out that eqs. 6.197 – 6.198, unlike their derivations, do not need any modifications at small θ_0.

Head wave

As was mentioned above, an integration along the path γ and the real axis of q gives the same result if singularities between them are absent. However, because of the term $(1-q^2)^{1/2}$ in eqs. 6.178 – 6.179, functions $f(q)$ and $F(q)$ have branch points at $q = 1$. In addition, $F(q)$ has a branch point $q = 0$ which is originated from the singularity of the Hankel function, eq. 6.174, and branch points $q = \pm n$ due to the term $\sqrt{n^2 - q^2}$ in the Fresnel reflection coefficient, V. Also $F(q)$ may have poles on the complex plane. However, in our case they are absent, since the imaginary parts of $\sqrt{1-q^2}$ and $\sqrt{n^2-q^2}$ are non-negative. In other words, the denominator of V has everywhere a value different from zero. As will be shown, the role of the branch points depends on the number of intersections between γ and the branch cuts originating at the branch points. It turns out that these lines, related to points $q = 0$ and $q = 1$, can be chosen in such a way that they are not intersected by the integration contour. Contributions to the integral resulting from branch points $q = \pm n$ will be considered in detail. From a mathematical point of view, they are responsible for an important component of the reflected field in the upper medium, which is known as the head wave, and to some extent it was studied earlier.

Position of branch cut

Physical considerations (decay of wave amplitude at large z), as well as the mathematical requirement of convergence of the integrals in eqs. 6.171 – 6.172, describing the incident waves, dictate that

$$\text{Im}\,\sqrt{1-q^2} \geq 0 \qquad \text{and} \qquad \text{Im}\,\sqrt{n^2-q^2} \geq 0$$

Let us assume that k_1 is real, that is in the upper half-space, where the source is located, absorption is neglected. Also

$$k_2 = k_2' + i\,k_2'' \qquad \text{or} \qquad n = n_1' + i\,n_2'',$$

where $n_1' > 0$ but $0 < n_2'' \ll 1$ and the latter characterizes a weak attenuation of acoustic energy in the lower half-space. Introduction of a very small imaginary part, n_2'', allows us to visualize a branch cut on the γ plane. It is obvious that the line

$$\text{Im}\,\sqrt{n^2-q^2} = 0$$

is defined by the equation

$$n^2 - q^2 = u \tag{6.201}$$

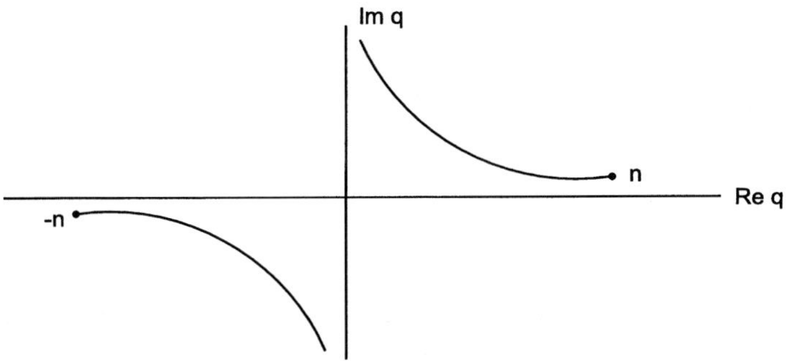

Figure 6.18: Branch cuts on the complex plane due to $\sqrt{n^2 - q^2}$.

Here u is a non-negative number $0 \leq u < \infty$, while n and q are complex. In terms of the real and imaginary parts of q and n, eq. 6.201 can be rewritten as

$$q_1^2 - q_2^2 = (n_1')^2 - (n_2'')^2 - u \qquad n_1' n_2'' = q_1 q_2 \tag{6.202}$$

This line is called a branch cut, C, and, in accordance with the second equation of the set 6.202, it is a part of a hyperbola. In the third quadrant of the complex plane its segment goes from $-n$ to $-i\infty$, while in the first quadrant the variable q changes along a symmetrical segment from n to $i\infty$, Fig. 6.18. The term $\sqrt{n^2 - q^2}$ is an analytical function of q away from the branch cut, but it varies discontinuously from a positive value, $u^{1/2}$, to a negative one, $-u^{1/2}$, on the branch cut.

Modification of the γ path

We recall that the saddle point method applies to integrals, eq. 6.189, where the function ϕ is varying slowly compared to the exponential. Besides it is supposed to be analytical, and hence, infinitely differentiable along the steepest descent path, γ. This condition is certainly violated when γ intersects a branch cut. Indeed, the function $F(q)$ in eq. 6.188 is not continuous. Hence the method of the steepest descent needs to be modified when branch cuts are involved. A complete analysis would require knowledge of Riemann surfaces and other concepts of the theory of multivalued analytical functions. We present here a much simpler and not mathematically rigorous version that still leads to correct results. First, in order to avoid the discontinuity of the integrand, we have to prevent an intersection with the cut C. It turns out that if the contour γ intersects the branch cut twice, we can deform this contour between the intersection points without a change of the integral value, Fig. 6.19a (dashed line). Therefore, in general, when the number of intersections is even, the presence of the branch cut can be essentially ignored. For instance, such a behavior takes place in the case of the branch line

$$\sqrt{1-q^2}=0\,,$$

and this is why the branch point $q=1$ is not considered. Inasmuch as the new element of the integration contour does not coincide with the steepest descent path, it may create the impression that this procedure gives rise to some errors. However, a more careful analysis, based on the use of Riemann surfaces, shows that by virtue of Cauchy's theorem this deformation is justified regardless of the distance between the saddle point and q_c, Fig. 6.19a.

On the contrary, if there is a single or odd number of intersections, a loop around the cut has to be added to connect segments of the contour γ on the opposite sides of C, Fig. 6.19b. Along such a contour the integrand remains an analytical function and the leading order contribution to the integral along γ is still given by eq. 6.197. However, in the case of an odd number of intersections between γ and C, we have to add to this result a contribution from the integral along the loop around the branch point n. Note that the steepest descent path γ does not enter the third quadrant of the complex plane q, and therefore an intersection with the other segment of the branch cut is absent. By definition, the path γ depends on the following parameters: n_1, n_2, and $\sin\theta_0$, while the position of C is determined by n_1 and n_2 only. Calculations of both curves show that if $n_2 \ll 1$, the number of intersections of the segment of the branch cut in the first quadrant and the path γ is

$$\begin{array}{ll} \text{one} & \text{if } n_1 < \sin\theta_0, \\ \text{two} & \text{if } \sin\theta_0 < n_1 < 1/\sin\theta_0, \\ \text{three} & \text{if } n_1 > (\sin\theta_0)^{-1}\,, \end{array}$$

here

$$n_1 = n = \frac{c_1}{c_2} \qquad \text{if} \qquad n_2 \to 0\,.$$

It is worthy to emphasize that this analysis is very important since it allows one to predict the area of existence of head waves.

Asymptotic expression for the head wave

Consequently, the reflected field has, aside from the specular component given by eq. 6.197, an additional one provided that the angle θ_0 is large enough. As we already know from section 6.4, it represents the head wave. Specifically, this wave is observed at

$$\theta_0 > \theta_c = \sin^{-1}\frac{c_1}{c_2}\,,$$

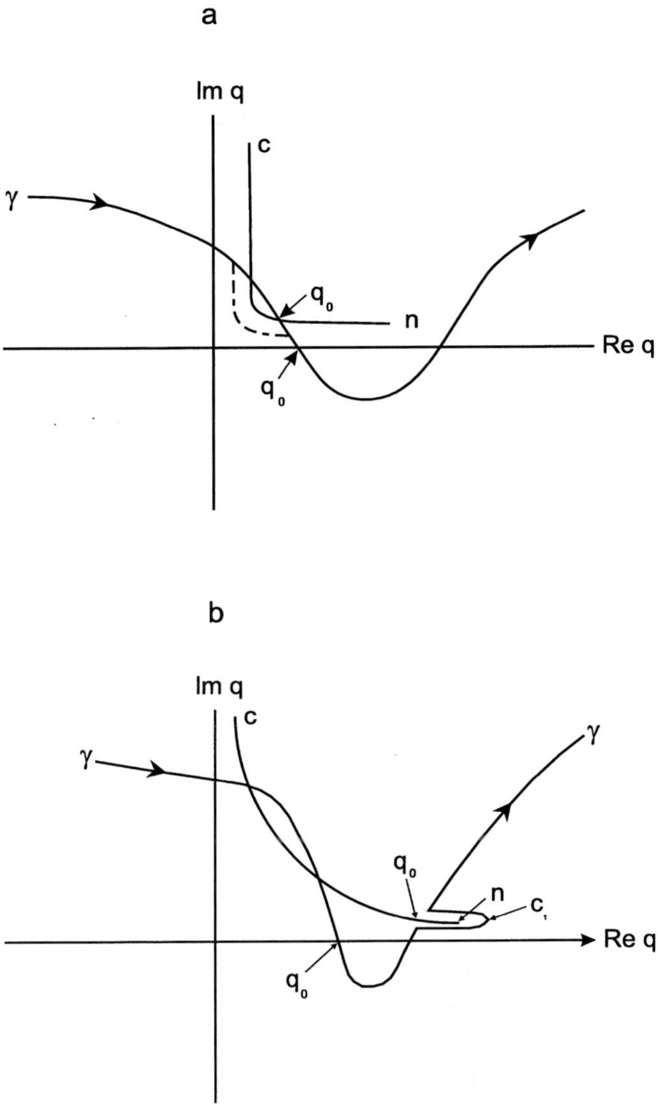

Figure 6.19: Deformation of an integration path with even (a) and odd (b) numbers of branch cut intersections.

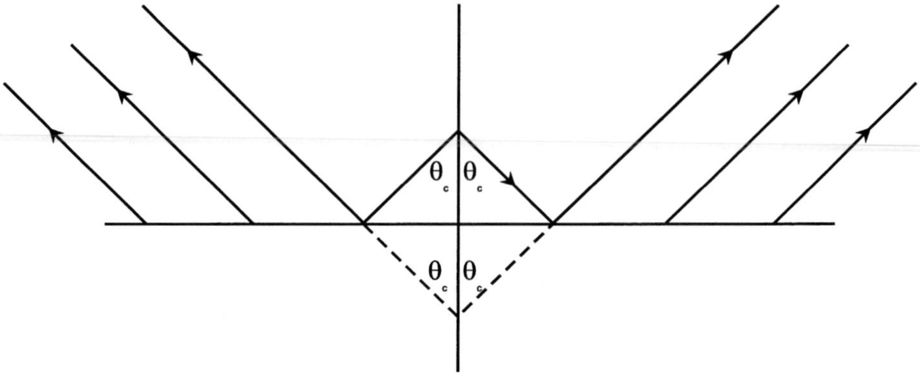

Figure 6.20: Rays corresponding to the head wave and its domain of existence.

when the velocity of propagation in the lower medium is greater than in the upper half space. Besides it arises if the angle θ_0 satisfies the condition

$$\theta_0 > \sin^{-1}\frac{1}{n_1} = \sin^{-1}\frac{c_2}{c_1} \qquad \text{and} \qquad c_2 < c_1$$

Unlike the specular component of the field, the head wave exists only in a certain part of the upper medium, adjoining the reflecting boundary, $z = 0$, Fig. 6.20.

Note that such a wave does not exist in the case $c_1 = c_2$, even if the densities of the upper and lower medium are not equal. As was mentioned in deriving the asymptotic for the head wave,we have to consider the integral along the path C_1, which is the portion of the branch cut C between the branch point $q = n$ and the point of intersection of γ and C, (q_c), closest to n. This integration is performed twice along C_1, namely, at both sides of the branch cut where the radical $\sqrt{n^2 - q^2}$ has opposite signs. Correspondingly, we arrive at an equation similar to eq. 6.176

$$P_l = \left(\frac{k_1}{2\pi r}\right)^{1/2} e^{i\pi/4} \int_{C_1} [F_+(q) - F_-(q)]\, \exp[ik_1 R_1 f(q)]\, dq$$

Here $F_+(q)$ and $F_-(q)$ are the values of the function $F(q)$ at the opposite sides of the branch cut. As follows from eq. 6.179,

$$F_+(q) - F_-(q) =$$

$$= \sqrt{\frac{q}{1-q^2}} \left(\frac{\sqrt{1-q^2} + b\sqrt{n^2-q^2}}{\sqrt{1-q^2} - b\sqrt{n^2-q^2}} - \frac{\sqrt{1-q^2} - b\sqrt{n^2-q^2}}{\sqrt{1-q^2} + b\sqrt{n^2-q^2}}\right) \left[1 + O\left(\frac{1}{k_1}\right)\right]$$

$$= \frac{4\, b\sqrt{n^2-q^2}\,\sqrt{q}}{1 - b^2 n^2 - (1-b^2)q^2} \left[1 + O\left(\frac{1}{k_1 R_1}\right)\right],$$

where $\sqrt{n^2 - q^2}$ is the value on the upper side of the cut. Respectively, we have

$$P_l = \left(\frac{8k_1}{\pi r}\right)^{1/2} e^{i\pi/4} \int\limits_{C_1} F_1(q)\,(n-q)^{1/2} \exp[ik_1 R_1 f(q)]\,dq \tag{6.203}$$

where

$$F_1(q) = \frac{b\sqrt{q\,(n+q)}}{1 - b^2 n^2 - (1-b^2)q^2}\left[1 + O\left(\frac{1}{k_1 R_1}\right)\right] \tag{6.204}$$

In evaluating this integral, we assume that branch point, $q = n$, is not close to either the stationary point, $(q_0 = \sin\theta_0)$, or the point of intersection, q_c. Also let us represent the function $f(q)$ as

$$f(q) = f(n) + i\,s \qquad \text{and} \qquad s > 0 \tag{6.205}$$

From eq. 6.178, it follows that

$$f(n) = n\,\sin\theta_0 + \sqrt{1-n^2}\,\cos\theta_0$$

or

$$f(n) = \cos(\theta_0 - \theta_c)\,, \tag{6.206}$$

since

$$\sin\theta_c = \frac{c_1}{c_2} \qquad \text{and} \qquad n_2 \to 0$$

Substitution of eq. 6.205 into eq. 6.203 yields

$$P_l = \left(\frac{8k_1}{\pi r}\right)^{1/2} e^{i\pi/4}\, e^{i\,k_1 R_1 f(n)} \int\limits_0^{s_c} \phi_1(s)\, s^{1/2} \exp(-ik_1 R_1 s)\,ds \tag{6.207}$$

Here

$$\phi_1(s) = \left(\frac{q-n}{s}\right)^{1/2} F_1(q)\,\frac{dq}{ds} \tag{6.208}$$

and

$$s_c = i\,[f(n) - f(q_c)]$$

The function $\phi_1(s)$ is regular and can be represented as a Taylor series in the vicinity of $s = 0$ (branch point). Inasmuch as the integrand in eq. 6.207 rapidly becomes smaller with the increase of s, the upper limit, s_c, can be replaced by $+\infty$. This introduces only an exponentially small error, when the parameter k_1R_1 is large.

Applying the integral representation of the gamma function

$$\Gamma\left(\frac{3}{2}\right) = \int_0^\infty s^{1/2}\, e^{-s}\, ds = \frac{\sqrt{\pi}}{2}\,, \tag{6.209}$$

we obtain

$$P_l = \left(\frac{2k_1}{r}\right)^{1/2} e^{i\pi/4}\, e^{i\,k_1R_1 f(n)}\, \phi_1(0)\left[1 + O\left(\frac{1}{k_1R_1}\right)\right] \tag{6.210}$$

By calculating $\phi_1(0)$ from eqs. 6.204 – 6.208 and discarding corrections $O\left(\dfrac{1}{k_1^3R_1^3}\right)$ we arrive at the following expressions for the pressure of the head wave

$$P_l = \frac{2\,i\,b\,\sin\theta_c\,\exp\left[i\,k_1R_1\cos(\theta_0-\theta_c)\right]}{k_1R_1^2\sqrt{\sin\theta_0\,\cos\theta_c\,\sin^3(\theta_0-\theta_c)}} \qquad \text{if} \quad c_1 < c_2\,,\ \theta_0 > \theta_c \tag{6.211}$$

and

$$P_l = \frac{-2inb\sqrt{n^2-1}\,\exp\left[i\,k_2r - k_1\sqrt{n^2-1}(z+d)\right]}{k_1r^2\left(\sqrt{n^2-1}+i\,n\tan\theta_0\right)^{3/2}} \tag{6.212}$$

$$\text{if} \quad c_1 > c_2\,,\ \theta_0 = \sin^{-1}\frac{c_2}{c_1}$$

Case one: $c_2 > c_1$ (Fig. 6.20). First, consider the wave when the lower medium has higher velocity, eq. 6.211. It is more instructive to express the trigonometric functions in terms of distances along which the wave travels. By definition,

$$R_1\sin\theta_0 = r\,, \qquad R_1\cos\theta_0 = z+d$$

Also, we have, Fig. 6.21

$$r = L_1 + (d+z)\tan\theta_c$$

and

$$k_1\sin\theta_c = k_1n = k_2$$

Taking into account these relationships, we obtain

$$k_1R_1\cos(\theta_0-\theta_c) = k_1\left[nr + \sqrt{1-n^2}(z+d)\right] = k_1(L_0+L) + k_2L_1$$

and

$$R_1 \sin(\theta_0 - \theta_c) = R_1(\sin\theta_0 \cos\theta_c - \cos\theta_0 \sin\theta_c) \tag{6.213}$$

$$= [r - (z+d)\tan\theta_c]\cos\theta_c = L_1 \cos\theta_c$$

As a result, eq. 6.211 becomes

$$P_l = \frac{2\,i\,n\,b}{k_1(n^2-1)\,\sqrt{r}\,L_1^{3/2}}\, \exp\left[ik_1(L_0+L) + ik_2 L_1\right] \tag{6.214}$$

The expression

$$k_1(L_0+L) + k_2 L_1$$

can be interpreted in the following way. This is a phase along the path, connecting the source and the observation point, Fig. 6.21, and it consists of three elements, namely, L_0, L_1, and L. Along L_0 the incidence wave propagates to the boundary with velocity c_1, and the incident angle is equal to θ_c. The segment L_1 is located at the interface and the head wave travels with the velocity of the lower medium, c_2. Finally, the third segment L forms the critical angle θ_c with the z-axis and this wave propagates with velocity c_1. As follows from eq. 6.214, its amplitude decreases inversely proportional to $\sqrt{r}$ along each ray and, correspondingly, it varies on the phase surface

$$k_1(L_0+L) + k_2 L_1 = \text{const}$$

Inasmuch as rays characterize the position of elementary tubes along which energy moves, we can conclude the following. This wave arises at the boundary near the points where $\theta_0 = \theta_c$, and propagates along it, while a portion of its energy penetrates into the upper medium. This is why such a wave is sometimes called the lateral one, Fig. 6.21. In accordance with eq. 6.214 at large distances, when $r \gg d+z$, we have $L_1 \approx r$ and

$$P_l = \frac{2\,i\,n\,b}{k_1\,(n^2-1)\,r^2}\, \exp i\left[k_1(L_0+L) + k_2 L_1\right]$$

This asymptotic formula for the head wave was derived earlier, using the method of stationary phase. Thus we see that the application of the steepest descent path allows us to obtain an asymptotic formula, which is more accurate and is valid at smaller distances from the z-axis. However, eq. 6.214 cannot be used when

$$n \to 1 \qquad \text{or} \qquad L_1 \to 0$$

Now let us make several comments:

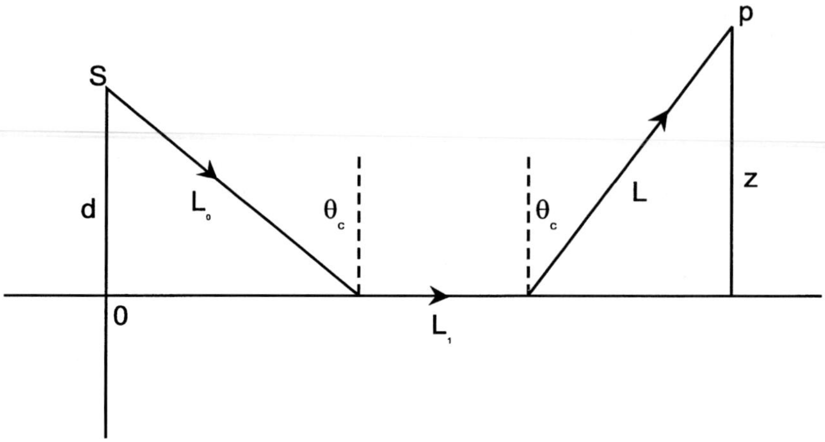

Figure 6.21: Ray presentation of the head wave.

(a) As we know, wave fields in the approximation of the geometrical acoustics are described by the Debye series and eq. 6.214 represents the leading term of the head wave. Its amplitude is inversely proportional to the wave number k_1, and therefore, unlike the reflected wave, this wave disappears when $\omega \rightarrow \infty$.

(b) In the frequency domain we observe simultaneously both waves (reflected and head) at each point. However, the behavior of the transient waves is different. Since the head wave propagates along the boundary with velocity c_2, $(c_2 > c_1)$, it arrives at the observation point earlier than the specular one and its delay becomes greater with the increase of the distance r.

Case two: $c_2 < c_1$ If the velocity of the lower medium is smaller, an evanescent wave arises in the upper half-space and propagates along the boundary with velocity c_2. At the same time it decays exponentially with the increase of either the source or receiver elevation. Its contribution is usually negligible unless they are located within a fraction of the wavelength from the reflector. Certainly, the wavefronts are cylindrical surfaces normal to the boundary

$$k_2 \, r = \text{const}$$

As in the previous case, we can observe this evanescent wave only at certain distances from the origin, when the specular angle satisfies the inequality

$$\theta_0 > \sin^{-1} \frac{c_2}{c_1} \tag{6.215}$$

Vicinity of the critical angle

Consider again the first case when $n = c_1/c_2 < 1$. Expressions for the specular component of the reflected wave, as well as for the head wave, diverge and have no physical meaning in the limit $\theta_0 \to \theta_c$. Geometrically, in the upper half space the equality $\theta_0 = \theta_c$ defines the ray of the reflected wave at the angle equal to that of the total reflection θ_c. These rays form the boundary of the region where the head wave is observed (Fig. 6.20). Mathematically, the asymptotics derived earlier diverge because the applicability of the steepest descent method is violated at $\theta_0 = \theta_c$. Indeed, in obtaining eq. 6.197 we assumed that the function F, eq. 6.179, varies slowly on the steepest descent path with respect to the exponent, eq. 6.189. However, the derivative $\partial F/\partial s$ is infinitely large at $q = n = \sin\theta_c$, that is F changes rapidly if $\theta_0 = \theta_c$. Also, in deriving eqs. 6.211 – 6.212 it was supposed that the stationary and branch points are not close to each other. At the same time if $\theta_0 = \theta_c$ these critical points coincide. In fact, we have

$$q_0 = \sin\theta_0 = \sin\theta_c = n$$

Now we present an elementary analysis that gives some insight into some features of the reflected field in the vicinity of the critical angle, and we consider the limiting case when $\theta_0 = \theta_c$. First, let us isolate the singularity in the Fresnel coefficient, eq. 6.173:

$$V(q) = \frac{n_1 - b\,n_2}{n_1 + b\,n_2} = \frac{n_1^2 + b\,n_2^2 - 2\,b\,n_1 n_2}{n_1^2 - b^2 n_2^2}$$

The latter can be written in the form

$$V(q) = V_1(q) + V_2(q)\sqrt{q-n}\,,$$

where

$$V_1(q) = \frac{1 + n^2b^2 - (1+b^2)q^2}{1 - n^2b^2 - (1-b^2)q^2} \qquad \text{and} \qquad V_2(q) = -\frac{2\,i\,b\sqrt{(1-q^2)(q+n)}}{1 - n^2b^2 - (1-b^2)q^2} \tag{6.216}$$

Unlike V, both functions $V_1(q)$ and $V_2(q)$ are regular in the vicinity of $q = n$. The contribution of the part $V_1(q)$ into the reflection field can be found by the method of the steepest descent and it does not encounter any difficulties. This portion of the pressure is given by eqs. 6.197 – 6.198, where V is replaced by V_1. From eqs. 6.197 and 6.216, it follows that

$$P_1 = \frac{\exp(ik_1R_1)}{R_1}\left[1 + O\left(\frac{1}{k_1R_1}\right)\right] \qquad \text{if} \quad \theta_0 = \theta_c \tag{6.217}$$

To determine the contribution of the reminder of the reflection coefficient, V_2, we again deform the integration path into the steepest descent one, eq. 6.182, and obtain from eqs. 6.176 – 6.179

$$P_2 = \left(\frac{k_1}{2\pi r}\right)^{1/2} e^{i\pi/4 + ik_1R_1} \int\limits_{-\infty}^{\infty} \phi_1(s)\sqrt{s}\exp(-k_1R_1s^2)\,ds \tag{6.218}$$

where

$$\phi_1(s) = \sqrt{\frac{q(q-n)}{s(1-q^2)}}\, V_2(q)\, \frac{dq}{ds}\left[1 + O\left(\frac{1}{k_1 R_1}\right)\right]$$

If $k_1 R_1 \gg 1$, the main contribution to P_2 comes from a small vicinity $|s| \leq (k_1 R_1)^{-1/2}$ of the point $s = 0$. The function $\phi_1(s)$ is regular in this vicinity and can be expanded in Taylor series in s. Each of the resulting integrals is found by using the equality

$$\Gamma(u) = \int_0^\infty s^{u-1} e^{-s}\, ds \tag{6.219}$$

and taking into account that $s^{1/2} = i|s|^{1/2}$ at $s < 0$. After doing some algebra, we find

$$P_2 = \frac{2^{5/4}\, b\, \Gamma(3/4)}{R_1 (k_1 R_1)^{1/4}} \sqrt{\frac{\tan\theta_c}{\pi}}\, \exp\left(ik_1 R_1 + i\frac{9\pi}{8}\right)\left[1 + O\left(\frac{1}{\sqrt{k_1 R_1}}\right)\right] \tag{6.220}$$

For the total reflected wave eqs. 6.217 and 6.220 give

$$P_r = \frac{\exp(ik_1 R_1)}{R_1}\left[1 - \frac{2^{5/4}\, e^{i\pi/8}\, b\, \Gamma(3/4)}{(k_1 R_1)^{1/4}} \sqrt{\frac{\tan\theta_c}{\pi}} + O\left(k_1 R_1\right)^{-3/4}\right] \tag{6.221}$$

It is interesting that eq. 6.197 with $N = 0$ still accurately predicts the leading term of the field asymptotics. We also see that the divergence of the asymptotics at $\theta_0 = \theta_c$ has indicated that the corrections to the leading term in the vicinity of the critical angle are much larger than away from θ_c. The difference is three orders of magnitude for $k_1 R_1 = 10^4$. Indeed, the relative corrections are $O\left[(k_1 R_1)^{-1/4}\right]$ at $\theta_0 = \theta_c$ rather than $O\left[(k_1 R_1)^{-1}\right]$ in the region $|\theta_0 - \theta_c| \geq (k_1 R_1)^{-1/2}$, where the specular component and the head wave can be considered as independent waves.

Transmitted wave

Next we begin to study the acoustic field in the lower medium, and, by analogy with the reflected wave, it can be represented as

$$P_t = \frac{i}{2} \int_{-\infty}^{\infty} W(m)\, H_0^{(1)}(mr)\, \exp(in_1 d - in_2 z)\, \frac{m}{n_1}\, dm \tag{6.222}$$

Here $z < 0$ and

$$W(m) = \frac{2\, n_1}{n_1 + b\, n_2} \tag{6.223}$$

Note that the expression

$$n_1 d - n_2 z$$

characterizes a phase that increases with $|z|$. In order to derive the asymptotic of the transmitted wave when $k_1 R \gg 1$, we again use an expansion of the Hankel function, and after replacement of variable $m = k_1 q$, we obtain

$$P_t = \left(\frac{k_1}{2\pi r}\right)^{1/2} e^{i\pi/4} \int_{-\infty}^{\infty} F_t(q) \exp\left[i\, k_1 R\, f_t(q)\right] dq \tag{6.224}$$

where

$$f_t(q) = q\,\frac{r}{R} + \sqrt{1-q^2}\,\frac{d}{R} + \sqrt{n^2-q^2}\,\frac{z_1}{R}, \tag{6.225}$$

$$\begin{aligned} F_t(q) &= \sqrt{\frac{q}{1-q^2}}\left[1 - \frac{i}{8k_1 r q} + O\left(\frac{1}{k_1^2 r^2}\right)\right] W(q) \\ &= \left(\frac{q}{1-q^2}\right)^{1/2}\left[1 - \frac{i}{8k_1 r q} + O\left(\frac{1}{k_1^2 r^2}\right)\right] \frac{2\sqrt{1-q^2}}{\sqrt{1-q^2} + b\sqrt{n^2-q^2}}, \end{aligned} \tag{6.226}$$

where $z_1 = -z > 0$ is the distance of the observation point from the boundary.

Inasmuch as the reflection and transmission coefficients are related by

$$W = V + 1,$$

the integrands in eq. 6.224 and 6.176 have exactly the same singularities (poles and branch points) in the complex plane.

To apply the steepest descent method, we have to find the stationary points of the function f_t in the exponent, eq. 6.224. Making use of eq. 6.225, the stationary point equation

$$\frac{d\,f_t}{d\,q} = 0$$

becomes

$$r - \frac{q\,d}{\sqrt{1-q^2}} - \frac{q\,z_1}{\sqrt{n^2-q^2}} = 0 \tag{6.227}$$

The latter has unique real-value solution

$$q = q_0 \qquad \text{and} \qquad 0 \leq q_0 \leq \min(1, n)$$

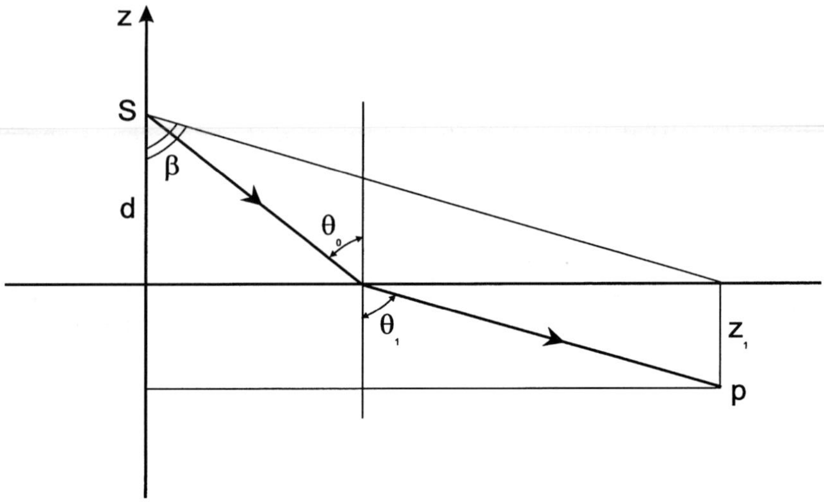

Figure 6.22: Rays of incident and transmitted waves.

Indeed, the left-hand side of eq. 6.227 decreases monotonically from r to $-\infty$ when q varies between zero and $\min(1, n)$, and thus, it has unique solution within this interval. When $q > \min(1, n)$, the imaginary part of the left hand side of the equation is positive, and therefore roots are absent. Similar reasoning shows that eq. 6.227 has no solution at $q < 0$. Note that stationary point q_0 is a simple, one and in terms of the incident angle

$$\theta_0 = \sin^{-1} q_0 \,,$$

eq. 6.227 can be written as

$$r = d \tan \theta_0 + z_1 \tan \theta_1 \,, \qquad \sin \theta_1 = \frac{\sin \theta_0}{n} \tag{6.228}$$

This is the equation of a ray that originates at the source and is refracted (transmitted) at the boundary according to Snell's law, where θ_1 is the refraction angle formed by the ray with the normal to the boundary in the lower half space, Fig. 6.22.

In addition to the real stationary point, q_0, there may be complex stationary points in the problem. We find the complex stationary point explicitly for the case $n < 1$ and small z_1 (that is, for observation points near the boundary in the lower half-space).

If $z_1 = 0$, the solution to eq. 6.227 is easily found to be $q = \sin \beta$, where $\beta = \tan^{-1}(r/d)$. At $z_1 > 0$, $z_1 \to 0$ we have either $q \to \sin \beta$ or $q \to \sin \theta_c$. In the latter case, z_1 is multiplied by a large factor in eq. 6.227. In the first case we obtain from eq. 6.227 $q = q_1$,

$$q_1 = \sin \beta - \frac{\sin \beta \cos^3 \beta \, z_1}{d \sqrt{\sin^2 \theta_c - \sin^2 \beta}} + O(z_1^3) \tag{6.229}$$

In the second case, we obtain from eq. 6.227 $q = q_2$,

$$q_2 = \sin\theta_c - \frac{\sin\theta_c\, z_1^2}{2\,d^2\,(\tan\beta - \tan\theta_c)^2} + O(z_1^3)\,, \qquad \beta > \theta_c \tag{6.230}$$

In the case $z_1 \to 0$, eq. 6.227 does not have a solution close to $\sin\theta_c$ when $\beta < \theta_c$. The solution q_2 is always real, while q_1 is real for $\beta < \theta_c$ and complex for $\beta > \theta_c$. We always have one real solution, as expected, but for $\beta > \theta_c$ there occurs an additional, complex stationary point. From eqs. 6.225 and 6.229, it follows that $\operatorname{Im} f_t(q_1) > 0$ at $\beta > \theta_c$. Therefore, at $z_1 > 0$ the complex stationary point makes an exponentially small contribution to the integral in eq. 6.224.

The steepest descent path

At the beginning we neglect the contribution due to the complex stationary point and deform the original integration path in eq. 6.224 to the steepest descent path, γ_t, passing through the real stationary point q_0, much like in the case of the reflected wave, eq. 6.182, the steepest descent path is defined by the equation

$$q + \sqrt{1-q^2}\,\frac{d}{r} + \sqrt{n^2-q^2}\,\frac{z_1}{r} = \frac{d}{r\,\cos\theta_0} + \frac{n\,z_1}{r\,\cos\theta_1} + i\,s^2 \tag{6.231}$$

and it can be analyzed in the same manner as eq. 6.182. It turns out that the path described by eq. 6.231 is geometrically rather similar to the path γ in the problem of the spherical wave reflection. Like γ, the path γ_t intersects the imaginary axis of the q plane once and has two intersections with the real axis, namely, at $q = q_0$ and at some point $q > 1$. For large q, the path approaches the lines

$$\arg q = \theta_0 \qquad \text{and} \qquad \arg q = \pi - \theta_0.$$

In order to connect γ_t to the original integration path, we use the same arcs in the first and second quadrants of the complex q plane. Again, as in the case of the reflected wave, the contributions of the integrals along the arcs tend to zero when their radius tends to infinity.

As was demonstrated earlier, q_0 is smaller than n, and, thus path γ_t intersects twice the branch cut associated with the branch point $q = n$. Therefore, this branch point does not contribute to the integral describing the refracted (transmitted) wave. The branch points $q = \pm 1$ do not contribute to the integral either. Hence, the leading terms of the integral asymptotic are related to the stationary point q_0 only.

Before we continue, let us point out that as in the case of the reflected wave, the solution of the stationary point equation allows us to predict the existence of different waves.

Asymptotic of the refracted wave

Applying eq. 6.196 to the integral representation, eqs. 6.224 – 6.226, we obtain after some algebra

$$P_t(x,y,z) = \sqrt{\frac{\sin\theta_0 \cos\theta_0}{r\left(d + z_1 \dfrac{\cos^3\theta_0}{n\cos^3\theta_1}\right)}} \exp\left[ik_1\left(\frac{d}{\cos\theta_0} + \frac{n\,z_1}{\cos\theta_1}\right)\right] \tag{6.232}$$

$$\left[W(\sin\theta_0) + \frac{i\cos^3\theta_0\, N_t}{2\,k_1\left(d + \dfrac{z_1\cos^3\theta_0}{n\cos^3\theta_1}\right)}\right]$$

where

$$N_t = \frac{3\sin\theta_0}{\cos^2\theta_0}\left(\frac{d + z_1\dfrac{\cos^5\theta_0}{n^3\cos^5\theta_1}}{d + z_1\dfrac{\cos^3\theta_0}{n\cos^3\theta_1}} - \frac{1+\sin^2\theta_0}{3\sin^2\theta_0\cos^2\theta_0}\right)\left.\frac{\partial W}{\partial q}\right|_{q_0=\sin\theta_0} - \left.\frac{\partial^2 W}{\partial q^2}\right|_{q_0=\sin\theta_0} \tag{6.233}$$

Here we neglected terms $O\left[(k_1R)^{-2}\right]$ compared to unity.

These expressions for the refracted wave are analogous to eqs. 6.197 – 6.198 for the specular component of the reflected wave. The first term in the square brackets in eq. 6.232 corresponds to the solution for the ray approximation, that is it is the leading term of the Debye series (Chapter 1). For sufficiently large distances from the source with respect to the wavelength it usually gives the main contribution to the refracted wave. The second term with N_t depends on the rate of the transmission coefficient variation and defines a correction to the ray approximation.

It is interesting to notice that in the case when a velocity contrast is absent, $n = 1$, W is independent of q and the correction vanishes. Correspondingly, the solution within the ray approximation coincides with the exact one

$$P_t = W\,\frac{e^{i\,k_1 R}}{R} = \frac{2}{1+b}\,\frac{e^{i\,k_1 R}}{R},$$

derived earlier.

Now assuming that $n \neq 1$ consider, as usual two different media.

Case one: $n < 1$, $(c_2 > c_1)$ First, suppose that $\beta > \theta_c$, that is, $r > d\tan\theta_c$. In accordance with eq. 6.230 we have

$$q_{02} \longrightarrow \sin\theta_c \,,$$

if z_1 approaches zero and, as follows from Snell's law, $\theta_1 \to \pi/2$. Indeed,

$$q_{02} = \sin\theta_0 = \sin\theta_c \,,$$

that is, $\theta_0 = \theta_c$ and

$$\sin\theta_1 = \frac{\sin\theta_c}{n} \longrightarrow 1$$

Whence

$$\frac{z_1}{\cos^3\theta_1} \longrightarrow \infty$$

regardless of how small the distance from the boundary. Correspondingly, the contribution of the first term in eq. 6.232 tends to zero if $z_1 \to 0$, and the second term in the brackets, which is proportional to N_t, dominates. Indeed, expanding the functions at the right side of eqs. 6.232 – 6.233 in powers of z_1, we find

$$\begin{aligned} P_t \; = \; & \frac{2\tan\theta_c}{\sqrt{r(r-d\tan\theta_c)}} \exp\left[ik_1\left(r\sin\theta_c + d\cos\theta_c + \frac{n\,z_1^2}{2\,(r-d\tan\theta_c)}\right)\right] \\ & \left[\frac{z_1}{r-d\tan\theta_c} + \frac{i\,b}{k_1\cos\theta_c\,(r-d\tan\theta_c)}\right] \qquad \text{if} \quad \beta > \theta_c \end{aligned} \tag{6.234}$$

Here we neglected terms $O(k_1 z_1^3)$ in the exponent and terms $O(z_1^2)$ elsewhere.

As in the general case, eq. 6.232, the first term in the second square brackets is the leading one, and the next represents a "correction" due to the transmission coefficient variation in the vicinity of the stationary point. Note that at the boundary, P_t is inversely proportional to wave frequency and for large r it is proportional to r^{-2}.

For any point of the boundary, the field can be equivalently represented as a sum of incident and reflected waves or, alternatively, as a refracted wave. In this light, it is useful to compare eq. 6.234 with eq. 6.211. Taking into account that $\theta_0 = \beta$ and $R = R_1$ at the interface, it is easy to verify that

$$P_l = P_t \qquad \text{if} \qquad z = 0$$

Mathematically, functions P_l and P_t have a rather different origin. It is interesting to note that the field originating from the branch cut, $(z \geq 0)$, matches exactly the field that is a second-order contribution from the stationary point of the integrand $(z \leq 0)$.

Physically, we see that the leading term of the geometric acoustic approximation predicts zero amplitude of the refracted wave at the interface for $\beta > \theta_c$. At the same time a more advanced analysis shows that the amplitude of this wave is, in fact, finite albeit small at high frequencies. In other words, eq. 6.234 takes into account the second term of the Debye series, and it defines the refracted wave at the boundary. Because of its finite amplitude at $z = 0$, the transmitted wave gives rise to the head wave in the upper half-space. It can be easily seen from eq. 6.234 that the main features of the head wave are determined by the properties of the refracted wave.

We note that with the increase of the distance z_1 from the boundary the influence of the second term of the Debye series becomes relatively small and the refracted wave is correctly described by its ray approximation which is independent of frequency, $(k_1 R \gg 1)$.

Evanescent wave

Until now we neglected the complex stationary point q_{01} for $\beta > \theta_c$, and its contribution to the field in the lower medium. According to eq. 6.229, the imaginary part of the stationary point, as well as Im $f_t(q)$, tends to zero when $z_1 \to 0$, and, correspondingly, this portion of the field ceases to be exponentially small. The asymptotic of this field can be derived using eq. 6.229 and eq. 6.196. However, we can avoid cumbersome transformations by noting that the only factor involving z_1 in eq. 6.224:

$$\exp\left[ik_1 z_1 (n^2 - q^2)^{1/2}\right]$$

varies slowly when z_1 is small. Then this term can be considered as a part of F_t in the integral representation, eq. 6.224. Respectively, the rapidly varying function becomes, (eq. 6.225):

$$\exp\left[ik_1 r q + ik_1 d(1 - q^2)^{1/2}\right]$$

Now let us compare the resulting form of the integral representation of the refracted and reflected waves. Inspection shows that P_t, eq. 6.232, can be obtained from P_r, eq. 6.176, by formally substituting

$$W \exp\left[ik_1 z_1 (n^2 - q^2)^{1/2}\right] \tag{6.235}$$

for V in eq. 6.179. Then, the contribution of the complex stationary point is given by eqs. 6.197 – 6.198, where V is described by eq. 6.235. In addition we should let

$$R_1 = (r^2 + d^2)^{1/2} \qquad \text{and} \qquad \theta_0 = \beta$$

in eqs. 6.197 – 6.198. This gives

$$P_d(x,y,z) = \frac{1}{\sqrt{r^2+d^2}} \left[\frac{2\cos\beta}{\cos\beta + i\,b\sqrt{\sin^2\beta - n^2}} - \frac{i\,N_d}{k_1\sqrt{r^2+d^2}} \right]$$

$$\exp\left[i\,k_1\sqrt{r^2+d^2} - k_1 z_1 \sqrt{\sin^2\beta - n^2} \right] \tag{6.236}$$

Here

$$N_d = N + \frac{2\,i\,k_1\,b\,z_1\,(1-n^2)\sin^2\beta\cos\beta}{(\sin^2\beta - n^2)\,(\cos\beta + i\,b\sqrt{\sin^2\beta - n^2})^2} \tag{6.237}$$

$$+ \frac{k_1 z_1 \cos\beta \left[2\,n^2 - \sin^2\beta\,(3\,n^2 + 1 - 2\,\sin^2\beta) + k_1 z_1 \sin^2\beta\,\cos^2\beta\,\sqrt{\sin^2\beta - n^2} \right]}{(\sin^2\beta - n^2)^{3/2}\,(\cos\beta + i\,b\sqrt{\sin^2\beta - n^2})}$$

and N is defined by eq. 6.199 with $\theta_0 = \beta$.

Thus, if $n < 1$ and $\beta > \theta_c$, the refracted wave is given by the sum of two components. One of them represents an approximation of the geometric acoustics, and the second one is a diffracted wave, that is, it is not described by the same Debye series. Also, if $n < 1$ but $\beta < \theta_c$, the latter is absent, and the total refracted wave is given by eq. 6.232. Both equations, 6.232 and 6.236 can diverge and are not applicable when $\beta = \theta_c$. This happens because neither the transmission coefficient, nor $\exp[ik_1 z_1 (n^2 - q^2)^{1/2}]$ is a slowly varying function when the stationary point approaches the branch one. A separate analysis is required to obtain the refracted wave asymptotics in such a case, much like as in the vicinity of the critical angle in the upper half space, described earlier.

Now let us discuss some of the properties of a diffracted wave. Its amplitude is mainly determined by the first term in square brackets in eq. 6.236. The second term provides a small correction with the exception of the case when $d/r \ll 1$, since the first term tends to zero. As a function of the distance from the interface, the amplitude of the pressure decreases mostly exponentially, and its rate of decay increases as the angle β increases. From eq. 6.236, we see that away from the boundary this wave is noticeable only at distances of the order of the wavelength. It is important to emphasize that at the boundary this wave coincides with the specular component of the reflected wave given by eqs. 6.197 – 6.198, when $\beta > \theta_c$.

The origin of the diffracted wave can be explained in the following way. This is an inhomogeneous wave generated in the vicinity of the boundary point M by the incident ray SM at an angle exceeding the critical one, Fig. 6.23a.

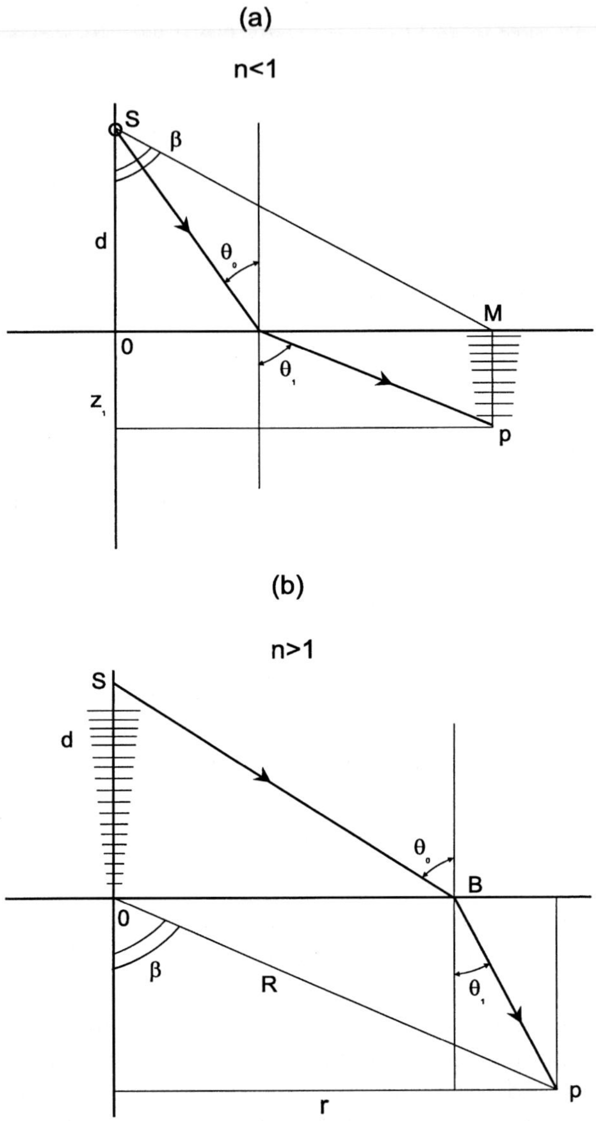

Figure 6.23: (a), (b) Evanescent waves, shown by systems of horizontal lines. The decrease of line length represents the decay of these waves.

Case $n > 1$, $(c_2 < c_1)$ Now we demonstrate that the equations already derived permit us to study wave behavior when the velocity of propagation in the lower medium is smaller in the upper one, $c_2 < c_1$. With this purpose in mind it is convenient to make use of the principle of reciprocity. First we write down eqs. 6.232 – 6.233 describing the transmitted wave in the lower half-space

$$P_t^{(1)} = \frac{i}{2}\int\limits_{-\infty}^{\infty} W(m)\, H_0^{(1)}(mr)\, \exp(in_1 d - in_2 z)\, \frac{m}{n_1}\, dm$$

and (6.238)

$$W(m) = \frac{2\,n_1}{n_1 + b\,n_2}$$

Of course, they are valid in both cases: $n > 1$ and $n < 1$. Next suppose that the positions of the source and the receiver are interchanged. Then, in place of eq. 6.238, we obtain

$$P_t^{(2)} = \frac{i}{2}\int\limits_{-\infty}^{\infty} W_1(m)\, H_0^{(1)}(mr)\, \exp(in_2 d - in_1 z)\, \frac{m}{n_2}\, dm \tag{6.239}$$

Here the z-axis is directed downward and

$$W_1 = \frac{2\,n_2}{n_2 + n_1/b} = \frac{2\,b\,n_2}{n_1 + b\,n_2} \tag{6.240}$$

Comparison with eq. 6.238 shows that if we now interchange notations $c_1 \leftrightarrow c_2$, then

$$P_t^{(2)} = b\,P_t^{(1)} \tag{6.241}$$

and the latter represents the principle of reciprocity, which is, of course, valid in a more general case. After calculation of the right-hand side of eq. 6.241, the following change of variables is made: $d \to -z\,,\ b \to 1/b\,,\ n \to 1/n\,.$

In our case, it is essential that if $P_t^{(1)}$ corresponds to the ratio $n < 1$, then the function $b\,P_t^{(1)}$ describes the refracted wave when $n > 1$. Taking into account this fact let us discuss the wave behavior in the lower medium when that $c_2 < c_1$. First of all, eqs. 6.232 – 6.233 hold regardless of the value of n. Note that with the increase of the distance r or a decrease of the height d, the amplitude corresponding to the ray approximation becomes relatively small, and the "correction" term in eq. 6.232 plays the dominant role. As in the previous case ($n < 1$), an additional (diffracted) wave is also present. After the above indicated interchange of notations in eqs. 6.237 – 6.238 and neglecting the term with N_d, we have:

$$P_d(x, y, z) = \frac{e^{i\, k_2 \sqrt{r^2 + z^2} - k_1 d \sqrt{n^2 \sin^2 \beta - 1}}}{\sqrt{r^2 + z^2}} \frac{2\, n \cos \beta}{n\, b\, \cos \beta + i\, \sqrt{n^2 \sin^2 \beta - 1}} \tag{6.242}$$

Here

$$\beta = \tan^{-1} \frac{r}{z}$$

As we see its amplitude decays rapidly with the increase of distance d between the source and the interface. This may suggest that the additional wave is generated at the boundary right below the source by elementary inhomogeneous waves which represent a portion of the incident wave. After refraction at the boundary these inhomogeneous waves give rise to usual (homogeneous) plane waves, propagating toward the observation point, Fig. 6.23b.

Chapter 7

Propagation of waves inside a layer

In this chapter we investigate the main features of waves caused by a spherical source, and we focus on the formation of normal modes in the frequency and time domain.

7.1 Acoustic potential of an elementary source located inside a layer

Derivation of the potential

Consider the more complicated problem of an elementary spherical source, located inside a layer overlying a half-space at depth d from its free surface, Fig. 7.1. The layer thickness is H, and c_1, ρ_1 and c_2, ρ_2 are the density and acoustic velocity in the layer and lower medium, respectively. In this section we derive formulae describing the waves propagating within the layer and the half-space. It is obvious that a spherical wave, traveling away from the source, experiences reflections from layer boundaries and, correspondingly, gives rise to the multiple reflections called the multiples in seismology. Let us assume at the beginning that the source generates a sinusoidal wave and then consider the transient waves.

Taking into account the axial symmetry of the problem it is natural to choose the cylindrical system of coordinates with the z-axis directed downward. The coordinate origin O is located at the free boundary. As usual, we formulate the boundary value problem in terms of the complex amplitude of potential $\mathcal{U}$, which is related to the pressure $\mathcal{P}$ and displacement $\mathcal{S}$ as

$$\mathcal{P} = \omega^2 \rho \mathcal{U} \qquad \text{and} \qquad \mathcal{S} = \text{grad}\,\mathcal{U} \tag{7.1}$$

The potential should satisfy the following conditions:

(a) At regular points, that is inside the layer and in the lower medium, it obeys the Helmholtz equation

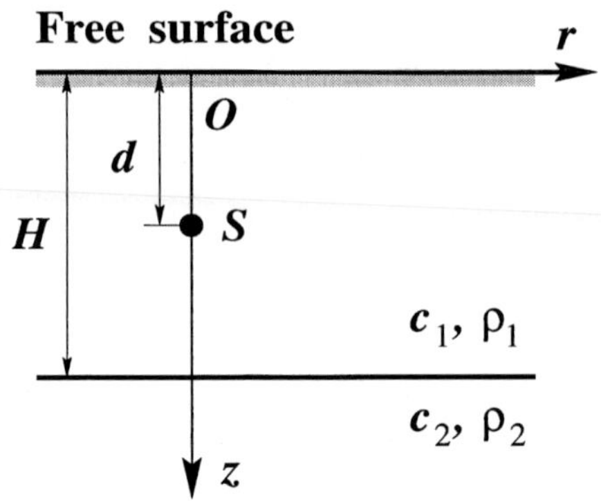

Figure 7.1: Elementary source inside a layer.

$$\nabla^2 \mathcal{U} + k^2 \mathcal{U} = 0\,, \tag{7.2}$$

where

$$k = \begin{cases} k_1 = \dfrac{\omega}{c_1} & \text{if} \quad 0 < z < H \\ k_2 = \dfrac{\omega}{c_2} & \text{if} \quad z > H \end{cases}$$

(b) At the free surface the pressure vanishes, and therefore

$$\mathcal{U} = 0 \qquad \text{if} \quad z = 0 \tag{7.3}$$

(c) At the boundary $z = H$ between the layer and the half-space, the pressure and the vertical component of the displacement are continuous functions. This means that

$$\rho_1 \mathcal{U}_1 = \rho_2 \mathcal{U}_2 \qquad \text{and} \qquad \frac{\partial \mathcal{U}_1}{\partial z} = \frac{\partial \mathcal{U}_2}{\partial z} \qquad \text{if} \qquad z = H \tag{7.4}$$

Here $\mathcal{U}_1$ and $\mathcal{U}_2$ are the potentials in the layer and the half-space, respectively.

(d) The potential of the direct wave generated by the source can be represented as

$$\mathcal{U}_0 = C\,\frac{e^{i\,k_1 R}}{R}\,, \tag{7.5}$$

where C is the known constant and

$$R = \sqrt{r^2 + (z - d)^2} \tag{7.6}$$

is the distance from the source to the observation point.

It is natural that the potential in the layer consists of two parts

$$\mathcal{U}_1 = \mathcal{U}_0 + \mathcal{U}_s \tag{7.7}$$

Here $\mathcal{U}_s$ is the potential of the secondary waves which has to be finite everywhere. Approaching the source, the total potential $\mathcal{U}_1$ tends to $\mathcal{U}_0$:

$$\mathcal{U}_1 \to \mathcal{U}_0 \qquad \text{if} \qquad R \to 0\,, \tag{7.8}$$

and the latter is the condition near the source.

(e) Since the spherical wave excited by the source has finite amplitude, the energy of the total wave field, which includes all reflected, transmitted, and head waves, is also finite. Therefore, the wave amplitude should decrease with the distance from the source. This leads to the condition at infinity

$$\mathcal{U} \to 0 \qquad \text{if} \qquad R \to \infty \tag{7.9}$$

Thus, we have formulated the boundary value problem. As follows from the theorem of uniqueness and from the physical point of view there is only one wave field that satisfies all five conditions, eqs. 7.2 - 7.4 and 7.8 - 7.9.

First we find the solution of the Helmholtz equation. Applying the method of separation of variables and using the results of the previous chapter, the solution of this equation can be written as

$$\mathcal{U}(r,\, z,\, \omega) = \int\limits_0^\infty \left(A_m\, e^{m_n z} + B_m\, e^{-m_n z}\right) J_0(mr)\, dm\,, \tag{7.10}$$

where

$$m_1 = \sqrt{m^2 - k_1^2}\,, \qquad m_2 = \sqrt{m^2 - k_2^2} \qquad \text{and} \qquad \operatorname{Im} m_n \le 0\,, \qquad n = 1,\, 2 \tag{7.11}$$

The coefficients A_m and B_m, which are independent of r and z, are the unknowns.

As we already know, the function $\mathcal{U}$ obeys eq. 7.2 regardless of the values of A_m and B_m. In other words, the Helmholtz equation has an infinite number of solutions, and we have to choose such A_m and B_m that $\mathcal{U}$ satisfies all other conditions of the boundary value problem. With this purpose in mind, we again make use of the Sommerfeld integral

$$\frac{e^{i\,k_1 R}}{R} = \int_0^\infty \frac{m}{m_1}\, e^{-m_1\,|z-d|}\, J_0(mr)\, dm \tag{7.12}$$

Then the expressions for the complex amplitude of the potential in the layer and the lower medium, satisfying conditions near the source and at infinity, are

$$\begin{aligned} \mathcal{U}_1 &= C \int_0^\infty \left(\frac{m}{m_1}\, e^{-m_1\,|z-d|} + A_{1m}\, e^{m_1 z} + B_{1m}\, e^{-m_1 z} \right) J_0(mr)\, dm \\ \mathcal{U}_2 &= C \int_0^\infty B_{2m}\, e^{-m_2 z}\, J_0(mr)\, dm \end{aligned} \tag{7.13}$$

Next we find coefficients A_{1m}, B_{1m} and B_{2m} such that conditions at both interfaces are satisfied. Substitution of eqs. 7.13 into eqs. 7.3–7.4 gives the following set of linear equations:

$$\frac{m}{m_1}\, e^{-m_1\,d} + A_{1m} + B_{1m} = 0\,, \tag{7.14}$$

$$b \left[\frac{m}{m_1}\, e^{-m_1\,(H-d)} + A_{1m}\, e^{m_1 H} + B_{1m}\, e^{-m_1 H} \right] = B_{2m}\, e^{-m_2 H}\,,$$

$$-m\, e^{-m_1\,(H-d)} + m_1\, A_{1m}\, e^{m_1 H} - m_1\, B_{1m}\, e^{-m_1 H} = -m_2\, B_{2m}\, e^{-m_2 H}\,,$$

where

$$b = \frac{\rho_1}{\rho_2} \tag{7.15}$$

Eliminating B_{2m} from the last two equations, we obtain

$$m \left(\frac{m_2}{m_1}\, b - 1 \right) e^{-m_1\,(H-d)} + (b\, m_2 + m_1)\, A_{1m}\, e^{m_1 H} + (b\, m_2 - m_1)\, B_{1m}\, e^{-m_1 H} = 0$$

or

$$A_{1m} + R_{12}\, B_{1m}\, e^{-2\,m_1 H} = -\frac{m}{m_1}\, R_{12}\, e^{-m_1\,(2\,H-d)}$$

where

$$R_{12} = \frac{b\, m_2 - m_1}{b\, m_2 + m_1} \tag{7.16}$$

Thus, in place of the system 7.14, we have

$$\begin{aligned} A_{1m} + B_{1m} &= -\frac{m}{m_1}\, e^{-m_1 d} \\ A_{1m} + R_{12}\, B_{1m}\, e^{-2\,m_1 H} &= -\frac{m}{m_1}\, R_{12}\, e^{-m_1\,(2\,H-d)} \end{aligned} \tag{7.17}$$

Whence

$$B_{1m} = -\frac{m}{m_1}\, e^{-m_1 d}\, \frac{1 - R_{12}\, e^{-2\,m_1\,(H-d)}}{1 - R_{12}\, e^{-2\,m_1 H}} \tag{7.18}$$

and

$$A_{1m} = \frac{m}{m_1}\, R_{12}\, e^{-m_1\,(2\,H+d)}\, \frac{1 - e^{2\,m_1 d}}{1 - R_{12}\, e^{-2\,m_1 H}} \tag{7.19}$$

The last two equations of the set 7.14 yield

$$B_{2m}\, e^{-m_2 H} = \frac{2\,b\,e^{-m_1 H}\left(m\,e^{m_1 d} + m_1\, B_{1m}\right)}{b\,m_2 + m_1}$$

Substitution of B_{1m} into the latter gives

$$\begin{aligned} m\,e^{m_1\,d} + m_1\,B_{1m} &= m\left[e^{m_1\,d} - e^{-m_1\,d}\,\frac{1 - R_{12}\,e^{-2\,m_1\,(H-d)}}{1 - R_{12}\,e^{-2\,m_1 H}}\right] \\ &= \frac{m}{1 - R_{12}\,e^{-2\,m_1 H}}\left(e^{m_1 d} - e^{-m_1 d}\right) \end{aligned}$$

Hence

$$B_{2m} = \frac{2\,b\,m\,e^{-m_1 H}\,e^{m_2 H}\left(e^{m_1 d} - e^{-m_1 d}\right)}{(b\,m_2 + m_1)\left(1 - R_{12}\,e^{-2\,m_1 H}\right)} \tag{7.20}$$

Thus, we have solved the boundary value problem and expressed the potentials in each medium in terms of the integrals. Correspondingly, in general, computation of the wave fields is reduced to numerical integration.

Consider several special cases.

(a) If $\rho_1 = \rho_2$ and $c_1 = c_2$, we arrive at the expressions for a homogeneous half-space with the free surface. As follows from eqs. 7.18–7.20, in this case

$$B_{1m} = -\frac{m}{m_1}\, e^{-m_1 d} \qquad \text{and} \qquad A_{1m} = 0\,,$$

since $R_{12} = 0$, and

$$B_{2m} = \frac{m}{m_1}\left(e^{m_1 d} - e^{-m_1 d}\right)$$

(b) If the boundary $z = H$ is also free

$$b = \frac{\rho_1}{\rho_2} \to \infty\,, \qquad m_2 = m \qquad \text{and} \qquad R_{12} = 1$$

Consequently,

$$\begin{aligned} A_{1m} &= \frac{m}{m_1}\, e^{-m_1\,(2\,H+d)}\, \frac{1 - e^{2\,m_1 d}}{1 - e^{-2\,m_1 H}}\,, \\ B_{1m} &= -\frac{m}{m_1}\, e^{-m_1 d}\, \frac{1 - e^{-2\,m_1\,(H-d)}}{1 - e^{-2\,m_1 H}} \end{aligned} \tag{7.21}$$

(c) In contrast, if the lower medium is rigid, then $R_{12} = -1$ and

$$\begin{aligned} A_{1m} &= -\frac{m}{m_1}\, e^{-m_1\,(2\,H+d)}\, \frac{1 - e^{2\,m_1 d}}{1 + e^{-2\,m_1 H}}\,, \\ B_{1m} &= -\frac{m}{m_1}\, e^{-m_1 d}\, \frac{1 + e^{-2\,m_1\,(H-d)}}{1 + e^{-2\,m_1 H}} \end{aligned} \tag{7.22}$$

(d) Suppose that the source approaches the free surface, that is $d \to 0$. Therefore,

$$B_{1m} \to -\frac{m}{m_1}\,, \qquad A_{1m} \to 0\,, \qquad B_{2m} \to 0$$

and

$$\mathcal{U}_1 \to 0\,, \qquad \mathcal{U}_2 \to 0$$

This means that the total wave field disappears due to the boundary condition at the free surface (eq. 7.3).

7.2 Expansion of integrands in series

As before, along with numerical integration, it is useful to derive asymptotic expressions for the wave field which are valid within a certain range of frequencies and observation distances. With this purpose in mind consider two methods, and the first one is based on the expansion of the integrands in eqs. 7.18–7.20 in a series. As is well known,

$$\frac{1}{1-R_{12}\,e^{-2m_1 H}} = \sum_{n=0}^{\infty} R_{12}^n e^{-2nm_1 H} \tag{7.23}$$

As $|R_{12}| \leq 1$, this series converges, except in the case when

$$|R_{12}| = 1 \qquad \text{and} \qquad \text{Re } m_1 = 0\,,$$

and it happens if the lower boundary is either free or rigid. Substituting eq. 7.23 into eqs. 7.18–7.20 and making use of the first equation of the set 7.13, we can represent the potential $\mathcal{U}_1$ as

$$\begin{aligned} \mathcal{U}_1 &= C \left[\frac{e^{i\,k_1 R}}{R} + \sum_{n=0}^{\infty} \int_0^{\infty} \frac{m}{m_1}\, L_n\, e^{m_1 z}\, J_0(mr)\, dm - \right. \\ &\qquad \left. - \sum_{n=0}^{\infty} \int_0^{\infty} \frac{m}{m_1}\, M_n\, e^{-m_1 z}\, J_0(mr)\, dm \right], \end{aligned} \tag{7.24}$$

where

$$\begin{aligned} L_n &= e^{-m_1 d}\, R_{12}^{n+1}\, e^{-2m_1 (n+1) H} \left(1 - e^{2m_1 d}\right) \\ M_n &= e^{-m_1 d}\, R_{12}^{n}\, e^{-2m_1 n H} \left[1 - R_{12}\, e^{-2m_1 (H-d)}\right] \end{aligned} \tag{7.25}$$

The latter shows that there is only one term with the zero power of R_{12}. Ignoring all other terms, we have

$$\mathcal{U}_{10} = C \left[\frac{e^{i\,k_1 R}}{R} - \int_0^{\infty} \frac{m}{m_1}\, e^{-m_1 (z+d)}\, J_0(mr)\, dm \right]$$

or

$$\mathcal{U}_{10} = C \left(\frac{e^{i\,k_1 R}}{R} - \frac{e^{i\,k_1 R_1}}{R_1} \right), \tag{7.26}$$

where

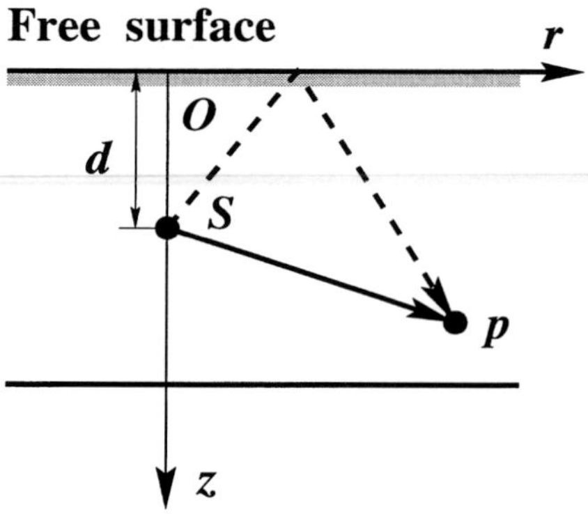

Figure 7.2: Reflection from free surface.

$$R_1 = \sqrt{r^2 + (z+d)^2}$$

The first term in the right-hand side of eq. 7.26 describes the direct wave while the second one characterizes the reflection from the free surface. It is clear that the reflected wave is also spherical.

Although there are two integrals in eq. 7.24 for each given $n \neq 0$ that contain the term R_{12}^n, they cannot be associated with any particular reflected wave. However, there is one exception, and in order to investigate it consider the integral of the following type:

$$\begin{aligned} I_n &= \int\limits_0^\infty \frac{m}{m_1} R_{12}^l\, e^{-m_1 z_n}\, J_0(mr)\, dm \\ &= \int\limits_0^\infty \frac{m}{m_1} \left(\frac{b\,m_2 - m_1}{b\,m_2 + m_1}\right)^l e^{-m_1 z_n}\, J_0(mr)\, dm\,, \end{aligned} \tag{7.27}$$

which is related to potentials $\mathcal{U}_1$ and $\mathcal{U}_2$. The parameters l and z_n in eq. 7.27 are related to n, z, H and d.

Assuming that the observations are performed relatively far away from the source, $k_1\, R > 1$, we can apply the method of the stationary phase in the same manner as it was previously done. In particular, we replace the Bessel function $J_0(mr)$ with its asymptotic,

$$J_0(mr) \approx \sqrt{\frac{2}{\pi\, m\, r}}\, \cos\left(m\, r - \frac{\pi}{4}\right) = \sqrt{\frac{1}{2\,\pi\, m\, r}}\, \left[e^{i\,(m\,r - \pi/4)} + e^{-i\,(m\,r-\pi/4)}\right]$$

Consequently, the stationary point m_{0n} is defined from the equality $\partial\,\varphi/\partial\, m = 0$. Here

$$\varphi(m) = -m\,r - n_1 z_n + \frac{\pi}{4}\,, \qquad n_1 = \sqrt{k_1^2 - m^2}$$

Whence

$$m_{0n} = \frac{k_1 r}{\sqrt{r^2 + z_n^2}} = \frac{k_1 r}{R_n} = k_1 \sin\theta_n \tag{7.28}$$

Note that the other exponent, describing the Bessel function, does not have a stationary point within the interval of integration. Correspondingly, the ratio

$$R_{12} = \frac{b\,m_2 - m_1}{b\,m_2 + m_1}$$

at the stationary point becomes

$$R_{12} = \frac{b\sqrt{k_1^2 \sin^2\theta_n - k_2^2} - \sqrt{k_1^2 \sin^2\theta_n - k_1^2}}{b\sqrt{k_1^2 \sin^2\theta_n - k_2^2} + \sqrt{k_1^2 \sin^2\theta_n - k_1^2}}$$

$$= \frac{b\sqrt{k_2^2 - k_1^2 \sin^2\theta_n} - k_1 \cos\theta_n}{b\sqrt{k_2^2 - k_1^2 \sin^2\theta_n} + k_1 \cos\theta_n}$$

or

$$R_{12} = \frac{Z_1\sqrt{1 - \dfrac{\sin^2\theta_n}{n^2}} - Z_2 \cos\theta_n}{Z_1\sqrt{1 - \dfrac{\sin^2\theta_n}{n^2}} + Z_2 \cos\theta_n} \tag{7.29}$$

The latter is known as the Rayleigh coefficient .

Assuming that the vicinity of the stationary point gives the main contribution to the integral, in place of eq. 7.27, we have

$$I_n \approx R_{12}^l(m_{0n}) \int_0^\infty \frac{m}{m_1}\, e^{-m_1 |z_n|}\, J_0(mr)\, dm \qquad \text{if} \quad k_1 r > 1 \tag{7.30}$$

Then, the use of the Sommerfeld integral gives

$$I_n \approx R_{12}^l(m_{0n}) \frac{e^{i\,k_1 R_n}}{R_n} = \beta_n \frac{e^{i\,k_1 R_n}}{R_n}\,, \tag{7.31}$$

where

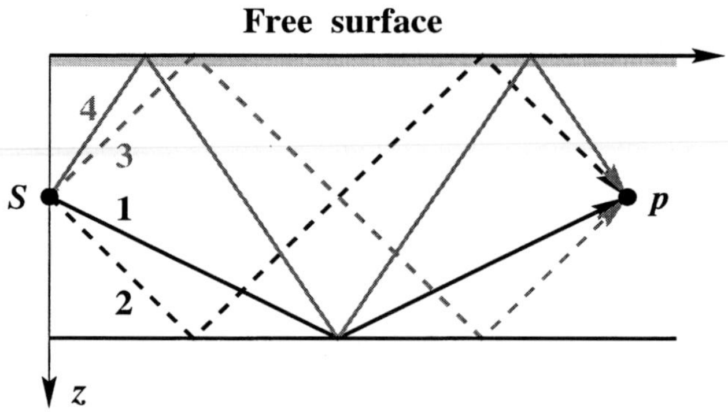

Figure 7.3: Rays reflected once from the lower boundary of the layer.

$$\beta_n = R_{12}^l(m_{0n}), \qquad R_n = (r^2 + z_n^2)^{1/2}$$

Substitution of latter into eq. 7.24 clearly shows that in the wave zone the total wave field is represented by superposition of spherical waves which experience multiple reflections at both layer boundaries and decay as $1/r$, provided that $r \gg z_n$. The center of phase surfaces of those waves is located on the z-axis ($r = 0$) and it moves away from the layer when n increases. Respectively, the apparent incidence angle θ_n becomes smaller. It is essential that the rays of each spherical wave obey Snell's law.

Consider the waves arriving at point p after a single reflection from the layer bottom. Therefore, their amplitudes are proportional to R_{12}. As we already know, there are four such waves and, in accordance with eqs. 7.25 and 7.30, the centers of their spherical fronts are located at points of the z-axis

$$z = \pm(2\,H - d) \qquad \text{and} \qquad z = \pm(2\,H + d)$$

Correspondingly, those waves can be described with the help of rays shown in Figures 7.2-7.3. In the same manner we can plot the ray trajectories that illustrate multiple reflections from the lower boundary. Inasmuch as those rays have different incidence angles θ_n, their apparent velocities differ from each other. Thus, the use of the concept of the stationary point and the Sommerfeld integral allows us to treat the total wave field as an infinite sum of spherical waves, provided that $r > \lambda_1$. In contrast, at relatively small distances, $r < \lambda_1$ diffusive harmonics may play a significant role, and the wave behavior becomes more complicated.

7.3 Integration along branch cuts and around poles

Our goal is to derive asymptotic expressions for the potential in the wave zone, that are simpler than the exact formulae 7.13. Expansion of the function $\mathcal{U}_1$ in an infinite sum of spherical waves, however, is not always convenient. For this reason, we make use of the method based on deformation of the path of integration in complex plane m. As in section 6.3, we proceed from Cauchy's theorem, according to which an integral of an analytic function along a closed path is determined by its singularities located inside the area surrounded by this path. Before we apply this approach, it is useful to represent the coefficients A_{1m}, B_{1m} and B_{2m}, eqs. 7.18–7.20 in a different form. Let us start from B_2, eq. 7.20. Since

$$\frac{1}{1-R_{12}\,e^{-2m_1H}} = \frac{b\,m_2+m_1}{(b\,m_2+m_1)-(b\,m_2-m_1)\,e^{-2m_1H}}$$

$$= \frac{b\,m_2+m_1}{b\,m_2\,(1-e^{-2m_1H})+m_1\,(1+e^{-2m_1H})},$$

we obtain

$$B_{2m} = \frac{2\,b\,m\,e^{m_2H}\,\sinh m_1 d}{b\,m_2\,\sinh m_1 H + m_1\,\cosh m_1 H} \tag{7.32}$$

It follows from eq. 7.19 that

$$A_{1m} = \frac{m}{m_1}\,e^{-m_1d}\,\frac{1-R_{12}\,e^{-2m_1(H-d)}}{1-R_{12}\,e^{-2m_1H}} - \frac{m}{m_1}\,e^{-m_1d}$$

$$= \frac{m}{m_1}\,e^{-m_1d}\,\frac{(b\,m_2+m_1)-(b\,m_2-m_1)\,e^{-2m_1(H-d)}}{(b\,m_2+m_1)-(b\,m_2-m_1)\,e^{-2m_1H}} - \frac{m}{m_1}\,e^{-m_1d}$$

$$= \frac{m}{m_1}\,e^{-m_1d}\,\frac{b\,m_2\left[1-e^{-2m_1(H-d)}\right]+m_1\left[1+e^{-2m_1(H-d)}\right]}{b\,m_2\,(1-e^{-2m_1H})+m_1\,(1+e^{-2m_1H})} - \frac{m}{m_1}\,e^{-m_1d}$$

Thus

$$A_{1m} = \frac{m}{m_1}\,\frac{b\,m_2\,\sinh m_1(H-d)+m_1\,\cosh m_1(H-d)}{b\,m_2\,\sinh m_1H+m_1\,\cosh m_1H} - \frac{m}{m_1}\,e^{-m_1d} \tag{7.33}$$

From eq. 7.18 we obtain

$$B_{1m} = -\frac{m}{m_1}\,\frac{b\,m_2\,\sinh m_1(H-d)+m_1\,\cosh m_1(H-d)}{b\,m_2\,\sinh m_1H+m_1\,\cosh m_1H} \tag{7.34}$$

Consider first the potential inside the layer. Introducing the notation

$$D = b\,m_2\,\sinh m_1 H + m_1\,\cosh m_1 H \tag{7.35}$$

and making use of eq. 7.13, we have

$$\mathcal{U}_1 = C\int_0^\infty \frac{m}{m_1}\left(e^{-m_1\,|z-d|} - e^{-m_1 d}\,e^{m_1 z}\right) J_0(mr)\,dm \tag{7.36}$$

$$+\,2C\int_0^\infty \frac{m}{m_1}\,\frac{b\,m_2\,\sinh m_1(H-d) + m_1\,\cosh m_1(H-d)}{D}\,\sinh m_1 z\,J_0(mr)\,dm$$

At the beginning suppose that $z < d$. Then $|z-d| = d - z$ and

$$e^{-m_1\,(d-z)} - e^{-m_1 d}\,e^{m_1 z} = 0$$

Therefore

$$\mathcal{U}_1 = 2C\int_0^\infty \frac{m}{m_1}\,\frac{b\,m_2\,\sinh m_1(H-d) + m_1\,\cosh m_1(H-d)}{D}\,\sinh m_1 z\,J_0(mr)\,dm \qquad \text{if} \quad z < d \tag{7.37}$$

In the opposite case when $z > d$, we have

$$e^{-m_1\,|z-d|} - e^{-m_1 d}\,e^{m_1 z} = e^{-m_1\,(z-d)} - e^{m_1\,(z-d)} = -2\,\sinh m_1(z-d)$$

Thus

$$\mathcal{U}_1 = 2C\int_0^\infty \frac{m}{m_1}\left[\frac{b\,m_2\,\sinh m_1(H-d)}{D}\,\sinh m_1 z + \frac{m_1\,\cosh m_1(H-d)}{D}\,\sinh m_1 z - \sinh m_1\,(z-d)\right] J_0(mr)\,dm$$

The expression in brackets can be represented as

$$[\ldots] = \frac{1}{D}\,\{b\,m_2\,[\sinh m_1(H-d)\,\sinh m_1 z - \sinh m_1 H\,\sinh m_1(z-d)] + m_1\,[\cosh m_1(H-d)\,\sinh m_1 z - \cosh m_1 H\,\sinh m_1(z-d)]\} \tag{7.38}$$

Using well-known expressions for $\sinh(x \pm y)$ and $\cosh(x \pm y)$ the right-hand side of eq. 7.38 can be greatly simplified:

$$b\,m_2\,[\ldots] = b\,m_2 \sinh m_1 d \sinh m_1\,(H - z)$$

Also

$$m_1\,[\ldots] = m_1 \sinh m_1 d \cosh m_1\,(H - z)$$

Therefore

$$\mathcal{U}_1 = 2\,C \int_0^\infty \frac{m}{m_1} \frac{b\,m_2 \sinh m_1(H - z) + m_1 \cosh m_1(H - z)}{D} \sinh m_1 d\, J_0(mr)\, dm \qquad \text{if} \quad z > d \tag{7.39}$$

After simple but rather cumbersome transformations, we obtain

$$\begin{aligned} \mathcal{U}_1 &= 2\,C \int_0^\infty m\, F_1(m^2)\, J_0(mr)\, dm\,, \\ \mathcal{U}_2 &= 2\,C \int_0^\infty m\, F_2(m^2)\, J_0(mr)\, dm\,, \end{aligned} \tag{7.40}$$

where $F_1(m^2)$ and $F_2(m^2)$ are the even functions of m given by

$$F_1(m^2) = \frac{1}{D} \begin{cases} \dfrac{\sinh m_1 z}{m_1}\,[b\,m_2 \sinh\, m_1(H - d) + m_1 \cosh m_1\,(H - d)] & \text{if} \quad z \le d \\[2ex] \dfrac{\sinh m_1 d}{m_1}\,[b\,m_2 \sinh\, m_1(H - z) + m_1 \cosh m_1\,(H - z)] & \text{if} \quad z \ge d \end{cases} \tag{7.41}$$

and, eq. 7.32,

$$F_2(m^2) = \frac{b}{D}\, e^{m_2(H - z)} \sinh\, m_1 d \tag{7.42}$$

Now we are ready to apply the Cauchy's theorem , so we again use the equality

$$2\,J_0(mr) = H_0^{(1)}(mr) + H_0^{(2)}(mr)$$

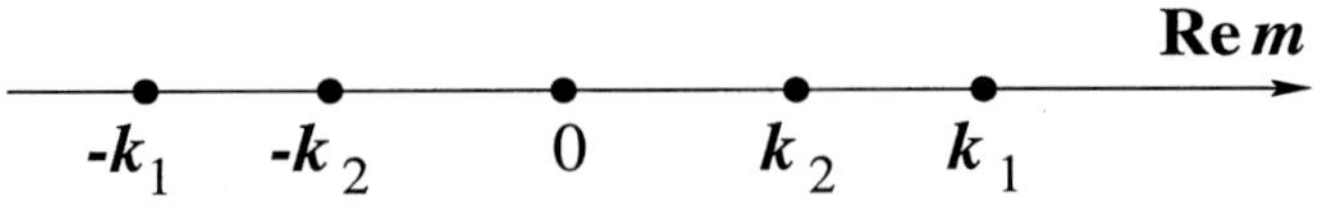

Figure 7.4: Singularities on the real m-axis.

Substitution of the latter into eq. 7.40 yields

$$\mathcal{U}_1 = C\left[\int\limits_0^\infty m\,F_1(m^2)\,H_0^{(1)}(mr)\,dm + \int\limits_0^\infty m\,F_1(m^2)\,H_0^{(2)}(mr)\,dm\right]$$

or

$$\mathcal{U}_1 = C\,(I_1 + I_2)\,, \tag{7.43}$$

where

$$I_1 = \int\limits_0^\infty m\,F_1(m^2)\,H_0^{(1)}(mr)\,dm\,,$$

$$I_2 = \int\limits_0^\infty m\,F_1(m^2)\,H_0^{(2)}(mr)\,dm \tag{7.44}$$

In order to choose the new integration paths properly and apply Cauchy's theorem, it is necessary to consider the singularities of the function $F_1(m^2)$. First, there are branch points defined by the equalities

$$m_1 = 0 \qquad \text{and} \qquad m_2 = 0 \tag{7.45}$$

and the latter give four such points located on the real m-axis:

$$k_1\,, \quad k_2\,, \quad -k_1\,, \quad -k_2 \tag{7.46}$$

In accordance with eq. 7.18, the function $F_1(m^2)$ may have poles, and their position is found from the condition $D = 0$, that is,

$$\frac{b\,m_2 - m_1}{b\,m_2 + m_1}\,e^{-2\,m_1 H} = 1 \tag{7.47}$$

Suppose that $c_2 > c_1$ or $k_1 > k_2$ and consider the axis Re m, Fig. 7.4. If $m > k_1$, coefficient R_{12} and exponent $e^{-m_1 H}$ are real quantities less than unity. This means that

$$R_{12}\,e^{-2\,m_1 H} < 1\,,$$

and therefore the poles are absent when $m > k_1$. Next, assume that $m < k_2$. Correspondingly,

$$m_1 = -i\sqrt{k_1^2 - m^2} \qquad \text{and} \qquad m_2 = -i\sqrt{k_2^2 - m^2}$$

are purely imaginary, but R_{12} still remains real. Since

$$\left|\exp\left(-2\,i\,H\sqrt{k_1^2 - m^2}\right)\right| = 1 \qquad \text{and} \qquad |R_{12}| < 1\,,$$

the condition 7.47 is not satisfied, that is the poles are also absent.

Now consider the interval between the branch points

$$k_2 < m < k_1$$

In this case, the condition for the existence of poles becomes

$$\frac{b\,m_2 + i\,|\,m_1\,|}{b\,m_2 - i\,|\,m_1\,|}\,e^{-2\,i\,|\,m_1\,|H} = 1\,, \tag{7.48}$$

where

$$m_2 = |\,m_2\,| \qquad \text{and} \qquad |\,m_1\,| = \sqrt{k_1^2 - m^2}$$

It is clear that regardless of the value of m, the magnitude of the complex number in the left-hand side of eq. 7.48 is equal to unity. Therefore, if its argument is zero, the poles are located between the branch points.

The same analysis shows that in the opposite case, when $c_2 < c_1$, the poles are also absent if $m > k_2$ and $m < k_1$. For the interval $k_1 < m < k_2$, the left-hand side of eq. 7.47 is

$$\frac{i\,b\,|\,m_2\,| - |\,m_1\,|}{i\,b\,|\,m_2\,| + |\,m_1\,|}\,e^{-2\,|\,m_1 H\,|} < 1$$

Correspondingly, there are no poles along the real axis of m. At the same time, the study of eq. 7.47 shows that its roots, that is the poles, are complex.

Next we analyze only the case $c_2 > c_1$ when the real poles are located between the branch points.

Layer has lower velocity, $c_2 > c_1$

It is convenient to assume, as was done in section 6.3, that all singularities are slightly moved away from the real m-axis and, then, after integration, they are placed back to their original positions. Making use of this approach, we change the integration path, eq. 7.40,

$$0 \le m < \infty$$

and first consider the integral

$$L_1 = \oint m\,F_1(m^2)\,H_0^{(2)}(mr)\,dm$$

along the closed line surrounding the forth quadrant, Fig. 7.5a. Since the singularities are absent inside this area, L_1 vanishes, and we have

$$\oint m\,F_1(m^2)\,H_0^{(2)}(mr)\,dm = 0$$

or

$$\int_0^{\infty} m\,F_1(m^2)\,H_0^{(2)}(mr)\,dm + \int_{C_\infty} m\,F_1(m^2)\,H_0^{(2)}(mr)\,dm$$
$$+ \int_{-i\infty}^{0} m\,F_1(m^2)\,H_0^{(2)}(mr)\,dm = 0 \qquad (7.49)$$

Taking into account that the path element C_∞ has an infinitely large radius, and function $H_0^{(2)}(mr) \to 0$ when $mr \to \infty$, the second integral in eq. 7.49 is zero. This gives

$$\int_0^{\infty} m\,F_1(m^2)\,H_0^{(2)}(mr)\,dm = -\int_{-i\infty}^{0} m\,F_1(m^2)\,H_0^{(2)}(mr)\,dm$$

From the second equation of the set 7.44, we have

$$I_2 = \int_0^{-i\infty} m\,F_1(m^2)\,H_0^{(2)}(mr)\,dm \qquad (7.50)$$

Thus, the integration along the real axis is replaced with that along the negative part of the imaginary axis.

Next, consider the integral

$$L_2 = \oint_C m\,F_1(m^2)\,H_0^{(1)}(mr)\,dm\,,$$

where C is also a closed path that surrounds the first quadrant in such a way that the singularities are situated outside it, Fig. 7.5b. Then, as follows from Cauchy's theorem,

$$L_2 = \oint_C m\,F_1(m^2)\,H_0^{(1)}(mr)\,dm = 0 \qquad (7.51)$$

Let us note that, for convenience, only two poles are shown in Fig. 7.5b. The path C, which is more complicated than the previous one, Fig. 7.5a, includes several elements, namely,

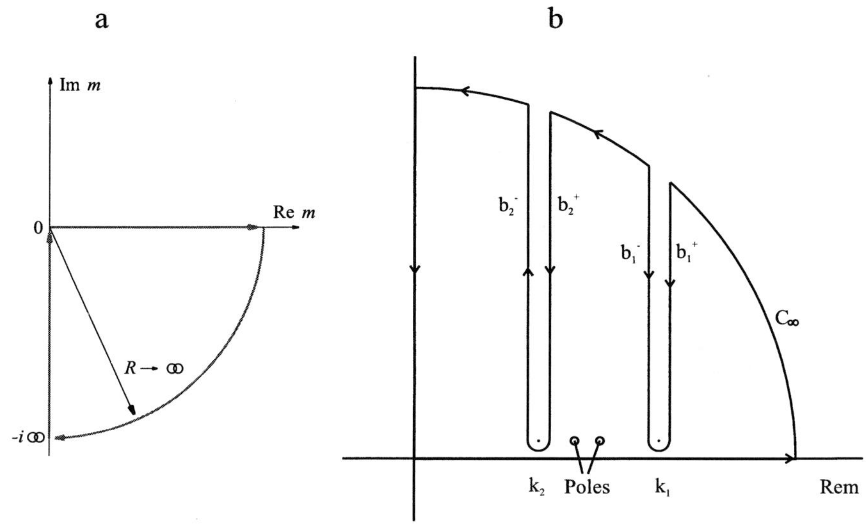

Figure 7.5: Integration paths.

(a) the real axis of m $0 \leq m < \infty$,

(b) branch lines b_1^+, b_1^- and b_2^+, b_2^-,

(c) circles around the poles,

(d) the portion of the circle C_∞ with infinitely large radius $R \to \infty$,

(e) the imaginary axis of m.

Because of the exponential decay of function $H_0^{(1)}(mr)$ for large values of its argument, the integral along the segment C_∞ is zero.

Therefore, in place of eq. 7.51 we can write

$$\int_0^\infty m\, F_1(m^2)\, H_0^{(1)}(mr)\, dm + \sum_b + \sum_p + \int_{i\infty}^0 m\, F_1(m^2)\, H_0^{(1)}(mr)\, dm = 0 \tag{7.52}$$

Here $\sum_b$ is the sum of integrals along the branch cuts, while $\sum_p$ is the sum of integrals around the poles.

In accordance with the first equation of set 7.44, we have

$$I_1 = -\sum_b - \sum_p - \int_{i\infty}^0 m\, F_1(m^2)\, H_0^{(1)}(mr)\, dm \tag{7.53}$$

Due to the equality

$$H_0^{(2)}(mr) = -H_0^{(1)}(-mr)\,,$$

the integral I_2, eq. 7.50, can be represented as

$$\begin{aligned} I_2 &= -\int\limits_0^{-i\infty} m\,F_1(m^2)\,H_0^{(1)}(-mr)\,dm \\ &= -\int\limits_0^{i\infty} m\,F_1(m^2)\,H_0^{(1)}(mr)\,dm \\ &= \int\limits_{i\infty}^{0} m\,F_1(m^2)\,H_0^{(1)}(mr)\,dm \end{aligned} \tag{7.54}$$

Thus,

$$I_1 + I_2 = -\sum_b - \sum_p \tag{7.55}$$

and, correspondingly, the expression for the potential inside the layer, eq. 7.43, becomes

$$\mathcal{U}_1 = -C\left(\sum_b + \sum_p\right) \tag{7.56}$$

It is essential that integration along the imaginary axis is eliminated and the function $\mathcal{U}_1$ is expressed in terms of integrals around the poles and along the branch cuts.

As follows from the theory of functions of a complex variable (Appendix A),

$$-\sum_p = 2\,\pi\,i\sum_{j=1}^{N} \mathrm{Res}_j\,, \tag{7.57}$$

where the right-hand side is the sum of residues of the function

$$m\,F_1(m^2)\,H_0^{(1)}(mr)$$

Finally, we have

$$\mathcal{U}_1 = \mathcal{U}_{1p} + \mathcal{U}_{1b} \tag{7.58}$$

Here

$$\mathcal{U}_{1p} = 2\,\pi\,i\sum_{j=1}^{N} \mathrm{Res}_j \qquad \text{and} \qquad \mathcal{U}_{1b} = -\sum_b \tag{7.59}$$

7.4 Normal modes

Numerical integration of eqs. 7.40 allows us to find the wave field at any distance r from the source. This task is relatively simple when separation between the source and the observation point is sufficiently small, and this case is of great practical importance for exploration seismology. With the increase of distance r, however, evaluation of infinite integrals (eq. 7.40) becomes more difficult because of rapid oscillations of the Bessel function $J_0(mr)$. For example, such a situation arises when we investigate propagation of waves in water layer, provided that the distance r is much greater than its thickness. Consequently, along with numerical integration, it is useful to derive asymptotic formulae describing these waves. Besides, this procedure allows us to understand wave behavior better. With this purpose in mind, we make use of eqs. 7.59 and, first, examine waves inside layers, which are related to the poles

$$\mathcal{U}_{1p} = 2\pi i C \sum \mathrm{Res}\,, \tag{7.60}$$

where Res is the residue of the integrand, eqs. 7.44,

$$m\, F_1(m^2)\, H_0^{(1)}(mr)$$

Such waves are usually called normal modes.

Suppose that the z-coordinate of the observation point satisfies the condition

$$d < z < H\,,$$

but later this restriction will be removed. Then, in accordance with eq. 7.41, we have

$$F_1(m^2) = \frac{\sinh m_1 d}{m_1} \frac{b\, m_2 \sinh m_1 (H-z) + m_1 \cosh m_1 (H-z)}{D}$$

and

$$D = b\, m_2 \sinh m_1 H + m_1 \cosh m_1 H.$$

Taking into account that

$$\sinh(-i\,x) = -i \sin x \qquad \text{and} \qquad \cosh(-i\,x) = \cos x\,,$$

the function $F_1(m^2)$ can be represented as

$$F_1(m^2) = \frac{\sin \tilde{m}_1 d}{\tilde{m}_1} \frac{b\, m_2 \sin \tilde{m}_1 (H-z) + \tilde{m}_1 \cos \tilde{m}_1 (H-z)}{b\, m_2 \sin \tilde{m}_1 H + \tilde{m}_1 \cos \tilde{m}_1 H}\,, \tag{7.61}$$

where

$$\tilde{m}_1 = \sqrt{k_1^2 - m^2} \qquad \text{and} \qquad m_2 = \sqrt{m_2 - k_2^2}$$

Correspondingly, the integrand in eq. 7.44 is

$$m\,F_1(m^2)\,H_0^{(1)}(mr) = \frac{\phi_1(m)}{\phi_2(m)} \tag{7.62}$$

Here

$$\begin{aligned} \phi_1(m) &= \frac{m}{\tilde{m}_1}\sin\tilde{m}_1 d\,\left[b\,m_2\,\sin\tilde{m}_1\,(H-z) + \tilde{m}_1\,\cos\tilde{m}_1\,(H-z)\right]\,H_0^{(1)}(mr)\,, \\ \phi_2(m) &= b\,m_2\,\sin\tilde{m}_1 H + \tilde{m}_1\,\cos\tilde{m}_1 H \end{aligned} \tag{7.63}$$

We assume throughout most of this section that velocity c_2 in the lower medium exceeds that in the layer:

$$c_2 > c_1$$

As was shown earlier, the real poles, x_n, of the integrand are located between the branch points

$$k_2 < x_n < k_1$$

and their positions are defined by the condition

$$\phi_2(x_n) = 0 \qquad \text{or} \qquad \tan\tilde{m}_1 H = -\frac{\tilde{m}_1}{b\,m_2} \tag{7.64}$$

Inasmuch as the poles are simple (Appendix A), their residue is determined as

$$\text{Res}\,\frac{\phi_1(m)}{\phi_2(m)} = \frac{\phi_1(m)}{\phi_2'(m)}\,, \tag{7.65}$$

where

$$\phi'(m) = \frac{d\,\phi(m)}{d\,m}$$

The last equation of set 7.63 gives

$$\begin{aligned} \frac{d\,\phi_2}{d\,m} &= \frac{b\,x_n}{m_2}\,\sin\tilde{m}_1 H - \frac{b\,m_2\,x_n H}{\tilde{m}_1}\,\cos\tilde{m}_1 H - \frac{x_n}{\tilde{m}_1}\,\cos\tilde{m}_1 H + x_n H\,\sin\tilde{m}_1 H \\ &= \frac{x_n}{\tilde{m}_1}\left[H\,(\tilde{m}_1\,\sin\tilde{m}_1 H - b\,m_2\,\cos\tilde{m}_1 H) - \cos\tilde{m}_1 H + \frac{b\,\tilde{m}_1}{m_2}\,\sin\tilde{m}_1 H\right] \end{aligned} \tag{7.66}$$

Substitution of eq. 7.64 into eq. 7.66 yields

$$\frac{d\,\phi_2}{d\,m} = \frac{x_n}{\tilde{m}_1}\left[H\left(\tilde{m}_1\,\sin\tilde{m}_1 H + \frac{\tilde{m}_1\,\cos\tilde{m}_1 H}{\tan\tilde{m}_1 H}\right)\right.$$

$$\left. -\cos\tilde{m}_1 H - b^2\,\sin\tilde{m}_1 H\,\tan\tilde{m}_1 H\right]$$

or

$$\frac{d\,\phi_2}{d\,m} = \frac{x_n}{\tilde{m}_1\,\sin\tilde{m}_1 H}\left[\tilde{m}_1\,H - \sin\tilde{m}_1 H\,\cos\tilde{m}_1 H - b^2\sin^2\tilde{m}_1 H\,\tan\tilde{m}_1 H\right] \tag{7.67}$$

Consider the function $\phi_1(m)$. The sum in brackets in eq. 7.63, can be simplified the following way:

$$[\ldots] = \cos\tilde{m}_1 z\,(b\,m_2\sin\tilde{m}_1 H + \tilde{m}_1\,\cos\tilde{m}_1 H)$$

$$+\sin\tilde{m}_1 z\,(\tilde{m}_1\,\sin\tilde{m}_1 H - b\,m_2\cos\tilde{m}_1 H)$$

Making use of eq. 7.64, the latter becomes

$$[\ldots] = \sin\tilde{m}_1 z\left(\tilde{m}_1\,\sin\tilde{m}_1 H + \tilde{m}_1\,\frac{\cos^2\tilde{m}_1 H}{\sin\tilde{m}_1 H}\right) = \frac{\tilde{m}_1\,\sin\tilde{m}_1 z}{\sin\tilde{m}_1 H}$$

Thus,

$$\phi_1(x_n) = \frac{x_n\,\sin\tilde{m}_1 z}{\sin\tilde{m}_1 H}\,\sin\tilde{m}_1 d\;H_0^{(1)}(x_n r) \tag{7.68}$$

Finally, from eqs. 7.60, 7.65, 7.67, and 7.68, we obtain

$$\mathcal{U}_{1p} = \frac{2\,\pi\,i}{H}\,C\sum_n \frac{\tilde{m}_1 H\,\sin\tilde{m}_1 z\,\sin\tilde{m}_1 d}{M_n}\,H_0^{(1)}(x_n r) \tag{7.69}$$

and

$$M_n = \tilde{m}_1 H - \sin\tilde{m}_1 H\,\cos\tilde{m}_1 H - b^2\tan m_1 H\,\sin^2 m_1 H\,,$$

where

$$\tilde{m}_1 = \sqrt{k_1^2 - x_n^2}$$

We note the following.

(a) To avoid confusion with m_1 and m_2, the notation x_n for poles is used.

(b) We have arrived at eq. 7.69 assuming that $z > d$.

However, similar derivations for the second equation of the set 7.41 show that eq. 7.69 remains valid for all points inside the layer except the source location.

Consider some features of the function $\mathcal{U}_{1p}$, eq. 7.69. When the argument $x_n r$ is small, $x_n r \ll 1$, as is usually the case in the near zone, none of the terms of the sum displays wave behavior. Rather they have the meaning of the diffusive harmonics. In contrast, if $x_n r > 1$, these terms can be interpreted as waves. In fact, taking into account the asymptotic expression for the Hankel function

$$H_0^{(1)}(x_n r) \approx \sqrt{\frac{2}{\pi\, x_n\, r}}\, e^{i\,(x_n r - \pi/4)}\,,$$

eq. 7.69 becomes

$$\mathcal{U}_{1p} = \frac{2\,C}{H}\sqrt{\frac{2\,\pi}{r}}\sum_n G_n\, e^{i\,(x_n r + \pi/4)}\,, \tag{7.70}$$

where

$$G_n = \frac{r_n\,\sin r_n \frac{d}{H}\,\sin r_n \frac{z}{H}}{\sqrt{x_n}\,(r_n - \sin r_n\,\cos r_n - b^2\,\tan r_n\,\sin^2 r_n)} \tag{7.71}$$

and

$$r_n = \tilde{m}_1 H \tag{7.72}$$

In accordance with the definition

$$U_{1p} = \text{Re}\,(\mathcal{U}_{1p}\, e^{-i\,\omega\, t})\,,$$

we have

$$U_{1p} = \frac{2\,C}{H}\sqrt{\frac{2\,\pi}{r}}\sum_n G_n\,\cos\,(\omega\, t - x_n r - \frac{\pi}{4}) \tag{7.73}$$

We see that x_n has a meaning of a wavenumber, and

$$c_{pn} = \frac{\omega}{x_n} \tag{7.74}$$

is a phase velocity of the nth propagating mode with frequency ω.

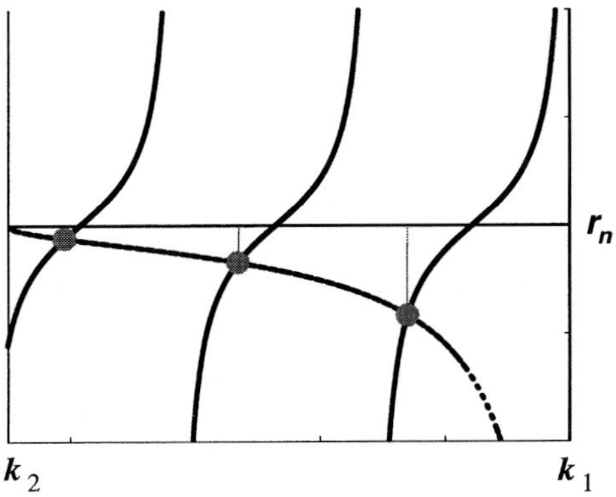

Figure 7.6: The left-hand side (solid), the right-hand side (dashed) of eq. 7.77 and its roots (gray dots).

Finding poles

To find positions of poles, we need to solve eq. 7.64. Let us introduce a new parameter p characterizing the relation between the wavelength λ and the thickness of the layer H

$$p = k_1 \mu H = 2\pi\mu \frac{H}{\lambda_1} = \frac{2\pi\mu H f}{c_1} \geq 0 \tag{7.75}$$

where f is a frequency and

$$\mu = \sqrt{1 - \left(\frac{c_1}{c_2}\right)^2} < 1 \tag{7.76}$$

Using eqs. 7.72 and 7.76, we transform eq. 7.64 into

$$\tan r_n = -\frac{r_n}{b(p^2 - r_n^2)} \tag{7.77}$$

and search for roots r_n of this equation for any given value of p. The left-hand side of this equation is independent of p and is a periodic function of r_n. In contrast, the right-hand side depends on the parameter p. The intersections of curves describing these two functions define the roots of eq. 7.77, as shown in Fig. 7.6.

We ignore the trivial root $r_0 = 0$, which does not correspond to traveling waves. Then, depending on the value of p, there are no roots, just one root, or several roots of eq. 7.77. It is clear that with the increase of parameter p, the number of roots also increases. Let us discuss this question in more detail. Analysis of eq. 7.77 for different values of p shows the following.

(a) The roots obeying eq. 7.77 are absent within the interval $0 \leq p < \pi/2$.

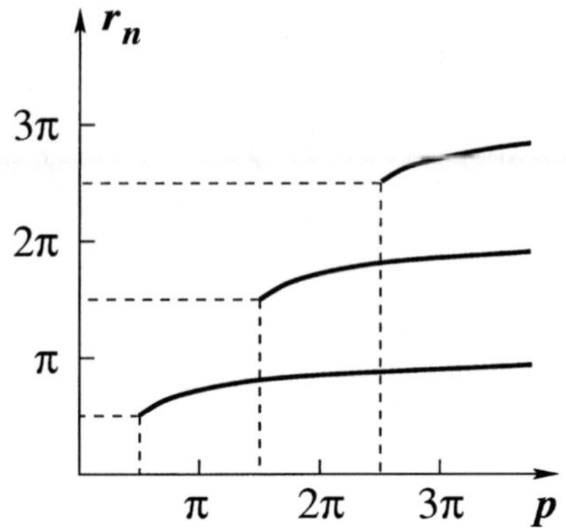

Figure 7.7: Roots r_n of eq. 7.77 as functions of p.

(b) The first real root r_1 appears when $p = \pi/2$. As is seen from eq. 7.77, it is equal to $r_1 = \pi/2$ if $p = \pi/2$.

(c) There is still one real root within the range $\pi/2 < p < 3\pi/2$, which becomes greater when p grows.

(d) The second real root r_2 arises when $p = 3\pi/2$. It is also equal to $r_2 = 3\pi/2$.

(e) There are two real roots within the interval, and they both gradually increase when the parameter p becomes larger.

Continuing this consideration, we obtain an infinite number of intervals of p which contain the roots of eq. 7.77. The boundaries between them are defined from the equality

$$p = r_n = (2\,n + 1)\,\frac{\pi}{2}\,, \tag{7.78}$$

and these pairs of p and r_n satisfy eq. 7.77. It is also useful to notice that each root approaches its limit as p increases. For instance,

$$r_1 \to \pi\,, \quad r_2 \to 2\,\pi\,, \quad r_3 \to 3\,\pi\,, \quad \ldots \quad \text{if} \quad p \to \infty\,,$$

that is

$$\frac{\pi}{2} < r_1 < \pi\,, \qquad \frac{3\,\pi}{2} < r_2 < 2\,\pi\,, \qquad \frac{5\,\pi}{2} < r_2 < 3\,\pi$$

and, in general,

$$(n-1)\,\frac{\pi}{2} < r_n < \pi\,n \tag{7.79}$$

The behavior of the first three roots is shown in Fig. 7.7. We see that the n-th pole exists

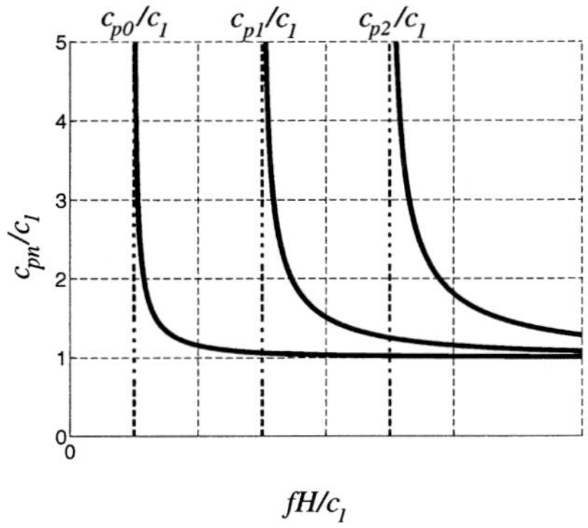

Figure 7.8: Phase velocity dispersion curves for first three normal modes.

in the range $p_n < p < \infty$ or $\pi(2n+1)/2 < p < \infty$, which corresponds to the range of frequencies $f_n < f < \infty$, where $f = \omega/2\pi$ and

$$f_n = \frac{(2n+1)\,c_1}{4H\sqrt{1-\left(\frac{c_1}{c_2}\right)^2}}, \tag{7.80}$$

f_n is called a cut-off frequency.

Dispersion

The analytical solution of eq. 7.77 for an arbitrary value of ω is absent. This means that x_n as well as the phase velocity

$$c_{pn} = \frac{\omega}{x_n} = \frac{\omega}{\sqrt{k_1^2 - \left(\frac{r_n}{H}\right)^2}} \tag{7.81}$$

should be found numerically. The relation 7.81 demonstrates that the phase velocity of each normal mode is frequency-dependent. Such dependence is called dispersion (Part I), and curves $c_{pn}(\omega)$ are called phase velocity dispersion curves. From analysis of behavior of roots x_n it is clear that $c_{pn}(\omega)$ is changing in the interval $c_2 \geq c_{pn} \geq c_1$ when ω changes in the range $\omega_n \leq \omega < \infty$ (Fig. 7.8).

The case $b = 0$

It is interesting to consider a special case when the lower medium is rigid and, therefore, $b = 0$. Then eq. 7.77 is greatly simplified and becomes

$$\cos r_n = 0 \tag{7.82}$$

Consequently, the roots are

$$r_n = \pi \left(n - \frac{1}{2} \right), \tag{7.83}$$

where $n = 1, 2, \ldots$ is an integer.

Unlike the general case, $b \neq 0$, the roots of eq. 7.77 are independent of the frequency and medium parameters. Therefore, the poles x_n, which characterize the behavior of normal modes, are defined as

$$x_n = \frac{1}{H} \sqrt{k_1^2 H^2 - \pi^2 \left(n - \frac{1}{2} \right)^2} \tag{7.84}$$

From eqs. 7.83 and 7.71, the expression for the phase velocity of the nth mode is

$$c_{pn} = \frac{\omega H}{\sqrt{k_1^2 H^2 - \pi^2 \left(n - \frac{1}{2} \right)^2}} \tag{7.85}$$

For the cut-off frequency of the same mode we have

$$f_n = \frac{c_1}{2H} (n - \frac{1}{2}) \tag{7.86}$$

Dependence of normal mode amplitude on depth

The expression 7.71 determines the change of the amplitude of a normal mode with the depth of the observation point and the source:

$$G_n \sim \sin r_n \frac{d}{H} \sin r_n \frac{z}{H}$$

We see that normal modes behave as standing waves along the z-axis, and with an increase of their order n, the number of nodes also increases, Fig. 7.9. The function G_n, characterizing the potential in the wave zone, is independent of the distance r from the source and, correspondingly, the relationship between amplitudes of different modes is defined by the frequency and the order n.

Waves in the lower medium

Let us briefly discuss the behavior of waves, which associate with normal modes, in the lower medium. As follows from eq. 7.42 and 7.60, the potential $\mathcal{U}_{2p}$ is

$$\mathcal{U}_{2p} = 2\pi i \sum \text{Res } m\, F_2(m^2)\, H_0^{(1)}(mr)\,, \tag{7.87}$$

where

$$F_2(m^2) = \frac{b}{D}\, e^{m_2(H - z)} \sinh m_1 d$$

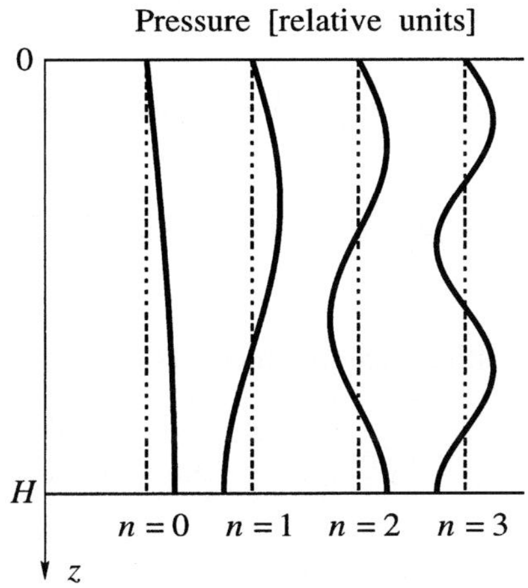

Figure 7.9: Standing waves.

and

$$z > H > d$$

Since Re $m_2 > 0$, each term in eq. 7.87 describes an evanescent wave propagating in the horizontal direction with phase velocity c_{pn} and exponentially decaying along the z-axis for $k_2 r > 1$.

The case $c_1 > c_2$

Until now we assumed that $c_2 > c_1$. In the opposite case, the wave behavior is different. As the internal reflection is absent, all waves inside the layer quickly attenuate as the distance from the source increases. The same result follows from eq. 7.64 because its roots become complex. Note that if $c_2 \ll c_1$, the reflection coefficient is close to unity and, correspondingly, attenuation is relatively small.

7.5 Asymptotic behavior of waves related to potential $\mathcal{U}_b$

In the previous section, we studied function $\mathcal{U}_p$, which describes the waves related to the poles of the integrand of eq. 7.56,

$$F(m_1, m_2) = \frac{\sinh m_1 d}{m_1} \frac{b\, m_2 \sinh m_1(H-z) + m_1 \cosh m_1(H-z)}{b\, m_2 \sinh m_1 H + m_1 \cosh m_1 H} \tag{7.88}$$

Since the function $F(m_1,\, m_2)$ is even with respect to m_1, the integrals along the lines of branch point $m = k_1$ cancel each other, and we have

$$\mathcal{U}_{1b} = -C \int\limits_{k_2+i\infty}^{k_2} m\, H_0^{(1)}(mr)\,[\,F(m_1,\, m_2) - F(m_1,\, -m_2)\,]\, dm \tag{7.89}$$

As follows from eq. 7.88, the difference of functions

$$\Delta F = F(m_1,\, m_2) - F(m_1,\, -m_2)$$

is equal to

$$\Delta F = \frac{\sinh m_1 d}{m_1}\left[\frac{b\, m_2\, \sinh m_1(H-z) + m_1\, \cosh m_1(H-z)}{b\, m_2\, \sinh m_1 H + m_1\, \cosh m_1 H} - \frac{m_1\, \cosh m_1(H-z) - b\, m_2\, \sinh m_1(H-z)}{m_1\, \cosh m_1 H - b\, m_2\, \sinh m_1 H}\right] \tag{7.90}$$

To simplify this expression, we perform some elementary algebra:

$$\begin{aligned}&[\,b\, m_2\, \sinh m_1(H-z) + m_1\, \cosh m_1(H-z)\,] \times [\,m_1\, \cosh m_1 H - b\, m_2\, \sinh m_1 H\,]\\ &= b\, m_1\, m_2\, \sinh m_1(H-z)\, \cosh m_1 H + m_1^2\, \cosh m_1(H-z)\, \cosh m_1 H\\ &\quad - b^2\, m_2^2\, \sinh m_1(H-z)\, \sinh m_1 H - b\, m_1\, m_2\, \sinh m_1 H\, \cosh m_1(H-z)\end{aligned} \tag{7.91}$$

Also,

$$\begin{aligned}&[\,b\, m_2\, \sinh m_1 H + m_1\, \cosh m_1 H\,] \times [\,m_1\, \cosh m_1(H-z) - b\, m_2\, \sinh m_1(H-z)\,]\\ &= b\, m_1\, m_2\, \sinh m_1 H\, \cosh m_1(H-z) + m_1^2\, \cosh m_1 H\, \cosh m_1(H-z)\\ &\quad - b^2\, m_2^2\, \sinh m_1 H\, \sinh m_1(H-z) - b\, m_1\, m_2\, \sinh m_1(H-z)\, \cosh m_1 H\end{aligned} \tag{7.92}$$

Subtraction of eq. 7.92 from eq. 7.91 gives

$$2\,b\,m_1\,m_2\,\sinh m_1(H-z)\,\cosh m_1 H\;-2\,b\,m_1\,m_2\,\sinh m_1 H\,\cosh m_1(H-z)$$

$$=-2\,b\,m_1\,m_2\,[\sinh m_1 H\,\cosh m_1(H-z)-\cosh m_1 H\,\sinh m_1(H-z)]$$

$$=-2\,b\,m_1\,m_2\,\sinh m_1 z \tag{7.93}$$

Certainly it is a great simplification. Thus, we have

$$\Delta F=-\frac{2\,b\,m_2\,\sinh m_1 d\,\sinh m_1 z}{m_1^2\,\cosh^2 m_1 H-b^2\,m_2^2\,\sinh^2 m_1 H}$$

and

$$\mathcal{U}_{1b}=2\,b\,C\int\limits_{k_2+i\infty}^{k_2}\frac{m_2\,\sinh m_1 d\,\sinh m_1 z}{m_1^2\,\cosh^2 m_1 H-b^2\,m_2^2\,\sinh^2 m_1 H}\,m\,H_0^{(1)}(mr)\,dm \tag{7.94}$$

Asymptotic expression for $\mathcal{U}_{1b}$

As before, we examine the case when $k_2\,r>1$. This means that the Hankel function can be replaced by its asymptotic expression

$$H_0^{(1)}(mr)\approx\sqrt{\frac{2}{\pi\,m\,r}}\,e^{i\,(m\,r\,-\,\frac{\pi}{4})}$$

In other words, it is assumed that the initial part of integration gives the main contribution due to the exponential decay of the integrand along the branch cut.

To evaluate this integral, eq. 7.94, let us introduce a new variable

$$m=k_1\left(\frac{c_1}{c_2}+i\,x\right) \tag{7.95}$$

Whence

$$dm=i\,k_1\,dx \tag{7.96}$$

Near the branch point, we approximately have

$$m\approx k_2\qquad\text{and}\qquad m_1=\sqrt{k_2^2-k_1^2}=i\,k_1\,\mu\,, \tag{7.97}$$

where

$$\mu=\sqrt{1-\frac{c_1^2}{c_2^2}}$$

The radical m_2 becomes

$$m_2 = \sqrt{\left(\frac{c_1}{c_2} + i\,x\right)^2 k_1^2 - k_2^2} \approx k_1 \sqrt{2\,i\,x\,\frac{c_1}{c_2}} = k_1 \sqrt{2\,x\,\frac{c_1}{c_2}}\,e^{i\,\pi/4} \tag{7.98}$$

We also have

$$\sinh m_1 d \,\sinh m_1 z = -\sin k_1\mu\, d\, \sin k_1 \mu z\,,$$

and the denominator of function ΔF is equal to

$$m_1^2 \cosh^2 m_1 H - b^2\, m_2^2 \sinh^2 m_1 H = \tag{7.99}$$

$$= -k_1^2\,\mu^2 \cos^2 k_1\mu H + 2\,k_1^2\, b^2\, x\,\frac{c_1}{c_2}\,e^{i\,\pi/2} \sin^2 k_1\mu H$$

Making use of eqs. 7.95–7.99, the function ΔF can be written as

$$\Delta F = \frac{2\,k_1\, b\,\sqrt{2\,x\,\frac{c_1}{c_2}}\;e^{i\,\pi/4}\,\sin k_1\mu\, d\,\sin k_1\mu z}{a_1 + a_2\, x}\,, \tag{7.100}$$

where

$$a_1 = -k_1^2\,\mu^2 \cos^2 k_1\mu H\,, \qquad a_2 = 2\,k_1^2\, b^2\,\frac{c_1}{c_2}\,e^{i\,\pi/2}\,\sin^2 k_1\mu H \tag{7.101}$$

Also

$$m\,dm = i\,k_1^2\,\frac{c_1}{c_2}\,dx$$

and

$$H_0^{(1)}(mr) \approx \sqrt{\frac{2}{\pi\,k_2\,r}}\,e^{i\,k_2\,r}\,e^{-k_1\,r\,x}\,e^{-i\,\pi/4} \tag{7.102}$$

Substitution of eqs. 7.100–7.102 into eq. 7.94 gives

$$\mathcal{U}_{1b} \approx -\frac{4\,i\,b\,C\,k_2\,k_1^{3/2}}{\sqrt{\pi\,r}}\,\sin k_1\mu d\,\sin k_1\mu z\,e^{i\,k_2\,r}\,I, \tag{7.103}$$

where

$$I = \int_0^\infty \frac{\sqrt{x}\,e^{-k_1\,r\,x}}{a_1 + a_2\,x}\,dx \tag{7.104}$$

Bearing in mind that the value of I is defined mainly by integration over small x, it can be approximately represented as

$$I \approx \frac{1}{a_1} \int_0^\infty \sqrt{x}\, e^{-k_1 r x}\, dx$$

As is well known, this integral is tabular, and it is equal to

$$I = \frac{\sqrt{\pi}}{2\, a_1\, (k_1 r)^{3/2}}$$

Correspondingly, in place of eq. 7.103 we obtain

$$\mathcal{U}_{1b} \approx \frac{2\, i\, b\, C\, k_2}{k_1^2\, r^2} \frac{\sin k_1 \mu d\, \sin k_1 \mu z}{\mu^2 \cos^2 k_1 \mu H}\, e^{i\, k_2\, r} \tag{7.105}$$

The latter shows that at sufficiently large distances from the source the potential $\mathcal{U}_{1b}$ characterizes a wave that propagates inside the layer with the phase velocity equal to c_2 and the amplitude decaying with the distance as $1/r^2$. This wave is the result of constructive interference of head waves and multiple reflections with the angle of incidence, equal to the critical angle $\theta = \sin^{-1}(c_1/c_2)$. Note that this wave behaves as the standing wave along the z-coordinate. It is obvious that we could obtain the same expression for $\mathcal{U}_{1b}$, eq. 7.105, if it is assumed in the beginning that

$$\sin k_1\, \mu\, H = 0 \qquad \text{or} \qquad k_1\, \mu\, H = \pi\, n \tag{7.106}$$

This condition can be rewritten as

$$\frac{\omega_n}{c_1}\, \mu\, H = \pi\, n \qquad \text{or} \qquad f_n = \frac{n\, c_1}{2H \sqrt{1 - \frac{c_1^2}{c_2^2}}} \tag{7.107}$$

Since the critical angle θ_c is defined as

$$\sin \theta_c = \frac{c_1}{c_2},$$

we also have

$$f_n = \frac{n\, c_1}{2H\, \cos \theta_c} \tag{7.108}$$

Thus, when the frequencies obey this condition, we observe a relatively strong decrease of the wave field $\mathcal{U}_{1b}$ with distance r.

Now suppose that parameter a_1 vanishes, that is

$$\cos k_1\,\mu\,H=0 \qquad \text{or} \qquad k_1\,\mu\,H=\frac{(2\,n-1)\,\pi}{2} \tag{7.109}$$

Correspondingly, eq. 7.104 becomes

$$I=\frac{1}{a_2}\int\limits_0^\infty \frac{e^{-k_1\,r\,x}}{\sqrt{x}}\,dx$$

Letting $y=\sqrt{x}$, we have

$$dy=\frac{1}{2}\frac{dx}{\sqrt{x}}=\frac{dx}{2\,y}$$

This yields

$$I=\frac{2}{a_2}\int\limits_0^\infty e^{-k_1\,r\,y^2}\,dy \qquad \text{or} \qquad I=\frac{1}{a_2}\sqrt{\frac{\pi}{k_1\,r}}$$

Substitution of the latter into eq. 7.103 gives

$$\mathcal{U}_{1b}\approx-\frac{2}{b\,r}\,\frac{\sin k_1\mu d\,\sin k_1\mu z}{\sin^2 k_1\mu H}\,e^{i\,k_2\,r}\,, \tag{7.110}$$

that is, the wave again propagates with the velocity c_2 but its amplitude decreases with the distance slower that in the previous case, $a_2 = 0$. Such behavior is observed if frequencies satisfy eq. 7.109:

$$f_n=\frac{(2\,n-1)\,c_1\,c_2}{4\,H\,\sqrt{c_2^2-c_1^2}} \tag{7.111}$$

As was shown earlier this set of frequencies coincides with the cut-off frequencies for the normal modes, and therefore constructive interference occurs. Comparison of waves related to both potentials $\mathcal{U}_{1p}$ and $\mathcal{U}_{1b}$ shows that normal modes play a dominant role in the wave zone.

7.6 Transient waves inside a layer

Now we discuss propagation of transient waves in the layer, and, as in the previous sections, the main attention is paid to the wave zone. As it is already known, the field of normal modes prevails in this case, that is $\mathcal{U}_p \gg \mathcal{U}_b$. Correspondingly, the complex amplitude of the potential is

$$\mathcal{U}(p,\,\omega)=\mathcal{U}_p(p,\,\omega) \tag{7.112}$$

Potential of transient waves

In accordance with eq. 7.70, the function $\mathcal{U}_p$, caused by a source of sinusoidal waves with the unit amplitude, $C = 1$, is

$$\mathcal{U} = \frac{2}{H}\sqrt{\frac{2\pi}{r}}\sum_{n=1}^{\infty} G_n\, e^{i\,(x_n\, r + \pi/4)}\,, \tag{7.113}$$

where

$$G_n = \frac{r_n \sin\left(r_n \frac{d}{H}\right)\,\sin\left(r_n \frac{z}{H}\right)}{\sqrt{x_n}\,(r_n - \sin r_n\, \cos r_n - b^2\, \tan r_n\, \sin^2 r_n)}, \tag{7.114}$$

and r_n are the roots of the dispersion equation

$$\tan r_n = -\frac{r_n}{(p^2 - r_n^2)}. \tag{7.115}$$

Here

$$b = \frac{\rho_1}{\rho_2}, \qquad p = k_1\,\mu\, H\,, \qquad \text{and} \qquad \mu = \sqrt{1 - \frac{c_1^2}{c_2^2}} \tag{7.116}$$

Finally, the wavenumber of the nth normal mode is

$$x_n = \frac{1}{H}\sqrt{k_1^2\, H^2 - r_n^2}\,, \tag{7.117}$$

where r_n is a known function of the wave frequency, layer thickness H, and velocities c_1 and c_2.

Suppose that the potential, caused by motion of the primary source, behaves as

$$U_0(t) = \begin{cases} 0\,, & t < 0 \\ U_0(t)\,, & t \geq 0 \end{cases} \tag{7.118}$$

Applying the Fourier transform and taking into account eq. 7.118, we have for the spectrum of the source

$$\mathcal{U}_0(p,\,\omega) = \int_0^{\infty} U_0(p,\,t)\, e^{i\,\omega\, t}\, dt \tag{7.119}$$

The latter allows us to assume that the source simultaneously generates an infinite number of sinusoidal waves with infinitely small amplitudes and different phases, and each of them may give rise to a normal mode. Then, making use of the principle of superposition (Fourier transform) we obtain the potential of total field in the wave zone,

$$U(p,\,t) = \frac{1}{2\pi}\int\limits_{-\infty}^{\infty} \mathcal{U}_0(\omega)\,\mathcal{U}(\omega)\,e^{-i\,\omega\,t}\,d\omega\,, \tag{7.120}$$

where $\mathcal{U}$ and $\mathcal{U}_0$ are the known functions given by eqs. 7.113 and 7.119, respectively.

Substitution of eqs. 7.113 into eq. 7.120 and change the order of integration and summation gives

$$U(p,\,t) = \frac{1}{H}\sqrt{\frac{2}{\pi\,r}}\sum_{n=1}^{\infty}\int\limits_{-\infty}^{\infty} G_n\,\mathcal{U}_0\,e^{-i\,(\omega\,t - x_n\,r - \pi/4)}\,d\omega \tag{7.121}$$

or

$$U(p,\,t) = \sum_{n=1}^{\infty} U_n(p,\,t)\,, \tag{7.122}$$

where

$$U_n(p,\,t) = \frac{1}{H}\sqrt{\frac{2}{\pi\,r}}\int\limits_{-\infty}^{\infty} G_n\,\mathcal{U}_0\,e^{-i\,(\omega\,t\,-\,x_n\,r\,-\,\pi/4)}\,d\omega \tag{7.123}$$

By analogy with the frequency domain the transient wave is now represented as a sum of normal modes. Since the functions $G_n(\omega)$, $\mathcal{U}_0(\omega)$ and $x_n(\omega)$ in the right-hand side of eq. 7.123 can be calculated, numerical integration allows us to find the wave field at any instant of time, provided that the observation point is located far away from the source.

Prior to discussion of results of calculation of the function $U(t)$, it is useful to describe qualitatively the main features of the transient modes. In accordance with eq. 7.123, each mode arises due to interference of an infinite number of sinusoidal waves with different frequencies, amplitudes, and phases

$$\frac{1}{H}\sqrt{\frac{2}{\pi\,r}}\,|\,G_n\,\mathcal{U}_0\,|\,\exp(-i\psi_n\,)d\omega\,,$$

where the phase $\psi_n = \omega t - x_n r - \pi/4$. The condition of constructive interference of these sinusoidal waves at given distance r and at time t is the closeness or stationarity of their phases, that is,

$$\frac{d\psi_n}{d\omega} = t - \frac{dx_n}{d\omega} = t - \frac{r}{c_{gn}(\omega)} = 0$$

As it was discussed in Part I, the function $c_{gn} = d\omega/dx_n$ is called a group velocity. This function defines the waveform of a transient (non-stationary) normal mode. In this light, note that we already considered several kinds of velocities. One of them is the physical velocity defined by the medium bulk modulus M and its density ρ. In this model,

$$c_1 = \sqrt{\frac{M_1}{\rho_1}} \qquad \text{and} \qquad c_2 = \sqrt{\frac{M_2}{\rho_2}}$$

The second is the phase velocity c_p which characterizes the propagation of phase surfaces of any single sinusoidal wave, for instance, a normal mode. The phase velocity may depend on frequency and usually differs from the physical velocity. Also, we often deal with the apparent velocity c_a, which describes the movement of phase surface of a sinusoidal wave along a given direction. It is related to phase velocity c_p as

$$c_a = \frac{c_p}{\cos\alpha},$$

where α is the angle between this direction and the normal to the phase surface.

Now we deal with a different kind of velocity which defines the speed of propagation of wave groups (Part I). To find the relation between the group and phase velocities we use the formulae

$$c_{pn} = \frac{\omega}{x_n}$$

Then,

$$c_{gn} = \frac{d\omega}{dx_n} = \frac{d\omega}{d\left(\dfrac{\omega}{c_{pn}(\omega)}\right)} = \frac{c_{pn}}{1 - \dfrac{\omega}{c_{pn}}\dfrac{dc_{pn}}{d\omega}} \tag{7.124}$$

It is possible to find an expression for the group velocity which does not contain $dc_{pn}/d\omega$ using the formula for the derivative of an implicit function known from calculus:

$$\frac{dy}{dx} = -\frac{\dfrac{\partial F(x,y)}{\partial x}}{\dfrac{\partial F(x,y)}{\partial y}}$$

In our case $x = \omega, y = c_{pn}$ and

$$F(\omega, c_{pn}) = \tan\left(\omega H\sqrt{c_1^{-2} - c_{pn}^{-2}}\right) + \frac{1}{b}\sqrt{\frac{c_1^{-2} - c_{pn}^{-2}}{c_{pn}^{-2} - c_{pn}^{-2}}}$$

After performing some cumbersome algebra, we arrive at the expression for c_{gn},

$$c_{gn} = \frac{c_{pn}(\omega)}{1 + (c_{pn}^2 c_1^{-2} - 1)\left[1 + \dfrac{c_1^{-2} - c_2^{-2}}{b(c_2^{-2} - c_{pn}^{-2})^{3/2}\omega H}\right]} \tag{7.125}$$

Knowing c_{pn} for a given mode, it is possible to calculate c_{gn} using eq. 7.125. The result

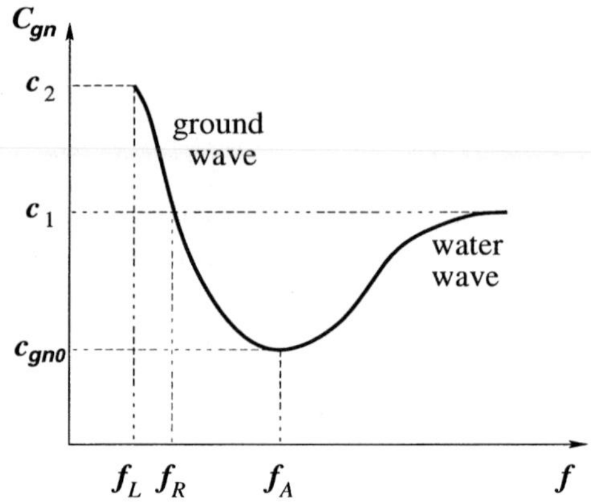

Figure 7.10: Group velocity as a function of frequency.

of such evaluation for the first normal mode, characterizing the dependence of the group velocity, c_{gn}, on frequency f, is shown in Fig. 7.10. As we see, the wave group with the cut-off frequency f_L propagates with velocity c_2 of the lower medium. This is the highest velocity of wave groups. Then with an increase of frequency the group velocity decreases and reaches its minimum at the frequency f_A. With further increase of f, the group velocity begins to grow and asymptotically approaches layer velocity c_1. Certainly this is an important information because it tells us about arrival times of different wave groups. As an illustration, consider the normal transient mode of the first order inside the layer, when $k_1 r > 1$. Note that the left and right parts of curve $c_{gn}(f)$ in Fig. 7.10 are called the ground and water waves, respectively.

It follows from the behavior $c_{gn}(f)$ that the ground wave with the lowest frequency arrives first, and its group velocity is equal to c_2. With an increase of time the dominant frequency of wave groups gradually increases, which takes place when f is less than f_R: $f_L \leq f < f_R$. Within this frequency range the group velocity changes in the limits $c_1 \leq c_{gn} < c_2$. Both water wave and ground wave behave as amplitude modulated and frequency modulated signals.

Then there is always an instant when both the ground and water waves arrive simultaneously at the observation point, and their group velocities are equal to c_1. The ground wave is characterized by a relatively low frequency f_R, while the water wave has a much higher frequency. Thus, starting from this moment, the transient mode is formed by wave groups with two distinct frequencies. In other words, we observe superposition of the ground and water waves. With the further increase of time the group velocities of these waves become smaller and their dominant frequencies approach each other. The second portion of the transient wave is terminated when the wave group starts to move with the velocity c_{gn0}, equal to its minimum value with dominant frequency f_A. This third and final stage of the transient wave is named the Airy phase. An example showing all three stages of the first normal mode is given in Fig. 7.11.

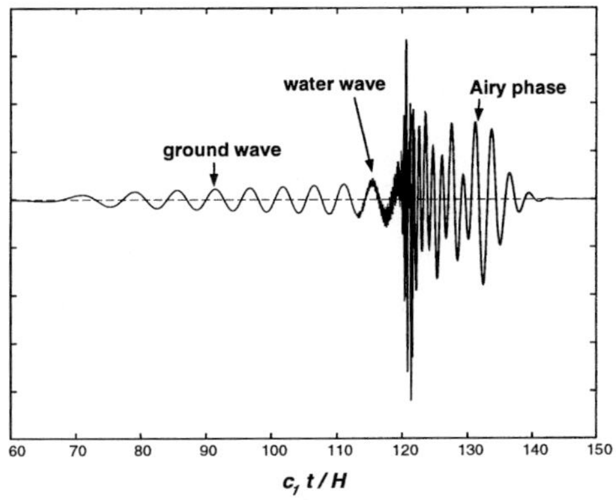

Figure 7.11: Behavior of the first transient mode.

7.7 Evaluation of normal modes

In the previous section we qualitatively described behavior of transient modes and, in particular, distinguished three different parts of the wave train, namely, the ground wave, the water wave, and the Airy phase. However, the dependence of their magnitudes on time and distance still remains unknown. In order to find those relationships, we can, in principle, perform numerical integration of each term in the right-hand side of eq. 7.121.

At the same time it is possible to arrive at sufficiently accurate values of the wave magnitude by applying approximate methods. Let us assume that the function $U_0(t)$ is

$$U_0(t) = \begin{cases} 0\,, & t < 0 \\ A\,e^{-\beta t}\,, & t \geq 0 \end{cases} \tag{7.126}$$

where β is the constant characterizing the rate of change of the source pressure. In fact, from the relationship

$$P = -\rho\,\frac{\partial^2 U}{\partial t^2}\,,$$

we have $P = -\rho\,\beta^2 A\,e^{-\beta t}$ if $t \geq 0$.

As follows from the inverse Fourier transform, the spectrum of function $U_0(t)$ is

$$\mathcal{U}_0(\omega) = A\int\limits_0^\infty e^{-\beta t}\,e^{i\omega t}\,dt \qquad \text{or} \qquad \mathcal{U}_0(\omega) = \frac{A}{\beta - i\,\omega}$$

Therefore, the transient normal mode of the nth order (eq. 7.121) can be written as

$$U_n(t) = \frac{A}{H}\sqrt{\frac{2}{\pi\,r}}\int\limits_{-\infty}^{\infty}\frac{G_n(\omega,\,r,\,t)}{\beta - i\,\omega}\,e^{-i\,(\omega\,t\,-\,x_n\,r\,-\,\pi/4)}\,d\omega \tag{7.127}$$

It is useful to make some transformations of the integrand in this expression. To the accuracy of a constant, the function

$$F(\omega) = \frac{G_n(\omega,\, r,\, t)}{\beta - i\omega}\, e^{-i\,(x_n\, r\, -\, \pi/4)} \tag{7.128}$$

represents the spectrum of $U(t)$. As follows from the properties of the Fourier transform, the real and imaginary parts of its spectrum are even and odd function of ω, respectively,

$$\text{Re } F(\omega) = \text{Re } F(-\omega) \qquad \text{and} \qquad \text{Im } F(\omega) = -\text{Im } F(-\omega)$$

Therefore,

$$\int\limits_{-\infty}^{\infty} F(\omega)\, e^{-i\,\omega\, t}\, d\omega \;=\; 2\,\text{Re} \int\limits_{0}^{\infty} F(\omega)\, e^{-i\,\omega\, t}\, d\omega \tag{7.129}$$

Taking into account that G_n is real and

$$\frac{1}{\beta - i\,\omega} = \frac{1}{\sqrt{\beta^2 + \omega^2}}\, \exp\left(i\, \tan^{-1}\frac{\omega}{\beta}\right)$$

we have

$$\text{Re}\, \left[F(\omega) e^{-i\,\omega\, t}\right] = \frac{G_n(\omega,\, r,\, t)}{\sqrt{\beta^2 + \omega^2}}\, \cos\left(\omega\, t - x_n\, r - \frac{\pi}{4} - \tan^{-1}\frac{\omega}{\beta}\right)$$

Whence

$$U_n(t) = \frac{2\,A}{H}\sqrt{\frac{2}{\pi\, r}}\, I_n \tag{7.130}$$

Here

$$I_n = \int\limits_{0}^{\infty} \frac{G_n(\omega,\, r,\, t)}{\sqrt{\beta^2 + \omega^2}}\, \cos\left(\omega\, t - x_n\, r - \frac{\pi}{4} - \tan^{-1}\frac{\omega}{\beta}\right)\, d\omega \tag{7.131}$$

The integrand represents a product of two functions. One of them

$$\frac{G_n(\omega,\, r,\, t)}{\sqrt{\beta^2 + \omega^2}}\,, \tag{7.132}$$

varies with ω relatively slowly. The same is true for the term $tan^{-1}(\omega/\beta)$. In contrast, the function

$$\cos\left(\omega\, t - x_n\, r - \frac{\pi}{4} - \tan^{-1}\frac{\omega}{\beta}\right)$$

rapidly changes even when the variations of ω are small, because the wave is considered at large distances from the source, $x_n\, r \gg 1$.

It is natural to treat the integrand as a sinusoidal wave whose phase depends on frequency. Applying the terminology of interference, we can say that if the phase changes rapidly in the vicinity of a certain frequency, this interval of integration does not make a noticeable contribution, due to destructive interference, and therefore it can be neglected. However, near such frequencies, where the phase variation is small, the constructive interference takes place and those ranges of ω determine the value of the integral. Respectively, our goal is to find such stationary points and evaluate the integrals around them. As we know, this is the essence of the method of stationary phase, and it is used in studying the ground and water stages of transient waves. In accordance with eq. 7.131, the phase φ_n is

$$\varphi_n(\omega) = \omega\, t - x_n\, r - \frac{\pi}{4} - \tan^{-1}\frac{\omega}{\beta} \tag{7.133}$$

Suppose that ω_0 is the stationary point. Taking into account that the interval of integration around ω_0 is small, it is proper to express the phase at its points in terms of $\varphi_n(\omega_0)$ and corresponding derivatives. This procedure greatly simplifies evaluation of $U(t)$. Expanding $\varphi_n(\omega)$ in the Taylor series, we obtain

$$\varphi_n(\omega) = \varphi_n(\omega_0) + \frac{d\varphi_n}{d\omega}\nu + \frac{d^2\varphi_n}{d\omega^2}\frac{\nu^2}{2} + \frac{d^3\varphi_n}{d\omega^3}\frac{\nu^3}{6} + \ldots, \tag{7.134}$$

where

$$\nu = \omega - \omega_0 \qquad \text{and} \qquad \frac{|\,\omega - \omega_0\,|}{\omega_0} \ll 1$$

Here all the derivatives are evaluated at point $\omega = \omega_0$.

Making use of eq. 7.133 and neglecting the change of the last term, we find

$$\frac{d\varphi_n}{d\omega} = t - r\,x_n', \qquad \frac{d^2\varphi_n}{d\omega^2} = -r\,x_n'', \qquad \frac{d^3\varphi_n}{d\omega^3} = -r\,x_n''', \qquad \ldots \tag{7.135}$$

Correspondingly eq. 7.134 becomes

$$\varphi_n(\omega) = \alpha_n + a_n\,\nu + b_n\,\nu^2 + c_n\,\nu^3 + \ldots, \tag{7.136}$$

where

$$\alpha_n = \omega_0\, t - x_n(\omega_0)\, r - \frac{\pi}{4} - \tan^{-1}\frac{\omega_0}{\beta} \tag{7.137}$$

and

$$a_n = t - x_n'\, r, \qquad b_n = -\frac{x_n''}{2}\, r, \qquad c_n = -\frac{x_n'''}{6}\, r,$$

$$x'_n = \frac{dx_n}{d\omega}, \qquad x''_n = \frac{d^2x_n}{d\omega^2}, \qquad x'''_n = \frac{d^3x_n}{d\omega^3}$$

By definition, in the vicinity of the stationary point a change of phase is minimal, and therefore the second term in the series, $a_n\,\nu$, which usually causes the main variation in $\varphi_n(\omega)$, vanishes. For this reason, in place of eq. 7.136, we have

$$\varphi_n(\omega) \approx \varphi_n(\omega_0) - \frac{r\,x''_n}{2}\,\nu^2 - \frac{r\,x'''_n}{6}\,\nu^3 \tag{7.138}$$

and

$$t - r\,x'_n = 0 \qquad \text{or} \qquad c_{gn}(\omega_0) = \frac{r}{t} \tag{7.139}$$

The latter is the known expression for the group velocity and it is of a great importance. In fact, eq. 7.139 allows us to determine the velocity of a group formed at a given instant t and at a certain distance r from the source. Since the dispersion curve $c_{gn}(\omega)$ is assumed to be known, the dominant frequency ω_0 (that is, the stationary point) corresponding to the group velocity $c_{gn}(\omega_0)$ can be easily found. The ratio

$$\frac{G_n(\omega,\,r,\,t)}{\sqrt{\beta^2+\omega^2}} \tag{7.140}$$

can, of course, be calculated, as well as the first term of the series, that is $\varphi_n(\omega_0)$.

The second-order derivative x''_n is represented as

$$x''_n = \frac{d}{d\omega}\left(\frac{dx_n}{d\omega}\right) = \frac{d}{d\omega}\left(\frac{1}{c_{gn}}\right) = -\frac{1}{c_{gn}^2}\,\frac{d\,c_{gn}}{d\omega} \tag{7.141}$$

Thus, knowing the group velocity as a function of the frequency ω, we can also calculate the second-order derivative of the wavenumber and, correspondingly, the second term in the Taylor series, eq. 7.138. The last term containing x'''_n is evaluated in a similar manner and makes it possible to approximate the phase $\varphi_n(\omega)$ around the stationary point.

Next, consider two cases:

$$x''_n \neq 0 \qquad \text{and} \qquad x''_n = 0$$

First, let us discuss the meaning of these conditions. With this purpose in mind, it is necessary to refer again to dispersion curve $c_{gn}(\omega)$ in Fig. 7.10. At the low-frequency part $\omega < \omega_A$, the group velocity decreases with an increase of the frequency, and therefore,

$$\frac{d\,c_{gn}}{d\omega} < 0$$

As follows from eq. 7.141, $x''_n > 0$ within this range. In contrast, at the high-frequency part of the curve, where $\omega > \omega_A$, the group velocity grows with the frequency and, correspondingly,

$$\frac{d\,c_{gn}}{d\,\omega} > 0 \qquad \text{and} \qquad x''_n < 0$$

Thus, the first case above includes all the frequencies except one, where the group, velocity reaches its minimum, and

$$\frac{d\,c_{gn}}{d\,\omega} = 0 \qquad \text{or} \qquad x''_n = 0$$

This is the second case which describes the Airy stage of the transient wave.

Before we consider the two cases, it is useful to simplify eq. 7.131. In view of the fact that the function, given by eq. 7.140, can be treated as a constant in the vicinity of ω_0, it can be taken out of the integral. This yields

$$I_n = \frac{G_n(\omega_0,\, r,\, t)}{\sqrt{\beta^2+\omega_0^2}} \int\limits_0^\infty \cos\left(\omega\, t - x_n\, r - \frac{\pi}{4} - \tan^{-1}\frac{\omega}{\beta}\right) d\omega$$

or, using Euler's formula

$$I_n = \frac{G_n(\omega_0,\, r,\, t)}{\sqrt{\beta^2+\omega_0^2}}\, \mathrm{Re}\, \mathcal{I}_n\,, \tag{7.142}$$

where

$$\mathcal{I}_n = \int\limits_0^\infty \exp\left[-i\left(\omega\, t - x_n\, r - \frac{\pi}{4} - \tan^{-1}\frac{\omega}{\beta}\right)\right] d\omega \tag{7.143}$$

We also assume that

$$\tan^{-1}\frac{\omega}{\beta} \approx \tan^{-1}\frac{\omega_0}{\beta}$$

Thus, we have the following expression for $U_n(t)$:

$$U_n(t) = \frac{2\,A}{H}\sqrt{\frac{2}{\pi\, r}}\, \frac{G_n(\omega_0,\, r,\, t)}{\sqrt{\beta^2+\omega_0^2}}\, \mathrm{Re}\, \mathcal{I}_n \tag{7.144}$$

Case one: $x''_n \neq 0$ Since the difference of frequencies $\nu = \omega - \omega_0$ is small the third term in series 7.138 can be ignored, and we have

$$\varphi_n(\omega) \approx \varphi_n(\omega_0) - \frac{r\, x''_n}{2}\,(\omega-\omega_0)^2 \tag{7.145}$$

Substituting the latter into eq. 7.143, we obtain

$$\mathcal{I}_n = e^{-i\,\varphi_n(\omega_0)} \int_0^{\infty} \exp\left[i\,\frac{r}{2}\,x_n''(\omega)\,(\omega-\omega_0)^2\right] d\omega \tag{7.146}$$

As we already know, only the vicinity of the stationary point gives the main contribution to the integral. For this reason, the integration limits can be changed, and in place of eq. 7.146, we can write

$$\mathcal{I}_n = e^{-i\,\varphi_n(\omega_0)} \int_{-\infty}^{\infty} \exp\left[i\,\frac{r}{2}\,x_n''(\omega)\,(\omega-\omega_0)^2\right] d\omega \tag{7.147}$$

Since this integral is tabular

$$\int_{-\infty}^{\infty} \exp\left[i\,\frac{r}{2}\,x_n''(\omega)\,(\omega-\omega_0)^2\right] d\omega = \sqrt{\frac{2\,\pi}{r\,|\,x_n''\,|}}\;e^{\pm i\frac{\pi}{4}}\,,$$

we have

$$\mathcal{I}_n = \sqrt{\frac{2\,\pi}{r\,|\,x_n''\,|}}\;e^{-i\,\varphi_n(\omega_0)}\,e^{\pm i\frac{\pi}{4}}\,, \tag{7.148}$$

where the plus sign is taken if $x_n'' < 0$ and the minus when $x_n'' > 0$.

Finally, making use of eqs. 7.144 and 7.148 we obtain

$$U_n(t) = \frac{2\,A}{H}\sqrt{\frac{2}{\pi\,r}}\,\frac{G_n(\omega_0,\,r,\,t)}{\sqrt{\beta^2+\omega_0^2}}\sqrt{\frac{2\,\pi}{r\,|\,x_n''\,|}}\,\cos\left(\omega\,t - x_n\,r - \frac{\pi}{4} \pm \frac{\pi}{4} - \tan^{-1}\frac{\omega_0}{\beta}\right)$$

or

$$\begin{aligned} U_n(t) &= \frac{4\,A}{H\,r}\,\frac{G_n(\omega_0,\,r,\,t)}{\sqrt{|\,x_n''\,|\,(\beta^2+\omega_0^2)}}\cos\left(\omega\,t - x_n\,r - \tan^{-1}\frac{\omega_0}{\beta}\right) \qquad \text{if} \quad x_n'' < 0 \\ U_n(t) &= \frac{4\,A}{H\,r}\,\frac{G_n(\omega_0,\,r,\,t)}{\sqrt{|\,x_n''\,|\,(\beta^2+\omega_0^2)}}\sin\left(\omega\,t - x_n\,r - \tan^{-1}\frac{\omega_0}{\beta}\right) \qquad \text{if} \quad x_n'' > 0 \end{aligned} \tag{7.149}$$

These equations describe transient modes corresponding to the ground and water waves. They allow us to calculate the waveform at any distance from the source, as long as the observation point is located in the wave zone.

Thus, the procedure of obtaining the wave field consists of the following steps:

(a) The group velocity $c_{gn} = r/t$ is found for any given values of r and t.

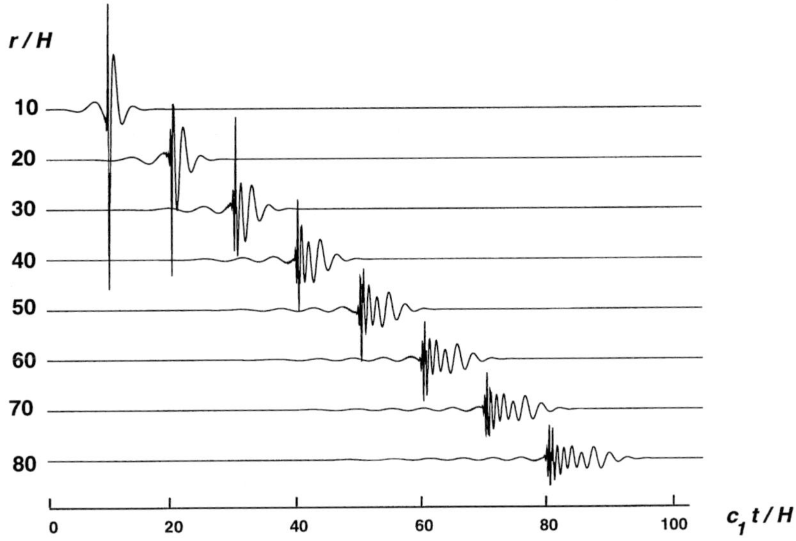

Figure 7.12: Transient mode at different distances from the source.

(b) The stationary point ω_0 is calculated from the dispersion curve $c_{gn}(\omega)$. It is proper to notice in this regard that if $c_1 < c_{gn} \leq c_2$, there is only one stationary point, and the second equation of the set 7.149 describes the ground wave. In contrast, when $c_{gn0} < c_{gn} \leq c_1$, there are two wave groups with different frequency content, Fig. 7.10, corresponding to the ground and water portions of the dispersion curve. Therefore, the total wave field is described by a sum of these two waves.

As follows from eqs. 7.149 the amplitude of the normal modes decreases with distance as r^{-1}, that is, by a factor of $\sqrt{r}$ faster than the amplitudes of stationary (sinusoidal) waves. This is due to the dispersion which causes the change of the waveform with distance. The waveform stretches in time and in accordance with energy conservation law the amplitudes decrease (Fig. 7.12).

Case two: $x''_n = 0$ Now we evaluate the last stage of the transient wave, which is characterized by dominant frequency ω_0 and the group velocity c_{gn0}. At this point, we have

$$\frac{d\,c_{gn}(\omega)}{d\,\omega} = 0 \qquad \text{and} \qquad \frac{d^2\varphi_n(\omega)}{d\,\omega^2} = 0\,, \tag{7.150}$$

which indicate that the group velocity varies relatively slowly near frequency ω_0, and, therefore, constructive interference is observed. Assuming, as before, that only a small range of frequencies close to ω_0 makes the most significant contribution, we use Taylor

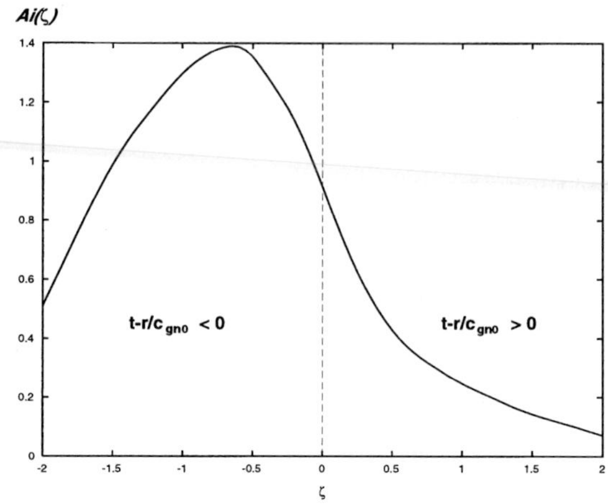

Figure 7.13: Airy function.

series 7.136 again:

$$\varphi_n(\omega) = \alpha_n + a_n\,\nu + c_n\,\nu^3 + \ldots \tag{7.151}$$

The term proportional to ν^2 is discarded, since the conditions 7.150 are met at ω_0. The coefficients α_n, a_n, and c_n are given by eq. 7.137. The second term of the series,

$$a_n\,\nu = \left(t - \frac{r}{c_{gn0}}\right)\,\nu$$

in general differs from zero except those pairs of r and t, when the equality $t = r/c_{gn0}$ takes place.

Substitution eq. 7.151 into eq. 7.143 gives

$$\mathcal{I}_n = \int_{-\infty}^{\infty} \exp\left[-i\left(\alpha_n + a_n\,\nu + c_n\,\nu^3\right)\right]\,d\omega\,,$$

because the value of the integral is defined by the vicinity of ω_0.

Bearing in mind that

$$\nu = \omega - \omega_0 \qquad \text{and} \qquad d\nu = d\omega\,,$$

we have

$$\mathcal{I}_n = \int_{-\infty}^{\infty} \exp\left[-i\left(\alpha_n + a_n\,\nu + c_n\,\nu^3\right)\right]\,d\nu \tag{7.152}$$

In accordance with eq. 7.142,

$$I_n = \frac{2\,G_n(\omega_0,\,r,\,t)}{\sqrt{\beta^2+\omega_0^2}} \int\limits_0^\infty \cos\left(\alpha_n + a_n\,\nu + c_n\,\nu^3\right) d\nu \tag{7.153}$$

The integral in eq. 7.153 can be presented as

$$\int\limits_0^\infty \cos\left(\alpha_n + a_n\,\nu + c_n\,\nu^3\right) d\nu \tag{7.154}$$

$$= \cos\alpha_n \int\limits_0^\infty \cos\left(a_n\,\nu + c_n\,\nu^3\right) d\nu - \sin\alpha_n \int\limits_0^\infty \sin\left(a_n\,\nu + c_n\,\nu^3\right) d\nu$$

Since ν is small, the second integral can be ignored and we have

$$\int\limits_0^\infty \cos\left(\alpha_n + a_n\,\nu + c_n\,\nu^3\right) d\nu \;=\; \cos\alpha_n \int\limits_0^\infty \cos\left(a_n\,\nu + c_n\,\nu^3\right) d\nu \tag{7.155}$$

The last integral is expressed in terms of the Airy function $Ai(\zeta)$ (Chapter 3) in the following way:

$$\int\limits_0^\infty \cos\left(a_n\,\nu + c_n\,\nu^3\right) d\nu \;=\; \frac{\pi}{(2\,c_n)^{1/3}}\,Ai(\zeta)\,, \tag{7.156}$$

where

$$\zeta = \frac{2}{3^{3/2}}\sqrt{\frac{|\,a_n\,|^3}{|\,c_n\,|}} \tag{7.157}$$

The behavior of the Airy function is different for negative and positive $a_n = t - r/c_{gn}$ (Fig. 7.13). In fact, for positive a_n $Ai(\zeta)$ gradually decreases with increase of its argument ζ and decays almost as an exponent when ζ is large. For negative a_n $Ai(\zeta)$ passes through the maximum and then slowly decreases.

Finally, from eqs. 7.130 and 7.153–7.156, we obtain

$$\begin{aligned} U_n(t) \;&=\; \frac{4\sqrt[3]{6}\,A}{H}\,\frac{\sqrt{2}\,\pi}{r^{\frac{5}{6}}}\,\frac{G_n(\omega_0,r,t)}{\sqrt{\beta^2+\omega_0^2}}\,\frac{Ai(\zeta)}{\sqrt[3]{-x_n'''(\omega_0)}} \\ &\times \cos\left(\omega_0\,t - x_n(\omega_0)\,r - \frac{\pi}{4} - \tan^{-1}\frac{\omega_0}{\beta}\right) \end{aligned} \tag{7.158}$$

The amplitude of the Airy phase decreases with distance r as $r^{-5/6}$, that is, slightly slower than amplitudes of water and ground waves. This means that far away from the source the Airy phase may be a dominant part of the signal carried out by a given mode. This phase represents the amplitude modulated signal, as the frequency of oscillations ω_0 remains constant.

Chapter 8

Acoustic potential in a medium with a cylindrical interface

In this chapter we study wave propagation in a medium with the cylindrical interface and mostly at relatively large distances from the source. As in the case of the horizontally layered medium elasticity is not taken into account, but such a model is nevertheless very useful for understanding some important features of longitudinal waves in a borehole.

8.1 Solution of the boundary value problem

Suppose that there is one cylindrical interface only and an elementary spherical source of a sinusoidal wave is located on its axis, Fig. 8.1a. The density and wave velocity inside and outside the cylinder (borehole) are ρ_1, c_1 and ρ_2, c_2, respectively. Taking into account the geometry of the model, we choose the cylindrical system of coordinates with the origin at point O, where the source is situated, and the z-axis coincides with the axis of the borehole. Because of the axial symmetry potential U depends on two coordinates only

$$U = U(r, z) \tag{8.1}$$

Correspondingly, the wave field

$$\mathbf{s} = \text{grad}\, U\,, \qquad P = \omega^2 \rho U \tag{8.2}$$

is independent of the azimuthal coordinate, ϕ, and vector $\mathbf{s}$ has two components: s_r and s_z.

Next we derive formulae for the potential inside and outside the borehole, and with this purpose in mind, we formulate the boundary value problem. First, the complex amplitude of the potential $\mathcal{U}$ satisfies the Helmholtz equation at the regular points

$$\nabla^2 \mathcal{U}_1 + k_1^2 \mathcal{U}_1 = 0 \qquad \text{if} \qquad r < a \tag{8.3}$$

and

$$\nabla^2 \mathcal{U}_2 + k_2^2 \mathcal{U}_2 = 0 \qquad \text{if} \qquad r > a \tag{8.4}$$

Here a is the borehole radius and

$$k_1 = \frac{\omega}{c_1}, \qquad k_2 = \frac{\omega}{c_2} \tag{8.5}$$

Of course, eq. 8.3 is invalid at the origin, where the source of the primary field is situated. Due to a continuity of the pressure and the normal component of the displacement at the boundary the function $\mathcal{U}$ has to obey the following conditions

$$\rho_1 \mathcal{U}_1 = \rho_2 \mathcal{U}_2 \qquad \text{and} \qquad \frac{\partial \mathcal{U}_1}{\partial r} = \frac{\partial \mathcal{U}_2}{\partial r} \qquad \text{if} \qquad r = a \tag{8.6}$$

It is natural to represent the wave field inside the borehole as the sum

$$\mathcal{U}_1 (r,\, z) = \mathcal{U}_0 (r,\, z) + \mathcal{U}_s\,, \tag{8.7}$$

where $\mathcal{U}_0$ is the potential of the primary field source in the homogeneous medium with parameters ρ_1 and c_1. As we know

$$\mathcal{U}_0 = C \frac{e^{i\,k_1\,R}}{R}$$

Here C is the constant and R is the distance from the origin:

$$R = (r^2 + z^2)^{1/2}$$

The function $\mathcal{U}_s$ describes the secondary wave, caused by the presence of the boundary and is finite everywhere. It is clear that in approaching the source the primary field becomes dominant:

$$\mathcal{U}_1 \rightarrow \mathcal{U}_0 = C \frac{e^{i\,k_1\,R}}{R} \tag{8.8}$$

Finally, assuming the presence of an attenuation, even if it is very small, we conclude that at very large distances from the source, the wave field vanishes:

$$\mathcal{U}_1 \rightarrow 0 \qquad \mathcal{U}_2 \rightarrow 0\,, \qquad \text{if} \qquad R \rightarrow \infty\,, \tag{8.9}$$

and the latter is a condition at infinity.

Thus, we have formulated the boundary value problem, and our goal is to find the potential $\mathcal{U}_1$, that satisfies all four conditions, eqs. 8.3–8.4, 8.6, 8.8–8.9.

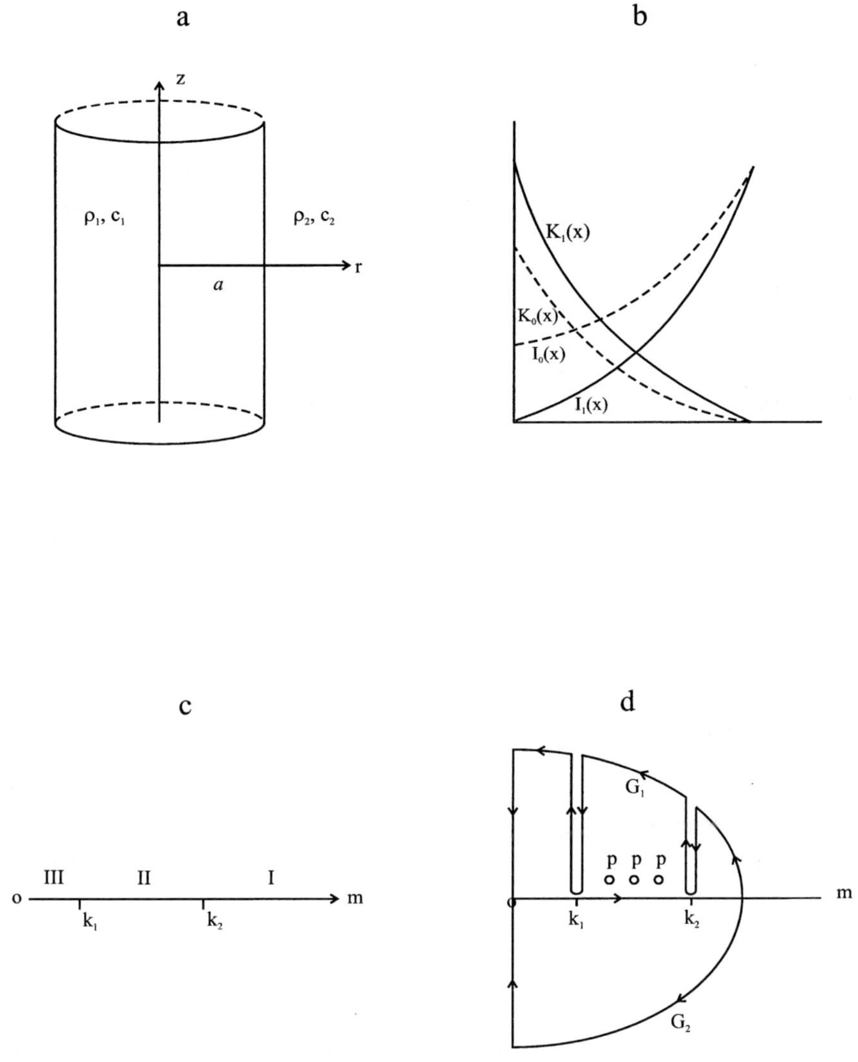

Figure 8.1: (a) Model of a medium (b) Modified Bessel functions (c) Three intervals of integration (d) Contours of integration.

Solution of the Helmholtz equation

In the cylindrical system of coordinates it has the following form, (Part I):

$$\frac{\partial^2 \mathcal{U}}{\partial r^2} + \frac{1}{r}\frac{\partial \mathcal{U}}{\partial r} + \frac{\partial^2 \mathcal{U}}{\partial z^2} + k^2 \mathcal{U} = 0\,, \tag{8.10}$$

since $\partial \mathcal{U}/\partial \varphi = 0$.

Applying again the method of separation of variables the function $\mathcal{U}$ can be written as

$$\mathcal{U}(r,\, z,\, \omega) = T(r,\, \omega)\, Z(z) \tag{8.11}$$

Substitution of eq. 8.11 into eq. 8.10 gives

$$\frac{1}{T}\frac{d^2 T}{dr^2} + \frac{1}{rT}\frac{dT}{dr} + \frac{1}{Z}\frac{d^2 Z}{dz^2} + k^2 = 0 \tag{8.12}$$

This shows that the sum of the two first terms, as well as others, are constants, and therefore in place of eq. 8.12, we obtain two ordinary differential equations:

$$\frac{1}{T}\frac{d^2 T}{dr^2} + \frac{1}{rT}\frac{dT}{dr} + k^2 = \pm m^2 \tag{8.13}$$

and

$$\frac{1}{Z}\frac{d^2 Z}{dz^2} = \mp m^2\,,$$

where m is an arbitrary variable of separation.

In choosing a sign on the right-hand side of eq. 8.13 we proceed from the fact that the potential $\mathcal{U}$ is the even function with respect to the z-coordinate

$$\mathcal{U}(r,\, z,\, \omega) = \mathcal{U}(r,\, -z,\, \omega)\,,$$

and this follows from the symmetry of the medium and the wave field. Taking into account this fact let us choose minus at the right hand side of the second equation of the set eq. 8.13 and it gives

$$\frac{d^2 Z}{dz^2} + m^2 Z = 0$$

Its partial solution is the sum of functions

$$C_m \sin mz \qquad \text{and} \qquad B_m \cos mz$$

and the latter is an even function of z. Correspondingly it is used to represent the complex amplitude $\mathcal{U}$.

Therefore, the first equation of set 8.13 becomes

$$\frac{d^2T}{dr^2} + \frac{1}{r}\frac{dT}{dr} - (m^2 - k^2)T = 0, \tag{8.14}$$

and its partial solutions are modified Bessel functions of the first and second type:

$$I_0(\sqrt{m^2 - k^2}\, r) \qquad \text{and} \qquad K_0(\sqrt{m^2 - k^2}\, r)$$

In particular, if the argument is real and positive the behavior of these functions is very simple, Fig 8.1b. The asymptotical expressions of these functions are

$$I_0(x) \to 1, \qquad K_0(x) \to -\ln x \qquad \text{if} \quad x \to 0 \tag{8.15}$$

and

$$I_0(x) \to \left(\frac{1}{2\pi x}\right)^{1/2} e^x, \qquad K_0(x) \to \left(\frac{\pi}{2x}\right)^{1/2} e^{-x} \qquad \text{if} \quad x \to \infty$$

Further, we also use the modified Bessel functions of the first order, $I_1(x)$ and $K_1(x)$. Their asymptotical formulae are

$$I_1(x) \to \frac{x}{2}, \qquad K_1(x) \to \frac{1}{x} \qquad \text{if} \quad x \to 0 \tag{8.16}$$

and

$$I_1(x) \to \left(\frac{1}{2\pi x}\right)^{1/2} e^x, \qquad K_1(x) \to \left(\frac{\pi}{2x}\right)^{1/2} e^{-x} \qquad \text{if} \quad x \to \infty$$

Expressions for potential

By definition, the sum

$$T_m(r) = A_m I_0(m_i r) + B_m K_0(m_i r) \tag{8.17}$$

satisfies eq. 8.14, and

$$m_i = \begin{cases} m_1 = \sqrt{m^2 - k_1^2} & \text{if} \quad r < a \\ \\ m_2 = \sqrt{m^2 - k_2^2} & \text{if} \quad r > a \end{cases}$$

As follows from eq. 8.11 the partial solution of the Helmholtz equation is

$$\mathcal{U}_m = [A_m I_0(m_i r) + B_m K_0(m_i r)] \cos mz$$

Inasmuch as the wave field is independent of the separation of a variable m we have to deal with the general solution

$$\mathcal{U}(r,\, z,\, \omega) = \int_0^\infty [A_m\, I_0\,(m_i r) + B_m\, K_0\,(m_i r)] \cos mz\, dm \tag{8.18}$$

As we pointed out earlier, the secondary field has everywhere a finite value, and therefore inside the cylinder,

$$\mathcal{U}_s\,(r,\, z,\, \omega) = \int_0^\infty A_m\, I_0\,(m_1 r)\, \cos mz\; dm \qquad \text{if} \qquad r < a \tag{8.19}$$

since function $K_0(m_1\, r)$ is infinitely large at points of the borehole axis ($r = 0$). In approaching the origin, $r \to 0$, $z \to 0$, the Bessel function in eq. 8.19 tends to a finite value, and therefore the function $\mathcal{U}_1$ satisfies the condition near the source. Bearing in mind that $I_0(m_2 r)$ unlimitedly increases with an increase of r, the potential in the surrounding medium is written as

$$\mathcal{U}_2\,(z,\, \omega) = \int_0^\infty B_m\, K_0\,(m_2 r)\, \cos mz\; dm \qquad \text{if} \qquad r > a \tag{8.20}$$

Thus, functions $\mathcal{U}_1$ and $\mathcal{U}_2$ satisfy the Helmholtz equation as well as conditions near the source and at infinity when $r \to \infty$. It is also proper to notice that they also vanish with an increase of z, because of the presence of the oscillating factor $\cos mz$ in integrands of eqs. 8.19 – 8.20.

Coefficients A_m and B_m

Next determine such coefficients A_m and B_m that the acoustic potential obeys boundary conditions at the cylindrical surface. With this purpose in mind, we make use of the Sommerfeld integral

$$\frac{e^{ik_1 R}}{R} = \frac{2}{\pi} \int_0^\infty K_0\,(m_1 r)\, \cos mz\; dm\,, \tag{8.21}$$

which allows us to represent eqs. 8.19 – 8.20 in the form

$$\mathcal{U}_1 = \frac{2}{\pi} C \int_0^\infty [K_0\,(m_1 r) + A_m\, I_0\,(m_1 r)]\, \cos mz\; dm \tag{8.22}$$

and

$$\mathcal{U}_2 = \frac{2}{\pi} C \int_0^\infty B_m\, K_0\,(m_2 r)\, \cos mz\; dm\,,$$

where A_m and B_m are still unknowns.

We have expressed functions $\mathcal{U}_1$ and $\mathcal{U}_2$ in terms of Fourier integrals. As is well known, from the equality of the two functions follows that their spectra are equal. Taking into account this fact and eqs. 8.6, we arrive at the system of equations with respect to A_m and B_m:

$$\rho_1 \left[K_0\,(m_1 a) + A_m\, I_0\,(m_1 a)\right] = \rho_2\, B_m\, K_0\,(m_2 a) \tag{8.23}$$

and

$$-m_1\, K_1(m_1 a) + m_1\, A_m\, I_1(m_1 a) = -m_2\, B_m\, K_1(m_2 a)$$

Here relationships

$$\frac{d}{dr} I_0\,(m_1 r) = m_1\, I_1\,(m_1 r)\,, \qquad \frac{d}{dr} K_0\,(m_1 r) = -m_1\, K_1\,(m_1 r)$$

were applied. Solution of this system gives

$$A_m = \frac{m_1\, \rho_2\, K_0(m_2 a)\, K_1(m_1 a) - m_2\, \rho_1\, K_0(m_1 a)\, K_1(m_2 a)}{m_1\, \rho_2\, K_0(m_2 a)\, I_1(m_1 a) + m_2\, \rho_1\, I_0(m_1 a)\, K_1(m_2 a)} \tag{8.24}$$

and

$$B_m = \frac{\rho_1}{a}\, \frac{1}{m_1\, \rho_2\, K_0(m_2 a)\, I_1(m_1 a) + m_2\, \rho_1\, I_0(m_1 a)\, K_1(m_2 a)} \tag{8.25}$$

In deriving the latter we have used the equality

$$I_0(x)\, K_1(x) + I_1(x)\, K_0(x) = \frac{1}{x}$$

Thus, the functions $\mathcal{U}_1$ and $\mathcal{U}_2$ with coefficients A_m and B_m, given by eqs. 8.24 – 8.25, satisfy all conditions of the boundary value problem. In other words, they describe the complex amplitudes of the acoustic potential due to the spherical source, located on the borehole axis, and their expressions are

$$\mathcal{U}_1 = C\,\frac{e^{ik_1 R}}{R} \tag{8.26}$$

$$+\frac{2}{\pi}\, C \int\limits_0^\infty \frac{m_1\, \rho_2\, K_0(m_2 a)\, K_1(m_1 a) - m_2\, \rho_1\, K_0(m_1 a)\, K_1(m_2 a)}{\mathcal{D}}\, I_0(m_1 r)\, \cos mz\; dm$$

if $\quad r < a$

and

$$\mathcal{U}_2 = \frac{2\,\rho_1}{\pi\, a}\, C \int\limits_0^\infty \frac{K_0(m_2 r)}{\mathcal{D}} \cos mz\, dm \qquad \text{if} \qquad r > a \tag{8.27}$$

Here

$$\mathcal{D} = m_1\,\rho_2\, K_0(m_2 a)\, I_1(m_1 a) + m_2\,\rho_1\, I_0(m_1 a)\, K_1(m_2 a) \tag{8.28}$$

Further we pay an attention to the case when $c_2 > c_1$ only. By analogy with a field in a horizontally layered medium it is useful to describe some general features of a non-oscillating term of the integrand in eqs. 8.26 - 8.27. For this purpose we distinguish three ranges of m, Fig. 8.1c.

Before we continue, it is necessary to point out that radicals have to satisfy the inequalities

$$\text{Re } m_1 > 0 \qquad \text{and} \qquad \text{Re } m_2 > 0 \tag{8.29}$$

Otherwise, the condition at infinity is not met.

Behavior of integrands

In the first interval, we have

$$k_1 \leq m < \infty\,,$$

where $m = k_1$ is the branch point. In this range, m_1 and m_2 are real and positive and, therefore, the modified Bessel functions are also positive numbers, Fig. 8.1b. This means that denominator $\mathcal{D}$, eq. 8.28, does not have zeroes, when $m > k_1$. Making use of asymptotical formulae of the Bessel functions, eqs. 8.15 – 8.16, consider integrands near $m = k_1$. In such a case, $m_1 \to 0$, and we have

$$\mathcal{D} \to m_2\,\rho_1\, K_1(m_2 a)$$

Whence

$$A_m\, I_0(m_1 r) \to -K_0(m_1 a) \qquad \text{and} \qquad \frac{K_0(m_2 r)}{\mathcal{D}} \to \frac{K_0(m_2 r)}{m_2\,\rho_1\, K_1(m_2 a)}$$

The logarithmic singularity of A_m is integrable, and it can be easily removed. In the opposite case, $m \to \infty$, we have

$$\mathcal{D} = (\rho_1 + \rho_2)\,\frac{1}{2a}\,,$$

while the numerator of A_m tends to

$$(\rho_2 - \rho_1)\,\frac{\pi}{2a}\, e^{-2ma}$$

and

$$A_m\, I_0(m_1 r) \to \frac{\rho_2 - \rho_1}{\rho_2 + \rho_1} \left(\frac{\pi}{2mr}\right)^{1/2} e^{(-2a + r)m} \to 0\,, \qquad \text{if} \quad m \to \infty$$

since $r < a$

It is obvious that the function

$$\frac{K_0(m_2 r)}{\mathcal{D}}$$

also tends to zero when $m \to \infty$.

Within the second interval,

$$k_2 < m < k_1$$

the radical m_2 still remains real and positive, while m_1 can be represented as

$$m_1 = \left(m^2 - k_1^2\right)^{1/2} = i\,\tilde{m}_1\,, \tag{8.30}$$

where

$$\tilde{m}_1 = \left(k_1^2 - m^2\right)^{1/2} \tag{8.31}$$

is the positive number. Correspondingly, function $\mathcal{D}$ has the form

$$\mathcal{D} = i\,\tilde{m}_1\,\rho_2\,K_0(m_2 a)\,I_1(i\,\tilde{m}_1 a) + m_2\,\rho_1\,I_0(i\,\tilde{m}_1 a)\,K_1(m_2 a)$$

Taking into account that

$$I_0(ix) = J_0(-x)\,, \qquad I_1(ix) = -i\,J_1(-x),$$

but

$$J_0(-x) = J_0(x)\,, \qquad J_1(-x) = -J_1(x),$$

we have

$$\mathcal{D} = -\tilde{m}_1\,\rho_2\,K_0(m_2 a)\,J_1(\tilde{m}_1 a) + m_2\,\rho_1\,J_0(\tilde{m}_1 a)\,K_1(m_2 a) \tag{8.32}$$

Therefore the denominator becomes equal to zero at points where

$$\tilde{m}_1\,\rho_2\,K_0(m_2 a)\,J_1(\tilde{m}_1 a) = m_2\,\rho_1\,J_0(\tilde{m}_1 a)\,K_1(m_2 a) \tag{8.33}$$

Finally, within the third interval, $m \le k_2$, we obtain

$$\mathcal{D} = -\tilde{m}_1\,\rho_2\,K_0(i\tilde{m}_2 a)\,J_1(\tilde{m}_1 a) + i\,\tilde{m}_2\,\rho_1\,J_0(\tilde{m}_1 a)\,K_1(i\tilde{m}_2 a) \tag{8.34}$$

Here

$$\tilde{m}_2 = \sqrt{k_2^2 - m^2}$$

Inasmuch as functions $K_0(i\tilde{m}_2 a)$ and $K_1(i\tilde{m}_2 a)$ have complex values, the function $\mathcal{D}$ is complex too. At the same time, the presence of zeroes requires that

$$\text{Re } \mathcal{D} = 0 \qquad \text{and} \qquad \text{Im } \mathcal{D} = 0$$

and calculations show that poles are absent inside the last interval of integration.

Deformation of the integration path

In the next two sections we consider asymptotical behavior of the field at relatively large distances from the source. With this purpose in mind let us again, as in the case of a horizontally layered medium, apply Cauchy's theorem and change a path of integration in eqs. 8.26 – 8.27. The potential $\mathcal{U}_1$ can be written as

$$\mathcal{U}_1 = C\,\frac{e^{i k_1 R}}{R} + \frac{2C}{\pi}\int\limits_0^\infty \frac{N}{\mathcal{D}}\left(e^{imz} + e^{-imz}\right) dm$$

or

$$\mathcal{U}_1 = C\left[\frac{e^{i k_1 R}}{R} + \frac{2}{\pi}\,(I_1 + I_2)\right] \tag{8.35}$$

where

$$N = [m_1\,\rho_2\,K_0(m_2 a)\,K_1(m_1 a) - m_2\,\rho_1\,K_0(m_1 a)K_1(m_2 a)]\;I_0(m_1 r) \tag{8.36}$$

and

$$I_1 = \int\limits_0^\infty \frac{N}{\mathcal{D}}\,e^{imz}\,dm\,, \qquad I_2 = \int\limits_0^\infty \frac{N}{\mathcal{D}}\,e^{-imz}\,dm$$

We assume that singularities are situated slightly above the real axis m. However, in evaluating the contribution of poles and integrals along branch cuts, they will be considered real. First, it is proper to change the path of integration of I_2. Since the function e^{-imz} vanishes when $\text{Im } m \to -\infty$, we have (Fig. 8.1d):

$$\oint\limits_{G_2} \frac{N}{\mathcal{D}}\,e^{-imz}\,dm = \int\limits_0^\infty \frac{N}{\mathcal{D}}\,e^{-imz}\,dm + \int\limits_{C_\infty} \frac{N}{\mathcal{D}}\,e^{-imz}\,dm + \int\limits_{-i\infty}^0 \frac{N}{\mathcal{D}}\,e^{-imz}\,dm = 0$$

or

$$I_2 = -\int\limits_{-i\infty}^0 \frac{N}{\mathcal{D}}\,e^{-imz}\,dm \tag{8.37}$$

Note that we disregard the integral along part of the circle C_∞ with an infinitely large radius. Bearing in mind that functions N and $\mathcal{D}$ are even with respect to m and changing the sign of the variable, eq. 8.37 becomes

$$I_2 = \int_{i\infty}^{0} \frac{N}{\mathcal{D}} e^{imz} \, dm \tag{8.38}$$

Now consider the integral I_1. Applying the Cauchy theorem to the integral along the closed path G_1, Fig. 8.1d, located in the first quadrant, we obtain

$$\oint_{G_1} \frac{N}{\mathcal{D}} e^{imz} \, dm = 0 ,$$

since poles and branch points are situated beyond the area surrounded by contour G_1. The latter can be written as

$$I_1 + M_p + M_b + \int_{i\infty}^{0} \frac{N}{\mathcal{D}} e^{imz} \, dm = 0$$

or

$$I_1 = -M_p - M_b - \int_{i\infty}^{0} \frac{N}{\mathcal{D}} e^{imz} \, dm \tag{8.39}$$

Here M_p and M_b are sums of integrals around poles and branch points, respectively, and the integral over C_∞ vanishes.

Thus, making use of eqs. 8.35, 8.38 – 8.39, we have for the potential inside the cylinder (borehole):

$$\mathcal{U}_1 = C \left[\frac{e^{ik_1 R}}{R} - \frac{2}{\pi} (M_p + M_b) \right] \tag{8.40}$$

In a similar manner potential $\mathcal{U}_2$ can be expressed in terms of integrals around poles and along branch cuts.

8.2 Normal modes inside the cylinder

In the previous section we represented the complex amplitude of the acoustic potential as the sum of three terms

$$\mathcal{U}_1 = \mathcal{U}_0 + \mathcal{U}_p + \mathcal{U}_b \tag{8.41}$$

First, consider the function $\mathcal{U}_p$. By analogy with the case of a horizontally layered medium we can expect that this part of the acoustic potential represents the normal modes inside the borehole if $c_2 > c_1$. In accordance with eq. 8.41, we have

$$\mathcal{U}_p = -\frac{2C}{\pi} M_p \tag{8.42}$$

or, taking into account the residual theorem,

$$\mathcal{U}_p = 4\, C\, i \sum_{n=0}^{\infty} \text{Res } F_n\, e^{i\, m_n\, z} \tag{8.43}$$

As follows from eqs. 8.32 – 8.36:

$$F_n = \frac{N}{\mathcal{D}} \tag{8.44}$$

where

$$N = [m_1 \rho_2\, K_0(m_2 a)\, K_1(m_1 a) - m_2 \rho_1\, K_0(m_1 a)\, K_1(m_2 a)]\, I_0(m_1 r) \tag{8.45}$$

and

$$\mathcal{D} = -\tilde{m}_1 \rho_2\, K_0(m_2 a)\, J_1(\tilde{m}_1 a) + m_2 \rho_1\, J_0(\tilde{m}_1 a)\, K_1(m_2 a)$$

Here

$$m_1 = \sqrt{m_n^2 - k_1^2}\,, \qquad \tilde{m}_1 = -i m_1\,, \qquad m_2 = \sqrt{m_n^2 - k_2^2}$$

and m_n are poles of function F_n which are defined from the dispersion equation

$$\tilde{m}_1 \rho_2\, K_0(m_2 a)\, J_1(\tilde{m}_1 a) = m_2 \rho_1\, J_0(\tilde{m}_1 a)\, K_1(m_2 a) \tag{8.46}$$

Bearing in mind that

$$U_p = \text{Re } \mathcal{U}_p\, e^{-i\omega t}$$

and making use of eq. 8.43, we have

$$\mathcal{U}_p = \text{Re} \sum_{n=0}^{\infty} G_n\, e^{-i(\omega t - m_n z)}\,, \tag{8.47}$$

where

$$G_n = 4C\, i \,\text{Res } F_n \tag{8.48}$$

The expression of $\mathcal{U}_p$ clearly shows that each term of the sum describes a sinusoidal wave traveling along the cylinder (borehole) with phase velocity

$$c_{pn} = \frac{\omega}{m_n}\,, \tag{8.49}$$

and the function G_n characterizes the complex amplitude of these waves.

It is natural to begin a study from the simplest case when the borehole is surrounded by the rigid medium.

Case one ($\rho_2 \to \infty$) Then the function F_n is greatly simplified, and we have

$$F_n = -i\,\frac{K_1(i\tilde{m}_1 a)}{J_1(\tilde{m}_1 a)}\,J_0(\tilde{m}_1 r), \tag{8.50}$$

since

$$I_0(i\tilde{m}_1 a) = J_0(\tilde{m}_1 a)$$

Correspondingly, the dispersion equation becomes

$$J_1(\tilde{m}_1 a) = 0 \tag{8.51}$$

and its first roots are given below.

n	0	1	2	3	4	5	6
$\tilde{m}_1 a$	0	3.83	7.02	10.17	13.32	16.47	19.62

Inasmuch as with an increase of the argument the behavior of the Bessel functions is almost sinusoidal (Chapter 6), the difference between roots tends to π. Letting

$$\tilde{m}_1 a = r_n \tag{8.52}$$

we obtain for the wave number, m_n, of each mode the following expression

$$m_n = \frac{1}{a}\left(k_1^2\,a^2 - r_n^2\right)^{1/2} \tag{8.53}$$

It is essential that, unlike a general case, when $0 < \rho_2 < \infty$, roots r_n are independent of the frequency.

In accordance with eq. 8.53 for all normal modes, except $n = 0$, there is a frequency when

$$k_1\,a = r_n\,,$$

and as we know, it is called the cut-off frequency. In such a case the wave number m_n is equal to zero and the phase velocity becomes infinitely large. With the increase of ω, the phase velocity, c_{pn}, gradually decreases, and in the limit it tends to the velocity of the borehole fluid.

As in the case of a horizontally layered medium, every normal mode is a result of the constructive interference of waves reflected from the cylindrical boundary. Then, with an increase of the frequency the angle between the direction of propagation of these waves and the z-axis decreases, and in the limit when $\omega \to \infty$, the normal mode travels with velocity c_1. Dispersion curves of the phase velocity $c_{pn}(\omega)$ are similar to those in the case of horizontally layered medium.

Unlike for other modes the phase velocity of the normal mode of the zero order, $n = 0$ is independent of the frequency, and it is equal to c_1. In fact, since $r_0 = 0$ we have $m_0 = k_1$ and $c_{p0} = c_1$. Thus, regardless of the frequency, this mode propagates with constant velocity c_1 along the borehole, and its cut-off frequency is equal to zero.

As follows from eq. 8.53 the modes, $n \neq 0$, are formed when the wavelength is greater than the borehole radius. For instance, in the case of the cut-off frequency and $n = 1$, we have

$$k_1\, a = r_1 \qquad \text{or} \qquad \frac{a}{\lambda} = \frac{r_1}{2\pi} \approx 0.6$$

In this regard we recall that an "ancient" telephone was based on the use of air pipes, and communication took place due to a propagation of the normal mode of the zero order.

In accordance with eq. 8.50, the residue of function F_n is

$$\text{Res } F_n = i\,\frac{\tilde{m}_1\, K_1(i\tilde{m}_1 a)\, J_0(\tilde{m}_1 r)}{m_n\, a\, J_1'(\tilde{m}_1 a)}\,, \tag{8.54}$$

since

$$\frac{\partial}{\partial m}\, J_1(\tilde{m}_1 a) = -\,\frac{m_n a}{\tilde{m}_1}\, J_1'(\tilde{m}_1 a)$$

Taking into account the equality

$$J_1'(x) = J_0(x) - \frac{J_1(x)}{x} \tag{8.55}$$

in place of eq 8.54 we obtain

$$\text{Res } F_n = i\,\frac{\tilde{m}_1\, K_1(i\tilde{m}_1 a)\, J_0(\tilde{m}_1 r)}{m_n\, a\,\left[J_0(\tilde{m}_1 a) - J_1(\tilde{m}_1 a)\,/\,\tilde{m}_1 a\right]} \tag{8.56}$$

Making use of the asymptotic behavior of the Bessel functions

$$J_0(x) \to 1\,, \qquad J_1(x) \to \frac{x}{2}\,, \qquad K_1(x) \to \frac{1}{x} \qquad \text{if} \quad x \to 0$$

we have for the normal mode of the zero order when $r_0 = 0$ and $m_0 = k_1$,

$$\text{Res } F_0 = \frac{2}{k_1 a^2} \qquad \text{if} \quad n = 0 \tag{8.57}$$

As follows from eq. 8.51, in place of eq. 8.54, we have ($n \neq 0$):

$$\text{Res } F_n = i\,\frac{\tilde{m}_1\, K_1(i\tilde{m}_1 a)\, J_0(\tilde{m}_1 r)}{m_n\, a\, J_0(\tilde{m}_1 a)} \tag{8.58}$$

In the case of the normal mode of zero order, the function G_0, eq. 8.48, is

$$G_O = \frac{8\,i\,C}{k_1\,a^2} \tag{8.59}$$

and correspondingly,

$$\mathcal{U}_{0p} = \frac{8\,i\,C}{k_1\,a^2}\,e^{-i\,(\omega t - k_1 z)} \tag{8.60}$$

Proceeding from the equality

$$\mathcal{P} = \omega^2\,\rho_1\,\mathcal{U}$$

the expression for additional pressure becomes

$$\mathcal{P}_0 = \frac{8\,C\,i\,\rho_1\,c_1\,\omega}{a^2}\,e^{-i\,(\omega t - k_1 z)} \tag{8.61}$$

Thus, it is uniformly distributed over the borehole cross-section and, as we may expect, it is inversely proportional to its area.

Inasmuch as

$$\mathcal{S}_z = \frac{\partial \mathcal{U}}{\partial z} \qquad \text{and} \qquad \mathcal{S}_r = \frac{\partial \mathcal{U}}{\partial r}$$

we have

$$\mathcal{S}_z = -\frac{8\,C}{a^2}\,e^{-i\,(\omega t - k_1 z)} \qquad \text{and} \qquad \mathcal{S}_r = 0$$

This means that inside the borehole the displacement has only the vertical component, which is independent of the coordinate r and frequency.

Next let us discuss general features of nonzero normal modes. From eq. 8.48 it follows

$$G_n = -4\,C\,\frac{\tilde{m}_1\,K_1(i\tilde{m}_1 a)\,J_0(\tilde{m}_1 r)}{m_n\,a\,J_0(\tilde{m}_1 a)} \tag{8.62}$$

and, unlike for the mode of the zero order, functions G_n depend on coordinate r. Since the complex amplitude of the potential is

$$\mathcal{U}_{np} = -4\,C\,\frac{\tilde{m}_1\,K_1(i\tilde{m}_1 a)\,J_0(\tilde{m}_1 r)}{m_n\,a\,J_0(\tilde{m}_1 a)}\,e^{-i\,(\omega t - m_n z)} \tag{8.63}$$

equations for the pressure and displacement components are

$$\mathcal{P} = -4\,C\,\rho_1\,\omega^2\,\frac{\tilde{m}_1\,K_1(i\tilde{m}_1 a)\,J_0(\tilde{m}_1 r)}{m_n\,a\,J_0(\tilde{m}_1 a)}\,e^{-i\,(\omega t - m_n z)},$$

$$\mathcal{S}_z = -4\,C\,i\,\frac{\tilde{m}_1\,K_1(i\tilde{m}_1 a)\,J_0(\tilde{m}_1 r)}{a\,J_0(\tilde{m}_1 a)}\,e^{-i\,(\omega t\;-\;m_n z)}\,, \tag{8.64}$$

$$\mathcal{S}_r = 4\,C\,\frac{\tilde{m}_1^2\,K_1(i\tilde{m}_1 a)\,J_1(\tilde{m}_1 r)}{m_n\,a\,J_0(\tilde{m}_1 a)}\,e^{-i\,(\omega t\;-\;m_n z)}$$

This clearly shows that with an increase of the order, n, the wave field varies more rapidly in the radial direction, forming a sequence of expansional and compressional zones. At each point of the borehole except its surface, there are usually two components of the displacement and the phase shift between them is equal to $\pi/2$. For this reason the vector $\mathbf{s}$ is elliptically polarized. It may be proper to emphasize that the radial component s_r vanishes at the boundary since

$$J_1(\tilde{m}_1 a) = 0\,,$$

which is obvious because the surrounding medium is rigid one.

Case two Consider the second limiting case, when $\rho_2 = 0$, that is around the cylinder there is a free space. In accordance with eq. 8.45, we have

$$F_n = -\frac{K_0(m_1 a)}{J_0(\tilde{m}_1 a)}\,J_0(\tilde{m}_1 r) \tag{8.65}$$

Respectively, the dispersion equation is

$$J_0(\tilde{m}_1 a) = J_0(r_n) = 0 \tag{8.66}$$

and its roots are shown below

n	1	2	3	4	5	6
r_n	2.40	5.52	8.65	11.79	14.93	18.07

As in the previous case, ($\rho_2 = \infty$), with an increase of the order of a normal mode a difference of roots approaches π but the normal mode of zero order is absent. The roots of eq. 8.66, r_n, are also independent of the frequency and at the cut-off one the phase velocity c_{nr}, as before, tends to an infinity but with an increase of the frequency it approaches c_1.

In the same manner as in the first case we find that the residual of the function F_n is

$$\text{Res}\;F_n = -\frac{\tilde{m}_1\,K_0(m_1 a)}{m_n a\,J_1(\tilde{m}_1 a)}\,J_0(\tilde{m}_1 r) \tag{8.67}$$

and therefore

$$G_n = -4\,C\,i\,\frac{\tilde{m}_1\,K_0(m_1 a)\,J_0(\tilde{m}_1 r)}{m_n a\,J_1(\tilde{m}_1 a)} \tag{8.68}$$

and

$$\mathcal{U}_{np} = -4\,C\,i\,\frac{\tilde{m}_1\,K_0(m_1 a)\,J_0(\tilde{m}_1 r)}{m_n\,a\,J_1(\tilde{m}_1 a)}\,e^{-i\,(\omega t\,-\,m_n z)}$$

Taking into account eq. 8.66 we see that boundary conditions at the borehole surface are satisfied, that is the pressure, as well as the vertical component of the displacement, are equal to zero. At the same time inside the borehole the vector **s** is the elliptically polarized and we can observe a regular change of the pressure sign in the radial direction.

General case

When both media have nonzero and finite values of the density, poles are defined from the dispersion equation, eq. 8.46, and correspondingly, the phase velocity, eq. 8.50, is

$$c_{pn} = \frac{\omega\,a}{(k_1^2\,a^2\,-\,r_n^2)^{1/2}}$$

Here r_n are roots of eq. 8.46. Unlike the limiting cases at the cut-off frequency the phase velocity of normal modes coincides with that of the surrounding medium

$$c_{pn} = c_2 \qquad \text{if} \qquad f = f_{cn} \tag{8.69}$$

and with an increase of frequency, the phase velocity gradually decreases and tends to that of the borehole, c_1.

By analogy with a horizontally layered medium, a study of frequency responses $c_{pn}(\omega)$ allows us to discuss general features of transient waves in a borehole. Suppose that a spectrum of the pressure, caused by the primary source, is a continuous function of the frequency. As an example, we pay attention to the first normal mode. By definition, the primary wave with frequency f, which is smaller than the critical one,

$$f < f_{c1}$$

does not generate the normal mode. The latter appears when $f = f_{c1}$, and with an increase of frequency its phase velocity c_{p1} decreases and varies within the range

$$c_1 < c_{p1} < c_2$$

The source produces the normal mode of the first order, and each frequency component in this mode propagates at a different phase velocity. As was shown in Chapter 7 an interference of this system of waves gives rise to different wave groups, and each of them is characterized by a certain value of the dominant frequency and the group velocity. As we know, the phase and group velocities are related to each other in the following way

$$\frac{1}{c_{qn}} = \frac{1}{c_{pn}} - \frac{\omega}{c_{pn}^2}\,\frac{dc_{pn}}{d\omega}$$

Because the function $c_{pn}(\omega)$ is known, the group velocity can be easily calculated, and the behavior of the group velocity as a function of the frequency in the borehole and in a horizontally layered medium is similar (Fig. 7.10). At the cut-off frequency, the group velocity coincides with physical velocity c_2 of the surrounding medium. Then with an increase of the frequency it becomes smaller and at some frequency $f = f_A$ the velocity c_{qn} reaches the minimum:

$$c_{qn} = c_{qA}$$

With further increase of the frequency velocity c_{qn} begins to grow and asymptotically approaches to c_1. Knowledge of function $c_{qn}(\omega)$ allows us to describe qualitatively a behavior of the transient wave, related to the potential $U_{1p}(t)$. Suppose that the source, located at the origin, is turned on at the instant $t = 0$, and at some distance z the receiver measures the wave.

During the time interval

$$t < \frac{z}{c_2},$$

the wave corresponding to the first mode is absent. Then, at the instant $t = z/c_2$ the wave group arrives, and its dominant frequency is equal to the cut-off one. With an increase of time, wave groups with higher frequencies begin to arrive. Within some interval

$$f_{1c} < f < f_1$$

the dominant frequency of the wave group gradually increases with time. At the moment, which is slightly greater than the ratio z/c_1, two wave groups arrive simultaneously at the observation point, and their group velocities are equal to each other and almost coincide with c_1.

The dominant frequency of the first group equals, f_1, while the second is characterized by a much higher frequency. Such superposition of two wave groups is observed within the time interval

$$\frac{z}{c_1} < t < \frac{z}{c_{qA}}$$

and with an increase of time the dominant frequencies of both groups approach each other. Finally, starting from the instant

$$t = \frac{z}{c_{qA}},$$

there is only one wave group, which propagates with velocity c_{qA}, and its dominant frequency is equal to f_A. It is natural to treat this last stage of a transient wave as the Airy phase.

8.3 Head wave in borehole

In order to evaluate a contribution to the potential caused by an integration along branch cuts it is convenient to represent the potential of the incident wave in the form (Chapter 6):

$$\mathcal{U}_i(r,z) = \frac{C\, e^{ik_1R}}{R} = C\,\frac{i}{2}\int\limits_{-\infty}^{\infty} H_0^{(1)}(\tilde{m}_1 r)\, e^{imz}\, dm \tag{8.70}$$

Here

$$\tilde{m}_1 = \sqrt{k_1^2 - m^2}\,, \qquad \text{Im}\; \tilde{m}_1 \geq 0\,, \qquad R^2 = r^2 + z^2$$

Correspondingly, the potential of the secondary wave inside the borehole has the form

$$\mathcal{U}_s = C\,\frac{i}{2}\int\limits_{-\infty}^{\infty} A(m)\, J_0(mr)\, e^{imz}\, dm\,, \qquad r < a \tag{8.71}$$

while outside of it

$$\mathcal{U}_2 = C\,\frac{i}{2}\int\limits_{-\infty}^{\infty} B(m)\, H_0^{(1)}(mr)\, e^{imz}\, dm \tag{8.72}$$

Coefficients A_m and B_m are determined from the boundary conditions at the borehole surface, $r = a$. In particular,

$$A = -\frac{\dfrac{\rho_1\tilde{m}_2}{\rho_2\tilde{m}_1}\, H_0^{(1)}(\tilde{m}_1 a) H_1^{(1)}(\tilde{m}_2 a) - H_0^{(1)}(\tilde{m}_2 a) H_1^{(1)}(\tilde{m}_1 a)}{\dfrac{\rho_1\tilde{m}_2}{\rho_2\tilde{m}_1}\, J_0(\tilde{m}_1 a) H_1^{(1)}(\tilde{m}_2 a) - H_0^{(1)}(\tilde{m}_2 a) J_1(\tilde{m}_1 a)}\,, \tag{8.73}$$

where

$$\tilde{m}_2 = \sqrt{k_2^2 - m^2}$$

Assuming a weak dissipation in both media, we have

$$\text{Im}\; k_1 > 0 \qquad \text{and} \qquad \text{Im}\; k_2 > 0$$

Then, as was mentioned earlier, two branch points in the upper half-plane are $m = k_1$ and $m = k_2$.

Consider the first contribution, $\mathcal{U}_{1s}$, that arises from the integral along the branch cut originating at $m = k_1$. Radical $\tilde{m}_1$ is discontinuous at the cut and takes values $\pm|\tilde{m}_1|$ on its "banks", while m, $\tilde{m}_2$ and $J_0(\tilde{m}_1 r)$ remain continuous. Then

$$\mathcal{U}_{1s} = -\frac{iC}{2}\int\limits_{k_1}^{i\infty} \left[A_{\tilde{m}_1=|\tilde{m}_1|} - A_{\tilde{m}_1=-|\tilde{m}_1|}\right] J_0(\tilde{m}_1 r)\, e^{imz}\, dm \tag{8.74}$$

Using the identities

$$H_0^{(1)}\left(e^{i\pi}u\right) = -\left[H_0^{(1)}(u)\right]^*, \qquad H_1^{(1)}\left(e^{i\pi}u\right) = -\left[H_1^{(1)}(u)\right]^*, \qquad u>0, \tag{8.75}$$

where $*$ indicate conjugate, eq. 8.73 gives

$$A_{\tilde m_1=|\tilde m_1|} - A_{\tilde m_1=-|\tilde m_1|} = -2 \tag{8.76}$$

Introducing a new variable,

$$u = \sqrt{k_1^2 - m^2}\, r\,,$$

eq. 8.74 becomes

$$\mathcal{U}_{1s} = \frac{iC}{r}\int_0^\infty J_0(u)\exp\left(i\frac{z}{r}\sqrt{u^2+k_1^2r^2}\right)\frac{u\,du}{\sqrt{u^2+k_1^2r^2}} \tag{8.77}$$

The latter is a tabular integral if $z>r$ and it can be written as

$$\mathcal{U}_{1s} = \frac{C}{\sqrt{z^2-r^2}}\exp\left(i\,k_1\sqrt{z^2-r^2}\right) \tag{8.78}$$

Thus, the contribution due to an incident wave and $\mathcal{U}_{1s}$ is equal to

$$C\left[\frac{1}{\sqrt{z^2+r^2}}\exp\left(i\,k_1\sqrt{z^2+r^2}\right) - \frac{1}{\sqrt{z^2-r^2}}\exp\left(i\,k_1\sqrt{z^2-r^2}\right)\right],$$

which describes a wave propagating along the borehole axis with velocity c_1 and rapidly decreasing when $k_1 r < 1$.

Next, we calculate the contribution that arises from the integral along the branch cut originating at $m = k_2$. Here $\tilde m_2$ is discontinuous at the cut and takes values $\pm|\tilde m_2|$ on its "banks", while m and $\tilde m_1$ are continuous. Then

$$\mathcal{U}_{2s} = -\frac{iC}{2}\int_{k_2}^{i\infty}\left[A_{\tilde m_2=|\tilde m_2|} - A_{\tilde m_2=-|\tilde m_2|}\right] J_0(\tilde m_2 r)\, e^{imz}\, dm$$

and introducing a new variable

$$m = i\sqrt{k_2^2 - s^2}\,,$$

we have

$$\mathcal{U}_{2s} = \frac{i}{2}C\int_0^\infty F\left(\sqrt{k_1^2-k_2^2+s^2}\,,\, s\right) J_0\left(r\sqrt{k_1^2-k_2^2+s^2}\right) e^{iz\sqrt{k_2^2-s^2}}\frac{s\,ds}{\sqrt{k_2^2-s^2}} \tag{8.79}$$

From the equality

$$H_1^{(1)}(u)J_0(u) - H_0^{(1)}(u)J_1(u) = \frac{2i}{\pi u}$$

and eqs. 8.75, we obtain for the function F:

$$\begin{aligned} F &= A_{\tilde{m}_2=|\tilde{m}_2|} - A_{\tilde{m}_2=-|\tilde{m}_2|} = \\ &= \frac{4\rho_1}{\pi^2\rho_2}\left[\tilde{m}_1 a\, J_1(\tilde{m}_1 a)\, H_0^{(1)}(\tilde{m}_2 a) - \frac{\rho_1}{\rho_2}\,\tilde{m}_2 a\, J_0(\tilde{m}_1 a)\, H_1^{(1)}(\tilde{m}_2 a)\right]^{-1} \\ &\quad \times \left[\tilde{m}_1 a\, J_1(\tilde{m}_1 a)\, H_0^{(2)}(\tilde{m}_2 a) - \frac{\rho_1}{\rho_2}\,\tilde{m}_2 a\, J_0(\tilde{m}_1 a)\, H_1^{(2)}(\tilde{m}_2 a)\right]^{-1}, \end{aligned} \tag{8.80}$$

where

$$H_\nu^{(1)}(u) = J_\nu(u), + i\, Y_\nu(u)\,, \qquad H_\nu^{(2)}(u) = J_\nu(u) - i\, Y_\nu(u)$$

and $J_\nu(u)$, $Y_\nu(u)$ are Bessel functions of the first and second kind.

As is well known, the Hankel functions have singularities at zero argument, and their replacement by an asymptotical expression when $0 < \tilde{m}_2 \leq 1$ gives

$$F = \frac{\rho_1}{\rho_2}\left[\tilde{m}_1\, a\, J_1(\tilde{m}_1 a)\left(\gamma + \ln\frac{\tilde{m}_2\, a}{2}\right) + \frac{\rho_1}{\rho_2}\, J_0(\tilde{m}_1 a)\right]^{-2}, \tag{8.81}$$

where $\gamma \approx 0.57722$ is Euler's constant.

To evaluate $\mathcal{U}_{2s}$, eq. 8.79, at large z we apply the method of the steepest descent. Assuming that $a/z \ll 1$, the steepest descent path is given by

$$m = k_2 + i\,\frac{q}{z}\,, \qquad 0 \leq q < \infty\,,$$

and the asymptotic value of the integral is determined by contributions from $q \sim 1$. For such values of q, we have

$$|m - k_2| \ll 1 \qquad \text{and} \qquad \tilde{m}_2 \approx \sqrt{-2\, i\, k_2\,\frac{q}{z}}$$

Respectively, the integral in eq. 8.79 can be approximately represented as

$$\mathcal{U}_{2s} = -\frac{i}{2}\, C \int\limits_0^\infty \frac{J_0\left(\sqrt{k_1^2 - k_2^2}\; r\right)\, e^{ik_2 z - q}\, dq}{\left(\alpha + \beta \ln\left(k_2 q a^2/z\right)\right)\left(\alpha^* + \beta^* \ln\left(k_2 q a^2/z\right)\right)}\,, \tag{8.82}$$

where α and β are constants. At large z, where $k_2 a/z \ll 1$, the function in the denominator is a rapidly varying function of q, which prevents use of the standard version of integral evaluation with the steepest descent method. To overcome this difficulty, we make the variable substitution $q = k_2 z e^w$. Then the integral 8.82 becomes

$$\mathcal{U}_{2s} = -\frac{i}{2} C J_0 \left(\sqrt{k_1^2 - k_2^2}\, r\right) e^{ik_2 z} \int_{-\infty}^{\infty} F\left(\sqrt{k_1^2 - k_2^2}, \sqrt{2}\, k_2\, e^{w/2 - i\pi/4}\right) \exp\left(w - k_2 z e^w\right) dw \tag{8.83}$$

Here

$$F\left(\sqrt{k_1^2 - k_2^2}, \sqrt{2}\, k_2\, e^{w/2 - i\pi/4}\right) \approx \left[(\alpha + 2\beta \ln k_2 a + \beta w)\,(\alpha^* + 2\beta^* \ln k_2 a + \beta^* w)\right]^{-1}$$

The main contribution to the integral comes from the vicinity of stationary point $w_s = -\ln k_2 z$ of the exponent in the integrand. Note that $w_s < 0$, $|w_s| \gg 1$, and the factor in front of the exponential in the integrand of eq. 8.83 is a slowly varying function of w in the vicinity of $w = w_s$. Then, the leading order of the head wave asymptotic expansion is obtained by taking the factor F out of the integral at $w = w_s$:

$$\mathcal{U}_{2s} = \frac{-i}{2k_2 z} J_0 \left(\sqrt{k_1^2 - k_2^2}\, r\right) e^{ik_2 z} F\left(\sqrt{k_1^2 - k_2^2}, \sqrt{2k_2/z}\, e^{-i\pi/4}\right) \left[1 + O\left(\frac{1}{\ln k_2 z}\right)\right] \tag{8.84}$$

According to eqs. 8.84 and 8.79, as a function of z, $\mathcal{U}_{2s}$ is inversely proportional to $k_2 z \ln^2 k_2 z$, unless $J_1\left(\sqrt{k_1^2 - k_2^2}\, a\right) = 0$, and to $k_2 z$ when $J_1\left(\sqrt{k_1^2 - k_2^2}\, a\right) = 0$. The dependence on r is given by the Bessel function in eq. 8.84. The field of the head wave has a rather complicated dependence on the borehole radius which enters through terms $J_0(\tilde{m}_1 a)$ and $\tilde{m}_1\, a\, J_1(\tilde{m}_1 a)$ in eq. 8.79.

Subsequent orders of the asymptotic expansion for large $k_2 z$ can be easily calculated from eq. 8.83 by developing F in powers of $w - w_s$. For each power the resulting integral is expressed in terms of the corresponding order of function $\Gamma(u)$ at $u = 1$.

It is worthy to note that the presence of the exponent $k_z z$ indicates that at the points of the cylindrical surface the angle between the rays of the head wave and the radius is equal to the critical angle.

Appendix A

Functions of complex variables

A.1 Single and multi-valued functions

Let us start by considering several elementary functions which were introduced in Part I when we described complex numbers.

Example one The function

$$w = z^2$$

can be written as

$$w = (x + i\,y)^2 = x^2 - y^2 + 2\,i\,x\,y \tag{A-1}$$

Making use of the notation

$$u = x^2 - y^2 \qquad \text{and} \qquad v = 2\,x\,y\,, \tag{A-2}$$

we write

$$w(z) \;=\; u(x,\,y) + i\,v(x,\,y)\,,$$

where u and v are the real and imaginary parts of the complex function w.

Example two The function

$$w = e^z = e^x\,e^{i\,y} \tag{A-3}$$

can be represented in the same form,

$$w(z) = u + i\,v$$

Here

$$u = e^x\,\cos y \quad \text{and} \quad v = e^x\,\sin y \tag{A-4}$$

Example three Our next function is

$$w = \sin z \tag{A-5}$$

As follows from Euler's formula, Part I,

$$\sin z = \frac{e^{iz} - e^{-iz}}{2i}$$

Thus, we have

$$w = \frac{e^{ix}\, e^{-y} - e^{-ix}\, e^{y}}{2\,i}$$

or

$$w = u + i\,v\,,$$

where

$$u = \frac{1}{2}\left(e^{y} + e^{-y}\right)\sin x\,, \qquad v = \frac{1}{2}\left(e^{y} - e^{-y}\right)\cos x \tag{A-6}$$

In all three examples, we dealt with single valued functions, since there is only one value of w which corresponds to any given argument z.

Example four Suppose that

$$w = z^{1/n}\,, \tag{A-7}$$

where n is an integer. Taking into account that

$$z = x + i\,y = r\,e^{i\,\phi}$$

and

$$r = \sqrt{x^2 + y^2}\,, \qquad \phi = \tan^{-1}\frac{y}{x}\,,$$

we have

$$w_k = \left(x^2 + y^2\right)^{1/2n} e^{i\,(\phi + 2\,n\,k)/n}\,, \qquad k = 0,\,1,\,2,\,\ldots,\,n-1 \tag{A-8}$$

Correspondingly, there are n different values of w for each value of z, which represents a classical example of a multi-valued function. At the same time for any given k the function w_k is single-valued, and it is usually called a branch of w.

To illustrate the relationship between z and w it is convenient to use two planes, as shown in Fig. A.1a. It is clear that the function $w(z)$ produces a translation of points of the z-plane into the w-plane, and this is characterized by two real functions:

$$u(x,\,y) \qquad \text{and} \qquad v(x,\,y)\,,$$

which, similarly to x and y, serve as coordinates of a point on the w-plane.

Further we consider only the single-valued functions :

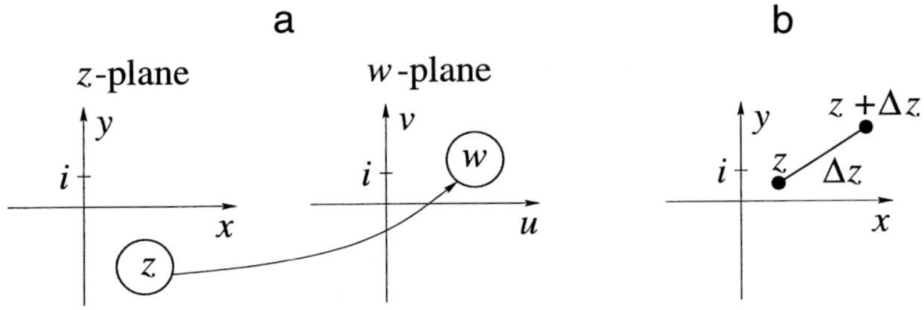

Figure A.1: (a) Relationship between points of the z- and w-planes (b) Illustration of eq. A-11.

$$f(z) = u(x, y) + i\,v(x, y) \tag{A-9}$$

Note that the notation $f(z)$ does not mean that the function depends on x and y only in the combination $x + i\,y$. For example,

$$f(z) = x + i\,n\,y$$

is still the function of the complex variable z.

By analogy with the scalar and vector fields we distinguish the regular points of the complex plane z, where the function $f(z)$ is continuous, and the singular points. As in the case of functions of a real argument, the continuity of $f(z)$ is one of the most important features, and it is defined in following way. The function $f(z)$ is continuous at the point $z = a$ if the absolute value of the difference

$$|f(z) - f(a)| \to 0 \qquad \text{when} \qquad |z - a| \to 0 \tag{A-10}$$

A.2 Analytical functions and Cauchy-Riemann conditions

To characterize the rate of change of function $f(z)$ at some point of the z-plane, consider the ratio

$$\frac{f(z + \Delta z) - f(z)}{\Delta z}$$

By definition, the derivative of $f(z)$ at point z is

$$f'(z) = \lim_{\Delta z \to 0} \frac{f(z + \Delta z) - f(z)}{\Delta z}, \tag{A-11}$$

where Δz is the distance between points z and $z + \Delta z$, located close to each other, Fig. A.1b. The coordinates of these points are

$$x,\, y \qquad \text{and} \qquad x + \Delta x,\, y + \Delta y,$$

respectively. Since

$$f(z) = u(x,\, y) + i\, v(x,\, y)$$

and

$$f(z + \Delta z) = u(x + \Delta x,\, y + \Delta y) + i\, v(x + \Delta x,\, y + \Delta y)\,,$$

the ratio in eq. A-11 can be written as

$$\frac{f(z + \Delta z) - f(z)}{\Delta z} = \frac{\Delta u + i\,\Delta v}{\Delta x + i\,\Delta y}\,, \tag{A-12}$$

where

$$\Delta u \;=\; u(x + \Delta x,\, y + \Delta y) - u(x,\, y)$$

and

$$\Delta v \;=\; v(x + \Delta x,\, y + \Delta y) - v(x,\, y)$$

Bearing in mind that the change in the real and imaginary parts of the function occurs because of change in both coordinates, we obtain in the limit $\Delta z \to 0$:

$$f'(z) \;=\; \frac{\dfrac{\partial u}{\partial x}\, dx + \dfrac{\partial u}{\partial y}\, dy + i\left(\dfrac{\partial v}{\partial x}\, dx + \dfrac{\partial v}{\partial y}\, dy\right)}{dx + i\, dy}$$

or

$$f'(z) \;=\; \frac{\left(\dfrac{\partial u}{\partial x} + i\,\dfrac{\partial v}{\partial x}\right) dx + \left(\dfrac{\partial u}{\partial y} + i\,\dfrac{\partial v}{\partial y}\right) dy}{dx + i\, dy} \tag{A-13}$$

In deriving the latter we assumed that the function is continuous and that all four of its derivatives

$$\frac{\partial u}{\partial x}\,, \qquad \frac{\partial u}{\partial y}\,, \qquad \frac{\partial v}{\partial x} \qquad \text{and} \qquad \frac{\partial v}{\partial y}$$

exist.

Earlier, in Part I, we studied behavior of a scalar field T and found that its rate of change essentially depends on the direction along which this variation is considered. In fact, such dependence is always observed except in the special case when $\operatorname{grad} T = 0$, and all directional derivatives are equal to zero. By analogy, we may expect that the derivative $f'(z)$ is, in general, also a function of orientation of element Δz. In other words, there is an infinite number of derivatives $f'(z)$ at the same point z. However, unlike for scalar fields, there is an extremely important class of functions of complex variable whose derivatives $f'(z)$ are independent of Δz, that is, $f'(z)$ has only one value. Such single-valued functions, which are called analytical, constitute the main subject of our study. To establish the conditions when a function of a complex variable becomes analytical, it is convenient to rewrite eq. A-13 as

$$f'(z) = \frac{\left(\frac{\partial u}{\partial x} + i\frac{\partial v}{\partial x}\right) + \left(\frac{\partial u}{\partial y} + i\frac{\partial v}{\partial y}\right)\frac{dy}{dx}}{1 + i\frac{dy}{dx}} \tag{A-14}$$

As is seen from Fig. A.1b, the ratio dy/dx defines the slope of the element dz, that is, its orientation.

Now suppose that the derivatives of $f(z)$ with respect to x and y,

$$\frac{\partial u}{\partial x} + i\frac{\partial v}{\partial x} \qquad \text{and} \qquad \frac{\partial u}{\partial y} + i\frac{\partial v}{\partial y}$$

are related as

$$\frac{\partial u}{\partial y} + i\frac{\partial v}{\partial y} = i\left(\frac{\partial u}{\partial x} + i\frac{\partial v}{\partial x}\right) \tag{A-15}$$

Its substitution into eq. A-14 gives

$$f'(z) = \frac{\partial u}{\partial x} + i\frac{\partial v}{\partial x} \qquad \text{or} \qquad f'(z) = -i\left(\frac{\partial u}{\partial y} + i\frac{\partial v}{\partial y}\right) = \frac{\partial v}{\partial y} - i\frac{\partial u}{\partial y}, \tag{A-16}$$

and the influence of the orientation of dz vanishes. Therefore, the function $f(z)$ is analytical, provided that equality A-15 takes place.

The latter can be represented in the form of two equations

$$\frac{\partial u}{\partial x} = \frac{\partial v}{\partial y} \qquad \frac{\partial u}{\partial y} = -\frac{\partial v}{\partial x} \tag{A-17}$$

which are called Cauchy-Riemann conditions .

For instance, applying eqs. A-17, it is easy to see that the function $f(z) = z^2$ is analytical at all points of the z-plane such that $z \neq \infty$. In fact, as follows from eq. A-2,

$$\frac{\partial u}{\partial x} = 2x, \qquad \frac{\partial v}{\partial y} = 2x,$$

and

$$\frac{\partial u}{\partial y} = -2y, \qquad \frac{\partial v}{\partial x} = 2y$$

It is a simple matter to verify that functions e^z and $\sin z$ are also analytical on the z-plane except for the points at infinity.

In contrast, for the function

$$f(z) = x + 2iy$$

we have

$$\frac{\partial u}{\partial x} = 1, \qquad \frac{\partial v}{\partial y} = 2, \qquad \text{and} \qquad \frac{\partial u}{\partial y} = \frac{\partial v}{\partial x} = 0,$$

that is, one of the Cauchy-Riemann conditions is not met, and therefore this function is not analytical. In this light let us formulate one more criterion of analyticity, namely, a function of the complex variable is analytical if it depends on the single argument z. In other words, the variables x and y appear only in the combination $x + iy$.

To prove this criterion, we treat the variable x as a function of z and y:

$$x = z - iy \tag{A-18}$$

Performing differentiation of the function

$$f(z) = u(x, y) + iv(x, y)$$

with respect to y, we obtain

$$\frac{\partial f}{\partial y} = \frac{\partial u}{\partial x}\frac{\partial x}{\partial y} + \frac{\partial u}{\partial y} + i\left(\frac{\partial v}{\partial x}\frac{\partial x}{\partial y} + \frac{\partial v}{\partial y}\right) \tag{A-19}$$

It follows from eq. A-18 that

$$\frac{\partial x}{\partial y} = -i$$

and correspondingly,

$$\frac{\partial f}{\partial y} = -i\frac{\partial u}{\partial x} + \frac{\partial u}{\partial y} + \frac{\partial v}{\partial x} + i\frac{\partial v}{\partial y}$$

or

$$\frac{\partial f}{\partial y} = \frac{\partial u}{\partial y} + \frac{\partial v}{\partial x} + i\left(\frac{\partial v}{\partial y} - \frac{\partial u}{\partial x}\right) \tag{A-20}$$

Taking into account the Cauchy-Riemann conditions, we see that if the function is analytical, then the right-hand side of eq. A-20 is zero:

$$\frac{\partial f}{\partial y} = 0$$

Thus, in reality analytical functions depend on the single variable z. This fact allows us to make one important generalization. Suppose that function $f(x)$ of the real argument exists along with its first derivative. Then, the function

$$f(z) = f(x + i\,y)$$

is analytical. For example, the function

$$f(z) = e^{z} \sin z$$

is analytical on the z-plane.

It is proper to notice that the Cauchy-Riemann conditions establish relationships between the real and imaginary parts of the analytical function, and this feature plays an important role in many applications.

A.3 Functions $u(x, y)$ and $v(x, y)$

In our study of single-valued and continuous functions $u(x, y)$ and $v(x, y)$ we proceed from Cauchy-Riemann conditions, eq. A-17. Their differentiation with respect to x and y, respectively, and then summation yields

$$\frac{\partial^2 u}{\partial x^2} + \frac{\partial^2 u}{\partial y^2} = 0 \tag{A-21}$$

In a similar manner, we obtain

$$\frac{\partial^2 v}{\partial x^2} + \frac{\partial^2 v}{\partial y^2} = 0 \tag{A-22}$$

Thus,

$$\nabla^2 u = 0 \qquad \text{and} \qquad \nabla^2 v = 0\,, \tag{A-23}$$

that is, both the real and imaginary parts of the analytical function are solutions of the Laplace equation. Respectively, they possess several interesting features. To demonstrate them, it is useful to treat these functions as scalar fields in the x-y plane, and we start from the function $u(x, y)$. In order to visualize its behavior, we imagine a system of level lines

$$u(x, y) = C\,, \tag{A-24}$$

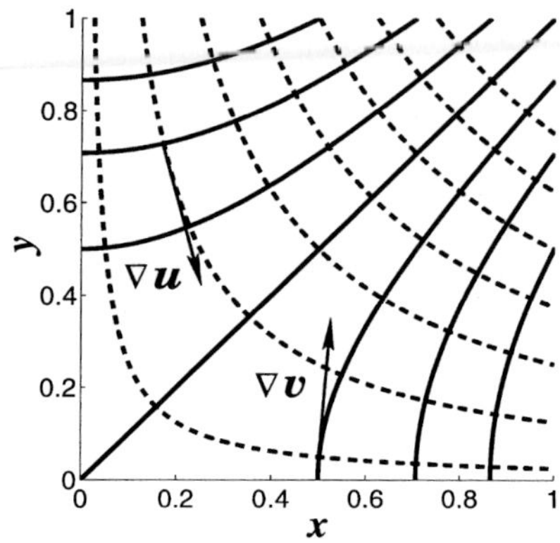

Figure A.2: Level lines of functions u (solid) and v (dashed) given by eq. A-2.

where C is some constant. By definition, function $u(x,\,y)$ remains constant along such lines. As was shown in Part I, the rate of change of the scalar field u in any direction is defined by the equality

$$\frac{\partial u}{\partial l} = \mathbf{i}_l \cdot \operatorname{grad} u \qquad \text{or} \qquad \frac{\partial u}{\partial l} = |\operatorname{grad} u|\,\cos(\mathbf{i}_l \cdot \nabla u) \tag{A-25}$$

Here dl is the elementary displacement, $\mathbf{i}_l$ is the unit vector specifying the direction along which the derivative is taken, and

$$\operatorname{grad} u = \frac{\partial u}{\partial x}\mathbf{i}_1 + \frac{\partial u}{\partial y}\mathbf{i}_2 \tag{A-26}$$

The unit vectors $\mathbf{i}_1$ and $\mathbf{i}_2$ point in the directions of the coordinate axes. At the same time the gradient defines the maximum rate of change of the function and the corresponding direction of such a change. Inasmuch as the directional derivative $\partial u/\partial l$ is zero along the lines $u(x,\,y)$ =const, it follows from eq. A-25 that the gradient is perpendicular to those lines, Fig. A.2.

In the same manner, we introduce the level lines of the scalar field $v(x,\,y)$:

$$v(x,\,y) = C\,, \tag{A-27}$$

and correspondingly,

$$\frac{\partial v}{\partial l} = \mathbf{i}_l \cdot \operatorname{grad} v\,, \tag{A-28}$$

where $\operatorname{grad} v$ is perpendicular to such lines.

As was already pointed out, Cauchy-Riemann conditions state that the functions $u(x,\,y)$ and $v(x,\,y)$ relate to each other. For this reason, it is natural to expect that the orientations of the level lines,

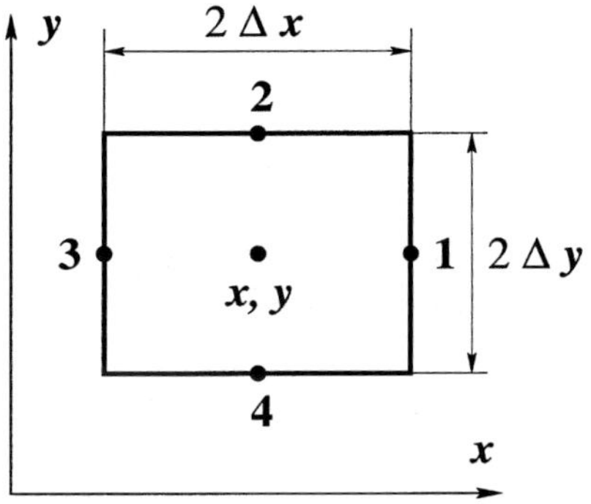

Figure A.3: Derivation of eq. A-30.

$$u = \text{const} \qquad \text{and} \qquad v = \text{const}$$

are also related at all points where the function $f(z)$ is analytical. In order to prove this, we make use of Cauchy-Riemann conditions and represent $\operatorname{grad} v$ as

$$\operatorname{grad} v = -\frac{\partial u}{\partial y}\mathbf{i}_1 + \frac{\partial u}{\partial x}\mathbf{i}_2$$

Then, taking a dot product of both gradients, we obtain

$$\nabla u \cdot \nabla v = \frac{\partial u}{\partial x}\frac{\partial v}{\partial x} + \frac{\partial u}{\partial y}\frac{\partial v}{\partial y}$$

or

$$\nabla u \cdot \nabla v = -\frac{\partial u}{\partial x}\frac{\partial u}{\partial y} + \frac{\partial u}{\partial y}\frac{\partial u}{\partial x} = 0 \tag{A-29}$$

Therefore, the directions of maximum rate of change of those two functions are orthogonal to each other. Correspondingly, the angle between the level lines

$$u(x,\, y) = C_1 \qquad \text{and} \qquad v(x,\, y) = C_2$$

is equal to $\pi/2$ at any point, Fig. A.2. This result can be formulated in a different way. Level line $v(x, y) = C$ coincides with the vector line of the field ∇u, while the line $u(x, y) = C$ points in the direction of ∇v.

Next consider one more interesting feature of those functions that follows from the Laplace equation. By analogy with the three dimensional case, Part I, we replace the derivatives by finite differences, Fig. A.3,

$$u'_x\left(x+\frac{\Delta x}{2},\, y\right)=\frac{1}{\Delta x}\left[u(x+\Delta x,\, y)-u(x,\, y)\right]$$

$$u'_x\left(x-\frac{\Delta x}{2},\, y\right)=\frac{1}{\Delta x}\left[u(x,\, y)-u(x-\Delta x,\, y)\right]$$

Then

$$\frac{\partial^2 u}{\partial x^2}\approx\frac{1}{(\Delta x)^2}\left[u(x+\Delta x,\, y)+u(x-\Delta x,\, y)-2\,u(x,\, y)\right]$$

In the same manner,

$$\frac{\partial^2 u}{\partial y^2}\approx\frac{1}{(\Delta y)^2}\left[u(x,\, y+\Delta y)+u(x,\, y-\Delta y)-2\,u(x,\, y)\right]$$

Assuming that $\Delta x=\Delta y$, the Laplace equation becomes

$$\frac{1}{(\Delta x)^2}\left[\sum_{j=1}^{4}u_j-4\,u(x,\, y)\right]=0\,,$$

where u_j are the values of function u at the middle points of the sides of a quadrant centered at the point with the coordinates x, y, Fig. A.3. The latter can be rewritten as

$$u(x,\, y)=\frac{1}{4}\sum_{j=1}^{4}u_j=u^{av}(x,\, y)\,, \tag{A-30}$$

that is $u(x,\, y)$ is equal to its average value at each point and it is called the harmonic function. If we suppose that the function $u(x,\, y)$ reaches its minimum or maximum at some point, the following inequalities have to be observed in its vicinity:

$$u(x,\, y)>u^{av}(x,\, y) \qquad \text{or} \qquad u(x,\, y)<u^{av}(x,\, y)$$

Since both of them contradict eq. A-30, we conclude that the real part of the analytical function has neither minima nor maxima. The same is true for the function $v(x,\, y)$. Note that only regular points were considered in both cases. This remarkable feature of analytical functions can be derived directly from eqs. A-21 or A-22. For instance, if

$$\frac{\partial^2 u}{\partial x^2}>0 \qquad \text{or} \qquad \frac{\partial^2 u}{\partial x^2}<0\,,$$

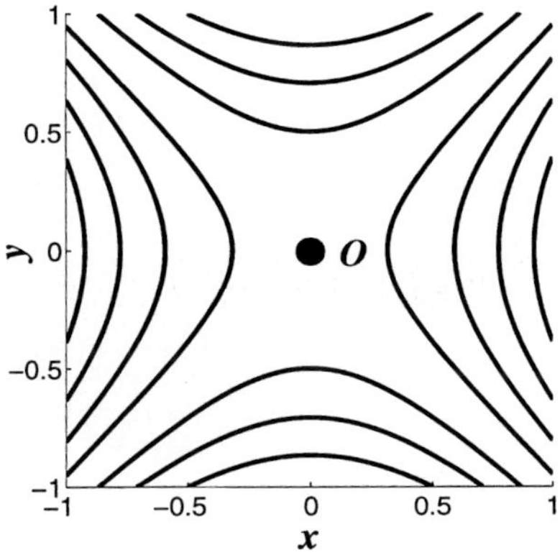

Figure A.4: Level lines of function $u = x^2 - y^2$ and its saddle point O.

then

$$\frac{\partial^2 u}{\partial y^2} < 0 \qquad \text{and} \qquad \frac{\partial^2 u}{\partial y^2} > 0\,,$$

respectively. Since the sum of the second derivatives is zero, the rate of increase of the first derivative in one direction has to be accompanied by the same rate of decrease in the perpendicular direction. An example of such a behavior is given in Fig. A.2.

A.4 Saddle points

Now let us assume that

$$\operatorname{grad} u = 0 \tag{A-31}$$

at some point O. In other words, the first derivative vanishes in all directions around this point. In particular,

$$\frac{\partial u}{\partial x} = 0 \qquad \text{and} \qquad \frac{\partial u}{\partial y} = 0$$

As follows from Cauchy-Riemann conditions,

$$\frac{\partial v}{\partial x} = 0 \qquad \text{and} \qquad \frac{\partial v}{\partial y} = 0\,, \tag{A-32}$$

that is,

$$\operatorname{grad} v = 0 \tag{A-33}$$

at O. Equations A-31–A-33 suggest that the geometry of level lines has some special features in the vicinity of O, for example, they do not pass through this point but rather asymptotically approach it, Fig. A.4. To visualize the behavior of functions near such

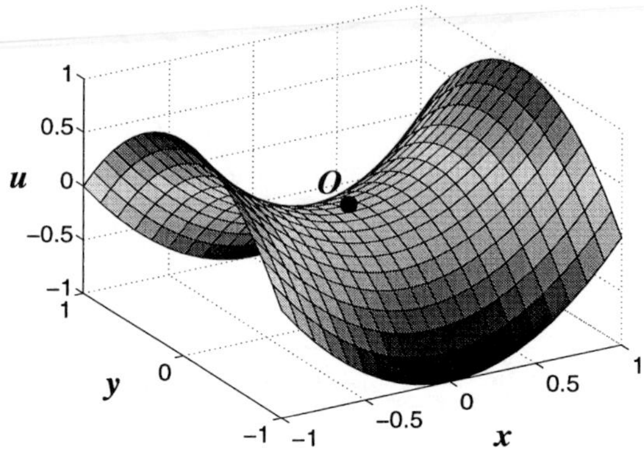

Figure A.5: The surface $u = x^2 - y^2$ and its saddle point O.

points, it is convenient to consider a surface of their values, as shown in Fig. A.5. It is clear that surface S has the shape of a saddle in the vicinity of O. Unlike other points, here we can distinguish two types of lines passing through O. Along each of them there is either a minimum or a maximum of function $u(x, y)$ at this point. The function $v(x, y)$ displays the same behavior. It is natural that such points are called the saddle points of the real and imaginary parts of the analytical function. Taking into account eq. A-14 and Cauchy-Riemann conditions, we conclude that the first derivative of $f(z)$ is equal to zero at the saddle point $z = z_0$:

$$f'(z_0) = 0 \tag{A-34}$$

Therefore, expansion of function $f(z)$ into Taylor series in the vicinity of this point yields

$$f(z) \approx f(z_0) + \frac{f''(z_0)}{2!}(z - z_0)^2 + \cdots \tag{A-35}$$

We have paid special attention on saddle points because the behavior of functions of the complex variable in the vicinity of those points is often used to derive various approximations of integrals that describe wave fields.

A.5 Curvilinear integrals of functions of the complex variable

Suppose that function $f(z)$ is given on the z-plane, and C is some path with the terminal points z_0 and z_n, as shown in Fig. A.6a. We divide this curve into an arbitrary number of small segments Δz_m and consider the sum

$$S_n = \sum_{m=1}^{n} f(a_m)\,\Delta z_m, \tag{A-36}$$

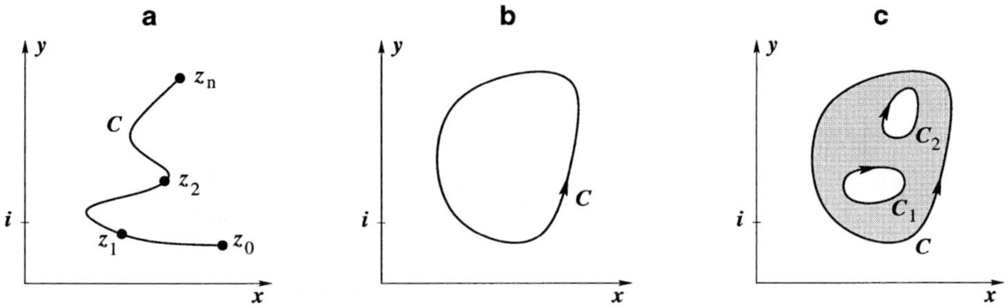

Figure A.6: Contours in the complex plane.

where a_m is an intermediate point of the element

$$\Delta z_m = z_m - z_{m-1}\,,$$

while $f(a_m)\,\Delta z_m$ is the product of two complex numbers.

Let us assume that n tends to infinity in such a way that the maximum interval Δz_m approaches zero and the sum A-36 has a limit that is independent of the choice of the intermediate points a_m. This limit is called the curvilinear integral of function $f(z)$ along the path C. It is denoted

$$\int_C f(z)\,dz$$

and in general, it depends on the function $f(z)$, the positions of terminal points z_0 and z_n, as well as the shape of curve C.

Taking into account that

$$f(z) = u(z) + i\,v(z) \qquad \text{and} \qquad dz = dx + i\,dy\,,$$

we have

$$\int_C f(z)\,dz = \int_C (u\,dx - v\,dy) + i\int_C (v\,dx + u\,dy) \tag{A-37}$$

A.6 Cauchy theorem

As was mentioned previously, we consider mainly analytical functions that, in particular, may be branches of multi-valued functions. Now let us establish one of the most remarkable feature of those functions, namely, the integral of analytical function $f(z)$ along an arbitrary but closed path C is zero, Fig. A.6b,

$$\oint_C f(z)\, dz = 0 \tag{A-38}$$

We prove this fundamental relationship making use of the Stokes theorem described in Part I. In the case when the integration path is located in the x-y plane, it can be written as

$$\oint_C (M_x\, dx + M_y\, dy) = \int_S \left(\frac{\partial M_y}{\partial x} - \frac{\partial M_x}{\partial y} \right) dx\, dy\,, \tag{A-39}$$

where S is the area enclosed by path C, M_x and M_y are any differentiable functions of the variables x and y, and integration along C is performed counterclockwise. First, in eq. A-39 we let

$$M_x = u \qquad \text{and} \qquad M_y = -v \tag{A-40}$$

and then

$$M_x = v \qquad \text{and} \qquad M_y = u\,, \tag{A-41}$$

we obtain either

$$\oint_C (u\, dx - v\, dy) = -\int_S \left(\frac{\partial v}{\partial x} + \frac{\partial u}{\partial y} \right) dS \tag{A-42}$$

or

$$\oint_C (v\, dx + u\, dy) = \int_S \left(\frac{\partial u}{\partial x} - \frac{\partial v}{\partial y} \right) dS \tag{A-43}$$

As follows from Cauchy-Riemann conditions, the surface integrals vanish in eqs. A-42 and A-43, and therefore the right-hand side of eq. A-37 is equal to zero, that is

$$\oint_C f(z)\, dz = 0$$

This result, obtained by Cauchy, plays an extremely important role in the theory of functions of complex variables and its numerous applications. To some extent, the Cauchy theorem is a consequence of the Stokes theorem, according to which the circulation of the vector field $\mathbf{M}$ is equal to zero:

$$\oint_C \mathbf{M} \cdot d\mathbf{l} = 0\,,$$

provided that the vortexes are absent inside closed path C. We proved the Cauchy theorem for the case when C is a single path, Fig. A.6b. However, eq. A-38 remains valid even when area S is surrounded by several paths, Fig. A.6c. This conclusion also follows

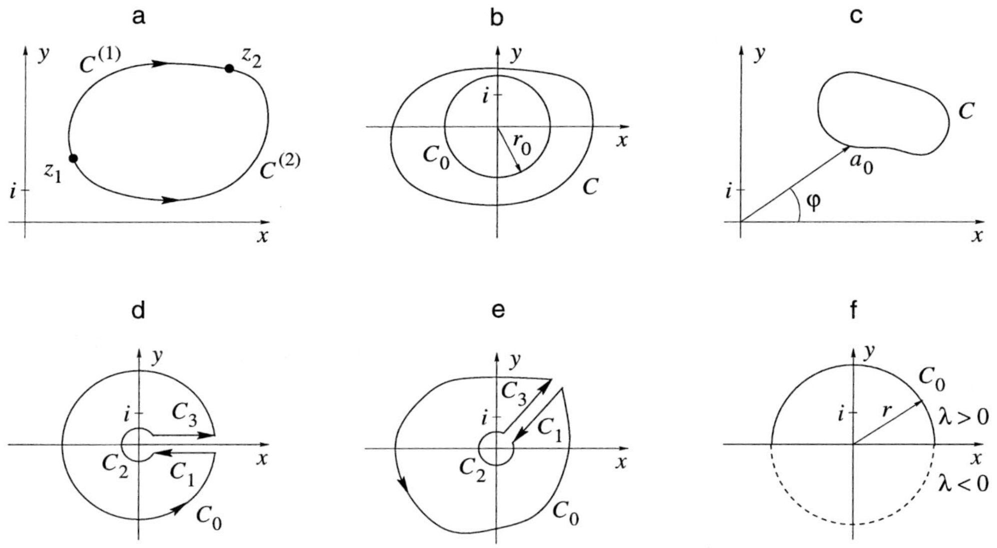

Figure A.7: Contours in the complex plane.

from the Stokes theorem and, unlike the external path C, the integration along C_1 and C_2 is performed clockwise. Thus, eq. A-38 becomes

$$\oint_C f(z)\,dz + \oint_{C_1} f(z)\,dz + \oint_{C_2} f(z)\,dz + \ldots = 0 \tag{A-44}$$

It is essential that function $f(z)$ has to be analytical within the area surrounded by those paths, whereas inside the area bounded by the internal paths (the holes) of analyticity is not necessary.

The change of direction of the integration leads to the change of sign

$$dz \approx \Delta z = z_{m+1} - z_m = -(z_m - z_{m+1})$$

For this reason eq. A-44 can be written as

$$\oint_C f(z)\,dz = \oint_{C_1} f(z)\,dz + \oint_{C_2} f(z)\,dz + \ldots, \tag{A-45}$$

where integration along all paths is performed in the positive direction, that is counterclockwise.

There is one more analogy between analytical function $f(z)$ and the vector field $\mathbf{M}$ caused by sources. As we know, Part I, the integral between two points

$$\int_{a_1}^{a_2} \mathbf{M} \cdot d\mathbf{l}$$

is path-independent in the absence of vortexes. It turns out the curvilinear integral of analytical function

$$\int_{z_1}^{z_2} f(z)\, dz$$

also does not depend on the shape of integration path. In fact, as follows from the Cauchy theorem, Fig. A.7a,

$$\oint_C f(z)\, dz = \oint_{C^{(1)}} f(z)\, dz + \oint_{C^{(2)}} f(z)\, dz = 0$$

Then, the change of direction along path $C^{(2)}$ gives

$$\oint_C f(z)\, dz = \oint_{C^{(1)}} f(z)\, dz - \oint_{C^{(2)}} f(z)\, dz = 0$$

or

$$\oint_{C^{(1)}} f(z)\, dz = \oint_{C^{(2)}} f(z)\, dz \tag{A-46}$$

Thus, we see that the integral

$$\int_{z_1}^{z_2} f(z)\, dz$$

of analytical function $f(z)$ depends only on the positions of terminal points z_1 and z_2. This fundamental result allows us to perform deformation of the path in such a way that evaluation of integrals is often greatly simplified. Of course, this procedure can be done if the function is analytical within the area of deformation.

A.7 Illustration of the Cauchy theorem

Consider several integrals along closed paths.

Example one To calculate the integral

$$\oint_C dz \tag{A-47}$$

along an arbitrary closed path C around the coordinate origin, Fig. A.7b, we make use of the Cauchy theorem and replace this path by circle C_0 with the radius r_0:

$$\oint_C dz = \oint_{C_0} dz \tag{A-48}$$

It is clear that function $f(z) = 1$ is analytical inside C. Using the exponential form of the complex number

$$z = r\,e^{i\phi},$$

we have

$$dz = i\,r\,e^{i\phi}\,d\phi$$

Then,

$$\oint_{C_0} dz = i\,r \int_0^{2\pi} e^{i\phi}\,d\phi = r\,e^{i\phi}\Big|_0^{2\pi} = 0$$

Thus, the summation of arbitrary elements Δz_m along any closed path is equal to zero. Similarity with the summation of vectors along a closed path is obvious.

Example two Suppose that

$$f(z) = (z - z_0)^n\,, \tag{A-49}$$

where n is an integer and z_0 is some point in the complex plane.

To evaluate the integral

$$\oint_C (z - z_0)^n\,dz$$

along the circle with the center at point z_0, we represent the difference $z - z_0$ as

$$z - z_0 = R\,e^{i\phi}\,, \tag{A-50}$$

where R is the radius of the circle.

The latter gives

$$dz = R\,i\,e^{i\phi}\,d\phi\,,$$

and correspondingly,

$$\oint_C f(z)\,dz = R^{n+1}\,i \int_0^{2\pi} e^{in\phi}\,e^{i\phi}\,d\phi$$

or

$$\oint_C f(z)\,dz = \frac{R^{n+1}\,e^{i\,(n+1)\,\phi}}{n+1}\Bigg|_0^{2\pi} = 0 \tag{A-51}$$

We can expect this result for any closed path, since $f(z)$ is an analytical function.

Example three Consider the integral

$$\oint_C \frac{dz}{z - z_0} \tag{A-52}$$

Here the function

$$f(z) = \frac{1}{z - z_0} \tag{A-53}$$

is analytical everywhere except the point z_0 which is the center of the circle C.

Again letting

$$z - z_0 = R\,e^{i\phi},$$

we obtain

$$\oint_C \frac{dz}{z - z_0} = \int_0^{2\pi} \frac{i\,R\,e^{i\phi}\,d\phi}{R\,e^{i\phi}} = 2\,\pi\,i$$

Thus

$$\oint_C \frac{dz}{z - z_0} = 2\,\pi\,i \tag{A-54}$$

Since there is the point $z = z_0$ inside closed path C where a function $f(z)$ is not analytical, it is natural that this integral is not zero. In contrast, if point z_0 is located outside the area, bounded by path C, the integral would vanish. It may be proper to notice that eq. A-54 is of great importance, and it will be used often to evaluate various integrals.

Example four The next integral is

$$\oint_C \frac{dz}{(z - z_0)^n} \tag{A-55}$$

Here $n > 1$ is an integer and C is an arbitrary path around z_0.

Proceeding from the Cauchy theorem, we can assume that C is the circle with radius R. Then we have

$$z - z_0 = R\,e^{i\phi}$$

and

$$\oint_C \frac{dz}{(z - z_0)^n} = \int_0^{2\pi} \frac{i\,R\,e^{i\phi}\,d\phi}{R^n\,e^{i\,n\,\phi}} = \frac{i}{R^{n-1}} \int_0^{2\pi} e^{i\,(1-n)\,\phi}\,d\phi$$

or

$$\oint_C \frac{dz}{(z - z_0)^n} = 0 \qquad \text{if} \qquad n > 1 \tag{A-56}$$

Thus, even though function $f(z)$ is not analytical at the point $z = z_0$, the integral is zero regardless of path C.

Example five Consider the integral

$$\oint_C \sqrt{z}\, dz \tag{A-57}$$

Here C is a circle with the radius r around the origin, Fig. A.7b.

The function

$$f(z) = \sqrt{z}$$

is not analytical at point $z = 0$ since its derivative $f'(0)$ does not exist. Moreover, each branch of this function is discontinuous at points of the interval

$$0 \le x \le r$$

For instance, let us choose the branch

$$f(z) = \sqrt{r}\, e^{i\,\phi/2}\,, \tag{A-58}$$

which has the positive values when the argument ϕ is equal to zero, that is

$$f(z) = \sqrt{r} \qquad \text{if} \qquad \phi = 0$$

Then, in the vicinity of the same points, provided that $\phi \to 2\pi$, we obtain

$$f(z) = \sqrt{r}\, e^{i\,\pi} = -\sqrt{r}$$

Thus, the discontinuity of this branch is equal to $2\sqrt{r}$ at the points $x > 0$ of the x-axis.

To perform integration, we select the initial point

$$z_0 = r\, e^{i\,\phi_0}$$

located on path C. Correspondingly, the final point has the same magnitude r, but its argument is

$$\phi = \phi_0 + 2\,\pi$$

Taking into account that

$$dz = i\, r\, e^{i\,\phi}\, d\phi\,,$$

we find

$$\oint_C \sqrt{z}\, dz = i\, r\, \sqrt{r} \int_{\phi_0}^{\phi_0+2\,\pi} e^{3\,i\,\phi/2}\, d\phi = \frac{2}{3}\, r^{3/2}\, e^{3\,i\,\phi/2} \Big|_{\phi_0}^{\phi_0+2\,\pi} = \frac{2}{3}\, r^{3/2}\, e^{3\,i\,\phi_0/2} \left(e^{3\,\pi\, i} - 1\right)$$

or

$$\oint_C \sqrt{z}\,dz = -\frac{4}{3}\,r^{3/2}\,e^{3\,i\,\phi_0/2} \tag{A-59}$$

Therefore, the value of the integral depends on the radius of the circle, the initial point z_0 of the integration, and, of course, on the branch of the function $f(z)$. Let us notice that the point $z = 0$ is usually called the branch point.

Consider a new path of integration which does not encircle the origin, Fig. A.7c. Unlike the previous case, each branch of $f(z)$ is continuous at every point of the area bounded by the new closed path C, and therefore $f(z)$ can be differentiated. Since any branch depends on z only, $f(z)$ is an analytical function, and as a consequence,

$$\oint_C \sqrt{z}\,dz = 0$$

It is obvious that counterclockwise motion from point z_0 along path C, Fig. A.7c, leads first to an increase of both the magnitude r and the argument ϕ. Then they start to decrease and finally approach r_0 and ϕ_0, that is function $f(z) = \sqrt{z}$ does not have discontinuities.

Example six It is useful to study a more general type of integrals:

$$\oint_{C_0} \sqrt{z}\,f(z)\,dz\,, \tag{A-60}$$

where $f(z)$ is the analytical function and C_0 is an arbitrary path around the point $z = 0$.

Bearing in mind that the integrand:

$$\varphi(z) = \sqrt{z}\,f(z) \tag{A-61}$$

is not differentiable at point $z = 0$ and is discontinuous in the vicinity of the x-axis when $x > 0$, the Cauchy theorem cannot be applied. However, it can be used in order to represent the integral A-60 in a different form that often allows us to simplify its evaluation. With this purpose in mind, we introduce a new integration path C, shown in Fig. A.7d, which includes the following four elements:

$$C = C_0 + C_1 + C_2 + C_3$$

Since it is assumed that function $\varphi(z)$ is analytical within the area surrounded by path C, we have

$$\oint_C \varphi(z)\,dz = 0$$

or

$$\oint_{C_0} \varphi(z)\,dz + \oint_{C_1} \varphi(z)\,dz + \oint_{C_2} \varphi(z)\,dz + \oint_{C_3} \varphi(z)\,dz = 0 \tag{A-62}$$

As follows from eq. A-59, the integral along the circle C_2 becomes smaller as the radius of the circle decreases and in the limit

$$\oint_{C_2} \varphi(z)\,dz \to 0 \qquad \text{if} \qquad r \to 0$$

At points of the path C_1 and C_3,

$$z = x\,e^{2\pi i} \qquad \text{and} \qquad z = x\,,$$

respectively.

Selecting the branch of $\sqrt{z}$ that has the negative values along path C_1, we obtain from eq. A-62

$$\oint_{C_0} \varphi(z)\,dz - \int_x^0 \sqrt{x}\,f(x)\,dx + \int_0^x \sqrt{x}\,f(x)\,dx = 0$$

or

$$\oint_{C_0} \sqrt{z}\,f(z)\,dz = 2\int_x^0 \sqrt{x}\,f(x)\,dx \tag{A-63}$$

Thus we have expressed the integral along a closed path on the complex plane in terms of the integral along the real axis z. Paths C_1 and C_3 are often called the branch cuts or the cut lines, and their position may vary, Fig. A.7d, e.

A.8 Jordan's lemma

Suppose we want to calculate the following integral

$$I = \int_{-\infty}^{\infty} e^{i\lambda x}\,f(x)\,dx\,, \tag{A-64}$$

where λ is either a positive or a negative number.

Sometimes this task can be accomplished if we consider a closed path and make use of the Cauchy theorem. With this purpose in mind consider an integral along a semicircle with an infinitely large radius, Fig. A.7f:

$$I_0 = \int_{C_0} e^{i\lambda z}\,f(z)\,dz \tag{A-65}$$

If the latter tends to zero when $r \to \infty$, we can write in place of eq. A-64

$$I = \int_{-\infty}^{\infty} e^{i\lambda x}\,f(x)\,dx + \int_{C_0} e^{i\lambda z}\,f(z)\,dz$$

or

$$I = \oint\limits_C e^{i\lambda z} f(z)\, dz \,, \tag{A-66}$$

where closed path C includes the x-axis and the semi-circle C_0.

Now we formulate a relatively general condition when

$$\lim_{r \to \infty} I_0 = 0$$

and correspondingly,

$$\int\limits_{-\infty}^{\infty} e^{i\lambda x} f(x)\, dx = \oint\limits_C e^{i\lambda z} f(z)\, dz$$

First, suppose that parameter λ is positive. Then if $f(z)$ is an analytical function everywhere in the upper half-space

$$0 \le \arg\, z \le \pi$$

except finite number of points, and it uniformly tends to zero with respect to $\arg z$ when $z \to \infty$, we have

$$\lim_{r \to \infty} \oint\limits_{C_0} e^{i\lambda z} f(z)\, dz = 0 \tag{A-67}$$

Since

$$z = r\, e^{i\phi} = r\, \cos\phi + i\, r\, \sin\phi \,,$$

it yields

$$\left| e^{i\,\lambda\, z} \right| = \left| e^{i\,\lambda\, r\, \cos\phi}\, e^{-\lambda\, r\, \sin\phi} \right| = e^{-\lambda\, r\, \sin\phi}$$

Taking into account that $f(z)$ uniformly tends to zero, we have for sufficiently large r:

$$|f(z)| < \epsilon \,,$$

where $\epsilon > 0$ is an arbitrarily small number.

Respectively, it gives

$$\left| \int\limits_{C_0} e^{i\,\lambda\, z} f(z)\, dz \right| < \epsilon r \int\limits_0^{\pi} e^{-\lambda\, r\, \sin\phi}\, d\phi \qquad \text{if} \qquad r \to \infty \tag{A-68}$$

To evaluate the right-hand side of this integral, consider the ratio $\sin\varphi/\varphi$ within the interval $0 \le \phi \le \pi/2$. It is obvious that at small ϕ we have

$$\frac{\sin\phi}{\phi} = 1 \,,$$

and this ratio tends to $2/\pi$ as ϕ goes to $\pi/2$.

Thus, the following inequality takes place:

$$\frac{\sin\phi}{\phi} \geq \frac{2}{\pi} \tag{A-69}$$

Its substitution into eq. A-68 gives

$$\left| \int\limits_{C_0} e^{i\lambda z} f(z)\,dz \right| < \epsilon r \int\limits_0^{\pi} e^{-2\lambda r\phi/\pi} d\phi < 2\epsilon r \int\limits_0^{\pi/2} e^{-2\lambda r\phi/\pi} d\phi =$$

$$= \frac{\pi\epsilon}{\lambda}\left(1 - e^{-r\lambda}\right) < \frac{\epsilon\pi}{\lambda}, \tag{A-70}$$

which proves the lemma. If parameter λ is negative, then semicircle C_0 is located in the lower half-space, Fig. A.7f. Jordan's lemma is often used in evaluating integrals that describe elastic and acoustic wave fields.

A.9 Cauchy formula

Now we derive one of the most important relationships of functions of complex variables. Suppose that $f(z)$ is an analytical function in the area bounded by closed path C and z_0 is one of the points of this area, Figure A.8a. Cauchy proved the following equality:

$$f(z_0) = \frac{1}{2\pi i} \oint\limits_C \frac{f(z)}{z - z_0}\, dz \tag{A-71}$$

In fact, the function

$$\varphi(z) = \frac{f(z)}{z - z_0}$$

is analytical everywhere inside the area except at the point $z = z_0$. Let us draw the circle γ with the radius ρ around this point. Since the function $\varphi(z)$ is analytical at all points of the area surrounded by both paths C and γ, we can apply the Cauchy theorem, which yields

$$\oint\limits_C \frac{f(z)}{z - z_0}\, dz = \oint\limits_\gamma \frac{f(z)}{z - z_0}\, dz \tag{A-72}$$

Here, the integration along both paths is performed in the positive direction.

It follows from eq. A-72 that the integral along circle γ is independent of its radius ρ. Taking into account this fact, assume that ρ tends to zero. First, we can write

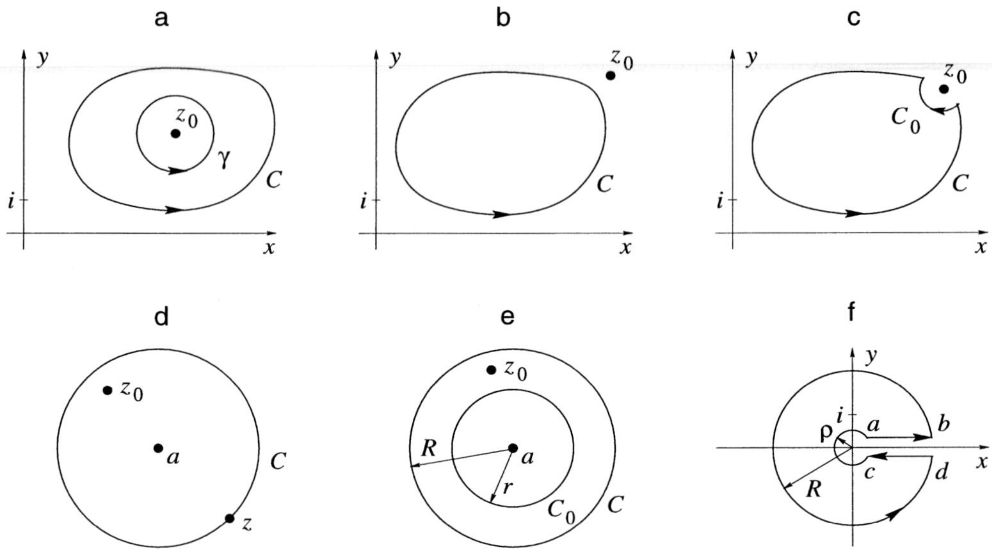

Figure A.8: (a), (b), (c) Illustration of the Cauchy formula (d), (e) Taylor and Laurent series (f) illustration of eq. A-108.

$$\oint_\gamma \frac{f(z)}{z-z_0}\,dz = \oint_\gamma \frac{f(z_0)}{z-z_0}\,dz + \oint_\gamma \frac{f(z)-f(z_0)}{z-z_0}\,dz \tag{A-73}$$

In accordance with eq. A-54, we obtain

$$\oint_\gamma \frac{f(z_0)}{z-z_0}\,dz = f(z_0)\oint_\gamma \frac{dz}{z-z_0} = 2\,\pi\,i\,f(z_0)$$

Bearing in mind that function $f(z)$ is continuous, we can choose such small ρ that

$$|f(z)-f(z_0)| < \epsilon$$

at all points of circle γ. Here $\epsilon > 0$ is an arbitrary small number. Thus,

$$\left| \oint_\gamma \frac{f(z)-f(z_0)}{z-z_0}\,dz \right| < \epsilon \int_\gamma \frac{|dz|}{|z-z_0|} = 2\,\pi\,\epsilon\,,$$

since the length of the path is equal to $2\,\pi\,\rho$ and $|z-z_0| = \rho$.

Correspondingly, the last integral at the right-hand side of eq. A-73 vanishes, and we have

$$\oint_C \frac{f(z)}{z - z_0}\, dz = 2\pi i\, f(z_0)\,,$$

that is, the Cauchy formula is proven.

The latter is one of the most remarkable results in mathematics and has many different applications. As follows from eq. A-71, we can determine the value of function $f(z)$ at any point z_0 if it is known at each point of an arbitrary path surrounding this point. In other words, eq. A-71 is the "bridge" between the values of analytical function at different points. Because of this feature Cauchy formula is the foundation for an analytical continuation which allows us to solve inverse problems in many important cases. Also, it helps us to establish the relationship between the amplitude and phase spectra. We note that there is a similarity between the Cauchy formula and various equations obtained from the Gauss theorem

$$\int_V \operatorname{div} \mathbf{M}\, dV = \oint_S \mathbf{M} \cdot \mathbf{dS}\,,$$

such as the Helmholtz and Kirchoff formulae describing the Huygens principle.

If a point z_0 is located outside the path C, as is shown in Fig. A.8b, function $f(z)/(z - z_0)$ is analytical inside the area bounded by C and, therefore

$$\oint_C \frac{f(z)}{z - z_0}\, dz = 0$$

Next suppose that point z_0 belongs to path C. In such a case we consider a slightly different contour, Fig. A.8c,

$$C_1 = C^\star + C_0\,,$$

where C_0 is a semicircle with some small radius ρ such that

$$C_0 \to 0 \qquad \text{and} \qquad C_1 \to C^\star = C$$

As follows from the Cauchy theorem

$$\oint_{C^\star} \frac{f(z)}{z - z_0}\, dz + \oint_{C_0} \frac{f(z)}{z - z_0}\, dz = 0 \tag{A-74}$$

Performing integration around point z_0, we find

$$P.V. \oint_{C_0} \frac{f(z)}{z - z_0}\, dz = -\pi i\, f(z_0)\,, \tag{A-75}$$

since $z - z_0 = \rho\, e^{i\phi}$ and the angle ϕ varies from 2π to π. Here "P.V." stands for the principal value of the Cauchy integral.

Thus, eq. A-74 becomes

$$P.V. \oint_{C^\star} \frac{f(z)}{z - z_0}\, dz - \pi\, i\, f(z_0) = 0\,,$$

and we obtain in the limit $\rho \to 0$:

$$P.V. \oint_{C} \frac{f(z)}{z - z_0}\, dz = \pi\, i\, f(z_0) \tag{A-76}$$

Summarizing, we can write

$$\oint_{C} \frac{f(z)}{z - z_0}\, dz = 2\,\pi\, i\, f(z_0) \begin{cases} 1 & z_0 \ \text{inside} \\ \frac{1}{2} & z_0 \ \text{on} \ \ C \qquad \text{(P.V.)} \\ 0 & z_0 \ \text{outside} \end{cases} \tag{A-77}$$

A.10 Taylor series

Now making use of the Cauchy formula, we derive a completely new relationship between the values of analytical functions at different points. Suppose that $f(z)$ is analytical inside circle C with its center at some point a, and z_0 is a point inside C, Fig. A.8d.

The term

$$\frac{1}{z - z_0}$$

of the integrand in eq. A-71 can be written as

$$\frac{1}{z - z_0} = \frac{1}{(z-a) - (z_0 - a)} = \frac{1}{(z-a)\left(1 - \dfrac{z_0 - a}{z - a}\right)}$$

This gives

$$\frac{f(z)}{z - z_0} = \frac{f(z)}{(z-a)\left(1 - \dfrac{z_0 - a}{z - a}\right)}$$

Bearing in mind that

$$|z_0 - a| < |z - a|$$

and that the ratio

$$\frac{1}{1 - \dfrac{z_0 - a}{z - a}}$$

is the sum of the geometrical progression, we can write the series

$$\frac{f(z)}{z-z_0} = \frac{f(z)}{z-a}\left[1+\frac{z_0-a}{z-a}+\ldots+\left(\frac{z_0-a}{z-a}\right)^n+\ldots\right]$$

Dividing the latter by $2\pi i$ and integrating along the path C, we obtain

$$\frac{1}{2\pi i}\oint\limits_C \frac{f(z)}{z-z_0}\,dz = f(z_0) = A_0 + A_1(z_0-a)+\ldots+A_n(z_0-a)^n+\ldots \tag{A-78}$$

Here

$$A_n = \frac{1}{2\pi i}\oint\limits_C \frac{f(z)}{(z-a)^{n+1}}\,dz \tag{A-79}$$

To understand the meaning of the coefficients A_n we use the Cauchy formula

$$f(a) = \frac{1}{2\pi i}\oint\limits_C \frac{f(z)}{z-a}\,dz \tag{A-80}$$

First, comparison of eqs. A-79 and A-80 shows that

$$A_0 = f(a) \tag{A-81}$$

Differentiation of eq. A-80 $n-1$ times in the vicinity of point a gives

$$f^{(n)}(a) = \frac{n!}{2\pi i}\oint\limits_C \frac{f(z)}{(z-a)^{n+1}}\,dz \tag{A-82}$$

As follows from eq. A-79,

$$A_n = \frac{f^{(n)}(a)}{n!} \tag{A-83}$$

Therefore, the right-hand side of eq. A-78 represents the Taylor series

$$f(z_0) = f(a) + \frac{f'(a)}{1!}(z_0-a) + \frac{f''(a)}{2!}(z_0-a)^2+\ldots+\frac{f^{(n)}(a)}{n!}(z_0-a)^n+\ldots \tag{A-84}$$

The latter, as well as the Cauchy formula, establishes the relationship between the values of analytical function at different points. Unlike the Cauchy formula, the Taylor series allows one to find the function $f(z_0)$, if its value is known at one point $z=a$, as well as all derivatives at this point. This consideration shows that $f(z)$ can be expanded in the Taylor series within the area of its analyticity. In other words, if the function can be represented as the Taylor series, then it is analytical and, therefore, satisfies Cauchy-Riemann conditions. Note that letting $a = 0$ in eq. A-84, we arrive at the McLaurin series

$$f(z_0) = f(0) + \frac{f'(0)}{1!}\, z_0 + \frac{f''(0)}{2!}\, z_0^2 + \ldots + \frac{f^{(n)}(0)}{n!}\, z_0^n + \ldots \qquad \text{(A-85)}$$

A.11 Singularities of analytical functions

Until now, we have considered the behavior of analytical functions at regular points. In other words, functions can be expanded in the Taylor series in the vicinity of such points. All other points, where eq. A-84 becomes invalid, are called irregular or singular points. There are three types of such points, namely,

(a) poles,

(b) essential singularities and

(c) branch points.

Poles are isolated irregular points in whose vicinity function $f(z)$ is single-valued. At the same time, those points are regular for function $1/f(z)$. For instance, the function

$$f(z) = \frac{1}{z^2}$$

has a pole of the second order at the origin $z = 0$.

The function

$$f(z) = \frac{1}{z^2 + a^2}$$

has two simple first-order poles,

$$z = i\,a \qquad \text{and} \qquad z = -i\,a\,,$$

which are located on the imaginary axis.

Essential singularities are also isolated points and function $f(z)$ is also single-valued in their vicinity. Unlike poles, however, these points are still irregular for function $1/f(z)$. For example, origin $z = 0$ is the essential singularity for function $\sin(1/z)$. Function $\exp\left[1/(z-a)\right]$ has essential singularity at $z = a$.

The critical or branch points are such irregular points such that $f(z)$ is multi-valued in their vicinity. For example, the function

$$f(z) = \sqrt{z^2 - a^2}$$

has two branch points

$$z_1 = a \qquad \text{and} \qquad z_2 = -a\,,$$

and each branch of this function is discontinuous in the vicinity of a line connecting the points a and $-a$.

As was shown in Part I the type of generators and their distribution define the behavior of vector fields. In a similar manner, the type of singularities and their locations characterize the analytical function between these points.

A.12 Laurent series

We have demonstrated that analytical functions can be expanded in the Taylor series in the vicinity of their regular points. Next, consider the behavior of the function $f(z)$ in the neighborhood of either a pole or an essential singularity, Fig. A.8e. Let us surround the singular point a by a small circle C_0 with radius r. Then function $f(z)$ is analytical within the area bounded by the two circles C_0 and C. The radius of the external circle is R. Applying the Cauchy formula to the closed path, $C_0 + C$, we obtain

$$f(z_0) = \frac{1}{2\pi i} \oint_{C_0+C} \frac{f(z)}{z - z_0}\, dz$$

or

$$f(z_0) = \frac{1}{2\pi i} \oint_{C} \frac{f(z)}{z - z_0}\, dz - \frac{1}{2\pi i} \oint_{C_0} \frac{f(z)}{z - z_0}\, dz\,, \tag{A-86}$$

where the direction of integration is positive for both terms.

In accordance with eq. A-78, the first integral in the right-hand side of eq. A-86 can be presented as

$$\frac{1}{2\pi i} \oint_{C} \frac{f(z)}{z - z_0}\, dz = A_0 + A_1\,(z-a) + A_2\,(z-a)^2 + \ldots + A_n\,(z-a)^n + \ldots\,, \tag{A-87}$$

where

$$A_n = \frac{1}{2\pi i} \oint_{C} \frac{f(z)}{(z - z_0)^{n+1}}\, dz \tag{A-88}$$

However, eq. A-82 is not valid in this case because the function $f(z)$ is not analytical at point $z = a$.

Consider the second integral in eq. A-86:

$$-\frac{1}{2\pi i} \oint_{C_0} \frac{f(z)}{z - z_0}\, dz = \frac{1}{2\pi i} \oint_{C_0} \frac{f(z)}{z_0 - z}\, dz \tag{A-89}$$

First, we have

$$\frac{1}{z_0 - z} = \frac{1}{(z_0 - a) - (z - a)} = \frac{1}{(z_0 - a)\left(1 - \dfrac{z - a}{z_0 - a}\right)}$$

Since

$$|z_0 - a| > |z - a|\,,$$

we again use the geometrical progression

$$\frac{1}{1 - \dfrac{z - a}{z_0 - a}} = 1 + \frac{z - a}{z_0 - a} + \ldots + \left(\frac{z - a}{z_0 - a}\right)^n + \ldots$$

Then, the integral along path C_0 can be represented as

$$-\frac{1}{2\pi i} \oint\limits_{C_0} \frac{f(z)}{z - z_0}\, dz = \frac{A_{-1}}{z_0 - a} + \frac{A_{-2}}{(z_0 - a)^2} + \ldots + \frac{A_{-n}}{(z_0 - a)^n} + \ldots\,, \tag{A-90}$$

where

$$A_{-n} = \frac{1}{2\pi i} \oint\limits_{C_0} f(z)\,(z - a)^{n-1}\, dz \tag{A-91}$$

Thus, as follows from eqs. A-86, A-87 and A-90, in the vicinity of the singularity the analytical function can be expanded in the Laurent series

$$\begin{aligned} f(z_0) = \ldots + &\frac{A_{-n}}{(z_0 - a)^n} + \ldots + \frac{A_{-1}}{z_0 - a} + A_0 \\ &+ A_1\,(z_0 - a) + \ldots + A_n\,(z_0 - a)^n + \ldots\,, \end{aligned} \tag{A-92}$$

where the coefficients are defined by eqs. A-88 and A-91.

The Laurent series allows us to classify the singularities. If the series has only a finite number of coefficients A_{-n}, then point a is the pole of the nth order. In fact, in this case we have

$$f(z_0) = \frac{A_{-n}}{(z_0 - a)^n}\,[1 + \varphi(z_0,\, a)]\,,$$

where $\varphi(z_0,\, a)$ is a polynomial. Correspondingly, function $1/f(z)$ is analytical at point a. If there is an infinite number of terms with the negative power of $(z_0 - a)$, then function $f(z)$ has the essential singularity at point a.

For illustration consider two examples. The first function

$$f(z) = \frac{1}{z\,(z-1)}$$

has two singularities at $z = 0$ and $z = 1$. In their vicinity

$$\frac{1}{z\,(z-1)} = -\frac{1}{z}\left(1 + z + z^2 + z^3 + \ldots\right) = -\frac{1}{z} - 1 - z - z^2 - \ldots$$

and

$$\begin{aligned} \frac{1}{(z-1)}\,\frac{1}{[1+(z-1)]} &= \frac{1}{z-1}\left[1 - (z-1) + (z-1)^2 + \ldots\right] \\ &= \frac{1}{z-1} - 1 + (z-1) - (z-1)^2 + \ldots \end{aligned}$$

Thus, both singularities are simple poles ($n = 1$).

The second example is the function

$$f(z) = e^{1/z}$$

Its expansion in the power series gives a Laurent series in the vicinity of $z = 0$,

$$e^{1/z} = 1 + \frac{1}{z} + \frac{1}{2!\,z^2} + \frac{1}{3!\,z^3} + \ldots$$

Since this series contains an infinite number of terms with the negative powers of z, the point $z = 0$ is the essential singularity.

Note that a function can be expanded in neither the Taylor nor the Laurent series in the vicinity of a branch point. This is related to the fact that the function itself or some its derivatives are not defined at such a point. Besides the function is multi-valued in the vicinity of a branch point.

A.13 Integration by the residual method

Suppose that the point a is the pole of function $f(z)$ and C is an arbitrary contour surrounding a. We also assume that $f(z)$ is analytical everywhere inside C except at point a. Now we demonstrate that

$$\oint_C f(z)\,dz = 2\,\pi\,i\,A_{-1}\,, \tag{A-93}$$

where A_{-1} is the coefficient of the term $1/(z-a)$ of the Laurent series, and it is called the residual of function $f(z)$ at point a:

$$\text{Res } f(a) = A_{-1} \tag{A-94}$$

As follows from eq. A-92, we have in the case of the pole of the nth order

$$f(z) = \frac{A_{-n}}{(z-a)^n} + \frac{A_{-n+1}}{(z-a)^{n-1}} + \ldots + \frac{A_{-1}}{z-a} + \varphi(z) \tag{A-95}$$

Here $\varphi(z)$ is described by the Taylor series, and correspondingly, it is an analytical function inside C including the point a.

Integration of eq. A-95 gives

$$\oint_C f(z)\,dz = A_{-n} \oint_C \frac{dz}{(z-a)^n} + A_{-n+1} \oint_C \frac{dz}{(z-a)^{n-1}} + \ldots + A_{-1} \oint_C \frac{dz}{z-a} + \oint_C \varphi(z)\,dz \tag{A-96}$$

In accordance with the Cauchy theorem the last integral is equal to zero. Now, making use of this theorem, we replace an arbitrary path C by a circle C_0 with the radius R and the origin at point a. This procedure allows us to apply eqs. A-54 and A-56, which show that

$$\oint_C \frac{dz}{z-a} = 2\pi i \qquad \text{and} \qquad \oint_C \frac{dz}{(z-a)^n} = 0$$

Here $n > 1$ is an integer. Thus, in place of A-96 we obtain eq. A-93. This shows that integration along a closed path is replaced by evaluating the residual A_{-1} of function $f(z)$. As we will see a calculation of the residual is much simpler than integration.

Let us further assume that integrand $f(z)$ can be represented as the ratio

$$f(z) = \frac{\varphi_1(z)}{\varphi_2(z)} \tag{A-97}$$

First, suppose that point a is a simple pole ($n = 1$). This means that function $f(z)$ behaves as

$$f(z) = \frac{A_{-1}}{z-a} + \varphi(z) \tag{A-98}$$

in the vicinity of this point. Here $\varphi(z)$ is an analytical function.

Multiplying both parts of eq. A-98 by $(z-a)$, we obtain

$$f(z)\,(z-a) = A_{-1} + \varphi(z)\,(z-a) \tag{A-99}$$

and therefore

$$\lim_{z\to a} \frac{\varphi_1(z)\,(z-a)}{\varphi_2(z)} = A_{-1} \tag{A-100}$$

The numerator and denominator tend to zero when $z \to a$, and, in order to resolve this uncertainty, we use the L'Hopital rule, which gives

$$\lim_{z\to a} \frac{\varphi_1(z) + (z-a)\,\varphi_1'(z)}{\varphi_2'(z)} = \frac{\varphi_1(a)}{\varphi_2'(a)}$$

Thus

$$\text{Res } f(a) = A_{-1} = \frac{\varphi_1(a)}{\varphi_2'(a)} \tag{A-101}$$

If there are several simple poles a_1, a_2, a_3, ... , a_n, then

$$\oint\limits_C f(z)\,dz = 2\pi i \sum_{i=1}^{n} \text{Res } f(a_i) = 2\pi i \sum_{i=1}^{n} \frac{\varphi_1(a_i)}{\varphi_2'(a_i)} \tag{A-102}$$

For example, if the integrand has the form

$$\frac{1}{z}\frac{\varphi_1(z)}{\varphi_2(z)} \qquad \text{and} \qquad \varphi_2(0) \neq 0\,,$$

then

$$\oint\limits_C \frac{1}{z}\frac{\varphi_1(z)}{\varphi_2(z)}\,dz = 2\pi i \left[\frac{\varphi_1(0)}{\varphi_2(0)} + \sum_{i=1}^{n} \frac{\varphi_1(a_i)}{a_i\,\varphi_2'(a_i)}\right] \tag{A-103}$$

Next, consider the case when function $f(z)$ has a pole of the nth order at point a. This means that

$$\frac{\varphi_1(z)}{\varphi_2(z)} = \frac{A_{-1}}{z-a} + \frac{A_{-2}}{(z-a)^2} + \ldots + \frac{A_{-n}}{(z-a)^n} + \varphi(z)$$

After multiplication by $(z-a)^n$, we obtain

$$(z-a)^n \frac{\varphi_1(z)}{\varphi_2(z)} = A_{-1}\,(z-a)^{n-1} + A_{-2}\,(z-a)^{n-2} + \ldots$$

$$+ A_{-n} + (z-a)^n\,\varphi(z) \tag{A-104}$$

Differentiating the latter $n-1$ times and letting $z=a$ gives

$$\lim_{z \to a} \frac{d^{n-1}}{dz^{n-1}} \left[(z-a)^n \frac{\varphi_1(z)}{\varphi_2(z)} \right] = (n-1)!\, A_{-1}$$

or

$$\text{Res } f(a) = \frac{1}{(n-1)!} \lim_{z \to a} \frac{d^{n-1}}{dz^{n-1}} \left[(z-a)^n \frac{\varphi_1(z)}{\varphi_2(z)} \right] \tag{A-105}$$

Correspondingly,

$$\oint_C f(z)\, dz = \frac{2\pi i}{(n-1)!} \lim_{z \to a} \frac{d^{n-1}}{dz^{n-1}} \left[(z-a)^n \frac{\varphi_1(z)}{\varphi_2(z)} \right]$$

Example To illustrate the application of the residual method, we consider the integral

$$I = \int_0^\infty x^\alpha f(x)\, dx \tag{A-106}$$

Here α is a non-integer number and function $f(x)$ does not have singularities within the interval

$$0 \le x \le \infty$$

Also, we have

$$\lim_{z \to 0} z^\alpha f(z) = 0 \tag{A-107}$$

It is convenient to introduce the integral

$$\oint_C z^\alpha f(z)\, dz\,,$$

where path C is shown in Fig. A.8f. Suppose that ρ and R are the small and large radii of two circles surrounding the area, where all poles of function $f(z)\, z^\alpha$ are located, Fig. A.8f. As follows from condition A-107, the integral along the small circle tends to zero when $\rho \to 0$. Assuming that the Jordan lemma can be applied, the integral along the large circle also vanishes. Therefore, in accordance with the residual theorem, eq. A-93, we obtain

$$\int_{\substack{0 \\ (ab)}}^{\infty} z^\alpha f(z)\, dz + \int_{\substack{\infty \\ (dc)}}^{0} z^\alpha f(z)\, dz = 2\pi i \sum \text{Res } f(z) \tag{A-108}$$

At the points of path ab,

$$z = x\,,$$

while along path *dc*

$$z = x\, e^{2\pi i}$$

Thus, in place of eq. A-108, we have

$$\int_0^\infty \left[x^\alpha f(x) - x^\alpha e^{2\pi i \alpha} f(x)\right] dx = 2\pi i \sum \mathrm{Res}\, f(z)$$

Then,

$$\int_0^\infty x^\alpha f(x)\, dx = \frac{2\pi i}{1 - e^{2\pi i \alpha}} \sum \mathrm{Res}\, f(z) \qquad \text{(A-109)}$$

Appendix B

Hilbert transform

In Appendix A we described the Cauchy-Riemann conditions that establish the relationship between the real and imaginary parts of the analytical function

$$f(z) = u(z) + i\, v(z)$$

Now, proceeding from the Cauchy formula we demonstrate a different relation between them. As an example consider the spectrum

$$F(\omega) = \int\limits_{-\infty}^{\infty} f(t)\, e^{i\omega t}\, dt\,, \tag{B-1}$$

which is usually an analytical function in the upper plane of the complex frequency

$$\omega = \operatorname{Re}\omega + i \operatorname{Im}\omega$$

Let us choose the integration contour shown in Fig. B.1a. Then, in accordance with the Cauchy formula

$$F(\omega_0) = \frac{1}{\pi\, i}\, P.V. \int\limits_{-\infty}^{\infty} \frac{F(\omega)}{\omega - \omega_0}\, d\omega \tag{B-2}$$

Here both ω and ω_0 are located on the real axis of ω. Of course, it is assumed that the Jordan lemma is valid, that is, the integral along the semicircle tends to zero when its radius r goes to infinity.

Introducing notations

$$F(\omega) = R(\omega) + i\, X(\omega), \tag{B-3}$$

eq. B-2 can be written as

$$R(\omega_0) + i\, X(\omega_0) = \frac{1}{\pi\, i}\, P.V. \int\limits_{-\infty}^{\infty} \frac{R(\omega) + i\, X(\omega)}{\omega - \omega_0}\, d\omega$$

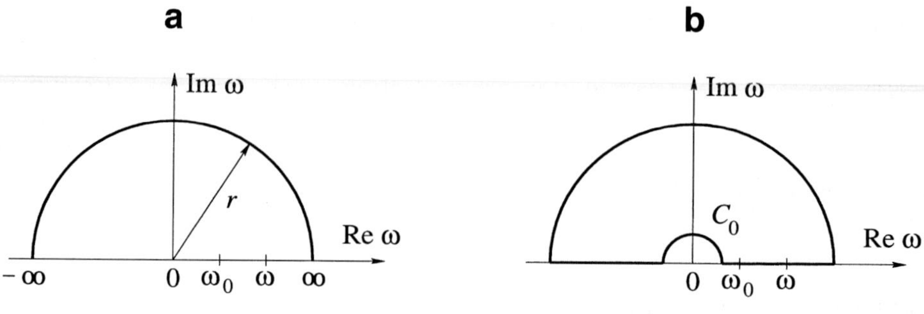

Figure B.1: (a) Illustration of eq. B-2 (b) Derivation of of eq. B-20.

or

$$R(\omega_0) = \frac{1}{\pi} P.V. \int_{-\infty}^{\infty} \frac{X(\omega)}{\omega - \omega_0} d\omega ,$$

$$X(\omega_0) = -\frac{1}{\pi} P.V. \int_{-\infty}^{\infty} \frac{R(\omega)}{\omega - \omega_0} d\omega \tag{B-4}$$

The integrands in eq. B-4 have singularities when ω approaches ω_0. In order to remove them we make use of the equality

$$\int_{-\infty}^{\infty} \frac{d\omega}{\omega - \omega_0} = 0 , \tag{B-5}$$

which follows from the fact that the integrand

$$\frac{1}{\omega - \omega_0}$$

is the odd function with respect to ω_0.

Respectively, eqs. B-4 can be rewritten as

$$R(\omega_0) = \frac{1}{\pi} \int_{-\infty}^{\infty} \frac{X(\omega) - X(\omega_0)}{\omega - \omega_0} d\omega \tag{B-6}$$

and

$$X(\omega_0) = -\frac{1}{\pi} \int_{-\infty}^{\infty} \frac{R(\omega) - R(\omega_0)}{\omega - \omega_0} d\omega \tag{B-7}$$

Since both numerators and denominators tend to zero in the same manner when $\omega \to \omega_0$, the integrands do not have singularities. The last two equations, called the Hilbert transforms, are of a great importance in many applications. For example, eqs. B-6 and B-7 allow us to calculate the real part of the spectrum when its imaginary part is known and vice versa. This means the spectra $R(\omega)$ and $X(\omega)$ contain the same information about the system they describe.

Next, consider the function

$$\Phi(\omega) = \ln F(\omega) \tag{B-8}$$

and assume that it is analytical and Jordan lemma is valid.

Taking into account that

$$F(\omega) = A(\omega)\, e^{i\,\varphi(\omega)},$$

we have

$$\Phi(\omega) = \ln A(\omega) + i\,\varphi(\omega) \tag{B-9}$$

Then, eqs. B-6 and B-7 yield

$$\ln A(\omega_0) = \frac{1}{\pi}\int\limits_{-\infty}^{\infty} \frac{\varphi(\omega) - \varphi(\omega_0)}{\omega - \omega_0}\, d\omega \tag{B-10}$$

and

$$\varphi(\omega_0) = -\frac{1}{\pi}\int\limits_{-\infty}^{\infty} \frac{\ln A(\omega) - \ln A(\omega_0)}{\omega - \omega_0}\, d\omega \tag{B-11}$$

For instance, knowing the amplitude spectrum, we can calculate its phase. In other words, both spectra contain the same information. This is a very useful result from the practical point of view, because in most cases the amplitude measurements are much simpler than those for the phase. We derived eqs. B-10 and B-11 assuming that function $F(\omega)$ does not have zeros in the upper half-space $\mathrm{Im}\,\omega > 0$. If it happens then $\ln F(\omega) \to \infty$, and these equations cannot be used. Now suppose that spectrum $F(\omega)$ is zero at a single point

$$\omega_1 = \mathrm{Re}\,\omega_1 + i\,\mathrm{Im}\,\omega_1 \qquad \text{and} \qquad \mathrm{Im}\,\omega_1 > 0$$

In such a case, it is useful to represent function $F(\omega)$ as

$$F(\omega) = F_0(\omega)\,\frac{\omega - \omega_1}{\omega - \omega_1^*}, \tag{B-12}$$

where $F_0(\omega)$ is an analytical function which does not have zeros in the upper half-plane, and ω_1^* is the complex conjugate of ω_1.

By definition, the magnitude of the ratio

$$\frac{\omega - \omega_1}{\omega - \omega_1^*}$$

is equal to unity, while its argument ϕ is

$$\phi(\omega) = -2 \tan^{-1} \frac{\text{Im } \omega_1}{\omega - \text{Re } \omega_1}, \tag{B-13}$$

because the integration is performed along the real axis of ω and $\text{Im } \omega = 0$.

Thus, functions $F(\omega)$ and $F_0(\omega)$ have the same magnitude but different phases. Correspondingly, eq. B-12 can be rewritten as

$$|F(\omega)| \, e^{i\,\varphi(\omega)} = |F_0(\omega)| \, e^{i\,\varphi_0(\omega)} \, e^{i\,\phi(\omega)}$$

or

$$\ln |F(\omega)| = \ln |F_0(\omega)|, \qquad \varphi(\omega) = \varphi_0(\omega) + \phi(\omega) \tag{B-14}$$

In particular,

$$\varphi(\omega_0) = \varphi_0(\omega_0) + \phi(\omega_0),$$

where

$$\phi(\omega_0) = -2 \tan^{-1} \frac{\text{Im } \omega_1}{\omega_0 - \text{Re } \omega_1}, \tag{B-15}$$

Applying the Hilbert transform, we can calculate the phase $\varphi_0(\omega_0)$ assuming that $\ln |F_0(\omega)|$ is known. Then, making use of eqs. B-14 and B-15, the phase of the spectrum $F(\omega)$ can be determined.

In general, if $F(\omega)$ is zero at n isolated points of the upper half-plane, the function $F_0(\omega)$ is introduced in the following way:

$$F(\omega) = F_0(\omega) \prod_{j=1}^{n} \frac{\omega - \omega_j}{\omega - \omega_j^*} \tag{B-16}$$

Here $\prod$ denotes the product of the fractions

$$\frac{\omega - \omega_j}{\omega - \omega_j^*}$$

As before, function $F_0(\omega)$ does not have zeros if $\text{Im } \omega > 0$, and therefore $\ln |F_0(\omega)|$ and $\varphi_0(\omega)$ are related to each other by eq. B-11. Also, the magnitude of each ratio in eq. B-16 is still equal to unity, and correspondingly,

$$\ln F(\omega) = \ln |F_0(\omega)| + i\,\varphi(\omega) = \ln |F_0(\omega)| + i \left[\varphi_0(\omega) + \sum_{j=1}^{\infty} \phi_j(\omega) \right] \tag{B-17}$$

Thus,

$$\varphi(\omega_0) = \varphi_0(\omega_0) - 2\sum_{j=1}^{\infty} \tan^{-1} \frac{\text{Im}\ \omega_j}{\omega_0 - \text{Re}\ \omega_j} \tag{B-18}$$

Consider one more example, in which it is necessary to introduce some modifications of eq. B-2 in order to find a relationship between $\ln F(\omega)$ and phase $\varphi(\omega)$. It may happen that the function $\Phi(\omega) = \ln F(\omega)$ does not satisfy the Jordan lemma. In this case we introduce the function

$$\Phi_1(\omega) = \frac{\ln F(\omega)}{\omega} \tag{B-19}$$

and assume that the integral of $\Phi_1(\omega)$ along the semicircle with an infinitely large radius can be neglected. Inasmuch as the function $\Phi_1(\omega)$ has a singularity at point $\omega = 0$, the path of integration in eq. B-2 should be slightly deformed, Fig. B.1b, and we have

$$\frac{\ln F(\omega_0)}{\omega_0} = \frac{1}{\pi i} \int_{-\infty}^{\infty} \frac{\ln F(\omega)}{\omega\,(\omega - \omega_0)}\, d\omega + \frac{1}{\pi i} \lim_{r\to 0} \int_{C_0} \frac{\ln F(\omega)}{\omega\,(\omega - \omega_0)}\, d\omega \tag{B-20}$$

Here C_0 is the semicircle around the point $\omega = 0$ when its radius r tends to zero.

Letting

$$\omega = r\, e^{i\theta}$$

and taking into account that the argument θ varies from π to 0, we obtain

$$\frac{1}{\pi i} \lim_{r\to 0} \int_{\pi}^{0} \frac{\ln F(\omega)}{\omega\,(\omega - \omega_0)}\, d\omega = \frac{\ln F(0)}{\omega_0}$$

Correspondingly, eq. B-20 becomes

$$\frac{\ln F(\omega_0)}{\omega_0} - \frac{\ln F(0)}{\omega_0} = \frac{1}{\pi i}\, P.V. \int_{-\infty}^{\infty} \frac{\ln F(\omega)}{\omega\,(\omega - \omega_0)}\, d\omega \tag{B-21}$$

In particular, we have

$$\frac{\varphi(\omega_0)}{\omega_0} - \frac{\varphi(0)}{\omega_0} = -\frac{1}{\pi}\, P.V. \int_{-\infty}^{\infty} \frac{\ln |F(\omega)|}{\omega\,(\omega - \omega_0)}\, d\omega \tag{B-22}$$

560

Appendix C

The saddle point method

Here we describe an approach for approximate evaluation of some integrals based on the behavior of analytical functions in the vicinity of saddle points. This method, as well as the method of stationary phase, is widely used to obtain various asymptotic formulae for wave fields. To illustrate some important features of the saddle point method it is useful to start with a relatively simple case in which integration is performed along the real axis x and the concept of the saddle point is not directly obvious.

Example one Consider the integral

$$I(p) = \int_0^\infty x^p e^{-x} dx \tag{C-1}$$

Here p is the parameter that strongly influences the integrand

$$F(x) = x^p e^{-x} \tag{C-2}$$

For example, its maximum is defined as

$$F'(x) = p x^{p-1} e^{-x} - x^p e^{-x} = 0$$

which yields

$$x = p \tag{C-3}$$

Therefore, the maximum shifts toward greater values of x as parameter p increases. Moreover, the width of the x-interval, where function $F(x)$ has relatively large values, decreases. In other words, the maximum becomes relatively sharper, Fig. C.1. For this reason, it is natural to expect that when parameter p is sufficiently large, integral $I(p)$ is defined mainly by the behavior of the integrand in the vicinity of its maximum $x = p$. It is essential that function $F(x)$ rapidly decreases with the increase of the distance $|x - p|$, thus making contributions of other intervals of x small.

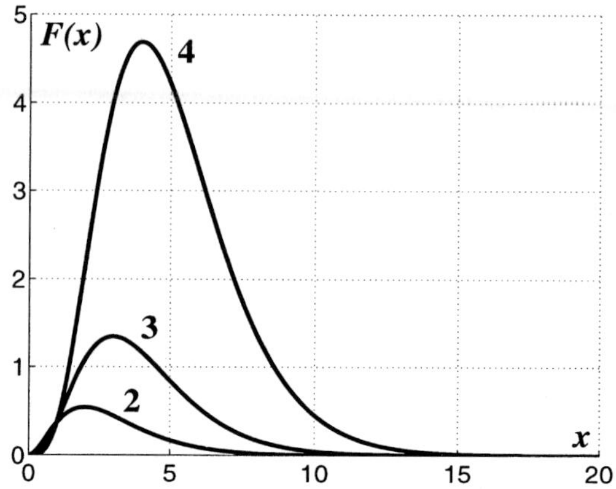

Figure C.1: Functions $F(x)$. The index of curves is p.

In order to find the asymptotic value of $I(p)$ when $p \gg 1$ we replace the integrand in the vicinity of its maximum with a simpler expression. With this in mind, it is convenient to represent function $F(x)$ in the exponential form. Since

$$x^p = e^{\ln x^p} = e^{p \ln x},$$

we have

$$F(x) = e^{f(x)}, \tag{C-4}$$

where

$$f(x) = p \ln x - x \tag{C-5}$$

Let us expand $f(x)$ into the Taylor series around its maximum $x = p$ and preserve only three first terms:

$$f(x) \approx f(p) + \frac{f'(p)}{1!}(x-p) + \frac{f''(p)}{2!}(x-p)^2 + \ldots \tag{C-6}$$

As follows from eq. C-5,

$$f(p) = p \ln p - p, \tag{C-7}$$

$$f'(x) = \frac{p}{x} - 1 \qquad \text{and} \qquad f'(p) = 0, \tag{C-8}$$

$$f''(x) = -\frac{p}{x^2} \qquad \text{and} \qquad f''(p) = -\frac{1}{p} \tag{C-9}$$

Then, eq. C-6 becomes

$$f(x) \approx p \ln p - p - \frac{1}{2p}(x-p)^2 \tag{C-10}$$

Before we continue it may be proper to notice the following. Both the functions $f'(x)$ and $F'(x)$ are equal to zero at the point $x = p$. This coincidence is not occasional, but the result of the equality

$$F'(x) = e^{f(x)} f'(x)$$

Now substitution of eq. C-10 into eq. C-1 yields

$$I(p) \approx \exp(p \ln p - p) \int_0^\infty \exp\left[-\frac{(x-p)^2}{2p}\right] dx \tag{C-11}$$

Certainly, our goal is achieved and integrand $F(x)$ is greatly simplified. Inasmuch as the contribution of the interval $-\infty < x < 0$ is exponentially small for positive p, the integral at the right-hand side of eq. C-11 can be rewritten as

$$\int_0^\infty \exp\left[-\frac{(x-p)^2}{2p}\right] dx = \int_{-\infty}^\infty \exp\left[-\frac{(x-p)^2}{2p}\right] dx \quad \text{if} \quad p \gg 1 \tag{C-12}$$

Introducing the new variable

$$m = \frac{x-p}{\sqrt{2p}}, \tag{C-13}$$

we have

$$dm = \frac{dx}{\sqrt{2p}}$$

and

$$\int_{-\infty}^\infty \exp\left[-\frac{(x-p)^2}{2p}\right] dx = \sqrt{2p} \int_{-\infty}^\infty e^{-m^2}\, dm \tag{C-14}$$

The last integral is tabular

$$\int_{-\infty}^\infty e^{-m^2}\, dm = \sqrt{\pi} \tag{C-15}$$

and therefore,

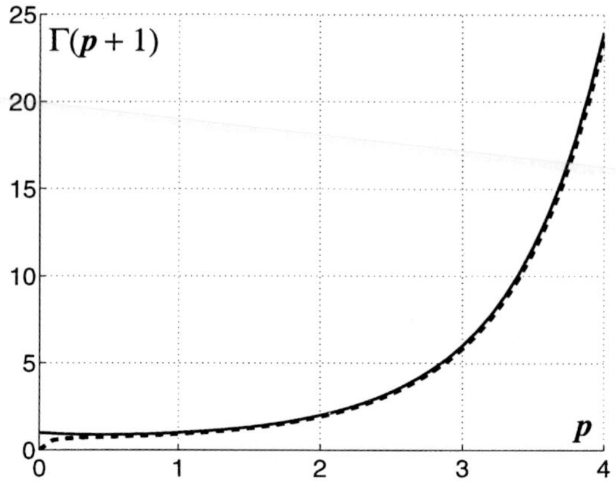

Figure C.2: Function $\Gamma(p+1)$ calculated from eq. C-1 (solid) and its asymptotic approximation given by eq. C-16 (dashed).

$$I(p) \approx \sqrt{2\,p\,\pi}\,\exp(p\,\ln p - p)$$

or

$$I(p) \approx \sqrt{2\,p\,\pi}\,p^p\,e^{-p} \qquad \text{(C-16)}$$

The latter represents the asymptotic expression for function $\Gamma(p+1)$, eq. C-1, and the comparison of numerical integration of eq. C-1 with eq. C-16 is shown in Fig. C.2.

Example two Suppose that

$$I(p) = \int_0^\infty x^{-p\,x}\,dx \qquad \text{(C-17)}$$

This integral can be written as

$$I(p) = \int_0^\infty e^{\,f(x)}\,dx\,, \qquad \text{(C-18)}$$

where

$$f(x) = -p\,x\,\ln x \qquad \text{(C-19)}$$

Correspondingly, we have

$$f'(x) = -p\,\ln x - p \qquad \text{(C-20)}$$

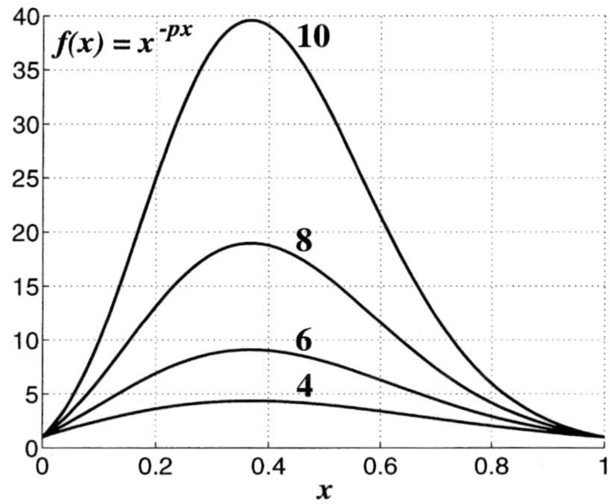

Figure C.3: Behavior of function $f(x) = x^{-p\,x}$. The indexes of the curves indicate the values of p.

and

$$f''(x) \;=\; -\frac{p}{x} \tag{C-21}$$

It follows from the condition $f'(x) = 0$ that the integrand has the extremum when

$$x = \frac{1}{e} \tag{C-22}$$

The second derivative is negative at this point, and it is equal to

$$f''\left(\frac{1}{e}\right) \;=\; -p\,e \tag{C-23}$$

This means that the maximum is observed at the point $x = 1/e$, Fig. C.3 and its value is

$$f\left(\frac{1}{e}\right) \;=\; \frac{p}{e} \tag{C-24}$$

Making use of the Taylor series, eq. C-6, we obtain

$$I(p) \;\approx\; e^{p/e} \int\limits_{-\infty}^{\infty} \exp\left[-\frac{p\,e}{2}\left(x - \frac{1}{e}\right)^2\right] dx \qquad \text{if} \qquad p \gg 1 \tag{C-25}$$

As in the first example, the low integration limit is replaced by $-\infty$.

Introducing the variable

$$m \;=\; \sqrt{\frac{p\,e}{2}}\left(x - \frac{1}{e}\right),$$

we have in place of eq. C-25

$$I(p) \approx e^{p/e} \sqrt{\frac{2}{p\,e}} \int_{-\infty}^{\infty} e^{-m^2}\, dm$$

or

$$I(p) \approx e^{p/e} \sqrt{\frac{2\,\pi}{p\,e}} \qquad \text{if} \qquad p \gg 1 \tag{C-26}$$

In essence we have demonstrated the use of the saddle point method, however, the reason for such a name still remains unclear. It is also proper to make comments:

(a) The simplicity of the examples above is related to the fact that in both cases the maximum of the integrand was located on the integration path. Next, however, we will deal with integrals of a different type where the original path of integration does not pass through the maximum of the integrand.

(b) Deriving the asymptotic formulae for $I(p)$ we have used only three terms of the Taylor series. To improve our approximation, let us take into account the forth term. For illustration consider the first example where

$$I(p) = \int_{0}^{\infty} \exp\left[f(x\,,p)\right]\, dx \qquad \text{when} \qquad p \gg 1 \qquad \text{and} \qquad f(x\,,p) = p\,\ln x - x$$

As follows from eq. C-9,

$$f'''(x) = \frac{2p}{x^3} \qquad \text{and} \qquad f'''(p) = \frac{2}{p^2}$$

Thus, in place of eq. C-10, we obtain

$$f(x) = p\,\ln p - p - \frac{1}{2\,p}\,(x-p)^2 + \frac{2}{3!\,p^2}\,(x-p)^3 + \ldots, \tag{C-27}$$

and the integrand has the new multiplier

$$\exp\frac{(x-p)^3}{3\,p^2} \tag{C-28}$$

Since the last term in eq. C-27 is small in the vicinity of the maximum, the exponent can be expanded in the Taylor series

$$\exp\frac{(x-p)^3}{3\,p^2} \approx 1 + \frac{(x-p)^3}{3\,p^2}$$

The correction factor is the odd function of x, therefore, the integral with the term

$$\frac{(x-p)^3}{3\,p^2}$$

vanishes. This means that it is necessary to find one more term of the Taylor series, namely,

$$\frac{f^{(IV)}(p)}{4!}\,(x-p)^4$$

Inasmuch as

$$f^{(IV)}(x) = -\frac{6p}{x^4} \qquad \text{and} \qquad f^{(IV)}(p) = -\frac{6}{p^3}\,,$$

the new correction factor follows from the equality

$$e^{n} \approx 1 + n + \frac{n^2}{2} + \ldots\,,$$

where

$$n = \frac{1}{3\,p^2}\,(x-p)^3 - \frac{1}{4\,p^3}\,(x-p)^4$$

or

$$e^{n} = 1 + \frac{(x-p)^3}{3\,p^2} - \frac{(x-p)^4}{4\,p^3} + \frac{(x-p)^6}{18\,p^4} + \ldots \tag{C-29}$$

Thus, the additional term, which corrects function $I(p)$, is

$$e^{f(p)} \int_{-\infty}^{\infty} \left[-\frac{(x-p)^4}{4\,p^3} + \frac{(x-p)^6}{18\,p^4} - \ldots \right] \exp\left(-\frac{(x-p)^2}{2\,p} \right)\, dx$$

$$= e^{-p + p\ln p} \int_{-\infty}^{\infty} e^{-m^2} \left(-\frac{m^4}{p} + \frac{4\,m^6}{9\,p} \right) \sqrt{2\,p}\, dm = \frac{\sqrt{2\,\pi}}{12}\,\frac{p^p}{\sqrt{p}}\,e^{-p} \tag{C-30}$$

Here we used the recursive relationship between integrals

$$\int_{-\infty}^{\infty} m^{2\,n}\, e^{-m^2}\, dm\,,$$

which can be established by applying integration by parts. Thus, a more accurate expression for $I(p)$ is

$$I(p) \approx \sqrt{2\pi p}\, e^{-p} p^p \left(1 + \frac{1}{12p}\right) \tag{C-31}$$

Saddle point method

Now we are prepared to describe the saddle point method when the integral $I(p)$ has a more general form

$$I(p) = \int\limits_C g(z)\, e^{p\, f(z)}\, dz \tag{C-32}$$

Here p is positive large number and C is an arbitrary contour of integration on the complex plane z.

We can imagine two situations. First, path C might pass through point z_0, where function $u(z) = \text{Re}\ f(z)$ has a maximum. Also $u(z)$ rapidly decreases with an increase of the distance from this point. Then, assuming that function $g(z)$ is almost constant in the vicinity of the maximum, eq. C-32 becomes

$$I(p) \approx g(z_0) \int\limits_C e^{p\, f(z)}\, dz$$

or

$$I(p) \approx g(z_0) \int\limits_C e^{p\, u(z)}\, e^{i\, p\, v(z)}\, dz \tag{C-33}$$

In addition, we suppose that the orientation of the path near the maximum is such that the oscillating factor $e^{i\, p\, v(z)}$ almost does not change. Correspondingly, the right-hand side of eq. C-33 is simplified to

$$I(p) \approx g(z_0)\, e^{\, i\, p\, v(z_0)} \int\limits_C e^{p\, u(z)}\, dz \tag{C-34}$$

Inasmuch as the last integral is similar to those considered earlier, we can use the asymptotic formulae that were already derived.

In the second case, when some of the above conditions are not met, this approach of evaluation of integrals becomes invalid. For this reason, let us try to deform path C in such a way as to satisfy these requirements. Below we explain how this goal might be achieved. First, assume that functions $f(z)$, $g(z)$ and, consequently,

$$F(z) = g(z)\, e^{p\, f(z)}$$

are analytical within the area of deformation. Then, in accordance with the Cauchy theorem, this procedure does not change integral $I(p)$. Of course, the terminal points of path C preserve their positions during the deformation.

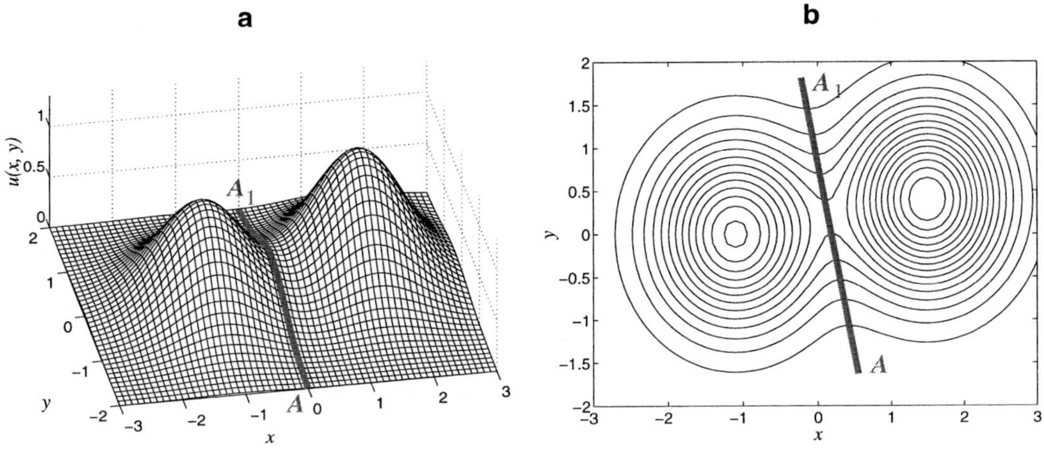

Figure C.4: Behavior of the function $u(x,\, y)$.

As was shown in Appendix A, functions $u(z)$ and $v(z)$ have no maxima or minima because they satisfy the Laplace equations

$$\nabla^2 u = 0 \qquad \text{and} \qquad \nabla^2 v = 0\,,$$

that if $\dfrac{\partial^2 u}{\partial x^2} < 0$, then $\dfrac{\partial^2 u}{\partial y^2} > 0$. Suppose that

$$\frac{\partial u}{\partial x} = \frac{\partial u}{\partial y} = 0 \tag{C-35}$$

at some point $z = z_0$. As we know, such a point is called a saddle point and the function $u(x,\, y)$ looks like a saddle or mountain passage in the vicinity of this point, Fig. C.4.

It also follows from the Cauchy-Riemann conditions and eq. C-35 that

$$\frac{\partial v}{\partial x} = \frac{\partial v}{\partial y} = 0\,, \tag{C-36}$$

and therefore,

$$f'(z_0) = 0 \tag{C-37}$$

Thus, the point $z = z_0$ is the saddle point for both functions $u(z)$ and $v(z)$.

Let us expand function $f(z)$ in the Taylor series near the saddle point. Taking into account eq. C-37 we have

$$f(z) \approx f(z_0) + \frac{1}{2} f''(z_0)\,(z - z_0)^2 + \ldots \tag{C-38}$$

In order to describe the behavior of $f(z)$ around the saddle point, it is convenient to represent the second term in eq. C-38 in trigonometric form. Suppose that

$$f''(z_0) = \rho\, e^{i\theta} \qquad \text{and} \qquad z - z_0 = s\, e^{i\phi} \tag{C-39}$$

Here ρ is the magnitude of the second derivative

$$\rho = |f''(z_0)| \tag{C-40}$$

Then, eq. C-38 takes the following form

$$u(x,\, y) \approx u(x_0,\, y_0) + \frac{1}{2}\rho\, s^2 \cos(\theta + 2\,\phi) \tag{C-41}$$

and

$$v(x,\, y) \approx v(x_0,\, y_0) + \frac{1}{2}\rho\, s^2 \sin(\theta + 2\,\phi)\,, \tag{C-42}$$

where s is the distance from the saddle point.

As is clearly seen from Fig. C.4a and C.4b, the surface $u(x,\, y)$ resembles a mountain range with a ridge and a valley. In accordance with eq. C-41, the inequality

$$u(x,\, y) < u(x_0,\, y_0) \tag{C-43}$$

occurs if

$$\cos(\theta + 2\,\phi) < 0$$

This means that either

$$\frac{\pi}{2} < \theta + 2\,\phi < \frac{3\,\pi}{2} \qquad \text{or} \qquad \frac{5\,\pi}{2} < \theta + 2\,\phi < \frac{7\,\pi}{2}$$

Thus, to satisfy condition C-43, angle ϕ should vary within the limits

$$\frac{\pi}{4} - \frac{\theta}{2} < \phi < \frac{3\,\pi}{4} - \frac{\theta}{2} \qquad \text{or} \qquad \frac{5\,\pi}{4} - \frac{\theta}{2} < \phi < \frac{7\,\pi}{4} - \frac{\theta}{2} \tag{C-44}$$

These inequalities define the points of the z-plane where the surface $u(x,\, y)$ has valleys. It is obvious that there are two such valleys at both sides of the saddle point, Fig. C.5. In a similar manner, the condition

$$\cos(\theta + 2\,\phi) > 1 \tag{C-45}$$

determines the ridges where

$$u(x,\, y) > u(x_0,\, y_0) \tag{C-46}$$

Therefore, there are four zones around the saddle point: two valleys and two ridges. Their boundaries are given by

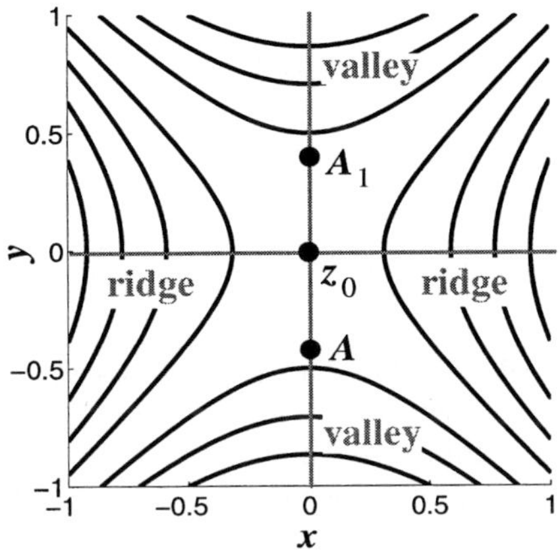

Figure C.5: Contours passing the saddle point z_0.

$$\cos(\theta + 2\phi) = 0 \tag{C-47}$$

Correspondingly, we can distinguish several paths of integration. For instance, if the contour C is located in both valleys, it can be deformed in such a way that the new path AA_1 passes through the saddle point, Fig. C.5 and C.6. Then, we can expect that the interval of integration near this point gives the main contribution to the integral. In fact, as was mentioned earlier, integrand $F(z)$ can be written as

$$F(z) = g(z)\, e^{p\,u(z)}\, e^{i\,p\,v(z)}$$

Since $p > 0$, magnitude of $F(z)$ reaches its maximum at the saddle point. Therefore, in this case one can apply the approximate method to evaluate the integral.

If both terminal points of contour C are located in the same zone (valley or ridge), the path can always be deformed in such a way that the integrand $F(z)$ changes monotonically. Then the interval where $|F(z)|$ is maximal gives the main contribution into the integral. The same situation occurs when one of the terminal points is located in the valley while the other is on the ridge. Of course, the minimum of $|F(z)|$ is observed when both these points are located at different zones of the ridges. Thus, evaluation of integral $I(p)$ implies that the integration path is located in both valleys and passes through the saddle point $z = z_0$. This, however, is not sufficient. As was pointed out earlier, it is also important to reduce the interval of integration as much as possible. In other words, our goal is to choose a path near z_0 along which the rate of decrease of the function $u(z)$ is maximal. Certainly, there is an infinite number of paths passing through the point z_0.

As follows from eq. C-41, its second term has maximum magnitude if

$$\cos(\theta + 2\phi) = -1, \tag{C-48}$$

and the maximum decrease of function $u(z)$ is observed. Thus, the orientation of the path of maximum descent is defined from the equality

$$\theta + 2\,\phi = \pi \qquad \text{or} \qquad \theta + 2\,\phi = 3\,\pi\,,$$

that is

$$\phi_1 = \frac{\pi}{2} - \frac{\theta}{2} \qquad \text{or} \qquad \phi_2 = \frac{3\,\pi}{2} - \frac{\theta}{2} \tag{C-49}$$

Comparison with eq. C-44 shows that the path given by eq. C-49 bisects the valley regions. For this reason, the saddle point method is sometimes called the method of the steepest descent. Both names emphasize the most essential features of this approach that allow one to find asymptotic formulae for integrals $I(p)$, eq. C-32.

In accordance with eqs. C-42 and C-48 the imaginary part of the function $f(z)$ remains constant along this path C:

$$v(x,\,y) = v(x_0,\,y_0) \tag{C-50}$$

The latter demonstrates one more special feature of the chosen contour. As we know from Appendix A, the level lines $u = \text{const}$ and $v = \text{const}$ are perpendicular to each other and, therefore, the steepest descent path follows the direction of $-\text{grad}\;u$, which is orthogonal to the lines $u = \text{const}$.

Inasmuch as $v = \text{const}$ at the path, eq. C-32 is greatly simplified:

$$I(p) \;=\; e^{i\,p\,v(x_0,\,y_0)} \int\limits_C g(z)\,e^{p\,u(z)}\,dz \tag{C-51}$$

This is a very important simplification. In fact, if the integrand would contain the term $\exp(i\,p\,v(z))$, its oscillations could decrease the contribution of the vicinity of the saddle point into the integral. At the same time the influence of the removed intervals might be relatively large and the saddle point method cannot be applied.

Next, bearing in mind that the interval of integration near point z_0 is small, function $g(z)$ can be treated as almost constant, and we again arrive at eq. C-34

$$I(p) \;\approx\; g(z_0)\,e^{i\,p\,v(z_0)} \int\limits_C e^{p\,u(z)}\,dz \tag{C-52}$$

Taking into account that along the steepest descent path

$$u(z) = u(z_0) - \frac{1}{2}\,\rho\,s^2 \qquad \text{and} \qquad z - z_0 = s\,e^{i\,\phi}\,,$$

we have

$$dz = e^{i\,\phi}\,ds \tag{C-53}$$

and

$$I(p) \;\approx\; g(z_0)\,e^{i\,p\,v(z_0)} \int\limits_C e^{p\,u(z_0)}\,e^{-p\,\rho\,s^2/2}\,e^{i\,\phi}\,ds$$

or

$$I(p) \approx g(z_0)\, e^{p\, f(z_0)}\, e^{i\phi} \int\limits_C e^{-p\,\rho\, s^2/2}\, ds \tag{C-54}$$

Deriving this equation, we assume that the path near the saddle point can be approximated by a straight line, so argument ϕ is constant, otherwise, eq. C-54 would be invalid. Since we neglect the contribution of those elements of path C that are relatively far from the saddle point, eq. C-54 can be rewritten as

$$I(p) \approx g(z_0)\, e^{p\, f(z_0)}\, e^{i\phi} \int\limits_{-\infty}^{\infty} e^{-p\,\rho\, s^2/2}\, ds$$

The last integral was considered in the first example, and therefore

$$I(p) \approx \sqrt{\frac{2\,\pi}{p\,|f''(z_0)|}}\, g(z_0)\, e^{p\, f(z_0)}\, e^{i\phi} \tag{C-55}$$

Here we used the fact that

$$\int\limits_{-\infty}^{\infty} e^{-m^2}\, dm = \sqrt{\pi} \qquad \text{and} \qquad \rho = |f''(z_0)|$$

As follows from eq. C-49, there are two possible values of ϕ. For instance, suppose that the second derivative $f''(z_0)$ is such that $\theta = \pi/2$. Then, the value of ϕ is either

$$\phi_1 = \frac{\pi}{4} \qquad \text{or} \qquad \phi_2 = \frac{5\,\pi}{4},$$

and it depends on the path direction at the saddle point. These two cases are shown in Fig. C.6a, b. First, suppose that the ridges are located as shown in Fig. C.6c. Then $\phi_1 = \pi/4$ because it characterizes the slope of the path in the vicinity of the saddle point. In the second case, Fig. C.6d, the direction along the path is opposite and, therefore, $\phi_2 = 5\,\pi/4 = \pi + \pi/4$.

Before we continue we note the following.

(1) At the beginning, eq. C-2, we studied integration of the real-valued function and found the maximum of integrand $F(x)$ at point $x = p$. It is a simple matter to verify that the latter is the saddle point of $F(z)$ on the complex plane.

(2) It may happen that the derivatives $f^{(m)}(z_0) = 0$ at the saddle point when $m < n$. Then the corresponding Taylor series has the form

$$f(z) = f(z_0) + \frac{f^{(n)}(z_0)}{n!}\,(z - z_0)^n + \dots$$

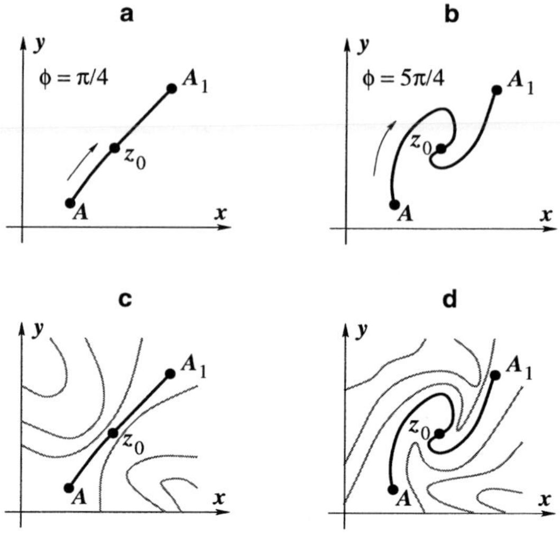

Figure C.6: Integration paths.

Therefore, we have in place of eqs. C-41–C-42:

$$u(x,\, y) = u(x_0,\, y_0) + \frac{\rho}{n!}\, s^n\, \cos(\theta + n\,\phi)$$

and

$$v(x,\, y) = v(x_0,\, y_0) + \frac{\rho}{n!}\, s^n\, \sin(\theta + n\,\phi) \tag{C-56}$$

Correspondingly, there are $2n$ zones of valleys and ridges. Assuming that path C passing through the saddle point is located in two zones of valleys, we can derive a formula similar to eq. C-55.

(3) If function $F(z)$ has several saddle points, then the same procedure can be applied to all of them.

(4) When one terminal point coincides with the saddle point, the interval of integration near this point becomes two times smaller. Correspondingly, eq. C-55 takes the form

$$I(p) \approx \sqrt{\frac{\pi}{2\,p\,|f''(z_0)|}}\, g(z_0)\, e^{p\, f(z_0)}\, e^{i\phi} \tag{C-57}$$

(5) As we know the method of stationary phase can be applied to evaluate integrals of the type

$$I(p) \,=\, \int\limits_C g(x)\, e^{i\,p\, f(x)}\, dx\,,$$

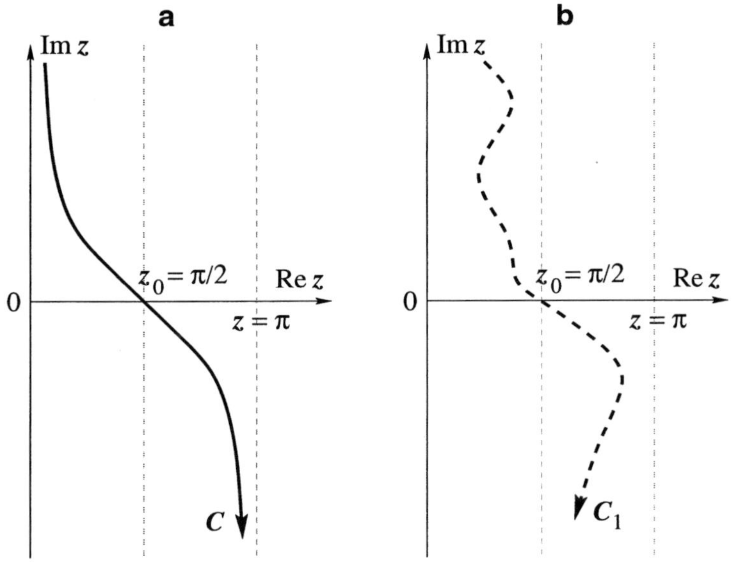

Figure C.7: Path of integration, given by eq. C-58.

where $f(x)$ is the real function along C and p is large. If the condition $f'(x) = 0$ is not met, the contribution of the interval close to x into $I(p)$ is almost zero due to strong oscillations of $exp(i\,p\,f(x))$. In contrast, the superposition of oscillations is constructive near the points where the derivative $f'(x)$ vanishes. Thus, both the saddle point and stationary phase methods are based on the fact that the interval of integration around the point where $f'(z) = 0$ or $f'(x) = 0$ gives the main contribution. This is their common feature. However, there is also an essential difference. In the case of the method of stationary phase the function $f(x)$ is real, and the contributions of other intervals cancel each other due to rapid oscillations of the integrand. In the case of the saddle point method when $f(z)$ is complex, the steepest descent path reduces the influence of intervals that are sufficiently far from the saddle point.

Next, consider an example illustrating the saddle point method.

Asymptotic formula for the Hankel function

As is well known the Hankel function of the first kind $H_\nu^{(1)}(p)$ can be represented by the integral

$$H_\nu^{(1)}(p) = \frac{1}{\pi} \int\limits_C \exp[i\,(p\,\sin z - \nu\, z)]\, dz \qquad \text{(C-58)}$$

Here ν is the order of the Hankel function and p is the positive number, such that

$$p > |\nu| \tag{C-59}$$

The integration path C in the complex plane z moves from the semi-strip

$$0 < \text{Re } z < \frac{\pi}{2}, \qquad \text{Im } z > 0$$

to the semi-strip

$$\frac{\pi}{2} < \text{Re } z < \pi\,, \qquad \text{Im } z < 0\,,$$

Fig. C.7a. From comparison with eq. C-32, we can write

$$g(z) = \frac{1}{\pi}\, e^{-i\,\nu\, z} \qquad \text{and} \qquad f(z) = i\,\sin z \tag{C-60}$$

Therefore,

$$f'(z) = i\,\cos z$$

and the saddle point is

$$z_0 = \frac{\pi}{2} \tag{C-61}$$

Thus, the path of integration passes through this point.

Now we demonstrate that both semi-strips correspond to valleys, that is

$$u(z_0) > u(z)$$

In fact, we have

$$f(z) \;=\; u(z) + i\,v(z) \;=\; i\,\sin(x + i\,y) \tag{C-62}$$

or

$$f(z) = i\,[\sin x\,\cos(i\,y) + \sin(i\,y)\,\cos x\,] = i\,(\sin x\,\cosh y + i\,\sinh y\,\cos x)$$

Thus,

$$u(z) = -\sinh y\,\cos x \tag{C-63}$$

In particular, at the saddle point,

$$u(z_0) = 0\,, \tag{C-64}$$

and $u < 0$ in the first and second semi-strips. Therefore, the real part of $f(z)$ decreases away from point z_0.

Making use of eq. C-60, we obtain

$$g(z_0) = \frac{1}{\pi}\, e^{-i\,\nu\,\pi/2}\,, \qquad f(z_0) = i \tag{C-65}$$

and

$$f''(z_0) = -i \sin z_0 = -i \qquad \text{or} \qquad f''(z_0) = e^{3\pi i/2},$$

that is,

$$|f''(z_0)| = 1 \qquad \text{and} \qquad \theta = \frac{3\pi}{2} \tag{C-66}$$

In accordance with eq. C-49, the angle ϕ is equal to

$$\phi_1 = -\frac{\pi}{4} \qquad \text{or} \qquad \phi_2 = \frac{3\pi}{4}$$

Taking into account the direction of path C, Fig. C.7a, we choose the value

$$\phi_1 = -\frac{\pi}{4} \tag{C-67}$$

and correspondingly, the path is the bisection of the semi-strips.

Substitution of eqs. C-65–C-67 into eq. C-55 gives the asymptotic expression for the Hankel function

$$H_\nu^{(1)}(p) \approx \sqrt{\frac{2}{\pi p}} \exp i \left[p - \frac{\nu\pi}{2} - \frac{\pi}{4}\right] \qquad \text{if} \qquad p \gg 1 \tag{C-68}$$

Let us notice that the path C, Fig. C.7a, can be treated as a result of deformation of an arbitrary contour C_1 as long as the latter is located within the semi-strips shown in Fig. C.7b.

Appendix D

Differential equations

In both volumes we often used various differential equations that play an important role in describing wave propagation. For instance, geometrical acoustics is based on solution of the system of linear differential equations, characterizing the ray behavior or on the partial differential equation (PDE) of the first order that describes the eikonal. For this reason, it seems proper to outline the main concepts related to solving those equations, starting from the simplest ones.

D.1 Ordinary differential equations of the first order

By definition, the general form of an ordinary differential equation (ODE) of the first order is

$$\Phi(x,\, y,\, y') = 0 \tag{D-1}$$

Here x is the argument, $y(x)$ is the unknown function and

$$y' = \frac{dy}{dx}$$

The left-hand side of eq. D-1, $\Phi(x,\, y,\, y')$, is the known relationship between the first derivative $y'(x)$, $y(x)$, and x. This is an example of an implicit form of differential equation. Resolving equality D-1 with respect to $y'(x)$, we may obtain its explicit form

$$\frac{dy}{dx} = y'(x) = f(x,\, y) \tag{D-2}$$

This equation clearly shows that our goal is to determine function $y(x)$, provided that its first derivative, $f(x,\, y)$, is known.

The function

$$y = \varphi(x) \tag{D-3}$$

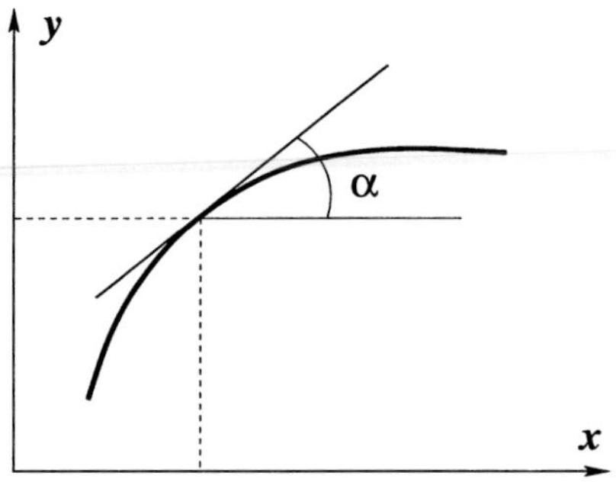

Figure D.1: Angle coefficient $y' = \tan\alpha$.

is called a solution of differential equation D-2 if a replacement of $y(x)$ and $y'(x)$ with $\varphi(x)$ and $\varphi'(x)$ in eq. D-2 leads to the equality

$$\varphi'(x) \equiv f\,[x,\, \varphi(x)\,] \tag{D-4}$$

The procedure of solving a differential equation is called its integration. Geometrical interpretation of a differential equation is very instructive. Let us assume that x and y are the coordinates of some point in the x-y plane. Then, a solution $y = \varphi(x)$ describes a curve on, more precisely, a family of curves located in this plane. Correspondingly, eq. D-2 is the relationship between the coordinates of a point on the curve and the angle coefficient

$$\tan\alpha \;=\; \frac{d\,y(x)}{d\,x} \;=\; f(x,\, y) \tag{D-5}$$

of the tangent to the curve at this point, Fig. D.1. In other words, a solution of a differential equation of the first order is a curve whose coordinates and the angle coefficient satisfy eq. D-1 or D-2. This curve is called the integral curve of the given differential equation. Note also that these equations are called the ordinary ones, because the unknown function y depends on the single argument x only.

Example For illustration, consider the simplest case, when the right-hand side of eq. D-2 is independent of function $y(x)$. Then we have

$$\frac{d\,y(x)}{d\,x} = f(x) \tag{D-6}$$

and its integration gives

$$y(x) = \int_{x_0}^{x} f(\xi)\, d\xi + C\,, \tag{D-7}$$

where C is an arbitrary constant and x_0 is the coordinate of some fixed point.

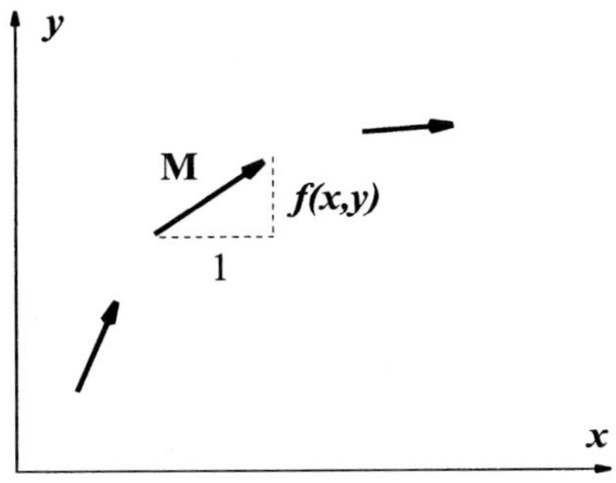

Figure D.2: Field of directions of a differential equation.

It is clear that the function $y(x)$ is the solution of eq. D-6 regardless of the value of C. For this reason it is called the general integral of this equation. Changing the constant C, we obtain particular solutions of the differential equation, which means that eq. D-6, as well as eqs. D-1 and D-2, has an infinite number of solutions. Correspondingly, there is a family of integral curves in the x-y plane. As follows from eq. D-7, we have

$$C = y(x_0) = y_0 \,,$$

that is the constant is equal to the y-coordinate at a fixed point.

Inasmuch as the integral in the right-hand side of eq. D-7 is a function of x, we can rewrite this equation as

$$\psi(x,\, y) = C \tag{D-8}$$

This important form of the solution will be discussed later. Of course, an arbitrary constant also appears in the general case, when the right-hand side of eq. D-2 is a function of both x and y.

Field of directions

As was pointed out earlier an ordinary differential equation of the first order establishes a link between coordinates of points on a plane curve $y(x)$ and the angle coefficients $y'(x)$ of tangents to this curve. Suppose that $f(x,\, y)$ is a single-valued and continuous function of x and y. Then we may say that eq. D-2 describes a field of directions, that is a vector field, and its orientation with respect to the x-axis is defined by the angle coefficient, Fig. D.2. Correspondingly, the integral curves of eq. D-2 are such that their tangents coincide with the field directions at each point.

We have two comments, related to this topic.

(1) The concept of a field of directions will be used for more complicated equations, in particular, for those with partial derivatives.

(2) Equation D-2 can be written as

$$\frac{dy}{f(x,\,y)} = \frac{dx}{1}$$

As follows from vector algebra, this equation shows that two vectors, Fig. D.2,

$$d\mathbf{l} = dx\,\mathbf{i} + dy\,\mathbf{j} \qquad \text{and} \qquad \mathbf{M} = \mathbf{i} + f(x,\,y)\,\mathbf{j}$$

are parallel to each other. Here, $d\mathbf{l}$ is the vector of an elementary displacement along the integral curve and $\mathbf{M}$ is the vector characterizing the field of directions. It has one important feature: its projection on the x-axis is equal to unity everywhere.

Construction of integral curves

As was already mentioned, differential equation D-2 describes a family of integral curves. To determine one of them uniquely, we also have to know a point through which this curve passes. For instance, it can be the point of intersection of the curve with the vertical line $x = x_0$, which is parallel to the y-axis. This means that we know coordinates of the point $(x_0,\,y_0)$, as well as the angle coefficient $f(x,\,y)$.

There are only a few types of differential equations which can be integrated explicitly. In most cases, it is necessary to apply some numerical methods designed to construct the integral curves, and some of them are described here. In essence, we will solve the initial-value Cauchy problem, when eq. D-2 is given along with the condition

$$y(x_0) = y_0 \tag{D-9}$$

Euler method

The simplest approach was suggested by Euler. Since the initial point $(x_0,\,y_0)$ is given, Fig. D.3, we can determine the direction of the tangent to the integral curve at this point:

$$y'(x_0) = f(x_0,\,y_0) \tag{D-10}$$

Let us also choose the points $x_1,\,x_2,\,\ldots,\,x_n$ and draw the vertical lines

$$x = x_i \tag{D-11}$$

In principle, the distances between them can be arbitrary. Then, the intersection of the first tangent with line $x = x_1$ gives the ordinate y_1 of the next point.

At this point, $(x_1,\,y_1)$, we calculate the angle coefficient

$$y'(x_1) = f(x_1,\,y_1) \tag{D-12}$$

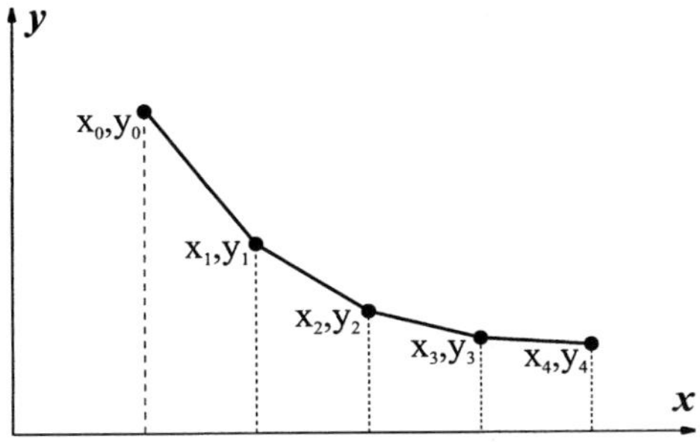

Figure D.3: Euler method of integration.

and draw the second tangent until it intersects the vertical line $x = x_2$. Repeating this procedure we obtain a line that consists of linear segments and approximately represents the integral curve passing through the initial point (x_0, y_0). Of course, with a decrease of intervals between points x_i the accuracy of solution of the Cauchy problem increases, but the number of required computations also becomes greater. In general, the angle coefficient $f(x, y)$ varies within each interval $[x_i, x_{i+1}]$. At the same time the Euler method implies that $f(x, y)$ is constant inside $[x_i, x_{i+1}]$, and its value is defined by that at the point x_i.

Now let us outline an approach that greatly improves the accuracy of the Euler method. As before we calculate the angle coefficient $y'(x_0, y_0)$ at the point (x_0, y_0) and draw the line until it intersects the vertical line $x = \xi_1$ located in the middle of the interval $[x_0, x_1]$ (half step). The intersection of those straight lines gives the point (ξ_1, η_1), where the angle coefficient

$$y' = f(\xi_1, \eta_1) \tag{D-13}$$

is calculated. Next, we return to the initial point (x_0, y_0) again and draw the tangent with the angle coefficient given by eq. D-13 until it intersects line $x = x_1$. This yields the value y_1 and, correspondingly, the angle coefficient $y' = f(x_1, y_1)$. Continuing this tangent to the middle of the next interval, $[x_1, x_2]$, we obtain the point (ξ_2, η_2) and the angle coefficient $f'(\xi_2, \eta_2)$. Then, the same procedure is repeated, and the line that consists of the constructed linear segments approaches the integral curve closer than that in the previous case, Fig. D.4. Unlike the Euler method, the angle coefficients are determined at the middle of each interval which allows us to achieve higher accuracy for the same interval lengths. The computational effort, however, also increases. An important feature of this modification of the Euler method is that the angle coefficient is calculated at the point located in front of the integral curve.

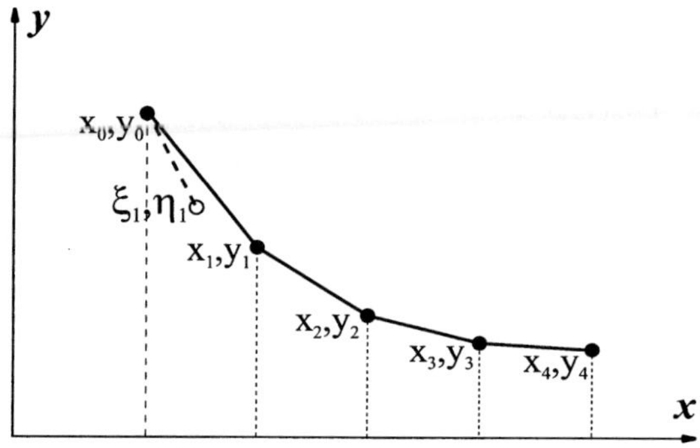

Figure D.4: Improvement to Euler method.

Runge-Kutta method

Among the variety of approaches for integrating differential equations, the Runge-Kutta method is one of the most efficient. It is also often used to solve systems of ordinary differential equations. The principle of this method may be described in the following way (Chapter 1). Consider an interval

$$[x_0,\, x_0 + h]\,,$$

where we specify the points $(\xi_\nu,\, \eta_\nu)$ and the constants R_ν $(\nu = 0,\, 1, \ldots,\, m)$. Those points and constants are selected in such a way that the ordinate $\varphi(x_0+h)$ of the integral curve $\varphi(x)$ passing through point $(x_0,\, y_0)$ is described by the expression

$$\varphi(x_0 + h) \;=\; y_0 + h\,[\,R_0\, f(\xi_0,\, \eta_0) + \ldots + R_m\, f(\xi_m,\, \eta_m)\,] \tag{D-14}$$

as accurately as possible. Therefore, it is required that when we expand in power series the equality

$$\varphi'(x_0 + h) \;=\; f\,[\,x_0 + h,\, \varphi(x_0 + h)\,]\,, \tag{D-15}$$

the maximum number of terms should coincide in its left- and right-hand sides.

Equation D-14 implies that

$$\xi_0 = x_0 \qquad \text{and} \qquad \eta_0 = y_0$$

and

$$\xi_1 = x_0 + \alpha_0\, h, \qquad \eta_1 = y_0 + \beta_0\, k_1, \qquad k_1 = h\, f(\xi_0,\, \eta_0);$$

$$\xi_2 = x_0 + \alpha_1\, h, \qquad \eta_2 = y_0 + \beta_1\, k_1 + \beta_2\, k_2, \qquad k_2 = h\, f(\xi_1,\, \eta_1); \quad \ldots$$

Here, the coefficients α_ν and β_ν are independent of function $f(x,\, y)$. Let us notice that there are formulae of the Runge-Kutta method of different orders, and in particular the first ones correspond to the Euler method and its modification.

We have discussed procedures for constructing the integral curves which show that only one such curve passes through a given point (x_0, y_0). This statement, however, is valid and can be proven only when function $f(x, y)$ has one important property: it is continuous with respect to its arguments and has a continuous derivative with respect to y at (x_0, y_0). In essence, we have just formulated the theorem of existence and uniqueness of solution of an ordinary differential equation of the first order with the initial conditions.

General solution

Let us now return to the differential equation

$$y' = f(x, y)$$

Changing the value of y_0 in the initial condition

$$y(x_0) = y_0 \,,$$

we obtain an infinite number of solutions of this equation, that is the family of integral curves corresponding to different y_0. As we know, the latter is the ordinate of the intersection of the integral curve with line $x = x_0$. An arbitrary constant C can enter the solution not only through the initial value of y, eq. D-7, but also as

$$y = \varphi(x, C) \tag{D-16}$$

The solution containing a constant C is called the general solution of the differential equation. It can also be written in implicit form

$$\psi(x, y, C) = 0 \tag{D-17}$$

Ascribing to C some numerical value we obtain a particular solution of eq. D-2. For instance, if solution is given by eq. D-17 and we want to distinguish the curve passing through the point (x_0, y_0), the constant C is found from the equality

$$\psi(x_0, y_0, C) = 0 \tag{D-18}$$

Note that the general integral D-17 could be resolved with respect to the constant C, which yields

$$\varpi(x, y) = C \tag{D-19}$$

Thus, we have three forms of the general integral of the differential equation D-2:

$$y = \varphi(x, C)\,, \qquad \psi(x, y, C) = 0 \qquad \text{and} \qquad \varpi(x, y) = C \tag{D-20}$$

and from a geometrical point of view, all of them represent the same family of integral curves.

Now it is useful to formulate the problem, that is the inverse of integration of the differential equation of the first order. Suppose that

$$\psi(x,\, y,\, C) = 0 \tag{D-21}$$

is the general solution of an unknown differential equation. In order to find the latter we differentiate eq. D-21 with respect to x and find

$$\frac{\partial\,\psi(x,\, y,\, C)}{\partial\, x} + \frac{\partial\,\psi(x,\, y,\, C)}{\partial\, y}\, y' = 0 \tag{D-22}$$

Eliminating constant C from eqs. D-21 and D-22, we obtain

$$\Phi(x,\, y,\, y') = 0 \tag{D-23}$$

which is a differential equation in implicit form.

Let us discuss two special cases.

Case one It may happen that when we resolve eq. D-23 with respect to y' and obtain eq. D-2, function $f(x,\, y)$ turns out to be multi-valued. Suppose it has m different values at some points, that is, there are m different directional fields at such places. For each field one integral curve goes through a given point, and therefore m integral curves of eq. D-23 may pass through the same point. For example if our equation describes ray behavior then m rays intersect each other. In this regard we note that the general integral of eq. D-23 contains one arbitrary constant, that is it has the form

$$\psi(x,\, y,\, C) = 0$$

but eq. D-18 may have m different values of C. As an illustration consider the differential equation

$$\Phi(x,\, y,\, y') \,\equiv\, (y')^2 - x\, y' \,=\, 0 \tag{D-24}$$

or

$$y'\,(y' - x) \,=\, 0$$

The latter splits into two differential equations

$$\frac{d\,y}{d\,x} = 0 \qquad \text{and} \qquad \frac{d\,y}{d\,x} = x$$

Their general integrals are

$$y(x) = C_1 \qquad \text{and} \qquad y(x) = \frac{x^2}{2} + C_2$$

The two equations can be represented as one:

$$(y - C_1)\left(y - \frac{x^2}{2} - C_2\right) = 0$$

which is the general form of eq. D-24. Therefore we obtained two families of integral curves: one consists of straight lines

$$y(x) = C_1\,,$$

while the other are parabolas,

$$y(x) = \frac{x^2}{2} + C_2$$

The curves of both families pass through any point of the x-y plane and, consequently, there are two fields of directions.

Case two It may turn out that equation

$$\frac{dy}{dx} = f(x,\, y) \qquad \text{or} \qquad \Phi(x,\, y,\, y') = 0 \tag{D-25}$$

have solutions that cannot be derived from the general integral, eq. D-20, regardless of value of C, and they called the singular ones. We will discuss the geometrical meaning of singular solutions later but now introduce one special type of curves.

Isocline lines

Replacing the derivative y' in eq. D-25 by an arbitrary constant C_1, we arrive at the family of curves

$$f(x,\, y) = C_1 \qquad \text{or} \qquad \Phi(x,\, y,\, C_1) = 0 \tag{D-26}$$

Thus, each curve of this family is a geometrical place of points on the x-y plane, where direction of the tangent, y', is the same. Respectively, we deal with the family of isoclines of the differential equation. For instance, considering a planar magnetic field, we can say that the field has the same direction at each point of its isocline. Of course, different isoclines are characterized by the different values of angle coefficient $y'(x)$. It may be proper to point out that unlike isoclines, this coefficient usually varies along the integral curve. It is useful to find a condition when the isocline coincides with the integral curve. Let us take some isocline,

$$\Phi(x,\, y,\, b) = 0$$

By definition the differential equation gives the same direction of tangents to the integral

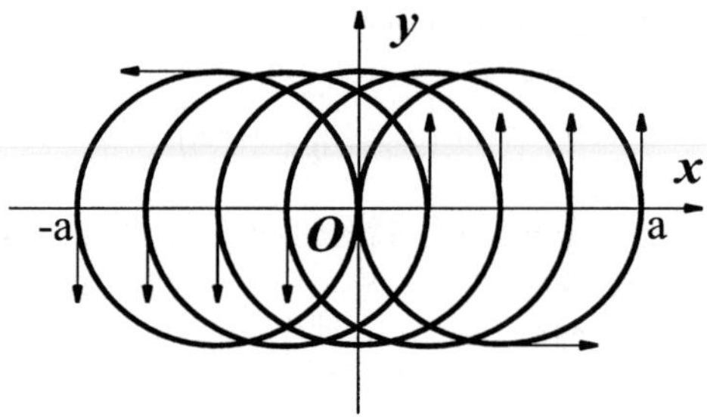

Figure D.5: Integral curves (D-30).

curves at points of this isocline. If the isocline coincides with the integral curve, the angle coefficient of the former has to be equal to b:

$$y' = b$$

Therefore, the isocline is a straight line.

Example Consider the following differential equation

$$\frac{dy}{dx} = \pm \frac{\sqrt{a^2 - y^2}}{y} \tag{D-27}$$

The right-hand side of this equation is defined within the strip

$$|\, y \,| \leq a$$

located between the straight lines

$$y = a \qquad \text{and} \qquad y = -a \tag{D-28}$$

and there are two values of angle coefficient $y'(x)$ at each point. Eq. D-27 can be written as

$$\frac{y\,dy}{\sqrt{a^2 - y^2}} = \pm\, dx \tag{D-29}$$

Its integration gives

$$(x - C)^2 + y^2 = a^2 \tag{D-30}$$

The latter describes a family of circles with centers at the x-axis and the radii equal to a, Fig. D.5. All of those circles are located inside the strip $|\, y \,| \leq a$, and two circles pass through always each of its points.

In deriving eq. D-29 we divided the left- and right-hand sides of eq. D-27 by $\sqrt{a^2 - y^2}$ and, therefore, it became possible to miss the solution

$$y = \pm a$$

In fact, direct substitution shows that the functions given by eqs. D-28 satisfy eq. D-27. They are straight lines that cannot be obtained from the general integral D-30 for any value of C, that is, eqs. D-28 represent singular solutions.

Suppose that derivative $y'(x)$ in eq. D-27 is replaced by some constant. Then we arrive at the family of isoclines,

$$\pm y \sqrt{1 + C_1^2} = a$$

They are also straight lines, which are parallel to the x-axis. In particular, lines $y = \pm a$ are also isoclines, since at their points the angle coefficient has a constant value equal to zero. Moreover, as we know, they are solutions of eq. D-27 too.

Envelope of family of integral curves and singular solutions

It is natural that the last example leads to a concept of an envelope of a family of integral curves. In our case the singular solutions are the straight lines $y = \pm a$ that touch the regular integral curves at each point. Now we assume that the family of integral curves is described by eq. D-21,

$$\psi(x, y, C) = 0 ,$$

where C is an arbitrary constant.

By definition, an envelope of this family is a line that touches each integral curve tangentially at a point. In other words, the tangent of the envelope coincides with that of the integral curve of the family passing through the same point. Note that if these curves are rays, the envelope is called caustic, (Chapter 1).

Equation for the envelope

To derive this equation we first find the angle coefficient of the integral curves from eq. D-21. Its differentiation with respect to x gives

$$\frac{\partial \psi(x, y, C)}{\partial x} + \frac{\partial \psi(x, y, C)}{\partial y} \frac{dy}{dx} = 0$$

Then

$$\frac{dy}{dx} = - \frac{\dfrac{\partial \psi(x, y, C)}{\partial x}}{\dfrac{\partial \psi(x, y, C)}{\partial y}} \tag{D-31}$$

Here we took into account the fact that y is a function of x, while C is a constant.

Suppose that the unknown equation for the envelope is

$$R(x,\,y) = 0 \tag{D-32}$$

We do not know the left-hand side yet, so we represent it in the same form as that for the integral lines:

$$R(x,\,y) \;=\; \psi(x,\,y,\,C) \;=\; 0 \tag{D-33}$$

However, there is one fundamental difference, it is now assumed that C is some function of x and y that has to be determined. Such an approach can be easily justified, since moving along the envelope, we touch the integral lines characterized by different values of C. Inasmuch as the function $\psi(x,\,y,\,C)$ is equal to zero, its differential, $d\psi$, vanishes too, and we have

$$d\psi \;=\; \frac{\partial\,\psi(x,\,y,\,C)}{\partial\,x}\,dx + \frac{\partial\,\psi(x,\,y,\,C)}{\partial\,y}\,dy + \frac{\partial\,\psi(x,\,y,\,C)}{\partial\,C}\,dC \;=\; 0 \tag{D-34}$$

It is clear that the angle coefficient of the tangent to the envelope should coincide with that of the integral curve passing through the same point. This means that eq. D-34 has to give the same expression for the derivative dy/dx as eq. D-31. This occurs only in the case when the third term in the left-hand side of eq. D-34 vanishes:

$$\frac{\partial\,\psi(x,\,y,\,C)}{\partial\,C}\,dC \;=\; 0$$

In principle, it is possible that

$$d\,C \;=\; 0 \qquad \text{or} \qquad C \;=\; \text{const}$$

However, such equalities indicate that in place of the envelope we deal with the integral curve. Correspondingly, in order to obtain the envelope, it is necessary to assume that

$$\frac{\partial\,\psi(x,\,y,\,C)}{\partial\,C} \;=\; 0 \tag{D-35}$$

This equality defines C as a function of x and y. Then, substituting $C(x,\,y)$ into eq. D-33, we arrive at the equation of the envelope of the family $\psi(x,\,y,\,C) = 0$. Thus, it is obtained by eliminating C from two equations:

$$\psi(x,\,y,\,C) = 0 \qquad \text{and} \qquad \frac{\partial\,\psi(x,\,y,\,C)}{\partial\,C} = 0 \tag{D-36}$$

As was demonstrated in the last example, an envelope is also a solution of the corresponding differential equation, even a singular one, and this result is not occasional. In fact, suppose that

$$\psi(x,\,y,\,C) = 0 \tag{D-37}$$

is the general integral of the differential equation

$$\Phi(x,\, y,\, y') = 0 \tag{D-38}$$

This means that the coordinates of a point $(x,\, y)$ and the angle coefficient y' satisfy eq. D-38 at any integral line, (eq. D-37). At the same time, at every point of the envelope the values of x, y and y' coincide with those of some integral line. In other words, x, y and y' of the envelope obey eq. D-38 too. Thus, the envelope of the family of integral curves is also a solution of the same differential equation. Now consider several examples of families of integral curves and their envelopes.

Example one Let us find the envelope of the family of circles

$$(x - C)^2 + y^2 = a^2$$

which was studied in detail earlier. In this case the function $\psi(x,\, y,\, C)$ is

$$\psi(x,\, y,\, C) = (x - C)^2 + y^2 - a^2$$

and eqs. D-36 can be written as

$$(x - C)^2 + y^2 - a^2 = 0 \qquad \text{and} \qquad -\,2\,(x - C) = 0$$

The second equation yields $x = C$, and after its substitution in the first one, we obtain

$$y^2 = a^2 \qquad \text{or} \qquad y = \pm a\,,$$

which coincides with eq. D-28. Thus, the solution of the differential equation D-27 consists of the family of integral curves within the strip $|\,y\,| \leq a$ and the two envelopes $y = \pm a$. There is no solution outside the strip $|\,y\,| \leq a$.

Considering the rays of acoustic waves in inhomogeneous media (Chapter 1), we often observe a similar behavior, namely, a regular distribution of rays, the appearance of their envelopes (the caustics), and the existence of regions where the rays are absent (shadow zones).

Example two It is a simple matter to see that the general integral of the equation:

$$y = x\,y' + \varphi(y')$$

is

$$y(x,\, C) = x\,C + \varphi(C)$$

Here φ is some function that has the first derivative. In such a case, as before, the envelope is obtained by eliminating C from two equations:

$$y = x\,C + \varphi(C) \qquad \text{and} \qquad x + \varphi'(C) = 0$$

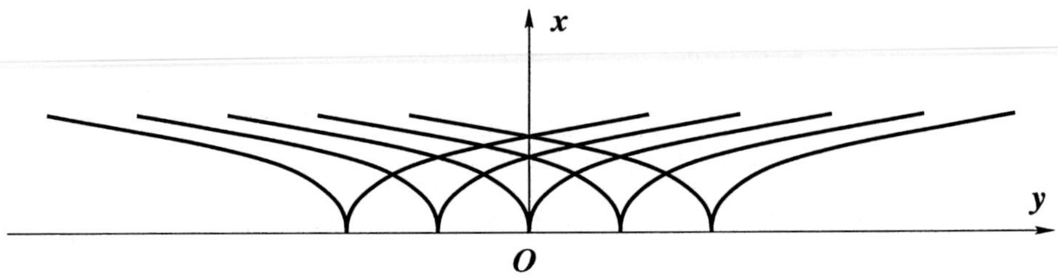

Figure D.6: Family of semi-cubic parabolas.

D.2 System of ordinary differential equations

To illustrate the main concepts related to solving a system of ordinary differential equations of the first order, suppose that three unknown functions of a single argument

$$y_1(x)\,, \qquad y_2(x) \qquad \text{and} \qquad y_3(x)$$

have to satisfy the following system:

$$\frac{d\,y_1}{d\,x} = f_1(x,\,y_1,\,y_2,\,y_3) \quad \frac{d\,y_2}{d\,x} = f_2(x,\,y_1,\,y_2,\,y_3) \quad \frac{d\,y_3}{d\,x} = f_3(x,\,y_1,\,y_2,\,y_3) \qquad \text{(D-39)}$$

Here f_1, f_2, and f_3 are given functions and each of them, in general, depends on the argument x and all three unknowns. Because of this the set D-39 forms the system of differential equations and its solution is

$$y_1 = \varpi_1(x)\,, \qquad y_2 = \varpi_2(x)\,, \qquad \text{and} \qquad y_3 = \varpi_3(x) \qquad \text{(D-40)}$$

Since integration of any equation of the system gives rise to an arbitrary constant, we may expect that each unknown function contains all three constants

$$y_i = \psi_i(x,\,C_1,\,C_2,\,C_3)\,, \qquad i = 1,\,2,\,3 \qquad \text{(D-41)}$$

Solving those equations with respect to the constants we obtain

$$\varphi_i(x,\,y_1,\,y_2,\,y_3) = C_i\,, \qquad i = 1,\,2,\,3 \qquad \text{(D-42)}$$

Each equation of this system is called the integral of set D-39. There are three such integrals in our case and it is assumed that they can be resolved for y_1, y_2 and y_3.

As before, we have the following three forms of solution:

$$y_i = \varpi_i(x)\,, \qquad y_i = \psi_i(x,\, C_1,\, C_2,\, C_3) \qquad \text{and} \qquad \varphi_i(x,\, y_1,\, y_2,\, y_3) = C_i\,,$$

and arbitrary constants C_i are defined from the initial conditions at $x = x_0$:

$$y_1(x_0) = y_{10}\,, \qquad y_2(x_0) = y_{20}\,, \qquad y_3(x_0) = y_{30} \tag{D-43}$$

It follows from the theory that as in the case of a single differential equation, if functions $f_i(x,\, y_1,\, y_2,\, y_3)$ have continuous derivatives with respect to y_i at $x = x_0$ and $y_i = y_{i0}$, then there is only one solution of the system D-39 that satisfies the initial conditions. In other words, we have formulated the theorem of existence and uniqueness of solution of a system of differential equations.

The second form of system D-39

It is useful to represent eqs. D-39 as

$$dx = \frac{dy_1}{f_1(x,\, y_1,\, y_2,\, y_3)} = \frac{dy_2}{f_2(x,\, y_1,\, y_2,\, y_3)} = \frac{dy_3}{f_3(x,\, y_1,\, y_2,\, y_3)} \tag{D-44}$$

Multiplying all denominators, including the unity in $dx/1$, by the same function, eq. D-44 becomes

$$\frac{dx_1}{X_1} = \frac{dx_2}{X_2} = \frac{dx_3}{X_3} = \frac{dx_4}{X_4}\,, \tag{D-45}$$

where

$$x_1 = x\,, \qquad x_2 = y_1\,, \qquad x_3 = y_2\,, \qquad x_4 = y_3$$

and X_i are given functions of the variables x_i.

Applying the same notation, we can write integrals of the system D-45 as

$$\varphi_i(x_1,\, x_2,\, x_3,\, x_4) = C_i\,, \qquad i = 1,\, 2,\, 3 \tag{D-46}$$

Here $x_1 = x$ is the argument of unknown functions x_2, x_3 and x_4. Sometimes only the left-hand side of eqs. D-46 is called the integrals of the system, provided that the latter becomes a constant after replacing x_2, x_3 and x_4 with any solution. In this light let us construct an arbitrary function F of the integrals φ_i:

$$F(\varphi_1,\, \varphi_2,\, \varphi_3) \tag{D-47}$$

Since substitution of any solution of system D-39 makes our new arguments φ_i constants, function F also becomes a constant, that is, it is independent of x. In other words, an arbitrary function of any integrals of the system is its integral too.

Example For illustration, consider the following system:

$$\frac{d\,x}{x\,z} = \frac{d\,y}{y\,z} = -\frac{d\,z}{x^2 + y^2}$$

It is equivalent to two ordinary differential equations,

$$\frac{d\,x}{x\,z} = \frac{d\,y}{y\,z} \tag{D-48}$$

and

$$\frac{d\,x}{x\,z} = -\frac{d\,z}{x^2 + y^2} \tag{D-49}$$

Removal of z at points where $z \neq 0$ from the first equation and its integration give

$$\ln x = \ln y - C \qquad \text{or} \qquad \ln \frac{y}{x} = C\,,$$

that is,

$$y = C_1\,x \tag{D-50}$$

This is the first integral of the system

$$\varphi_1 = \frac{y}{x} = C_1 \qquad \text{at} \qquad x \neq 0$$

Its substitution into eq. D-49 yields

$$\frac{d\,x}{z} = -\frac{d\,z}{x\,(1 + C_1^2)} \qquad \text{or} \qquad (1 + C_1^2)\,x\,dx + z\,dz = 0$$

Integrating this equation we have

$$(1 + C_1^2)\,x^2 + z^2 = C_2$$

Then, replacement of C_1 by the ratio y/x gives the second integral of the system,

$$\varphi_2 = x^2 + y^2 + z^2 = C_2 \tag{D-51}$$

Thus, the solution of sets D-48, D-49 consists of two integrals

$$\frac{y}{x} = C_1 \qquad \text{and} \qquad x^2 + y^2 + z^2 = C_2 \tag{D-52}$$

Comparison with eqs. D-45 shows that in our case there are two unknown functions $y_1(x) = y$ and $y_2(x) = z$. As follows from eqs. D-52

$$y_1(x) = C_1\,x \qquad \text{and} \qquad y_2(x) = \sqrt{C_2 - (1 + C_1^2)\,x^2} \tag{D-53}$$

We have described the solution of system of three differential equations of the first order and have illustrated it in the case when there are two unknowns. Of course the same approach can be applied for an arbitrary number of equations.

D.3 Linear homogeneous partial differential equations of the first order

Until now we have studied ordinary differential equations when all unknowns are functions of a single argument. Our next subject is the differential equations of the first order with partial derivatives. This means that an unknown function depends on several arguments. As usual, we start from the simplest case, when an equation is linear and homogeneous. For instance, if the unknown function φ depends on three arguments, the equation is

$$X_1 \frac{\partial \varphi}{\partial x_1} + X_2 \frac{\partial \varphi}{\partial x_2} + X_3 \frac{\partial \varphi}{\partial x_3} = 0 \tag{D-54}$$

It is essential that X_1, X_2 and X_3 are given functions of x_1, x_2 and x_3, but they are independent of $\varphi(x_1,\, x_2,\, x_3)$. For this reason eq. D-54 is linear, and it is also homogeneous, since its right-hand side is zero.

Equivalence of a partial differential equation to the system of ordinary differential equations

To demonstrate this very important relationship between a partial differential equation and a system of ordinary differential equations we again consider the simplest system

$$\frac{d\,x_1}{X_1} = \frac{d\,x_2}{X_2} = \frac{d\,x_3}{X_3} \tag{D-55}$$

Here $x_1 = x$ is an independent variable (an argument), but $x_2 = y_1(x)$ and $x_3 = y_2(x)$ are two unknown functions. Suppose that

$$\varphi(x_1,\, x_2,\, x_3) = C \tag{D-56}$$

represents an integral of the system, that is, after substitution of any solution of eqs. D-55 (x_2 and x_3) into eq. D-56, the latter becomes constant. In other words, it is independent of x, and correspondingly, its derivative with respect to x vanishes. This means that

$$\frac{\partial \varphi}{\partial x_1} + \frac{\partial \varphi}{\partial x_2}\frac{\partial x_2}{\partial x_1} + \frac{\partial \varphi}{\partial x_3}\frac{\partial x_3}{\partial x_1} = 0$$

Multiplying this equation by dx_1, we have

$$\frac{\partial \varphi}{\partial x_1}\, dx_1 + \frac{\partial \varphi}{\partial x_2}\, dx_2 + \frac{\partial \varphi}{\partial x_3}\, dx_3 = 0 \tag{D-57}$$

In accordance with eq. D-55, differentials dx_i are proportional to X_i, and their ratios

dx_i/X_i are independent of the index i. This allows us to replace dx_i with X_i in the last equality, and it gives the linear homogeneous differential equation with partial derivatives:

$$X_1 \frac{\partial \varphi}{\partial x_1} + X_2 \frac{\partial \varphi}{\partial x_2} + X_3 \frac{\partial \varphi}{\partial x_3} = 0 \qquad \text{(D-58)}$$

By definition, function $\varphi(x_1, x_2, x_3)$ satisfies eq. D-57 and, correspondingly, also the partial differential equation. Thus, we have demonstrated that if

$$\varphi(x_1, x_2, x_3) = C$$

is an integral of system D-55, then function

$$\varphi(x_1, x_2, x_3)$$

is a solution of eq. D-58. In other words, we have replaced the solution of a differential equation with partial derivatives with that of a system of linear ordinary differential equations. Note that the concept of an integral of a system played a vital role in deriving this equivalence. Moreover, if, for example,

$$\varphi_1(x_1, x_2, x_3) = C_1 \qquad \text{and} \qquad \varphi_2(x_1, x_2, x_3) = C_2 \qquad \text{(D-59)}$$

are two independent integrals of the system, then, as we know, an arbitrary function $F(\varphi_1, \varphi_2)$ is also an integral. Respectively, we can formulate the rule of integration of eq. D-58 in the following way. In order to find a solution of linear partial differential equation D-58, it is necessary to construct the corresponding system D-55 of ordinary differential equations and its independent integrals, D-59. Then, the solution of eq. D-58 is

$$\varphi = F(\varphi_1, \varphi_2)\,, \qquad \text{(D-60)}$$

where F is an arbitrary function of its arguments.

D.4 Linear inhomogeneous partial differential equations

We remember that eq. D-58 is linear and homogeneous, since functions X_i are independent of solution φ and its right-hand side is zero. The word "linear" is related to the fact that the equation does not contain the terms with the product of function φ and its derivatives. A more general case is the following equation

$$Y_1 \frac{\partial \varphi}{\partial x_1} + Y_2 \frac{\partial \varphi}{\partial x_2} + Y_3 \frac{\partial \varphi}{\partial x_3} = Y_4\,, \qquad \text{(D-61)}$$

that is, the term Y_4 is nonzero but Y_i are still the functions of x_1, x_2, and x_3.

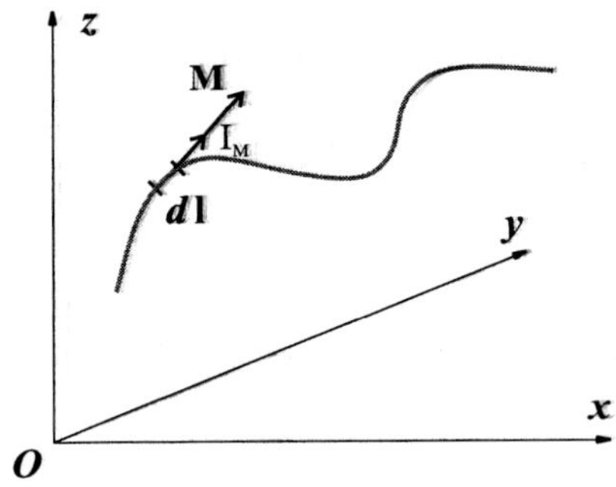

Figure D.7: Vector field $\mathbf{M}(x, y, z)$.

In order to reduce this equation to the previous one, eq. D-58, it is convenient to look for its solution in the form

$$\varpi(x_1, x_2, x_3, \varphi) = C\,, \tag{D-62}$$

where C is an arbitrary constant. In fact, the latter is an integral of eq. D-61. Taking the derivative with respect to x_i from both sides of this equality results in

$$\frac{\partial \varpi}{\partial x_i} + \frac{\partial \varpi}{\partial \varphi}\frac{\partial \varphi}{\partial x_i} = 0 \qquad \text{or} \qquad \frac{\partial \varphi}{\partial x_i} = -\frac{\partial \varpi / \partial x_i}{\partial \varpi / \partial \varphi} \tag{D-63}$$

Its substitution into eq. D-61 yields

$$Y_1 \frac{\partial \varpi}{\partial x_1} + Y_2 \frac{\partial \varpi}{\partial x_2} + Y_3 \frac{\partial \varpi}{\partial x_3} + Y_4 \frac{\partial \varpi}{\partial \varphi} = 0 \tag{D-64}$$

Thus, we have again obtained the linear homogeneous equation where function ϖ is unknown. If the latter is found, then, resolving eq. D-62, we can determine the function $\varphi(x_1, x_2, x_3)$. It is clear that the solution of eq. D-62 depends on some function, while solution of ordinary differential equations contains arbitrary constants.

Integral (characteristic) curves

Now we discuss the linkage between a system of ordinary differential equations and a partial differential equation from a geometrical point of view. With this purpose in mind, we introduce the Cartesian coordinate system x, y, z and consider some vector field $\mathbf{M}(x, y, z)$, Fig. D.7. As we know from vector algebra (Part I), the vector can be represented in two forms

$$\mathbf{M} = M\,\mathbf{i}_M \qquad \text{and} \qquad \mathbf{M} = u(x, y, z)\,\mathbf{i} + v(x, y, z)\,\mathbf{j} + w(x, y, z)\,\mathbf{k} \tag{D-65}$$

Here M is the vector magnitude and $\mathbf{i}_M$ is its unit vector; u, v and w are the scalar components of the vector. By definition, we have

$$\mathbf{i}_M = \cos(\mathbf{i}_M,\, \mathbf{i})\,\mathbf{i} + \cos(\mathbf{i}_M,\, \mathbf{j})\,\mathbf{j} + \cos(\mathbf{i}_M,\, \mathbf{k})\,\mathbf{k}\,, \qquad M = \sqrt{u^2 + v^2 + w^2}\,, \tag{D-66}$$

where $\cos(\mathbf{i}_M,\, \mathbf{i})$, $\cos(\mathbf{i}_M,\, \mathbf{j})$, and $\cos(\mathbf{i}_M,\, \mathbf{k})$ are the directional cosines of the vector and

$$\cos(\mathbf{i}_M,\, \mathbf{i}) = \frac{u(x,\, y,\, z)}{M}\,, \qquad \cos(\mathbf{i}_M,\, \mathbf{j}) = \frac{v(x,\, y,\, z)}{M}\,, \qquad \cos(\mathbf{i}_M,\, \mathbf{k}) = \frac{w(x,\, y,\, z)}{M} \tag{D-67}$$

We usually deal with regular points, where the magnitude and direction of vector $\mathbf{M}$ are uniquely defined. At the same time, it may happen that there are places (points, lines, surfaces), where the field displays various types of singularities.

It is conventional to characterize the direction of the vector field $\mathbf{M}$ with the vector lines that clearly show the orientation of the field at regular points, Fig. D.7. In order to derive equations for those lines, we recall that their tangents and the field have the same direction. Consider an element of vector line $d\mathbf{l}$. As is seen from Fig. D.7, it can be written as

$$d\mathbf{l} = \mathbf{i}\,dx + \mathbf{j}\,dy + \mathbf{k}\,dz\,, \tag{D-68}$$

where dx, dy and dz are the scalar components of $d\mathbf{l}$ along the coordinate axes. It is obvious that the directional cosines of this vector are

$$\frac{dx}{dl}\,, \qquad \frac{dy}{dl} \qquad \text{and} \qquad \frac{dz}{dl} \tag{D-69}$$

Inasmuch as vectors $d\mathbf{l}$ and $\mathbf{M}$ are parallel to each other, their corresponding directional cosines coincide and, in accordance with eqs. D-67 and D-69, we can write

$$\frac{u}{M} = \frac{dx}{dl}\,, \qquad \frac{v}{M} = \frac{dy}{dl} \qquad \text{and} \qquad \frac{w}{M} = \frac{dz}{dl}$$

or

$$\frac{dx}{u} = \frac{dl}{M}\,, \qquad \frac{dy}{v} = \frac{dl}{M} \qquad \text{and} \qquad \frac{dz}{w} = \frac{dl}{M}$$

Thus, we have obtained a very useful relationship between components of two parallel vectors:

$$\frac{dx}{u(x,\, y,\, z)} = \frac{dy}{v(x,\, y,\, z)} = \frac{dz}{w(x,\, y,\, z)} \tag{D-70}$$

At the same time the latter represents the system of differential equations describing the vector lines of field **M**. This means that determination of the orientation of the given field **M** can be based on solving the system of ordinary differential equations. In our case, eq. D-70, there are two such equations, and their integration results in two independent integrals,

$$\varphi_1(x,\, y,\, z) = C_1 \qquad \text{and} \qquad \varphi_2(x,\, y,\, z) = C_2 \tag{D-71}$$

Each of these describes a surface in three-dimensional space and, correspondingly, the intersection of two surfaces defines a line. Changing the values of C_1 and C_2, we obtain a family of vector lines, which are often called the characteristic curves. The position of a particular curve is determined by coordinates x_0, y_0, z_0 of a point (the initial conditions) through which the curve passes. Making use of eqs. D-71, these initial conditions allow us to find the constants C_1 and C_2.

Note that in solving system D-70, we may treat two coordinates, for example, y and z as unknown functions of x: $y = y_1(x)$ and $z = y_2(x)$. Summarizing, we can see that the system of ordinary differential equations D-70 gives a family of characteristic curves, or the vector lines, which defines a certain field of directions.

Geometrical interpretation of differential equations with partial derivatives

In this light it is important to solve a problem of determination of a surface formed by a given set of characteristic curves. Suppose that a family of such surfaces is described by equation

$$\varphi(x,\, y,\, z) = C \tag{D-72}$$

By definition, function $\varphi(x, y, z)$ remains constant at each surface of this family and, correspondingly, the gradient of φ,

$$\mathbf{N} = \operatorname{grad}\varphi\,, \tag{D-73}$$

is perpendicular to them. Otherwise, function φ would vary along any surface of this set (Part I). In the Cartesian coordinate system, for instance, we have

$$\mathbf{N} = \frac{\partial\varphi}{\partial x}\mathbf{i} + \frac{\partial\varphi}{\partial y}\mathbf{j} + \frac{\partial\varphi}{\partial z}\mathbf{k} \tag{D-74}$$

Inasmuch as vector **M**, eq. D-65, is tangent to the characteristic curve, it is orthogonal to **N**, and therefore their dot product is zero:

$$\mathbf{M}\cdot\mathbf{N} = 0$$

or

$$u(x,\, y,\, z)\,\frac{\partial\varphi}{\partial x} + v(x,\, y,\, z)\,\frac{\partial\varphi}{\partial y} + w(x,\, y,\, z)\,\frac{\partial\varphi}{\partial z} = 0 \tag{D-75}$$

Thus, to find the surface formed by the characteristic lines, we have to solve the partial

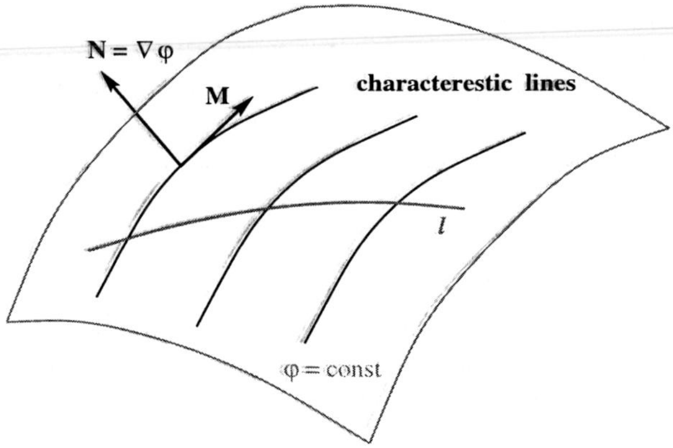

Figure D.8: Surface formed by characteristic lines.

differential equation D-75. Each term of this equation is the product of corresponding components of grad φ and vector $\mathbf{M}$. Their mutual orientation is shown in Fig. D.8.

The system of ordinary differential equation, corresponding to eq. D-75, as well as their integrals, are given by eqs. D-70 and D-71. Thus, the general solution of eq. D-75 is

$$\varphi(x,\, y,\, z) = F(\varphi_1,\, \varphi_2) \tag{D-76}$$

Here F is an arbitrary function of two independent integrals, eqs. D-71. Respectively, the equation of surfaces, eq. D-72, formed by characteristic curves is

$$\varphi(x,\, y,\, z) = F(\varphi_1,\, \varphi_2) = C$$

Bearing in mind that F is an arbitrary function, let us choose it in such a way that $C = 0$, and therefore the equation for the surface becomes

$$F(\varphi_1,\, \varphi_2) = 0 \tag{D-77}$$

This means that each surface is a geometrical place of characteristic curves of system D-70, provided that the constants in the right-hand side of eqs. D-71 are related as

$$F(C_1,\, C_2) = 0 \tag{D-78}$$

Initial conditions

By analogy with a solution of ordinary differential equations, some surface can be distinguished from others in the family by requiring that it passes through a given line l,

Fig. D.8. This is the initial condition for the differential equation with partial derivatives. It is obvious that in such a case the surface is formed by those characteristic curves which intersect the line l. In other words, the coordinates of points on l are the initial conditions for the characteristic curves, and, in accordance with the theorem of existence and uniqueness, we obtain certain surface. An exception are the lines l that are also characteristic curves of eqs. D-70. Suppose that line l is given as

$$\psi_1(x,\, y,\, z) = 0\,, \qquad \psi_2(x,\, y,\, z) = 0 \tag{D-79}$$

Eliminating variables x, y and z from eqs. D-71 and D-79, we arrive at the relationship between C_1 and C_2 that allows us to determine function F such that eq. D-77 gives a surface passing through line l. We will give several examples that illustrate this subject.

Example one Consider the partial differential equation

$$x\,z\,\frac{\partial\varphi}{\partial x} + y\,z\,\frac{\partial\varphi}{\partial y} - (x^2 + y^2)\,\frac{\partial\varphi}{\partial z} = 0 \tag{D-80}$$

The corresponding system of ordinary differential equations is

$$\frac{dx}{x\,z} = \frac{dy}{y\,z} = -\frac{dz}{x^2 + y^2} \tag{D-81}$$

We have already found two independent integrals of system D-81 earlier:

$$\varphi_1 = \frac{y}{x} = C_1 \qquad \text{and} \qquad \varphi_2 = x^2 + y^2 + z^2 = C_2 \tag{D-82}$$

The first one gives a family of planes passing through the z-axis, while the second represents a family of spheres with centers at the coordinate origin. Their intersection produces the characteristic curves, which are circles located in the vertical planes with the centers at the origin.

In accordance with eq. D-76, the general solution of eq. D-80 is

$$F = F(\varphi_1,\, \varphi_2) \qquad \text{or} \qquad F = F(C_1,\, C_2)\,,$$

that is

$$\varphi = F\left(\frac{y}{x},\; x^2 + y^2 + z^2\right), \tag{D-83}$$

where F is a differentiable function of its arguments.

For example, taking

$$\varphi = \left(\frac{y}{x}\right)^m + \left(x^2 + y^2 + z^2\right)^n\,,$$

we have

$$\begin{aligned}
\frac{\partial \varphi}{\partial x} &= -m\, y^m\, x^{-m-1} + 2\, n\, x\left(x^2+y^2+z^2\right)^{n-1} \\
\frac{\partial \varphi}{\partial y} &= m\, y^{m-1}\, x^{-m} + 2\, n\, y\left(x^2+y^2+z^2\right)^{n-1} \\
\frac{\partial \varphi}{\partial z} &= 2\, n\, z\left(x^2+y^2+z^2\right)^{n-1}
\end{aligned}$$

Their substitution into eq. D-80 yields

$$-\,m\, y^m\, z\, x^{-m} + m\, y^m\, z\, x^{-m} + 2\, x^2\, z\, n\left(x^2+y^2+z^2\right)^{n-1} +$$

$$+2\, y^2\, z\, n\left(x^2+y^2+z^2\right)^{n-1} - (x^2+y^2)\, 2\, z\, n\left(x^2+y^2+z^2\right)^{n-1} =$$

$$= 2\, x^2\, z\, n + 2\, y^2\, z\, n - 2\, x^2\, z\, n - 2\, y^2\, z\, n \equiv 0$$

This, of course, it is only an illustration of the remarkable fact established earlier.

Now let us return back to our example and suppose that the initial condition is

$$x = 1 \qquad \text{and} \qquad y = z \tag{D-84}$$

This means that l is the straight line which is located in the plane $x = 1$ and which makes 45° angle with the y- and z-axes, Fig. D.9. The first integral of set D-82 and the initial condition give

$$x = 1\,, \qquad y = C_1 \qquad \text{and} \qquad z = C_1 \tag{D-85}$$

After substitution of eqs. D-85 into the second integral, we obtain

$$1 + 2\, C_1^2 - C_2 = 0 \tag{D-86}$$

The latter is the relationship between arbitrary constants which, in accordance with eq. D-78 describes the surface

$$F(C_1,\, C_2) = 0$$

passing through line l, eq. D-84. In fact, taking into account eqs. D-82, we have

$$1 + 2\,\frac{y^2}{x^2} - \left(x^2+y^2+z^2\right) = 0$$

or

$$x^2 + 2\, y^2 - x^2\left(x^2+y^2+z^2\right) = 0 \tag{D-87}$$

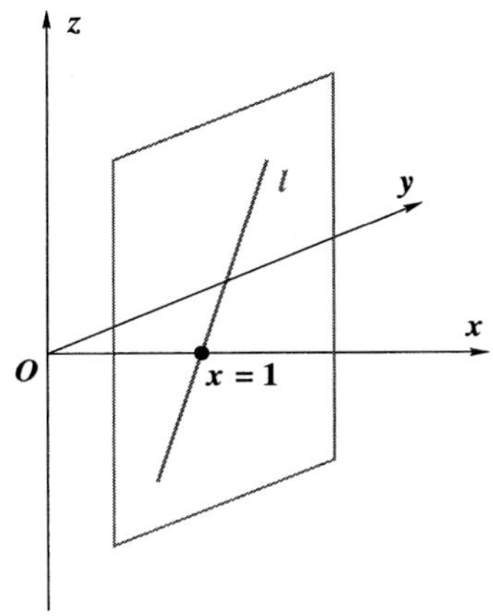

Figure D.9: Line l representing the initial condition.

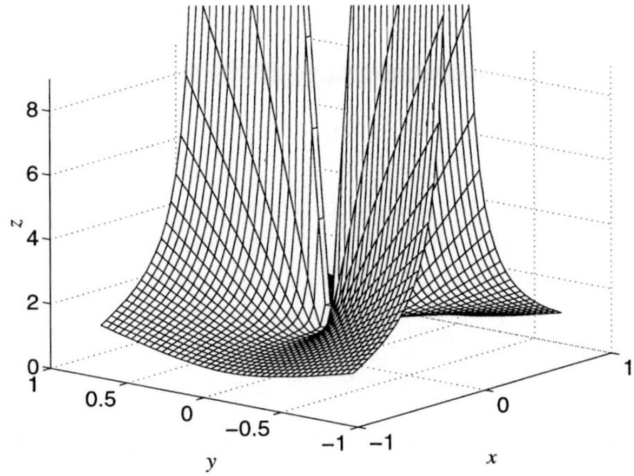

Figure D.10: Surface $z(x, y)$ given by eq. D-87.

The part of this surface corresponding to the real and positive values of z is shown in Fig. D.10.

Example two Suppose that the field of directions, defined by three numbers a, b, and c, is constant:

$$\mathbf{M} = a\,\mathbf{i} + b\,\mathbf{j} + c\,\mathbf{k}$$

Correspondingly, the system of differential equations is

$$\frac{dx}{a} = \frac{dy}{b} = \frac{dz}{c} \tag{D-88}$$

or

$$c\,dx - a\,dz = 0 \qquad \text{and} \qquad c\,dy - b\,dz = 0 \tag{D-89}$$

Each of these equations describes a family of planes passing through either the x- or y-axis. Therefore, their intersection forms a family of straight characteristic lines. The corresponding differential equation with partial derivatives is

$$a\frac{\partial \varphi}{\partial x} + b\frac{\partial \varphi}{\partial y} + c\frac{\partial \varphi}{\partial z} = 0 \tag{D-90}$$

and its general solution is given by

$$\varphi = F(cx - a\,z\,,\; cy - b\,z)\,, \tag{D-91}$$

where F is a differentiable function.

Thus, the characteristic lines form the cylindrical surface

$$F(cx - a\,z\,,\; cy - b\,z) = 0 \tag{D-92}$$

Example three In this case, the field of directions is governed by the vector pointing from the fixed point $A(a,\, b,\, c)$ to the point p with coordinates x, y, z. Then the system of ordinary differential equations is

$$\frac{dx}{x-a} = \frac{dy}{y-b} = \frac{dz}{z-c}$$

and they have two independent integrals

$$\frac{x-a}{z-c} = C_1 \qquad \text{and} \qquad \frac{y-b}{z-c} = C_2\,, \tag{D-93}$$

which describe two families of planes. Their intersection produces a family of straight characteristic lines passing through point $A(a, b, c)$. Correspondingly, the partial differential equation is

$$(x-a)\frac{\partial \varphi}{\partial x} + (y-b)\frac{\partial \varphi}{\partial y} + (z-c)\frac{\partial \varphi}{\partial z} = 0 \tag{D-94}$$

and it defines conical surfaces with the apex at A. The general solution for such surfaces is

$$F\left(\frac{x-a}{z-c}, \frac{y-b}{z-c}\right),$$

where F is a differentiable function.

Example four Consider the following system of differential equations:

$$\frac{dx}{cy-bz} = \frac{dy}{az-cx} = \frac{dz}{bx-ay} \tag{D-95}$$

We introduce a new variable t so that all three ratios are equal to dt. Then we can write

$$dx = (cy-bz)\,dt \qquad dy = (az-cx)\,dt, \qquad dz = (bx-ay)\,dt \tag{D-96}$$

The latter allows us to construct two equations that can be easily integrated. Multiplying each equation of set D-96 by a, b and c, respectively, and adding them together, we obtain the first equation

$$a\,dx + b\,dy + c\,dz = 0 \tag{D-97}$$

The same procedure with x, y, and z gives the second equation,

$$x\,dx + y\,dy + z\,dz = 0 \tag{D-98}$$

Performing integration, we obtain two integrals of the system

$$\varphi_1 = ax + by + cz = C_1 \qquad \text{and} \qquad \varphi_2 = x^2 + y^2 + z^2 = C_2 \tag{D-99}$$

The first integral describes a family of parallel planes. The directional cosines of their normals are proportional to a, b and c, as follows from the definition of grad φ_1. The second integral gives a family of spheres with the center at the coordinate origin. Their intersection produces a family of characteristic lines, which are the circles located within those planes. The centers of the circles are situated on the straight line

$$\frac{x}{a} = \frac{y}{b} = \frac{z}{c} \tag{D-100}$$

which passes through the coordinate origin and is perpendicular to the planes.

It is clear that the corresponding equation with partial derivatives is

$$(c\,y - b\,z)\,\frac{\partial\,\varphi}{\partial\,x} + (a\,z - c\,x)\,\frac{\partial\,\varphi}{\partial\,y} + (b\,x - a\,y)\,\frac{\partial\,\varphi}{\partial\,z} = 0$$

Its general solution is given by

$$F(a\,x + b\,y + c\,z\,,\; x^2 + y^2 + z^2) \;=\; 0$$

This describes the surfaces of rotation, when the straight line, given by eq. D-100, is the axis of rotation.

D.5 Quasi-linear equation with partial derivatives

So far, we have dealt with linear equations. Next consider a quasi-linear differential equation with partial derivatives

$$a(x,\,y,\,u)\,u_x \,+\, b(x,\,y,\,u)\,u_y \;=\; c(x,\,y,\,u) \tag{D-101}$$

Here, we assume that the unknown function $u(x,\,y)$ depends on two arguments, x and y, only. This simplification allows us to represent a solution as a surface in three-dimensional space or as a family of level curves on the x-y plane. We also use the following notation:

$$u_x = \frac{\partial\,u(x,\,y)}{\partial\,x} \qquad \text{and} \qquad u_y = \frac{\partial\,u(x,\,y)}{\partial\,y}$$

Unlike the previous cases, coefficients a, b and c contain the unknown function u, and this is why eq. D-101 is called the quasi-linear one.

Consider a surface that describes a solution

$$u = u(x,\,y)$$

In general, function u depends on both arguments x and y, and its differential is equal to

$$u_x\,dx + u_y\,dy = du$$

or

$$u_x\,dx + u_y\,dy - 1\,du = 0 \tag{D-102}$$

The latter can be interpreted as a dot product of two vectors,

$$d\,\mathbf{l}\cdot\mathbf{N} = 0 \tag{D-103}$$

One of them,

$$d\,\mathbf{l} = \mathbf{i}\,dx + \mathbf{j}\,dy + \mathbf{k}\,du\,, \tag{D-104}$$

characterizes the magnitude and orientation of any displacement along the solution surface $u(x, y)$. In other words, $d\mathbf{l}$ is tangent to the surface at each point. Therefore, vector

$$\mathbf{N} = \mathbf{i}\, u_x + \mathbf{j}\, u_y - \mathbf{k} \tag{D-105}$$

is normal to surface $u(x, y)$. Note that vectors $\mathbf{N}$ and $\operatorname{grad} u$ differ from each other. In particular, the latter is situated in the x-y plane, whereas $\mathbf{N}$ is located in three-dimensional space. As long as we know the direction of $\mathbf{N}$, let us represent partial differential equation D-101 in the form

$$\mathbf{N} \cdot \mathbf{M} = 0\,, \tag{D-106}$$

where

$$\mathbf{M} = a(x, y, u)\,\mathbf{i} + b(x, y, u)\,\mathbf{j} + c(x, y, u)\,\mathbf{k} \tag{D-107}$$

Thus, we have arrived at important conclusion: vector $\mathbf{M}$ is tangential to the solution surface $u(x, y)$ at each point. Now we direct vector $d\mathbf{l}$, eq. D-104, along $\mathbf{M}$ and introduce the vector lines of field $\mathbf{M}$. Inasmuch as vectors $d\mathbf{l}$ and $\mathbf{M}$ are both situated on the solution surface and parallel to each other, the equation for those vector lines is

$$\frac{dx}{a(x, y, u)} = \frac{dy}{b(x, y, u)} = \frac{dz}{c(x, y, u)} \tag{D-108}$$

As in the case of the linear equations we have reduced the solution of a quasi-linear differential equation to that of system of the ordinary ones, eq. D-108.

Suppose that

$$\varphi_1(x, y, u) = C_1 \qquad \text{and} \qquad \varphi_2(x, y, u) = C_2 \tag{D-109}$$

are its independent integrals. Each of them describes a surface, and correspondingly, their intersection gives the vector line of field $\mathbf{M}$ or the characteristic curve. Changing the constants C_1 and C_2, we can obtain an infinite number of solution surfaces made up of the characteristic curves. The equation for these surfaces is

$$F(\varphi_1, \varphi_2) = 0\,, \tag{D-110}$$

where F is a differentiable function.

To select a particular surface in the family we have to add the initial condition. As before, assume that such a surface passes through a given line l:

$$x = x_0(\tau)\,, \qquad y = y_0(\tau)\,, \qquad z = z_0(\tau) \tag{D-111}$$

Here τ is the parameter that characterizes a position on the prescribed curve. This means that the solution surface consists of characteristic curves intersecting a given line l. It is also convenient to introduce the second parameter σ along the characteristic curves. For example, it can be their length. Then, the derivative of the radius vector,

$$\frac{d\mathbf{R}}{d\sigma}$$

is tangential to the curve, and therefore $\mathbf{M}$ is parallel to $d\mathbf{R}/d\sigma$. Correspondingly,

$$\frac{dx}{d\sigma} = \lambda a, \qquad \frac{dy}{d\sigma} = \lambda b, \qquad \frac{dz}{d\sigma} = \lambda c, \tag{D-112}$$

where λ is an arbitrary constant.

Letting $\lambda = 1$, we obtain a different form of eq. D-108:

$$\frac{dx}{d\sigma} = a, \qquad \frac{dy}{d\sigma} = b, \qquad \frac{dz}{d\sigma} = c, \tag{D-113}$$

and the solution surface consists of points defined by two parameters, τ and σ:

$$x = x(\sigma, \tau), \qquad y = y(\sigma, \tau), \qquad u = u(\sigma, \tau) \tag{D-114}$$

We still need to find function $u(x, y)$. With this purpose the first two equations of the set D-114 are used to express σ and τ in terms of x and y. Except in some special cases, this procedure can be performed. Then, substituting these formulae into the last equality, $u = u(\sigma, \tau)$, we obtain $u(x, y)$.

It may be proper to make two comments.

(a) The method of transforming the partial differential equation into the system of the ordinary differential equations is called the method of characteristics.

(b) We have represented the solution surface as a continuous system of characteristic curves. There is another way to describe the solution. Let us imagine a system of level curves

$$u(x, y) = \text{const} ,$$

that are projections of the characteristic curves in the x-y plane. The latter are called the base characteristic curves.

D.6 First-order nonlinear differential equation

Now we discuss the solution of more complicated equations, the nonlinear ones. For instance, it can be the eikonal equation

$$\left(\frac{\partial u}{\partial x}\right)^2 + \left(\frac{\partial u}{\partial y}\right)^2 = n^2 \tag{D-115}$$

The general form of such a first-order equation with partial derivatives is

$$F(x,\, y,\, u,\, p,\, q) = 0\,, \tag{D-116}$$

where unknown function u depends on x and y. Here F is a given function, and

$$u = u(x,\, y)\,, \qquad p = \frac{\partial\, u}{\partial\, x} = u_x\,, \qquad q = \frac{\partial\, u}{\partial\, y} = u_y \tag{D-117}$$

Inasmuch as eq. D-116 is a differential equation with partial derivatives, it has to contain p or q or both of them, that is,

$$F_p \neq 0 \qquad \text{or} \qquad F_q \neq 0\,, \tag{D-118}$$

where

$$F_p = \frac{\partial\, F}{\partial\, p} \qquad \text{and} \qquad F_q = \frac{\partial\, F}{\partial\, q}$$

It turns out that solution of eq. D-116 is based on reducing it to quasi-linear equations with partial derivatives. With this in mind, let us find a small increment, ΔF, of function F due to elementary variation in x and y. Taking into account that, regardless the values of Δx and Δy, the change in F is zero, we have

$$\Delta F \equiv 0 \tag{D-119}$$

A variation of x and y causes a change in u, p and q, and therefore eq. D-119 becomes

$$0 = \Delta F = \left(\frac{\partial\, F}{\partial\, x} + \frac{\partial\, F}{\partial\, u}\frac{\partial\, u}{\partial\, x} + \frac{\partial\, F}{\partial\, p}\frac{\partial\, p}{\partial\, x} + \frac{\partial\, F}{\partial\, q}\frac{\partial\, q}{\partial\, x} \right) \Delta x$$

$$+ \left(\frac{\partial\, F}{\partial\, y} + \frac{\partial\, F}{\partial\, u}\frac{\partial\, u}{\partial\, y} + \frac{\partial\, F}{\partial\, p}\frac{\partial\, p}{\partial\, y} + \frac{\partial\, F}{\partial\, q}\frac{\partial\, q}{\partial\, y} \right) \Delta y$$

or

$$0 = (F_x + p\, F_u + p_x\, F_p + p_y\, F_q)\, \Delta x + (F_y + q\, F_u + q_x\, F_p + q_y\, F_q)\, \Delta y\,, \tag{D-120}$$

where we use the equality

$$p_y = q_x$$

Since the increments Δx and Δy are independent of each other, their coefficients should both be zero:

$$p_x\, F_p + p_y\, F_q = -F_x - p\, F_u \tag{D-121}$$

and

$$q_x\, F_p + q_y\, F_q = -F_y - q\, F_u \tag{D-122}$$

The latter are differential equations with respect to p and q. It is essential that eqs. D-121 and D-122 are quasi-linear, and the method of their solution was described earlier. As we know, the corresponding system of linear equations is

$$\frac{d\,x}{F_p} = \frac{d\,y}{F_q} = -\frac{d\,p}{F_x + p\,F_u} \quad \text{and} \quad \frac{d\,x}{F_p} = \frac{d\,y}{F_q} = -\frac{d\,q}{F_y + q\,F_u} \tag{D-123}$$

By definition, eqs. D-123 describe the base characteristic curves for functions p and q, and they are located in the x-y plane. Comparison of these equations shows that the two families of curves coincide with each other, and correspondingly we can combine eqs. D-123. This gives

$$\frac{d\,x}{F_p} = \frac{d\,y}{F_q} = -\frac{d\,p}{F_x + p\,F_u} = -\frac{d\,q}{F_y + q\,F_u} \tag{D-124}$$

In order to determine unknown function $u(x,\, y)$, consider its differential

$$d\,u = \frac{\partial\,u}{\partial\,x}\,dx + \frac{\partial\,u}{\partial\,y}\,dy$$

or

$$d\,u = p\,dx + q\,dy$$

Then, making use of eqs. D-124, we have

$$d\,u = p\,dx + q\,\frac{F_q}{F_p}\,dx = (p\,F_p + q\,F_q)\,\frac{d\,x}{F_p}$$

or

$$\frac{d\,u}{p\,F_p + q\,F_q} = \frac{d\,x}{F_p}$$

Therefore, we can add one more ratio to the set D-124:

$$\frac{d\,x}{F_p} = \frac{d\,y}{F_q} = \frac{d\,u}{p\,F_p + q\,F_q} = -\frac{d\,p}{F_x + p\,F_u} = -\frac{d\,q}{F_y + q\,F_u} \tag{D-125}$$

This contains four independent ordinary differential equations and allows us to determine $u(x,\, y)$, as well as its partial derivatives $\partial\,u/\partial\,x$ and $\partial\,u/\partial\,y$. Thus, our problem is solved and the nonlinear equation D-116 is replaced by th system D-125.

As in the case of quasi-linear equations, it is useful to introduce parameter σ along the characteristic curves. Then, eqs. D-125 become

$$\frac{dx}{d\sigma} = \lambda F_p, \qquad \frac{dy}{d\sigma} = \lambda F_q, \qquad \frac{du}{d\sigma} = \lambda (p F_p + q F_q),$$

$$\frac{dp}{d\sigma} = -\lambda (F_x + p F_u), \qquad \frac{dq}{d\sigma} = -\lambda (F_y + q F_u),$$

where λ is an arbitrary constant.

Letting $\lambda = 1$, we obtain

$$\frac{dx}{d\sigma} = F_p, \qquad \frac{dy}{d\sigma} = F_q, \qquad \frac{du}{d\sigma} = p F_p + q F_q,$$

$$\frac{dp}{d\sigma} = -F_x - p F_u, \qquad \frac{dq}{d\sigma} = -F_y - q F_u \tag{D-126}$$

Example For illustration, consider the eikonal equation and, first, assume that the medium is homogeneous and the eikonal u depends on x and y only. Then we have

$$p^2 + q^2 = 1 \tag{D-127}$$

Here

$$p = u_x \qquad \text{and} \qquad q = u_y$$

Comparison with eq. D-116 gives

$$F(x, y, u, p, q) = p^2 + q^2 - 1 \tag{D-128}$$

and, therefore

$$F_p = 2p, \qquad F_q = 2q, \qquad F_x = F_y = F_u = 0 \tag{D-129}$$

In accordance with set D-126, the equations for the characteristics are

$$\frac{dx}{d\sigma} = 2p, \qquad \frac{dy}{d\sigma} = 2q, \qquad \frac{du}{d\sigma} = 2, \qquad \frac{dp}{d\sigma} = 0, \qquad \frac{dq}{d\sigma} = 0 \tag{D-130}$$

As well known, these represent ray equations for homogeneous media.

By definition, eq. D-124, the vector

$$\mathbf{M} = F_x \mathbf{i} + F_y \mathbf{j}$$

is tangential to the base characteristic curves, and as follows from eq. D-129, it is equal to

$$\mathbf{M} = 2\,p\,\mathbf{i} + 2\,q\,\mathbf{j} \tag{D-131}$$

On the other hand, the level curves

$$u(x,\, y) = \text{const}$$

represent the eikonal lines, and

$$\operatorname{grad} u = \frac{\partial\, u}{\partial\, x}\,\mathbf{i} + \frac{\partial\, u}{\partial\, y}\,\mathbf{j} = p\,\mathbf{i} + q\,\mathbf{j} \tag{D-132}$$

is perpendicular to them. Comparison of eqs. D-131 and D-132 leads to the equality

$$\mathbf{M} = 2 \operatorname{grad} u\,,$$

that is, the characteristic curves are normal to the eikonal lines. In other words, they describe the rays.

Next, suppose that the medium is inhomogeneous. Then, in place of eq. D-127 we have

$$p^2 + q^2 = n^2\,, \tag{D-133}$$

where n is the function of x and y. Inasmuch as derivatives F_p and F_q are still defined by eqs. D-129, we conclude that the characteristic curves remain perpendicular to the eikonal lines. It is obvious that the same result remains valid in the three-dimensional case, when

$$u = u(x,\, y,\, z) \qquad \text{and} \qquad p^2 + q^2 + r^2 = n^2$$

Here

$$r = \frac{\partial\, u}{\partial\, z}$$

In fact, we have

$$\operatorname{grad} u \;=\; \frac{\partial\, u}{\partial\, x}\,\mathbf{i} + \frac{\partial\, u}{\partial\, y}\,\mathbf{j} + \frac{\partial\, u}{\partial\, z}\,\mathbf{k} \;=\; p\,\mathbf{i} + q\,\mathbf{j} + r\,\mathbf{k}$$

and

$$F_p = 2\,p\,, \qquad F_q = 2\,q\,, \qquad F_r = 2\,r\,,$$

that is,

$$\mathbf{M} \;=\; 2\,p\,\mathbf{i} + 2\,q\,\mathbf{j} + 2\,r\,\mathbf{k} \;=\; 2 \operatorname{grad} u$$

Therefore, the characteristic curves are again perpendicular to the eikonal surfaces

$$u(x,\, y,\, z) = C$$

References

K. Aki and P. Richards. Quantitative seismology: theory and methods. W.H. Freeman and Co., 1980.

M. A. Biot. "Propagation of elastic waves in a cylindrical bore containing a fluid." J. Appl. Phys. 23, 997, 1952.

M. Born and E. Wolf. Principles of optics. Pergamon Press, 1993.

N. Bleinstein. Mathematical methods for wave phenomena. Academic Press, 1984.

L. M. Brekhovskikh. Waves in layered media (2nd edition). Academic Press, 1980.

L. M. Brekhovskikh and O. A. Godin. Acoustics of layered media, I, Plane wave theories (Springer Ser. Wave Phen. Vol. 5). Springer-Verlag, 1990.

H. Dix. Seismic prospecting for oil. Harper Brothers, 1952.

W. Ewing, W. Jardetzky, and F. Press. Elastic waves in layered media. McGraw-Hill Book Co., 1957.

F. Grant and G. West. Interpretation theory in applied geophysics. McGraw-Hill Book Co., 1965.

I. I. Gurvich. Seismic prospecting. Nedra, 1970 (in Russian).

I. G. Hagedorn. "A process of seismic reflection interpretation." Geophys. Prospecting, 2, pp. 85-127, 1954.

W. J. Jacobi. "Propagation of sound waves along liquid cylinders." J. Acoust. Soc. Am., 21, 120, 1949.

A. Kaufman and A. Levshin. Acoustic and elastic wave fields in geophysics, I. Elsevier, 2000.

Yu. A. Kravtsov and Yu. I Orlov. Geometrical optics of inhomogeneous media. Springer-Verlag, 1990.

L. Landay and E. Lifshitz. The classical theory of fields. Pergamon Press, 1975.

J. Matthews and R. L. Walker. Mathematical methods of physics. W.A. Benjamin, Inc., 1964.

C. B. Officer. Introduction to the theory of sound transmission. McGraw-Hill Inc., 1958.

F. L. Paillet and C. Cheng. Acoustic waves in boreholes. CRC Press, 1991.

C. L. Perekis. Theory of propagation of explosive sound in shallow water, in J. L. Worzel et al., Propagation of sound in the ocean. Geol. Soc. Amer. Mem. 27, 1948 (reprinted 1963).

V. I. Smirnov. A course of higher mathematics. Pergamon Press, 1964.

D. H. Towne. Wave Phenomena. Dover Publications, Inc., 1988.

Index